Archaea

New Models for Prokaryotic Biology

Edited by Paul Blum

Caister Academic Press

Caister Academic Press
Norfolk, UK

www.caister.com

British Library Cataloguing-in-Publication Data
A catalogue record for this book is available from the British Library

ISBN: 978-1-904455-27-1

Printed and bound in Great Britain

Contents

Contributors

Brian R. Berquist
Center of Marine Biotechnology
University of Maryland Biotechnology Institute
Baltimore, MD
USA

berquistbr@nia.nih.gov

Elisabetta Bini
Department of Biochemistry and Microbiology
Rutgers, the State University of New Jersey
New Brunswick, NJ
USA

bini@aesop.rutgers.edu

Stan J.J. Brouns
Laboratory of Microbiology
Wageningen University
Wageningen
the Netherlands

stan.brouns@wur.nl

Shannon B. Conners
Department of Chemical and Biomolecular Engineering
North Carolina State University
Raleigh, NC
USA

Shannon.Conners@jmp.com

Edward J. Crane III
Department of Chemistry
Pomona College
Claremont, CA
USA

ej.crane@pomona.edu

Shiladitya DasSarma
Center of Marine Biotechnology
University of Maryland Biotechnology Institute
Baltimore, MD
USA

dassarma@umbi.umd.edu

Henri Grosjean
Institute of Genetics and Microbiology
University of Paris-Sud
Orsay, Essonne
France

henri.grosjean@igmors.u-psud.fr

Petr Gruz
Division of Genetics and Mutagenesis
National Institute of Health Sciences
Tokyo
Japan

gruz@nihs.go.jp

Ramesh Gupta
Department of Biochemistry and Molecular Biology
Southern Illinois University School of Medicine
Springfield, IL
USA

rgupta@siumed.edu

Evan T. Hall
Department of Chemistry
Pomona College
Claremont, CA
USA

evanthomashall@gmail.com

Cynthia A. Haseltine
School of Molecular Biosciences
Washington State University
Pullman, WA
USA

chaseltine@wsu.edu

Charles S. Hummel
Department of Chemistry
Pomona College
Claremont, CA
USA

charles.hummel@gmail.com

Robert M. Kelly
Department of Chemical and Biomolecular Engineering
North Carolina State University
Raleigh, NC
USA

rmkelly@eos.ncsu.edu

Peter J. Kennelly
Department of Biochemistry
Virginia Polytechnic Institute and State University
Blacksburg, VA
USA

pjkennel@vt.edu

Kathryn J. Leyva
Department of Microbiology
Arizona College of Osteopathic Medicine
Midwestern University
Glendale, AZ
USA

kleyva@midwestern.edu

E. Stuart Maxwell
Department of Molecular and Structural Biochemistry
North Carolina State University
Raleigh, NC
USA

stu_maxwell@ncsu.edu

Takehiko Nohmi
Division of Genetics and Mutagenesis
National Institute of Health Sciences
Tokyo
Japan

nohmi@nihs.go.jp

John van der Oost
Laboratory of Microbiology
Wageningen University
Wageningen
the Netherlands

john.vanderoost@wur.nl

Michael L. Rolfsmeier
School of Molecular Biosciences
Washington State University
Pullman, WA
USA

mrolfsmeier@wsu.edu

Richard F. Shand
Department of Biological Sciences
Northern Arizona University
Flagstaff, AZ
USA

richard.shand@nau.edu

Imke Schröder
Department of Microbiology, Immunology and Molecular Genetics
University of California, Los Angeles
Los Angeles, CA
USA

imkes@microbio.ucla.edu

Keith R. Shockley
Department of Chemical and Biomolecular Engineering
North Carolina State University
Raleigh, NC
USA

keith.shockley@jax.org

Sabrina Tachdjian
Department of Chemical and Biomolecular Engineering
North Carolina State University
Raleigh, NC
USA

stachdj@ncsu.ed

Simon de Vries
Department of Biotechnology
Delft University of Technology
Delft
the Netherlands

s.devries@tudelft.nl

Harmen J.G. van de Werken
Laboratory of Microbiology
Wageningen University
Wageningen
the Netherlands

harmen.vandewerken@wur.nl

Masami Yamada
Division of Genetics and Mutagenesis
National Institute of Health Sciences
Tokyo
Japan

myamada@nihs.go.jp

Preface

The 'concept' of Archaea arose over 30 years ago when Woese and Fox (1977) proposed that prokaryotes were not a monophyletic group (single root) because of differences between their small subunit ribosomal RNA sequences. Instead, they defined two distinct evolutionary lineages represented by the Bacteria and the Archaea (formerly called archaebacteria). This distinction has since received considerable support from diverse sources. A compelling example comes from whole genome sequencing studies that reveal extensive examples of genetic conservation common to the Archaea but absent from the Bacteria and the Eukarya (eukaryotes). Archaea are subdivided into four phyla of which two, the Crenarchaeota and the Euryarchaeota, are most intensively studied. The identity and function of the conserved features of the Archaea remain enigmatic and are worthy of research endeavour.

Research studies on the Archaea place strong emphasis on taxa from extreme habitats notably environments having high temperature, extreme salinity or a lack of oxygen. For this reason, selected taxa derived from these environment have emerged as models for the overall study of archaeal biology. This emphasis prompted development of 'Archaea: New Models for Prokaryotic Biology' and the opportunity to focus on research topics of particular relevance to the biology of these organisms.

Recent developments in genomics and proteomics have accelerated comparative biochemical studies on the Archaea. This topic plays a critical role as archaeal biodiversity grows. Information that clarifies lifestyle and growth habit for selected taxa become essential when used to infer the physiology of related but uncultivated organisms. Among these we can include the low temperature Crenarchaeota that are of growing importance in soil and marine nitrogen geochemical cycling. For this reason a particular emphasis in this book (Chapters 1–4) concerns topics addressing metallobiochemistry (Bini), redox chemistry (Crane), respiration (Schroeder and de Vris), and sugar catabolism (van de Werken, Brouns and van der Oost).

A second thematic emphasis concerns information processing pathways (Chapters 5–8). These topics remain of considerable interest to archaeal researchers since they broaden the foundation for model system development. The increasing use of Archaea to investigate subjects of shared relevance with Eukarya cannot be underestimated; many successful examples are evident in the field of structural biology. Topics presented here include nucleic acid modification (Grosjean, Gupta and Maxwell), DNA replication (Berquist and Das Sarma), DNA repair (Nohmi, Yamada and Gruz), and DNA recombination (Rolfsmeier and Haseltine).

In the final third of this book, three additional topics are included that move away from the bounds of single molecules and their mechanistic detail (Chapters 9–11). Post-transcriptional regulatory processes that govern cell functions are largely unexplored in the Archaea. In some Euryarchaeota there is conservation of known mechanisms with those of Bacteria (e.g. sensor units of two component regulatory systems) yet for the Crenarchaeota there is a lack of such conserved elements. In Chapter 9, Kennelly addresses the cell biology of archaeal signal transduction. In response to alterations in signal transduction, gene expression can change. Functional genomics has become a critical tool for studies that investigate behaviour of the entire transcriptome and can link these patterns with cell physiology. In Chapter 10 (Tachdjian, Shockley, Conners and Kelly) employ whole genome expression array analyses to examine stress physiology of selected archaeal taxa. Such studies can provide insight on niche adaptation and nicely complement the final book topic that addresses true competition in a community setting. In Chapter 10, Shand and Leyva, present ongoing studies on the role of archaeal antimicrobial compounds notably peptide antibiotics.

The future of archaeal research promises to be exciting and rewarding. Many of the book contributors have provided their thoughts about this subject. Continued insight into basic research mechanisms will continue while applications will expand creating new opportunities for research and employment. The emergence of model archaeal systems and their contribution to prokaryotic biology is now established.

I would like to acknowledge the important and creative contributions of many young and gifted scientists that have worked in my laboratory as well as the National Science Foundation for its support of our research efforts. I also thank my wife and children for their continued encouragement and Mark Ettlinger for inspiration and insight into the scientific process.

Paul Blum
University of Nebraska
USA

Reference

Woese, C.R. and Fox, G.E. (1977). Phylogenetic structure of the prokaryotic domain: the primary kingdoms. Proc. Natl. Acad. Sci. USA *74*, 5088–5090.

Respiratory Pathways in Archaea

1

Imke Schröder and Simon de Vries

Abstract

Respiratory diversity allows Archaea to adapt and thrive even in the most extreme places significantly impacting the geochemistry of this planet. This chapter will summarize the current advances in research on archaeal oxygen, nitrate and Fe(III) respiration. While much is known about the bioenergetics and biochemical properties of the enzymes involved in oxygen and nitrate respiration our understanding of the mechanisms of Fe(III) respiration in the Archaea is in its infancy. In general, respiratory enzymes are greatly conserved amongst Archaea and Bacteria but certain archaeal enzymes and bioenergetic pathways exhibit features that may reflect early evolutionary variants or adaptations to extreme environments. Unique archaeal enzymes in oxygen respiration include the oxidase supercomplexes present in *Sulfolobus* species that allow for efficient electron transfer, and quinone linked N-oxide reductases with several cofactor modifications in *Pyrobaculum* species representing a simple electron transport chain. Thus, the archaeal respiratory repertoire diverges from bacterial energy-generating pathways and future discoveries of other Archaea-specific traits are likely.

Introduction

Respiratory pathways are of profound importance for most Archaea as they generate the bulk of the energy to support growth. Archaea not only respire with O_2, a trait shared with the vast majority of the Eukarya, but can also reduce diverse alternative electron acceptors such as CO_2, elemental sulphur, sulphate, nitrate, or metal ions such as ferric iron, Mn(IV), arsenate, selenate and others to generate energy for cell propagation (Table 1.1). The versatility of archaeal respiratory capability resembles closely that of the bacterial kingdom with the exception of the ability to generate methane from reduction of CO_2 with H_2 or acetate. This latter respiratory pathway is exclusively found in the methanogens of the archaeal domain (Deppenmeier, 2002; Deppenmeier, 2004; Ferry, 1999). One is left to wonder whether respiratory enzymes may have evolved in Archaea given that these organisms are thought to be most ancient (Gribaldo and Brochier-Armanet, 2006). Through lateral gene transfer genes for respiratory enzymes may have found new hosts in the bacterial domain advancing rapid adaptation and providing competitive advantage (Doolittle, 1999). On the other hand, oxidized energy substrates such as sulphate, nitrate, selenate and arsenate, which are thought to be more abundant in modern times of O_2 evolution, could have given rise to the advancement of respiratory pathways in the bacterial domain that may have been transferred to Archaea via lateral gene transfer. Thus, the origin of respiratory pathways remains a much debated topic that will not be the subject of this chapter (Castresana, 2001; Castresana and Moreira, 1999; Castresana and Saraste, 1995; Schäfer *et al.*, 1996).

The extraordinary respiratory capability of Archaea was not readily recognized when Carl Woese and Ralph Wolfe discovered that the methanogens constitute members of the distinct archaeal kingdom (Balch *et al.*, 1979; Fox *et al.*,

Table 1.1 Energy-yielding reactions in Archaea

Energy-generating reaction	Representative genera
O_2 respiration	
$2H_2 + O_2 \rightarrow 2H_2O$	*Pyrobaculum, Aeropyrum*, Pyrolobus, Thermoplasma*, Picrophilus*, Sulfolobus*, Haloferax*, Halobacterium*, Halococcus*, Haloarcula*, Natronomonas, Natronococcus*, Natronobacterium**
$2S + 3O_2 + 2H_2O \rightarrow 2H_2SO_4$	*Sulfolobus, Acidianus, Metallosphaera*
$2FeS_2 + 7O_2 + 2H_2O \rightarrow 2FeSO_4 + 2H_2SO_4$	*Metallosphaera*
$NH_4^+ + 1.5O_2 \rightarrow NO_2^- + H_2O + H^+$	Ammonium oxidizing Archaea (AOA), Strain SCM1
Nitrate respiration	
$5H_2 + 2NO_3^- + 2H^+ \rightarrow N_2 + 6H_2O$	*Pyrobaculum, Haloferax*, Haloarcula*, Ferroglobus*
$4H_2 + NO_3^- + 2H^+ \rightarrow NH_4^+ + 3H_2O$	*Pyrolobus*
Sulfate/thiosulphate respiration	
$4H_2 + H_2SO_4 \rightarrow H_2S + 4H_2O$	*Archaeoglobus*
$4H_2 + H_2S_2O_3 \rightarrow 2H_2S + 3H_2O$	*Ferroglobus, Pyrolobus*, Pyrobaculum, Pyrodictium, Archaeoglobus**
S respiration	
$H_2 + S \rightarrow H_2S$	*Pyrodictium, Thermoproteus, Thermofilum*, Pyrobaculum, Thermoplasma*, Acidianus, Stygiolobus, Igneococcus, Paleococcus*, Thermodiscus*, Desulphurococcus*, Staphylothermus**
Metal ion respiration	
$H_2 + 2Fe(III)citrate \rightarrow 2Fe(II)citrate + 2H^+$	*Pyrobaculum*
$H_2 + 3Fe_2O_3 \rightarrow 2Fe_3O_4$ (magnetite) $+ H_2O$	*Pyrobaculum, Ferroplasma, Geoglobus, Geogemma, Pyrobaculum, Paleococcus**
$H_2 + AsO_4^{2-} \rightarrow AsO_3^- + H_2O$	*Pyrobaculum*
$3H_2 + 2H^+ + SeO_4^{2-} \rightarrow Se + 4H_2O$	*Pyrobaculum*
Methanogenesis	
$4H_2 + CO_2 \rightarrow CH_4 + 2H_2O$	*Methanococcus, Methanocaldococcus, Methanothermococcus, Methanotorris, Methanothermus, Methanosphaera, Methanothermobacter, Methanopyrus, Methanoculleus, Methanosarcina,*

1977; Woese *et al.*, 1990). The mechanism of methane production from H_2 and CO_2 was intensively studied by several groups and it was initially unclear whether energy generation by methanogenesis occurred by a fermentative pathway or whether ATP was generated via a chemiosmotic mechanism. However, work by Gottschalk and Thauer's groups established that important pathway intermediates are membrane bound, generating an electrochemical gradient that consists of protons and sodium ions (Blaut and Gottschalk, 1984; Blaut *et al.*, 1992; Bott and Thauer, 1987; Lienard *et al.*, 1996; Peinemann *et al.*, 1988). Methanogenic Archaea have been isolated from very diverse environments that include rice paddy fields, deep sea hydrothermal vents and termite guts (Chaban *et al.*, 2006). Since the bioenergetics of methanogenesis has been extensively reviewed by Deppenmeier and Schäfer *et al.* (Deppenmeier, 2004; Schäfer *et al.*, 1999) it will not be discussed in this chapter.

Aerobic respiration with elemental sulphur as electron donor was recognized as early as 1970 when Brock and Darland isolated the first *Sulfolobus* and *Thermoplasma* species from acidic hot springs in Yellowstone Park and from a coal refuse pile, respectively (Brock *et al.*, 1972; Darland *et al.*, 1970). The first glimpse into the mechanism of sulphur oxidation was provided in 1986 by Emmel *et al.*, who demonstrated the direct incorporation of $^{18}O_2$ into the final product $^{18}SO_4^{2-}$, suggesting the existence of a sulphur oxygenase (Emmel *et al.*, 1986). Much of the biochemistry of sulphur oxidation was elucidated by Kletzin's group using *Acidianus ambivalens* as model organism and is reviewed in (Kletzin *et al.*, 2004). The archaeal sulphur respiratory chain significantly differs in enzyme components and chemistry from those found in the bacterial kingdom (Friedrich *et al.*, 2005; Friedrich *et al.*, 2001).

In contrast to sulphur oxidation, which is confined to a relatively closely related group of the crenarchaeaota, the ability to grow anaerobically by reduction of elemental sulphur is widespread throughout the archaeal kingdom. Because this mode of respiration appears to be so prevalent in the hyperthermophiles it was suggested that sulphur respiration could be one of the earliest energy yielding pathways, which had evolved (Fischer *et al.*, 1983). Besides Brock's laboratory, the group of Stetter pioneered the isolation of many very diverse archaeal sulphur reducers including the *Thermoproteus*, *Pyrobaculum*, *Pyrodictium*, *Stygiolobus*, *Igniococus*, *Acidianus*, and many other species (Blöchl *et al.*, 1997; Dirmeier *et al.*, 1998; Fischer *et al.*, 1983; Keller *et al.*, 1995; Segerer *et al.*, 1993; Stetter, 1988; Stetter, 2006a; Stetter, 2006b; Stetter, 1987; Völkl *et al.*, 1993). Noticeably, all sulphur-reducing isolates are hyperthermophiles, and it was not until 2004 that the first non-thermophilic sulphur-reducing archaeon, *Haloferax sulfurifontis*, was isolated (Elshahed *et al.*, 2004). Sulphate respiration is confined to the hyperthermophilic *Archaeoglobus* species with *Archaeoglobus fulgidus* as the first isolate in 1988 by Stetter (Stetter, 1988). Sulphur chemistry is complicated, resulting in various inorganic sulphur compounds that may form, dependent on the environmental pH, temperature and the presence of HS^- or sulphur oxyanions (Kletzin *et al.*, 2004; Schauder and Müller, 1993). Consequently, sulphur compound respiration is complex involving distinct enzymes specific for the terminal sulphur electron acceptor. Respiration with inorganic sulphur compounds is excellently reviewed by Kletzin and will therefore not be covered in this chapter (Kletzin *et al.*, 2004).

Only very recently, it became obvious that ammonia-oxidizing Archaea (AOA) are abundantly present in terrestrial and aqueous environments (Francis *et al.*, 2005; Leininger *et al.*, 2006; Mincer *et al.*, 2007). Using the *amoA* gene encoding a subunit of the key enzyme ammonia monooxygenase as a probe, Leininger *et al.* demonstrated that uncultivated and abundant archaeal species vastly exceed bacterial ammonia oxidizing species (Leininger *et al.*, 2006). Only few ammonia-oxidizing Archaea have been isolated to date and virtually nothing is known about the biochemistry of this process in the Archaea (Könneke *et al.*, 2005).

Brock and Gustafson discovered in 1976 that *Sulfolobus acidocaldarius* was able to grow by reducing ferric iron using elemental sulphur as electron donor (Brock and Gustafson, 1976). However, only recently, through the work of Lovley and his coworkers it became clear that iron respiration is widespread amongst the Archaea and may represent another ancient form of energy acquisition (Vargas *et al.*, 1998).

In contrast, nitrate respiration appears to be limited to a few archaeal species with *Pyrobaculum aerophilum* representing the first denitrifying archaeon that was isolated by Stetter and his students in 1993 (Völkl *et al.*, 1993). It is unclear to date why nitrate is such a commonly used electron acceptor in the bacterial domain while it appears to play only a minor role as electron acceptor for the majority of the archaeal isolates.

This chapter will mainly focus on the O_2, nitrate, and Fe(III) respiratory pathways and summarize the current state of the research in these areas. While examining all three respiratory pathways we will highlight the biochemistry of the enzymes involved and ask whether the archaeal respiratory repertoire is distinct or similar to the pathways found in Bacteria. To date, much is known about the molecular basis and biochemistry of oxygen and nitrate respiration; however, our understanding of Fe(III) respiration is still very poor.

Oxygen respiration

Diverse Archaea that include the acidophilic *Sulfolobus* and *Acidianus* species and the neutrophilic *Aeropyrum pernix* and *Pyrobaculum* species from the crenarchaeota branch have the ability to respire with oxygen (Table 1.1). The Haloarchaeota are photosynthetic members of the euryarchaeota. Bacteriorhodopsin-dependent photosynthesis occurs under O_2-limitting environmental conditions and is repressed when the O_2 tension is high (Shand and Betlach, 1991; Yang and DasSarma, 1990).

Diversity of terminal oxidases

Archaeal O_2 respiration proceeds via a membrane-bound electron transport chain that terminates in a cytochrome oxidase. Membrane-bound terminal oxidases of all organisms can be divided into two groups: (1) the superfamily of haem–copper oxidases, which also include the NO reductases (Hendriks *et al.*, 1998; Saraste and Castresana, 1994) and (2) the cytochrome *bd* quinol oxidases (Junemann, 1997; Osborne and Gennis, 1999; Yang *et al.*, 2007). Enzymes from both groups are present in the Archaea and Bacteria. Phylogenetic analyses of both groups of oxidases reveals considerable deviation from the 16sRNA tree, suggesting that a large amount of horizontal gene transfer has occurred between the Archaea and the Bacteria (Pereira *et al.*, 2001).

The physiological function of the haem–copper and cytochrome *bd* terminal oxidases is to generate an electrochemical proton gradient upon O_2 reduction, which is utilized to drive phosphorylation of ADP to ATP by the membrane-bound ATP synthase, for transport of nutrients and to energize motility. Haem–copper and cytochrome *bd* oxidases are distinguished by their bioenergetic efficiency (charges, i.e. protons, translocated across the cytoplasmic membrane per electron transferred; q/e); upon O_2 reduction the haem–copper oxidases translocate twice as many protons to the exterior of the cell than the cytochrome *bd* oxidases (Bloch *et al.*, 2004; Faxen *et al.*, 2005; Garcia-Horsman *et al.*, 1995; Gomes *et al.*, 2001; Miller and Gennis, 1985; Puustinen *et al.*, 1991; Toledo-Cuevas *et al.*, 1998; Wikström and Verkhovsky, 2002):

$$O_2 + 4e + 8H^+_{in} \rightarrow 2H_2O + 4H^+_{out}$$

q/e = 2, haem–copper oxidases

$$O_2 + 4e + 4H^+_{in} \rightarrow 2H_2O$$

q/e = 1, cytochrome *bd* oxidases

The reason for this bioenergetic discrepancy is that haem–copper oxidases couple the catalysis of O_2 reduction to the pumping of four protons from the cell interior through the enzyme complex into the small space between the cytoplasmic membrane and the archaeal cell wall (König *et al.*, 2007). In contrast, cytochrome *bd* oxidases are not proton pumps *per se*; they generate a membrane potential exclusively by consuming four protons when reducing O_2 to form water (chemical protons), a feature that is shared with the proton-pumping haem–copper oxidases.

Hydroquinones are the sole substrates for the cytochrome *bd* oxidases while hydroquinones or one-electron carriers such as *c*-type cytochromes or small blue-copper proteins might serve as electron donors to members of the haem–copper oxidases. In Archaea and Gram-positive Bacteria, the one-electron carriers appear to be membrane-bound (Mattar *et al.*, 1994; Nunoura *et al.*, 2003; Nunoura *et al.*, 2005; Schäfer *et al.*, 1999; Scharf and Engelhard, 1993). In Gram-negative Bacteria, the electron mediator can be membrane-bound or soluble, periplasmic proteins (Richardson, 2000).

Archaeal cytochrome *bd* oxidases

Cytochrome *bd* oxidase consists of two subunits encoded by the *cydAB* genes and typically have a high affinity for O_2. Genes encoding putative cytochrome *bd* oxidases are present in many Euryarchaeota including the strict aerobic *Picrophilus torridus*, the facultative anaerobic *Halobacterium* (sp. NRC-1), *Halobacterium salinarum*, *Thermoplasma volcanium*, *Thermoplasma acidophilum*, *Thermoplasma elongates*, and *Ferroplasma acidarmanus* (Fer1), and, also in the strict anaerobic *A. fulgidus*, *Methanosarcina barkeri*, and *Methanosarcina acetivorans* C2A. Amongst the Crenarchaeota several, mostly strict anaerobic thermoproteales contain putative *bd* oxidases that include, *Hyperthermus butylicus*, *Thermofilum pendens* (strain Hrk 5), *Pyrobaculum arsenaticum*, *Pyrobaculum islandicum* and *Caldivirga maquilingensis* (see Boucher *et al.* for phylogenetic analysis (Boucher *et al.*, 2003)). To date, presence of a cytochrome *bd* oxidase on the protein level has been confirmed by optical or proteomic analyses only for *H. salinarum* (Hubmacher *et al.*, 2003; Klein *et al.*, 2005; Scharf *et al.*, 1997) but no archaeal *bd* oxidase has yet been purified.

Classification and phylogeny of haem–copper oxidases

Archaea can harbour a multitude of cytochrome oxidases. In addition to the *bd* oxidases, haem–copper oxidases can be present in the same organism. Genes for haem–copper oxidases have been identified in the genomes of several Archaea and their properties will be summarized below.

Haem–copper oxidases are classified according to their haem type as cytochrome aa_3, ba_3, bo_3 or cbb_3 oxidase. *Halobacterium* sp. NRC-1, *Haloarcula marismortui* and *A. pernix* contain genes for both the aa_3 and the ba_3-type cytochrome oxidase. In contrast, *S. solfataricus* P2, *S. acidocaldarius*, *H. salinarum*, and *Haloferax volcanii* contain only the low O_2 affinity aa_3-type cytochrome oxidase. Cytochrome oxidases of the ba_3-type are present in *A. pernix*, *Natronomonas pharaonis*, *Pyrobaculum calidifontis*, *P. aerophilum* and *Pyrobaculum oguniense*. Archaeal genomes analysed to date seem to lack the high O_2 affinity cbb_3-type cytochrome oxidases.

Haem–copper oxidases consist of two to five subunits. The large subunit (SUI or Cox1) is highly hydrophobic and contains 12–15 transmembrane α-helices dependent on the enzyme. The subunit harbours three redox centres, two haems and one copper atom centre called Cu_B (Fig. 1.1). The three redox centres are coordinated by a total of six conserved histidine residues. The low spin haem is coordinated by two His, the high-spin haem by one and Cu_B by three His residues. O_2 binds to the high-spin haem–Cu_B binuclear centre where it is reduced to water (Babcock and WikstrAppl. Environ Microbiolm,

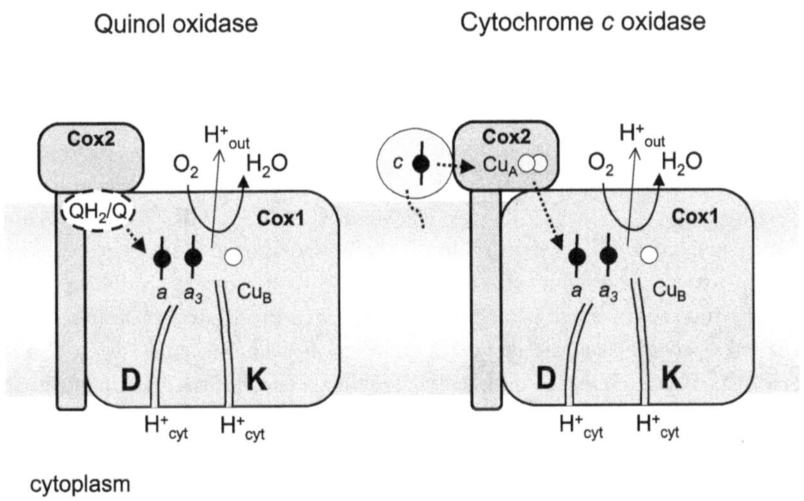

Figure 1.1 Schematic illustration of the subunit composition of quinol and cytochrome *c* oxidases. Cox1 and Cox2 refer to the conserved subunits of haem–copper oxidases carrying redox centres. The two types of oxidases differ mainly by the absence or presence of Cu_A in Cox2. The *c* cytochrome can be replaced by blue-copper proteins such as sulphocyanin or halocyanin located at the exterior face of the cytoplasmic membrane. These electron mediators are membrane associated as indicated by the wavy, dotted line for cytochrome *c*. Cu_A is indicated by two adjacent open circles indicating its di-copper structure, Cu_B by a single open circle. Haem groups are represented by a closed circle dissected by a vertical line. Proton translocation from the cytoplasm (H^+_{cyt}) occurs via well-defined D- and K-proton pathways in type A oxidases. Type B oxidases appear to lack the D-channel. The exit pathway for protons (H^+_{out}) and the precise route of oxygen to the binuclear haem a_3-Cu_B centre are not known. Water generated by the binuclear centre is expelled from the cell. Dotted arrows indicate electron-transfer pathways. Electron transfer from haem *a* to the binuclear a_3/CuB centre is not shown.

1992; Brzezinski and Larsson, 2003; Ferguson-Miller and Babcock, 1996; Ogura and Kitagawa, 2004). The high-spin haem a_3 receives electrons from the low-spin haem *b or a*, in ba_3 or aa_3 oxidases, respectively. Cox2 (SUII) mediates electrons from the electron donor to reduce the low-spin haem. Cox2 assumes a characteristic cupredoxin fold (Abramson *et al.*, 2000; Iwata *et al.*, 1995; Kelly *et al.*, 1993; Ostermeier *et al.*, 1997; Soulimane *et al.*, 2000; Wilmanns *et al.*, 1995; Yoshikawa *et al.*, 2006). When designated conserved Cys, His and Met ligands are present they coordinate the di-copper Cu_A redox centre (Fig. 1.1) (Blackburn *et al.*, 1994). The presence of these conserved amino acids in putative Cox2 polypeptides can be used to infer that the subunits contains the Cu_A site, which in turn means that this particular oxidase can be reduced by one-electron carriers such as cytochrome *c* or blue-copper protein. Conversely, when the conserved Cys, His and Met residues in Cox2 are absent, the enzyme functions as a quinol oxidase (Fig. 1.1). The precise binding site for hydroquinone in the quinol oxidases is not known but is proposed to be at the crevice of Cox1 and Cox2 (Abramson *et al.*, 2000). In fact, two quinone binding sites were postulated to reside in Cox2, a high-affinity site where quinone is permanently bound and a low-affinity site exchanging quinone with the Q-pool (Grimaldi *et al.*, 2003; Sato-Watanabe *et al.*, 1994).

While a high-resolution 3D structure for an archaeal haem oxidase has not yet been generated, the 3D structures of bacterial and mitochondrial enzymes provide accurate distances between the electron-transfer centres and have further identified two proton entry channels (H^+_{cyt}; D- and K-channels) and one proton-exit pathway in Cox1 (H^+_{out}) (Fig. 1.1) (Brzezinski and Adelroth, 1998; Iwata *et al.*, 1995; Tsukihara *et al.*, 1996). Sequence comparison of all haem–copper oxidases Cox1 subunits showed that the amino-acid residues characterizing the D- and K-proton entry channels in the crystallized enzymes are not completely conserved. This has led to a revised classification of the haem–copper oxidases in type A (subdivided into A1 and A2), type B and type C (the cbb_3 oxidases only) (Pereira *et al.*, 2001) (see also Boucher *et al.* for phylogenetic analysis of Cox1 (Boucher *et al.*, 2003)). Archaeal oxidases appear to belong either to type A1 or to type B. Both types include quinol and cytochrome *c*/blue-copper protein oxidases. The primary sequence of the D- and K-channels of the type A1 oxidases is well defined. In contrast, type B oxidases appear to lack consensus D-channel residues even though they are proton pumps and thus may have 'alternative proton channels' (Gilderson *et al.*, 2001).

Bacterial cytochrome *c* oxidases consist of four subunits even though the enzyme purified as a two-subunit complex (Cox1 and Cox2) is fully active as a proton pump (Solioz *et al.*, 1982). The function of Cox3 and Cox4 remains unclear. Cox3 is an integral membrane subunit tightly binding several phospholipids. Cox4 consists of a single transmembrane α-helix (Abramson *et al.*, 2000; Iwata *et al.*, 1995). While Cox1 and Cox2 are conserved in the Archaea, Cox3 and Cox4 can be absent or replaced by archaeal subunits. In the type A1 archaeal and (some) bacterial oxidases, the large subunit is a fusion protein of Cox1 and Cox3 peptides (Schäfer *et al.*, 1999). A phylogenetic tree of this chimeric subunit, called SoxM (see below) can be found in (Nunoura *et al.*, 2003).

Biochemical properties of *Sulfolobus* SoxM and SoxABCD oxidases

The best characterized archaeal oxidases are the SoxM and SoxABCD enzyme complexes from *S. acidocaldarius*. Both enzymes were purified in Schäfer's laboratory and have been extensively reviewed by Schäfer *et al.* (Schäfer *et al.*, 1999). *S. acidocaldarius* and other Sulfolobales are obligate aerobes that thrive optimally at a pH of 2.0–3.5 (Brock *et al.*, 1972). Moll and Schäfer demonstrated that *S. acidocaldarius* maintains a cytoplasmic pH of about 5.8–6.5 by respiration-driven proton extrusion (Lübben and Schäfer, 1989; Moll and Schäfer, 1988). Both the SoxM and SoxABCD enzyme complexes play an essential role in pH homeostasis in the Sulfolobales. While sharing some common features with bacterial cytochrome aa_3 oxidases both archaeal enzymes are uniquely different in that they form supercomplexes containing subunits reminiscent of the bacterial-type bc_1 complex (Figs 1.2 and 1.3). The bacterial bc_1 complex, which is also present in the electron transport chain of

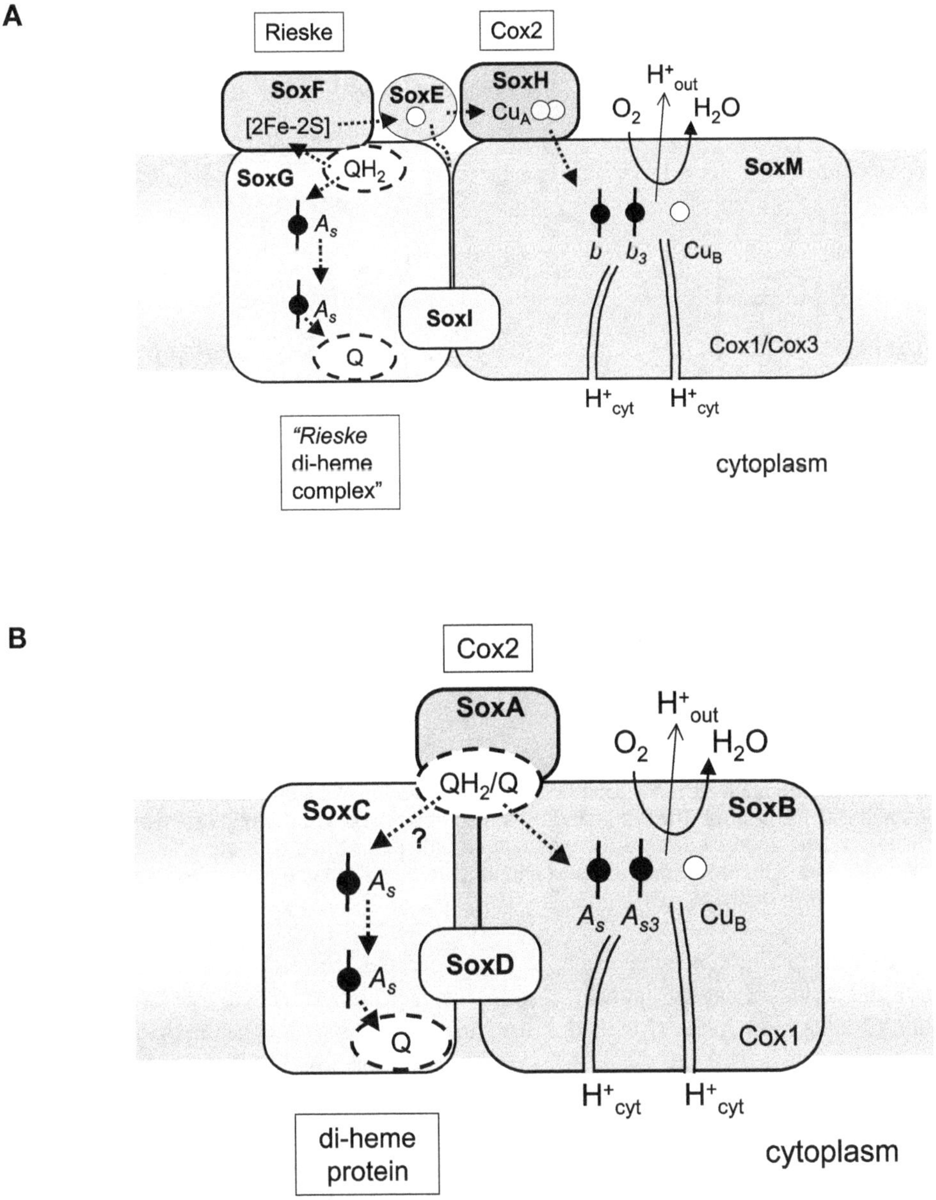

Figure 1.2 Subunit composition of the SoxM and SoxB supercomplexes from *S. acidocaldarius*. Cox1/Cox3 and Cox2 refer to the conserved subunits of haem–copper oxidases carrying redox centres. A: The SoxM supercomplex consists of an oxidase (SoxMH), the Rieske di-haem A_s subcomplex (SoxFG) and SoxE, which mediates electrons between SoxMH and SoxFG. The location and function of SoxI is not known. Operation of a Q-cycle mechanism by the Rieske di-haem A_s subcomplex requires QH_2- and Q-binding sites on opposite sides of the membrane as indicated. Electron transfer between these two quinone-binding sites is accomplished by the transmembrane arrangement of the two haems A_s in SoxG, analogous to the structure of the di-haem *b* subunit of the bacterial and mitochondrial cytochrome bc_1-complex. Upon oxidation of QH_2, the electron transfer pathway branches, one electron reducing haem A_s while the second electron reduces the Rieske [2Fe–2S] centre. The Rieske [2Fe–2S] centre donates its electrons to the membrane-bound blue-copper protein SoxE. SoxE reduces the Cu_A centre in the oxidase part of the supercomplex from which electrons flow to the binuclear b_3/Cu_B center. B: The SoxB oxidase supercomplex consists of the SoxB and SoxA oxidase component, the di-haem SoxB protein and SoxD with unknown function. A Rieske protein is absent. SoxAB functions as a quinol oxidase consistent with the absence of a Cu_A redox centre in SoxA. Both, the SoxM and the SoxB oxidase complexes can potentially generate an electrochemical proton gradient by a combination of Q-cycle and oxidase pumping mechanism as discussed in the text.

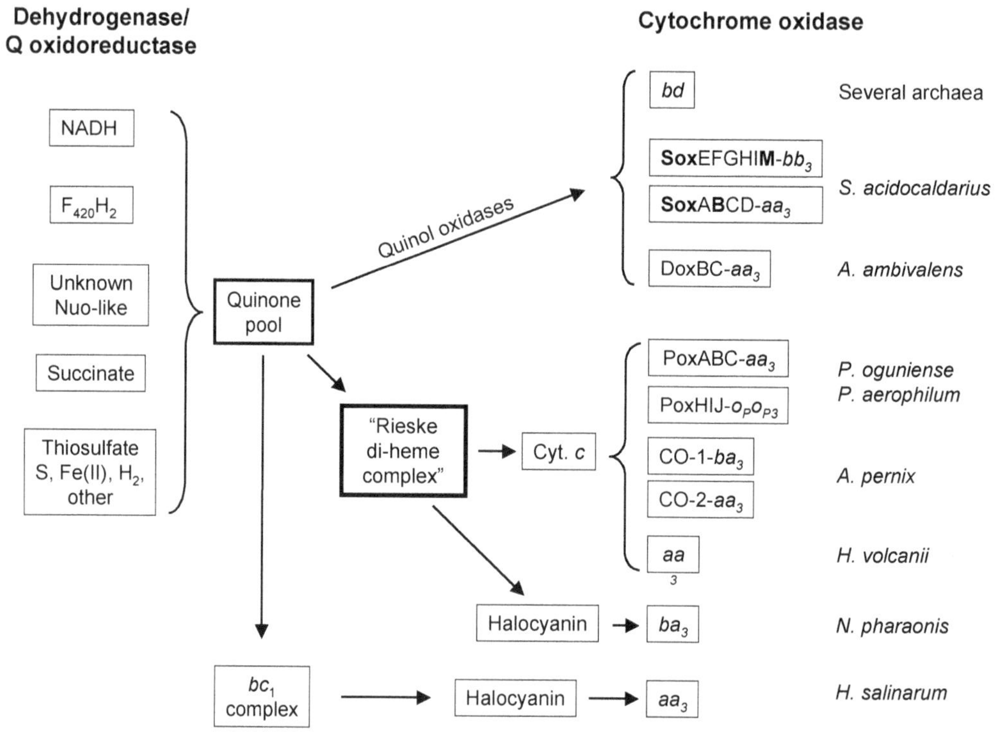

Figure 1.3 Diversity of aerobic respiratory pathways in the Archaea. The dehydrogenase indicated by 'Unknown Nuo-like' is the putative eleven-subunit complex homologous to Ndh1 of unknown function but suggested to interact with the Q-pool. The SoxM and SoxB supercomplexes and DoxBC are quinol oxidases. The archaeal 'Rieske di-haem complex' lacks a *c*-type cytochrome but is the physiological equivalent to the bacterial and mitochondrial cytochrome bc_1 complex and mediates electrons between the quinone pool and the cytochrome c/halocyanin oxidases.

mitochondria, functions as a quinol:cytochrome *c* oxidoreductase and acts as an energy transduction site via the Q-cycle mechanism (Mitchell, 1976).

In contrast to bacterial cytochrome aa_3 oxidases that receive their electrons from periplasmic cytochrome *c*, the *S. acidocaldarius* SoxM complex uses the *Sulfolobus*-specific caldariella hydroquinone, a thiophenobenzoquinone as physiological electron donor and is as such a quinol oxidase (Komorowski *et al.*, 2002). Thus, SoxM activity is not only inhibited by cyanide, a common cytochrome aa_3 oxidase inhibitor, but also by aurachins, quinone analogues shown to be inhibitors of bacterial quinol oxidases (Komorowski *et al.*, 2002). The genes for the SoxM enzyme complex are not organized in an operon but form a gene cluster comprising six genes. One of the genes encodes the large catalytic subunit SoxM that constitutes a fusion of bacterial-type Cox1 and Cox3 subunits and gives the entire enzyme complex its name (Fig. 1.2A). SoxM harbours the redox centres important for the enzyme's catalytic activity, the Cu_B redox centre and two haems. However, in place of the bacterial aa_3 haems SoxM binds haems bb_3 and is as such a bb_3-type oxidase. Nevertheless, based on amino acid sequence and COG analysis SoxM is more similar to bacterial aa_3 oxidases than to the distantly related bacterial cbb_3 oxidases. Subunit 2 (Cox2) encoded by *soxH* has been shown to coordinate the di-copper Cu_A centre. SoxH readily dissociates from the enzyme complex during purification resulting in loss of oxidase activity (Komorowski and Schäfer, 2001; Komorowski *et al.*, 2002). In place of the bacterial-type Cox4, the SoxM enzyme complex contains a SoxI subunit with yet unknown function.

The most remarkable feature of the SoxM oxidase is that it forms a supercomplex with three additional subunits, one harbouring a blue-copper centre (SoxE) and two that are homologous to the cytochrome *b* (SoxG) and Rieske [2Fe–2S] protein (SoxF) of the bacterial bc_1

complex (Fig. 1.2A). Archaea except *H. salinarum* (see below) lack a bacterial-type cytochrome c_1 homologue. SoxG contains two transmembrane arranged haems of the *a*-type. The quinol oxidation site is formed by one of the haems and the Rieske [2Fe–2S] centre (Fig. 1.2A). Upon oxidation of hydroquinone, one electron is diverted towards the Rieske [2Fe–2S] centre, the other to the nearest *a*-type haem according to the Q-cycle mechanism (de Vries *et al.*, 1981; Mitchell, 1976; Trumpower and Gennis, 1994). After this initial oxidation step, the electron travels from the *a*-type haem transmembrane to the second *a*-type haem, the energy conserving step, and is then returned to the quinone pool, hence the name Q-cycle. The Rieske [2Fe–2S] centre receives the second electron and donates this to the blue copper sulphocyanin, SoxE, which mediates the electron towards the Cu_A centre in SoxH. With respect to the bacterial and mitochondrial systems, the SoxM complex lacks both cytochromes *c* and c_1. SoxE acts as their functional replacement. Thus, this supercomplex is predicted to contain two energy conservation sites, one at the quinol oxidation site formed by the Rieske [2Fe–2S] centre and the di-haem SoxG subunit and a second via the proton pumping function of the SoxH/SoxM subunits. The overall reaction of the SoxM complex would yield the same q/e = 3 as obtained by the action of the bacterial cytochrome bc_1 complex and cytochrome aa_3 oxidase together. Unfortunately, neither kinetic nor H^+-pumping data are presently available to substantiate the bioenergetic efficiency of the SoxM complex.

The second cytochrome oxidase present in the Sulfolobales is the SoxABCD complex, based on its catalytic subunit also called SoxB (Fig. 1.2B). Like the SoxM complex, the SoxB enzyme is a quinol oxidase using caldariella hydroquinone as physiological electron donor (Lübben *et al.*, 1992). The four-subunit complex purified from *S. acidocaldarius* is encoded by the *soxABCD* operon and resembles more closely the canonical cytochrome aa_3 oxidase. However, the enzyme complex is unusual in that it lacks the Cu_A redox centre typically located in Cox2, here SoxA, and contains the di-haem SoxG-type bc_1 complex subunit, here SoxC. The large catalytic subunit I, SoxB, harbours two modified haems aa_3 and the Cu_B active site redox centre. The haems contain isoprenoid side chains that are longer by one isoprenoid unit than found in canonical *a* haems (Lübben and Morand, 1994). The modified haem *a* with a hydroxyethylgeranylgeranyl side chain was termed A_S to indicate that it was originally found in Sulfolobus but it is present also in other archaeal cytochrome oxidases (see below). Anemüller *et al.* demonstrated that the SoxB subunit alone is capable of catalysing the hydroquinone-dependent reduction of O_2 (Anemüller and Schäfer, 1990). In this light the function of SoxA is not clear but it might be to increase the binding affinity for caldariella hydroquinone and it is likely that the quinone-binding site is located at the interface between SoxB and SoxA (Fig. 1.2B). The function of SoxD remains to be determined.

In an elegant series of experiments, the SoxABCD complex was shown to be a proton pump with a stoichiometry of $1.2H^+/e$ (Gleissner *et al.*, 1997). Based on the presence of the di-haem SoxC subunit reminiscent of the bc_1 complex Gleissner *et al.* suggested the possibility of proton translocation via a Q cycle rather than a proton-pumping mechanism typically present in aa_3-type oxidases. The SoxABCD oxidase is classified as a type B oxidase lacking conserved amino acids critical for proton pumping and is thus not predicted to have a this type of proton translocation mechanism. In a Q cycle mechanism a $1H^+/e$ bioenergetic efficiency is in close agreement with the experimentally determined value. A stoichiometry of $> 2H^+/e$ would be expected if SoxB would act as a proton pump, (Schäfer *et al.*, 2002). However, this interpretation is debatable since proteoliposomes are notoriously leaky for protons and generally underestimate the H^+/e value. In addition, the *A. ambivalens* aa_3-quinol oxidase, also classified as a type B enzyme was shown by Gomes *et al.* to function as a proton pump (see below) (Gomes *et al.*, 2001). Thus, the SoxABCD oxidase could operate like the SoxM supercomplex as a combined quinone cycle/proton pump.

The *Sulfolobus* SoxM and SoxB oxidases serve as paradigms for the classification of cytochrome oxidases and an extensive phylogenetic tree illustrating the relationship of archaeal and bacterial haem–copper oxidases is presented in the paper by Nunoura *et al.* (2003). A more commonly used representation of haem–copper oxidase phylogeny is based on divergent proton translocation pathways and is presented by Pereira and Teixeira (2004).

Properties of the *A. ambivalens* DoxBC oxidase

A much simpler haem–copper oxidase is present in the facultative anaerobic *A. ambivalens*, a hyperthermophilic member of the Sulfolobales order. The purified enzyme consists of only two subunits, DoxB and DoxC, and acts as a quinol oxidase using the Sulfolobales-specific caldariella hydroquinone as physiological electron donor (Fig. 1.3) (Anemüller *et al.*, 1994; Giuffre *et al.*, 1997; Müller *et al.*, 2004; Purschke *et al.*, 1997). The catalytic subunit, DoxB, binds haems *a* and a_3 and the Cu_B redox centre and is distantly related to other haem–copper Cox1 subunits. DoxC bears no sequence homology to any oxidase subunit and its function remains unclear. Initial reports indicated a four- to five-subunit enzyme encoded by the bicistronic operon containing the *doxA* and *doxD* genes and the *doxBCEF* operon (Anemüller *et al.*, 1994; Purschke *et al.*, 1997). However, Müller *et al.* showed that DoxBC constitute the aa_3 oxidase while DoxAD functions as a thiosulphate:quinone oxidoreductase (Müller *et al.*, 2004).

Biochemical and biophysical analyses of the *A. ambivalens* cytochrome aa_3 oxidase indicate a similar turnover value (~1 ms), and similar electron transfer pathways and reaction intermediates as in bacterial and mitochondrial enzymes (Gilderson *et al.*, 2001; Pereira *et al.*, 2004). The *A. ambivalens* cytochrome aa_3 oxidase has a relatively low pH optimum and the pK_a values of the active site residues are shifted to lower pH values compared with oxidases from neutrophilic microorganisms. Although the reduced enzyme binds CO, addition of CO to the oxidized enzyme did not yield a mixed-valence enzyme, which implies that the enzyme cannot oxidize CO to CO_2, a characteristic reaction of all other oxidases studied to date (Das *et al.*, 2004).

Using reconstituted proteoliposomes, Gomes et al demonstrated that the DoxBC oxidase pumps protons with a stoichiometry of $1H^+$/electron across the lipid membrane (Gomes *et al.*, 2001). Thus, this simple oxidase that was classified originally as a non-proton pumping type B enzyme is as efficient in energy transduction as the canonical type A aa_3 oxidases. Using comparative modelling Victor *et al.* proposed alternative proton channels for the DoxBC oxidases that are also conserved in other type B enzymes (Victor *et al.*, 2004).

Properties of the *Pyrobaculum* and *Aeropyrum* cytochrome oxidases

The facultative anaerobic hyperthermophile *P. oguniense* contains two distinct cytochrome aa_3 oxidases, PoxC and PoxI, named after their respective catalytic subunits (Nunoura *et al.*, 2003; Nunoura *et al.*, 2005; Sako *et al.*, 2001). The PoxC oxidase groups to the *S. acidocaldarius* SoxM-type oxidases while the PoxI oxidase resembles more the SoxABCD-type enzyme complex (Fig. 1.3). However, while the *Sulfolobus* oxidases are quinol oxidases, the *P. oguniense* oxidases receive their electrons from a cytochrome *c* protein. The PoxC aa_3 oxidase was purified by Nunoura *et al.* and consists of three subunits encoded by the *poxABC genes.* The genes are also present in *P. aerophilum* and are part of a larger putative transcriptional unit. The catalytic subunit encoded by the *poxC* gene is, like SoxM, a chimeric protein of Cox1 and Cox3. PoxC binds the modified A_s-type haem. The *poxB* gene encodes the Cu_A-containing subunit II. Biophysical analysis of the purified PoxABC complex suggests that this enzyme does indeed contain the Cu_A redox centre, which is in contrast to the SoxM complex, which lacks Cu_A (Nunoura *et al.*, 2005). The *poxA* gene product is homologous to SoxI and its function is unclear. In contrast to the *S. acidocaldarius* SoxM oxidase complex, the *P. oguniense* PoxABC oxidase lacks the SoxEFG-type subunits and thus does not form a supercomplex with a bc_1-type protein assembly.

The *P. oguniense poxABC* genes are regulated by oxygen and are expressed optimally under atmospheric air growth conditions (Nunoura *et al.*, 2003). In contrast, the second cytochrome oxidase in *P. oguniense* encoded by *poxHIJ* is expressed constitutively. The PoxI oxidase has not yet been purified, however, haem analysis suggests the presence of modified haem *o* of the O_{P1}

type that are attributed to this enzyme (Lübben and Morand, 1994; Nunoura *et al.*, 2003).

The strictly aerobic hyperthermophile *A. pernix* contains two haem–copper oxidases that were purified by Ishikawa *et al.* (Fig. 1.3) (Ishikawa *et al.*, 2002). The CO-1 oxidase is a SoxB-type enzyme and closely related to the *P. oguniense* PoxHIJ complex. CO-1 consists of at least two subunits encoded by two independent gene loci, i.e. APE1623 and APE1720 for subunits I and II, respectively. The enzyme has the spectral properties of haems *a* and *b* and is as such classified as a cytochrome *ba*$_3$ oxidase. CO-2, the second *A. pernix* oxidase, is encoded by the *aoxABC* operon and groups phylogenetically to the SoxM-type oxidases. CO-2 exhibits its highest similarity to the *P. oguniense* PoxABC complex. AoxA (Cox2) coordinates the Cu$_A$ redox centre, AoxB is the chimeric subunit I/ subunit III peptide with the Cu$_B$ redox centre and haems *a* and *a*$_3$, and AoxC is a putative Cox4 analog. Interestingly, the activity of the CO-1 *ba*$_3$ oxidase was stimulated six-fold by 1M NaCl while CO-2 *aa*$_3$ oxidase activity increased only twofold. This led the authors to suggest that the *ba*$_3$ oxidase might act as a Na^+-pump rather than a H^+ pump. It is not clear whether this conclusion is justified as the enzyme assay was performed 60°C below the likely temperature optimum of *A. pernix* enzymes and NaCl may stabilize the enzyme to protect against cold shock.

Haloarchaeal cytochrome oxidases

Haloarchaea are facultative anaerobes that thrive in hypersaline environments. With rising temperatures and increasing salt concentrations oxygen can become readily limiting. Haloarchaeal genomes contain several cytochrome oxidase genes, however, only few enzymes have been purified thus far and all are mostly inactive. The oxidase from *H. volcanii* is a cytochrome *aa*$_3$ oxidase likely using cytochrome *c* as electron source (Fig. 1.3) (Tanaka *et al.*, 2002). The genome of *N. pharaonis* contains genes encoding three copper-haem oxidases all of which were suggested to function as halocyanin oxidases (Mattar and Engelhard, 1997; Scharf *et al.*, 1997). Halocyanins are small, blue copper electron mediator proteins that are predicted to reside in the extracellular compartment but remain tethered to the cytoplasmic membrane (Scharf and Engelhard, 1993). One of the *N. pharaonis* oxidases was reported to be a *ba*$_3$ oxidase containing the modified haem A$_s$ and its amino acid sequence suggests the presence of a Cu$_A$ centre.

H. salinarum contains three distinct oxidases that are expressed dependent on available iron concentration (Hubmacher *et al.*, 2003; Klein *et al.*, 2005). Under iron limitation, the prevalent condition at high salinity, the high-affinity cytochrome *bd* oxidase is mainly expressed. At high iron concentration cytochrome *bd* oxidase expression is decreased, and cytochrome *aa*$_3$ oxidase expression increased.

Dehydrogenases of aerobic respiratory pathways

Archaea utilize a wide variety of organic and inorganic compounds as electron sources for aerobic respiration. The following section will provide a short description of a subset of these dehydrogenases (Fig. 1.3).

NADH dehydrogenase

NADH dehydrogenases are membrane-bound enzymes and catalyse the two-electron transfer from NADH to quinone (Fig. 1.3). Three types of NADH dehydrogenases are distinguished (reviewed in (Kerscher *et al.*, 2007)): the rotenone-sensitive NADH dehydrogenase, also called complex I or Ndh1 constitutes type I and consists of 14 subunits. Ndh1 enzymes are energy transducing and pump protons across the cytoplasmic membrane. The type II class encompasses the rotenone-insensitive NADH dehydrogenase (Ndh2). They consist of one subunit and do not pump protons. The so-called Na^+-translocating NADH:quinone oxidoreductases compose type III. Archaeal genomes contain genes for all three types of NADH dehydrogenases (Melo *et al.*, 2004). However, the putative archaeal Ndh1 enzymes lack the flavoprotein subunit containing the NADH-binding site and the iron–sulphur homologues encoded by the *nqo*1–3 genes (also annotated as *nuoEFG*). This is consistent with the absence of rotenone-sensitive NADH dehydrogenase activity in Archaea (Falb *et al.*, 2005; Schäfer *et al.*, 1999). Nevertheless, homologues of *nqo*4–14 are present in many archaeal genomes and it is possible that these eleven genes encode an energy-conserving complex with yet unknown electron donor (Fig. 1.3).

Ndh2-type enzymes were purified from *A. ambivalens* and *Sulfolobus metallicus* (Bandeiras *et al.*, 2002; Bandeiras *et al.*, 2003). The archaeal Ndh2 enzymes contain FMN that is covalently bound to a conserved His residue. Both the *A. ambivalens* and *S. metallicus* NADH dehydrogenase use the *Sulfolobus*-specific caldariella quinone as electron acceptor. The genomes of *S. tokodaii* and *S. solfataricus* contain genes for two Ndh enzymes, while one enzyme is predicted to be present in *A. ambivalens, S. metallicus, Th. volcanium, Th. acidophilum* and *Halobacterium* NRC1. Ndh-2 genes are present in the *P. calidifontis*, but absent in the closely related *P. aerophilum*, consistent with the absence of NADH dehydrogenase activity in the latter organism (de Vries and Schröder, 2002).

$F_{420}H_2$ dehydrogenase

$F_{420}H_2$ dehydrogenase is exclusively found in Archaea and catalyses the oxidation of reduced cofactor F_{420} (Fig. 1.3). The enzyme uses either saturated menaquinone or methanophenazine from *A. fulgidus* and methanogens as electron acceptor, respectively (Bäumer *et al.*, 2000; Kunow *et al.*, 1994). $F_{420}H_2$ dehydrogenases have been purified from *A. fulgidus, Methanolobus tindarius* and *Methanosarcina mazei* (Bäumer *et al.*, 2000; Brüggemann *et al.*, 2000; Haase *et al.*, 1992; Kunow *et al.*, 1994). The purified enzymes form a multimeric complex of eleven subunits encoded by the *fqo* operon. The subunits form two subcomplexes, a membrane component (FqoAHJKLMN) proposed to be involved in proton pumping and a membrane-associated component (FqoBCDI) harbouring several iron–sulphur centres that mediate electron transfer between the catalytic subunit FqoF and quinone or methanophenazine. FqoF contains the $F_{420}H_2$ binding site, two distinct [4Fe–4S] centres and FAD.

$F_{420}H_2$ dehydrogenases are homologous to the FMN- and iron–sulphur-containing proton-pumping type I NADH dehydrogenases and are reviewed in (Deppenmeier, 2002; Deppenmeier, 2004; Friedrich, 2001; Friedrich and Scheide, 2000). In $F_{420}H_2$ dehydrogenases FqoF replaces the NADH-binding subunit present in Ndh-1.

The physiological roles of $F_{420}H_2$ dehydrogenases and Ndh-1 are also most likely conserved; they function to reoxidize reducing equivalents from central metabolic pathways while generating an electrochemical proton gradient. In the methanogens, the enzyme reoxidizes $F_{420}H_2$, which is the main reducing equivalent in the methanogensis pathway (Deppenmeier, 2004). Using coupled vesicles, Bäumer *et al.* demonstrated that the *M. mazei* $F_{420}H_2$ dehydrogenase functions as a proton pump with an energetic efficiency of about two H^+ translocated per two electrons transferred, which is about half of the efficiency generated by Ndh1 enzyme complexes (Bäumer *et al.*, 2000). Studies by Deppenmeier *et al.* suggested that $F_{420}H_2$ dehydrogenase forms a supercomplex with heterodisulphide reductase, an enzyme complex that reduces the heterodisulphide CoM-S-S-CoB in a dihydrophenazine-dependent manner and, thus, acts as the terminal reductase in this methanogenic electron transport chain (Deppenmeier *et al.*, 1990).

Succinate dehydrogenase

Membrane-bound succinate dehydrogenase catalyses the quinone-dependent oxidation of the citric acid cycle intermediate succinate to fumarate. Archaeal succinate dehydrogenases have been extensively reviewed by Schäfer *et al.* and will be summarized here briefly (Fig. 1.3) (Schäfer *et al.*, 2002). Based on their hydrophobic subunits and haem content succinate dehydrogenases are divided into five groups (types A–E), of which types A, C and E are present in Archaea (Hederstedt, 2002; Lancaster and Kröger, 2000; Pereira *et al.*, 2004). All succinate dehydrogenases consist of a large, catalytic subunit, SdhA, with covalently bound FAD, and a smaller subunit, SdhB, coordinating three distinct iron–sulphur centres. Succinate is oxidized at the FAD-containing active site in SdhA from where electrons are mediated by the iron–sulphur centres in SdhB to the quinone. The quinone-binding sites are located in the hydrophobic subunits, SdhC and SdhD that anchor the flavin and iron–sulphur subunits to the cytoplasmic membrane. SdhC and SdhD can coordinate either two haems (types A, B), one (type C) or no haem (type D).

The enzyme purified from *T. acidophilum* contains two haems in a single hydrophobic subunit (type C) (Bach *et al.*, 1993; Gärtner, 1991). However, the succinate:caldariella quinone oxidoreductases purified from *S. acidocaldarius*, *S. tokodaii*, and *A. ambivalens* lack the typical membrane-anchoring subunits and with that any haem (Iwasaki *et al.*, 2002; Iwasaki *et al.*, 1995; Janssen *et al.*, 1997). These type E enzymes contain two novel subunits, also called SdhC and SdhD although no sequence similarity exists to the hydrophobic C and D subunits of the canonical succinate dehydrogenases. In fact, both *Sulfolobus* SdhC and SdhD are hydrophilic in nature and it remains an open question as to how the Sulfolobales succinate dehydrogenase complex associates with the cytoplasmic membrane. *Sulfolobus* SdhC homologues are found in other archaeal and bacterial enzyme complexes including heterodisulphide reductase of methanogens (Schäfer *et al.*, 2002). The *Sulfolobus* SdhC peptide is interesting in that it contains an unusual Cys cluster arranged in tandem that was shown by Iwasaki *et al.* to coordinate a novel type of [2Fe–2S] centre (Iwasaki *et al.*, 2002). The new redox centre can be reduced by succinate but it is unclear, to date, in what way the [2Fe–2S] centre participates in the overall electron flow from succinate to quinone.

Organization of aerobic respiratory chains

The respiratory chain of *A. ambivalens* is composed of a NADH dehydrogenase of the Ndh2-type or a thiosulphate:caldariella quinone oxidoreductase (reviewed by Kletzin, 2004) and a cytochrome oxidase. In this simple electron transfer chain the dehydrogenases are linked to the terminal oxidase by the caldariella quinone pool (Fig. 1.3).

Oxidases can either be directly reduced by the Q-pool or utilize single-electron carriers such as a membrane bound *c*-type cytochrome or the blue copper proteins, sulphocyanin or halocyanin as electron donors (Fig. 1.3). Since the electron carrier proteins cannot be reduced directly by hydroquinone they require an additional hydroquinone:electron carrier oxidoreductase such as the archaeal Rieske di-haem complex or the canonical bc_1 complex. In the Sulfolobales, the Rieske di-haem complex subunits are part of the SoxM and SoxABCD supercomplexes, however, it is not clear how far this extends to other Archaea. It should be noted that the di-haem subunit can ligand haems other than of the *b*-type such as haems A_s. In contrast to most Archaea, which are not affected by myxothiazol and antimycin, respiration in *H. salinarum* is sensitive to these inhibitors, suggesting the presence of an operating canonical bc_1 complex (Sreeramulu *et al.*, 1998). *H. salinarum* contains genes for all three subunits of the bc_1 complex (*petABC*) and is, in this respect, similar to the Bacteria. A genomic survey indicates that *A. pernix*, *P. oguniense* and *P. aerophilum* contain putative genes for di-haem and the Rieske [2Fe–2S] protein. Whether these genes are expressed and assemble with the terminal oxidases to form SoxM and SoxABCD-like supercomplexes or constitute an oxidase independent Rieske di-haem complex remains to be investigated.

Nitrate respiration

To date, only few nitrate-respiring Archaea have been isolated, providing limited insight into the respiratory repertoire of these microbes. Nitrate respiration can proceed by two physiological pathways generating either ammonium (NH_4^+) or dinitrogen gas (N_2) as end products.

Pyrolobus fumarii, a deep sea hydrothermal vent isolate with optimal growth temperature of 106°C, is the sole ammonifying archaeon isolated thus far (Blöchl *et al.*, 1997). While its genome has been sequenced, the sequence is not publicly available. In Bacteria, dissimilatory ammonification is catalysed by two distinct membrane-bound enzymes, a nitrate reductase of the NarGHI type and a multihaem nitrite reductase, Nrf (Richardson, 2000; Richardson *et al.*, 2001). Both membrane-bound enzymes catalyse the following reactions:

$$NO_3^- \rightarrow NO_2^- \rightarrow NH_4^+$$

It is likely that these enzymes are also present in *P. fumarii* and they may exhibit very similar molecular properties. However, differences between bacterial and archaeal enzymes may provide important clues as to the evolution of these enzymes.

Nitrate respiration resulting in N_2 as the end product is called denitrification and is the mode of energy acquisition by several archaeal species including *P. aerophilum* and *Ferroglobus placidus* (Völkl *et al.*, 1993; Vorholt *et al.*, 1997). *P. aerophilum* is a facultative anaerobic crenarchaeote with optimum growth temperature of 100°C, and *F. placidus* is a strict anaerobic Fe^{2+} oxidizing crenarchaeote that grows optimally at 85°C. Euryarchaeal denitrifiers include the halophilic *Haloferax mediterranei, Haloarcula marismortui, Haloarcula vallismortui* and *Haloferax denitrificans* (Lledo *et al.*, 2004; Oren *et al.*, 1990; Tindall, 1989).

The denitrification pathway involves four distinct enzymes that catalyse the reduction of nitrate (NO_3^-) to dinitrogen (N_2) in a stepwise fashion (Fig. 1.4):

$$NO_3^- \rightarrow NO_2^- \rightarrow NO \rightarrow N_2O \rightarrow N_2$$

Nitrate reductase (Nar), the first pathway enzyme, is identical to the NarGHI complex of the nitrate ammonification pathway. The second pathway enzyme, nitrite reductase (Nir), is distinct from the enzyme in the ammonification pathway and reduces nitrite (NO_2^-) to nitric oxide (NO). The reduction of NO to nitrous

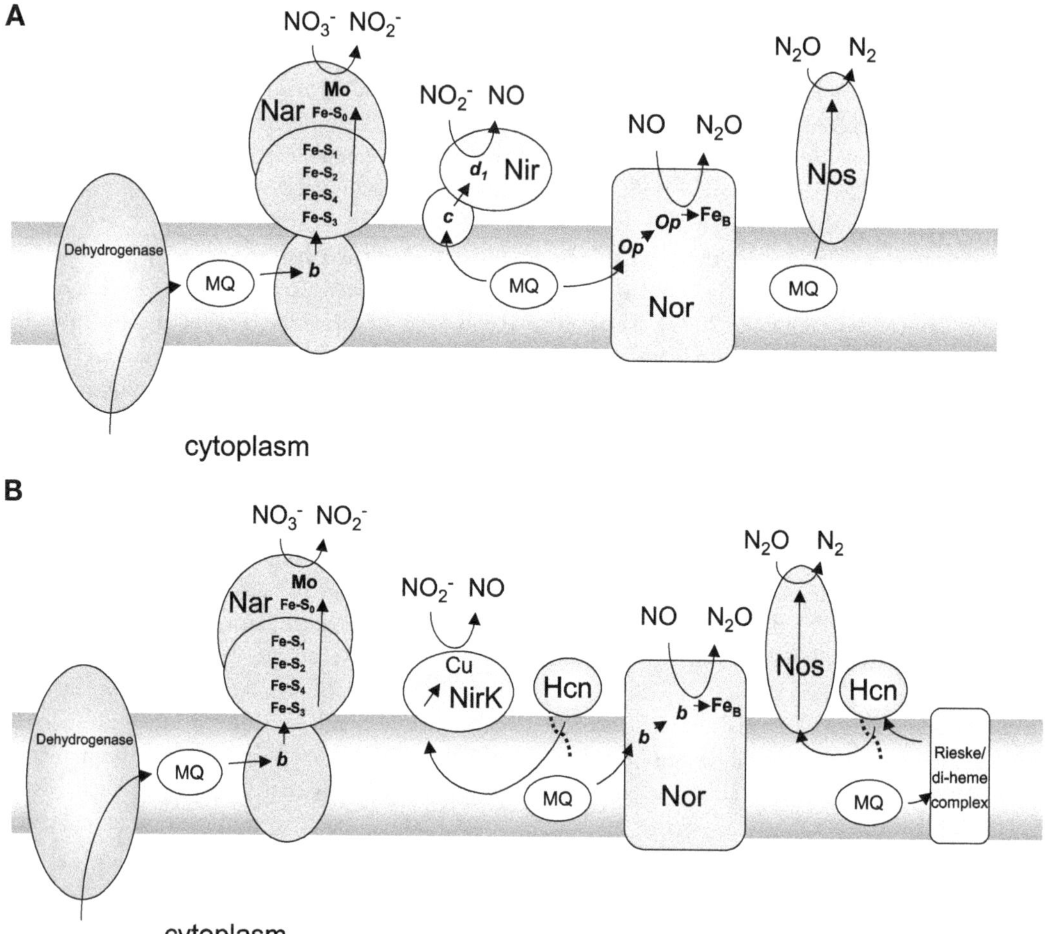

Figure 1.4 Denitrification pathway enzymes in *P. aerophilum* (A) and halophilic nitrate reducing Archaea (B). In *P. aerophilum*, all reductases are membrane-bound and interact with the quinone pool. The quinone pool is reduced by any of the dehydrogenases operating in *P. aerophilum*. In halophiles, the electron mediator to Nir, Nor and N_2OR, halocyanin (Hcn), is putative and based on the co-localization of the *hcn* gene with denitrification pathway genes. Halocyanin is predicted to be reduced by a putative the Rieske/di-haem complex as it is present in aerobic respiration. The wavy, dotted line indicates a membrane association of halocyanins in the halophiles (Mattar *et al.*, 1994).
*Use organic electron donors.

oxide (N_2O) is catalysed by NO reductase (Nor) and N_2O reductase (Nos) completes the redox pathway by generating the final product dinitrogen (N_2) (Cabello *et al.*, 2004; Philippot, 2002; Richardson and Watmough, 1999; Wasser *et al.*, 2002; Zumft, 1997).

The denitrification pathway plays an important role in the global N-cycle as it is responsible for converting the majority of the fixed nitrogen to N_2 gas. The pathway is prevalent in diverse bacterial species (for review, see Zumft, 1997). However, the bacterial pathway enzymes can be uncoupled releasing significant amounts of intermediate gasses such as the highly reactive NO and the greenhouse gas N_2O into the atmosphere (http://www.eia.doe.gov/oiaf/1605/ggrpt/nitrous.html). Thus, denitrifying Bacteria contribute considerably to greenhouse gas generation. To what extent denitrifying Archaea participate in this process is not yet understood.

Denitrification pathway enzymes

While bacterial denitrification pathway enzymes have been studied in great detail the properties of archaeal N-oxide reductases are limited to the enzymes mainly isolated from the hyperthermophile *P. aerophilum* and from the halophilic *H. marismortui*. The following section summarizes the current state of research on the archaeal denitrification pathway enzymes.

Nitrate reductase

Nitrate reductase catalyses the reduction of nitrate to nitrite using hydro(mena)quinone as the electron donor (Fig. 1.4):

$$(M)QH_2 + NO_3^- \rightarrow (M)Q + NO_2^- + H_2O$$

The enzymes are membrane-bound and are highly conserved in Archaea with respect to number of subunits and cofactors. To date, three archaeal Nars have been purified from *P. aerophilum*, *H. marismortui* and *H. mediterranei* (Afshar *et al.*, 2001; Lledo *et al.*, 2004; Yoshimatsu *et al.*, 2002). The *P. aerophilum* enzyme consists of three subunits encoded by the *narGHM* operon. The catalytic NarG subunit and the Fe–S containing NarH subunit are highly conserved with other archaeal and bacterial enzymes sharing 40–60% sequence identity. In contrast, the membrane anchor subunit NarM is conserved only amongst the archaeal Nars and has no similarity to the bacterial NarI membrane anchor. The different membrane anchors likely reflect the distinct lipid properties of the archaeal versus bacterial cytoplasmic membrane. The nitrate reductases from *H. marismortui* and *H. mediterranei* were purified as soluble, two-subunit enzymes with a $(NarGH)_2$ quarternary structure (Lledo *et al.*, 2004; Yoshimatsu *et al.*, 2002; Yoshimatsu *et al.*, 2000). It is likely that the soluble subunits were separated from the membrane anchor during the purification process.

The large catalytic subunit, NarG, harbours a molybdopterin cofactor (bis-molybdopterin guanine dinucleotide or Mo-bisMGD) where reduction of nitrate to nitrite takes place (Afshar *et al.*, 2001; Cabello *et al.*, 2004; Martinez-Espinosa *et al.*, 2001; Richardson *et al.*, 2001; Yoshimatsu *et al.*, 2002). Tightly associated with Mo-bisMGD is a modified [4Fe–4S] cluster (Fe–S0) in which three cysteine residues and one histidine serve as ligands to the four iron atoms. The Fe–S0 cluster is the direct electron donor to the Mo-bisMGD (Bertero *et al.*, 2003; Jormakka *et al.*, 2004). NarH contains three [4Fe–4S] centres and one [3Fe–4S] centre, all coordinated by conserved cysteine clusters that serve to transfer electrons from the membrane anchoring subunit to the NarG active site (Bertero *et al.*, 2003; Jormakka *et al.*, 2004).

All archaeal NarG subunits contain an N-terminal twin-arginine motive indicative for protein translocation by the TAT protein translocation pathway (Pohlschröder *et al.*, 2004). Using an enzyme assay with a membrane-impermeable dye as artificial substrate an exterior orientation of this enzyme was confirmed for the *H. marismortui* Nar (Yoshimatsu *et al.*, 2002). This orientation is distinct to all bacterial Nars whose active site faces the cytoplasm. Interestingly, all archaeal and bacterial Nars are associated with a private chaperone, NarJ (Turner *et al.*, 2004). NarJ belongs to a redox enzyme maturation protein (REMP) family of chaperones that function in folding, maturation and Mo-bisMGD cofactor assembly of Nar and other redox enzymes such as dimethyl sulphoxide (DMSO) reductase and trimethylamine (TMAO) reductase (Lanciano *et al.*, 2007; Turner *et al.*, 2004). Recent work by several groups have demonstrated that the bacterial NarJ homologues DmsD and TorD bind at the twin-Arg signal peptide of their target redox enzymes and escorts these for export to the TAT protein translocation machinery (Genest *et al.*, 2006a; Genest *et al.*, 2006b; Hatzixanthis *et al.*, 2005; Winstone *et al.*, 2006). Some bacterial NarGs still contain a rudimentary N-terminal twin-Arg motif region that is also bound by NarJ; however, this association no longer results in protein export (Chan *et al.*, 2006). Archaeal and bacterial NarJ chaperones are distantly related sharing less than 25% amino acid sequence identity. The low sequence similarity of NarJ may suggest a kingdom-specific adaptation of this chaperone. In Archaea, NarJ likely serves as an escort to the TAT translocation system whereas in Bacteria, NarJ function is restricted to NarG maturation. It seems that archaeal Nars may have been precursors to bacterial enzymes, which adapted to an interior cellular orientation of their active site. The arrangement of the Nar active site has important bioenergetic implications that are discussed below.

The *P. aerophilum* NarM membrane anchor not only differs in sequence but also in the amount of haem *b* from the bacterial membrane anchor subunit, NarI (Afshar *et al.*, 2001). While bacterial NarIs coordinate 2 haems *b* per enzyme, approximately 1 haem *b* per enzyme was determined for the *P. aerophilum* Nar. The haem *b* is presumably located at the exterior face of the cytoplasmic membrane where it is reduced by archaeal hydroquinone and mediates electrons to the Fe–S3 centre located in NarH (Fig. 1.4).

Archaeal Nars are remarkably stable. The half-life of the *P. aerophilum* nitrate reductase was 1.5 h at 100°C when the enzyme was detergent-extracted but 6 h when contained in its native membrane environment (Afshar *et al.*, 2001). Haloarchaeal Nars retained their activity in the absence of NaCl, which is unlike many other proteins from halophiles that require high salt concentrations for a properly folded state. While the *H. mediterranei* Nar activity is salt independent, the activity of the *H. marismortui* Nar is stimulated twofold in the presence of 2 M NaCl (Lledo *et al.*, 2004; Yoshimatsu *et al.*, 2002).

Archaeal Nars are similar to the bacterial Nars in that they also reduce chlorate. Reduced haem *b* of the *P. aerophilum* Nar can be oxidized by nitrate as well as by chlorate (Afshar *et al.*, 2001) The *P. aerophilum* Nar has the highest reported activity of 1130/s for nitrate and 1300/s for chlorate, both at 75°C. This activity increases two-fold at 95°C, the highest temperature experimentally accessible and close to the optimum growth temperature of the archaeon.

Nitrite reductase

Denitrifying Archaea contain one of two types of nitrite reductases. A copper-containing nitrite reductase encoded by the *nirK* gene is found in halophilic Archaea while *P. aerophilum* contains a cytochrome cd_1 nitrite reductase encoded by *nirS*. Both types of nitrite reductases are also found in the bacterial domain (Zumft, 1997).

A NirK type nitrite reductase was purified from *H. denitrificans* and *H. marismortui*. as soluble protein (Ichiki *et al.*, 2001; Inatomi and Hochstein, 1996). The *H. marismortui* enzyme constitutes a homotrimer harbouring two copper atoms per monomer. The type I blue copper is the site of electron entry and transfers the electrons to the type II (non-blue) copper site in the enzyme where reduction of nitrite to nitric oxide takes place. Both halophilic Nirs exhibit their optimal activity in the presence of 2 M NaCl, which corresponds to approximately 600 NO_2^-/s for the *H. marismortui* enzyme.

The *nirK* gene is highly conserved amongst halophilic Archaea including *H. marismortui*, *H. denitrificans*, *Haloferax* sp. (strain Aa2.2) and *N. pharaonis* displaying about 60% amino acid sequence identity. The two copper atoms are coordinated by two conserved His, one Cys and one Met ligand for the (blue) type I Cu, and three conserved His and one solvent ligand for the type II Cu. This type of copper coordination is shared with bacterial Nirs of the NirK type.

To date, the physiological electron donor for the halophilic Nirs is not known. However, all halophilic genomes contain blue copper halocyanin genes, *hcp*, that, in some instances, cluster with the *nirK* gene. In Bacteria, these types of proteins are soluble and periplasmic, and mediate electrons to periplasmic, soluble Nirs and it is likely that halocyanin mediates electrons to the denitrification pathways enzymes in halophilic Archaea (Fig. 1.4A).

$$Hcp_{red} + NO_2^- + 2H^+ \rightarrow Hcp_{ox} + NO + H_2O$$

All halocyanins and archaeal NirK enzymes contain the signature sequence for TAT-dependent protein translocation. The *H. marismortui* NirK enzyme was purified as a soluble protein lacking the N-terminal portion with the twin-Arg motif (Ichiki *et al.*, 2001). Mattar *et al.* showed that *N. pharonis* halocyanin is cleaved at its N-terminus and subsequently lipidated with a diphytanyl glycerol diether (Mattar *et al.*, 1994). This lipid serves to anchor halocyanin to the archaeal membrane and it is feasible that other translocated enzymes such as NirK may receive similar modifications.

The cytochrome cd_1 type NirS present in *P. aerophilum* is bound to the cytoplasmic membrane and utilizes menahydroquinone as the physiological electron donor (Fig. 1.4A) (de Vries and Schröder, 2002).

$$MQH_2 + 2NO_2^- + 2H^+ \rightarrow MQ + 2NO + 2H_2O$$

This property greatly distinguishes this enzyme from all other archaeal and bacterial Nirs that depend on protein electron donors. An N-terminal twin-Arg motif predicts an exterior location for this enzyme. Unlike bacterial *nirS* genes that encode a single polypeptide with an N-terminal cytochrome *c* and a C-terminal cytochrome d_1 domain, the *P. aerophilum nirS* gene is encoded by two genes, *nirSc* for the cytochrome *c* subunit and *nirSd* for the cytochrome d_1 subunit. Both NirSc and NirSd are only distantly related to the corresponding domains of bacterial NirS enzymes sharing approximately 25% amino acid sequence identity.

Nitric oxide reductase

To date, three distinct nitric oxide reductases (Nors) are distinguished, a two subunit cytochrome *bc*-containing cNor that is commonly found in Gram-negative denitrifiers, a one subunit quinol-dependent qNor present in few Bacteria including pathogens and a quinol- and cytochrome c-dependent two subunit qCuANor, thus far only found in the Gram-positive *Bacillus azotoformans* (Cramm *et al.*, 1999; de Vries *et al.*, 2003; Girsch and deVries, 1997; Suharti *et al.*, 2001). Of these, only the qNor-type NO reductase is present in Archaea and may represent the archaic precursor to this family of enzymes.

Archaeal and bacterial Nors are distant members of the haem–copper oxidase superfamily and most closely related to the bacterial *cbb*$_3$-type oxidases. The overall structure of the main catalytic subunit NorB is similar to Cox1 of the oxidases in terms of the number transmembrane helices (12) and ligation of the haem and non-haem iron redox cofactors. Three His coordinate the two haems and three ligand the non-haem Fe atom (de Vries and Schröder, 2002; de Vries *et al.*, 2003; Hendriks *et al.*, 1998; Hendriks *et al.*, 2000; Richardson and Watmough, 1999; Wasser *et al.*, 2002). In contrast to the oxidases, Nors do not pump protons or generate a proton electrochemical gradient. In the reaction catalysed by the *P. aerophilum* Nor periplasmic protons produced by MQH_2 oxidation are consumed in the formation of water:

$$MQH_2 + 2NO \rightarrow MQ + N_2O + H_2O$$

The menahydroquinone-dependent qNor purified from *P. aerophilum* is a monomeric, integral membrane protein encoded by the *norB* gene (de Vries *et al.*, 2003).

The genes for qNor-type enzymes are also present in *H. marismortui*, and *S. solfataricus* although the latter archaeon is a strict aerobic sulphur oxidizer and not known to reduce N-oxides. The *P. aerophilum* qNOR is distinct from bacterial Nors in that the enzyme harbours two modified haems of the *o*-type, Op1 and Op2, in place of the haems *b* in bacterial Nors (de Vries *et al.*, 2003; Lübben and Morand, 1994). These haems were also attributed to the *P. oguniense* PoxHIJ oxidase. One of the *P. aerophilum* Nor haems is low spin and involved in electron transfer from menaquinol to the second, high-spin haem. The latter haem, together with the non-haem iron centre (Fe_B), comprises the enzyme's active site. The Fe_B centre and the high-spin haem *b* form a functional binuclear iron–iron centre where NO binding and reduction occurs (Wasser *et al.*, 2002) (Moenne-Loccoz, 2007; Wasser *et al.*, 2002). This arrangement of a haem *b*/Fe_B binuclear centre is the functional equivalent to the haem *a*/Cu_B centre in the cytochrome oxidases (Fig. 1.1), where O_2 binding and reduction occurs.

The discovery that Nors catalyse oxygen reduction (Butland *et al.*, 2001; Fujiwara and Fukumori, 1996) and some oxidases have rudimentary NO reduction activity (Giuffre *et al.*, 1999) extends the structural similarity to the functional level and suggests that both types of enzymes may have evolved from a common ancestor. This ancestral protein may have been an anaerobic, NO reducing enzyme, which could have functioned to detoxify environmental NO, a component of volcanic gas emissions (Hendriks *et al.*, 1998; Hendriks *et al.*, 2000; Richardson, 2000; Wasser *et al.*, 2002).

Nitrous oxide reductases

Information of archaeal nitrous oxide reductases, Nos, is scarce. To date, no archaeal Nos enzyme has been purified; however, a bacterial type *nosZ* gene is present in the *H. marismortui* genome. The putative *H. marismortui* Nos exhibits 47% sequence identity with bacterial Nos and contains the conserved His, Cys and Met residues required to coordinate the Cu_A and Cu_Z redox sites present in Nos enzymes (Charnock *et al.*, 2000; Haltia *et al.*, 2003; Wasser *et al.*, 2002; Zumft and Kroneck, 2007). Bacterial Nos contain multiple copper atoms, two are organized in the Cu_A centre and transfer electrons from the physiological electron donor to the Cu_Z centre at the active site. The Cu_Z centre consists of four copper atoms and one inorganic sulphur atom and is responsible for the reduction of N_2O to N_2. The *H. marismortui* Nos is predicted to be a soluble protein with a twin-Arg motif targeting the protein for TAT-dependent protein translocation. Whether lipid modification tethers the enzyme to the exterior of the cytoplasmic membrane has yet to be established.

The *P. aerophilum* genome, in contrast, lacks a *nosZ*-type gene, suggesting yet a different class of Nos enzyme not present in Bacteria. Unlike bacterial Nos, the *P. aerophilum* enzyme has been shown to be membrane bound and reduces N_2O with menahydroquinone as physiological electron donor (de Vries and Schröder, 2002). However, the identity of this enzyme is still elusive.

P. aerophilum Nos: $MQH_2 + N_2O \rightarrow MQ + N_2 + H_2O$

Archaeal denitrification bioenergetics

In contrast to other respiratory pathways that employ a single terminal reductase, the denitrification pathway consists of four reductase complexes that each function as electron acceptor in N-oxide respiration (Fig. 1.4). While all enzymes are distinct from one another they have to be kinetically coupled as the product of one enzyme's activity is the substrate for the subsequent reaction. To date, very little is known about the kinetic properties and coupling of archaeal denitrification enzymes.

Denitrifying Archaea are less well explored, with *P. aerophilum* and *H. marismortui* serving as the sole model organisms for understanding the biochemistry of denitrification electron transport. All archaeal denitrification pathway enzymes are suggested to be located on the exterior face of the cytoplasmic membrane and are either tightly associated with or modified and tethered to the membrane surface (Fig. 1.4). This is unlike bacterial denitrification pathway enzymes that include an interior location of Nar and soluble Nir and Nos enzymes. Membrane localization was confirmed experimentally for all *P. aerophilum* denitrification enzymes (Afshar *et al.*, 1998; de Vries and Schröder, 2002). The *P. aerophilum* electron transport chain is greatly simplified as it lacks the bc_1 complex and periplasmic, electron-mediating proteins such as cytochrome *c* or copper-containing (pseudo) azurin, which are obligatory components of Gram-negative denitrification pathway (Zumft, 1997). Instead, all reductases in *P. aerophilum* receive their electrons from the quinone pool (Fig. 1.4A) (de Vries and Schröder, 2002).

In contrast, halophilic denitrifiers may share features more commonly found in bacterial denitrifiers (Fig. 1.4B). The *H. marismortui* genome contains genes for the di-haem *b* and Rieske [2Fe–2S] proteins that cluster with one of the copies for halocyanin. Whether halocyanin is indeed the physiological electron donor to Nir and Nos remains to be established. However, the possible involvement of the di-haem *b* and Rieske [2Fe–2S] proteins may provide haloarchaea with the bioenergetical advantage of an additional proton translocation site.

Interestingly, the active sites of all four *P. aerophilum* reductases are predicted to face the outside of the cell. This topological arrangement eliminates the need for nitrate uptake in Archaea, which is accomplished by dedicated secondary transporters of the NarK family in Bacteria (Moir and Wood, 2001). However, with an exterior positioning of the archaeal Nar active site the enzyme joins the other electroneutral denitrification pathway reductases in that it receives its electrons and protons at the same face of the cytoplasmic membrane. Thus, none the denitrification pathway enzymes contribute to the generation of the proton motive force. Instead, the reductases function as electron sink for various dehydrogenases. To date, little is known about the nature of these dehydrogenases in the denitrifying Archaea. Their diversity and abundance may depend on the environmental substrate availability including oxygen.

Fe(III) respiration

With the discovery that the ability to reduce Fe(III) is widespread among hyperthermophilic Archaea the significance of archaeal Fe(III) respiration has been realized only in the recent years (Vargas *et al.*, 1998). Not all Fe(III) reducing Archaea are capable of coupling this process to the generation of energy, however, diverse euryarchaeota and crenarchaeota were isolated that are either strictly dependent on Fe(III) for growth or utilize Fe(III) as alternative electron acceptor (Table 1.1) (reviewed in Weber *et al.*, 2006).

Fe(III) respiration combines many fascinating and challenging attributes. One of its interesting aspects is the fact that albeit its abundance, iron rapidly forms insoluble precipitates under aerobic and neutral conditions and is, thus, not readily available as energy substrate. Fe(III) respiring Archaea have to actively move towards and/or attach themselves to the ferric iron precipitates. Alternatively, for *P. aerophilum* it has been shown that direct contact to the oxidized metal was not required to support Fe(III)-dependent growth (Feinberg and Holden, 2006). Electron transfer in this archaeon may be mediated via a secreted yet unidentified compound that is produced more abundantly by *P. aerophilum* when grown with poorly crystalline Fe(III) oxide. Similar strategies have been observed for Fe(III) respiring Bacteria that have been examined in much more detail (Lovley *et al.*, 2004; Weber *et al.*, 2006).

Yet another interesting feature of Fe(III) respiration is its potentially broad impact on the chemistry of other compounds including metals in the environment. Once Fe(III) is reduced to the Fe(II) form, the highly reactive ferrous iron can react abiotically with organic compounds or metals in the anaerobic environment. For example, Fe(III) reducers can rapidly reduce Mn(IV) to Mn(II) abiotically while recycling the iron from the ferrous back to the ferric state (Lovley *et al.*, 2004).

Fe(III) reduction is an anaerobic respiratory mode, yet Fe(III) oxides are largely formed in aerobic environments. Ferric oxides may distribute over time into anaerobic environments. Alternatively, archaeal Fe(III) reducers may thrive at the oxic/anoxic interface to gain access to ferric oxides. In addition, many Archaea have flexible catabolic capacities that allow them to switch to Fe(III) respiration once more favourable electron acceptors such as oxygen are depleted from the environment. Indeed, *P. aerophilum* can respire with oxygen and nitrate; however, in the absence of other electron acceptors the organism switches to Fe(III) reduction when ferric citrate is present (Feinberg and Holden, 2006). On the other hand, Fe(III) sulphides (Fe_2S_3) are readily abundant in sulphur-rich anaerobic environments with pyrite (FeS_2) being only one of its various forms. Hydrogen sulphide emitted from deep-sea hydrothermal vents rapidly reduces various metals including Fe(II) to form vent chimney precipitates.

In general, iron is present only in traces in oceanic environments; however, it is highly abundant in soils and sediments. Here, it has the potential to serve as electron acceptor for the oxidation of organic compounds and thus may contribute significantly to the turnover of organic carbon material. To what respect archaeal Fe(III) reducers contribute to this process is not yet understood.

To date the mechanism of archaeal Fe(III) reduction remains elusive. Electron transport activity with Fe(III) citrate and H_2 as electron donor was demonstrated using washed cell suspensions of *P. islandicum* and *P. aerophilum* (Vargas *et al.*, 1998). However, this activity was lost upon cell breakage. The identification of Fe(III) reductase is especially difficult as the only assay currently employed uses NAD(P)H as electron donor and may not distinguish between the respiratory and an assimilatory type Fe(III) reductase. Thus, *P. islandicum* and *P. aerophilum* exhibit mostly soluble NADPH-dependent Fe(III) citrate reductase activity with specific activities of 1.5 and 0.7 U/mg (determined 25°C below the optimum growth temperature for both Archaea to minimize abiotic Fe(III) reduction) (Childers and Lovley, 2001; Feinberg and Holden, 2006). The NAD(P)H-dependent *Archaeoglobus fulgidus* Fe(III) reductase exhibits comparable Fe(III) reductase activity (1.2 U/mg assayed 8°C below the optimum growth temperature), and constitutes close to 1% of the soluble *A. fulgidus* protein (Vadas, 1999). In *A. fulgidus*, the close proximity of the Fe(III) reductase gene to the gene for iron storage protein, ferritin, suggests an assimilatory function for this enzyme. Whether the *Pyrobaculum* enzymes serve an assimilatory or dissimilatory function has yet to be established.

Conclusion

Research over the past decades has shown that archaeal respiration closely matches that of the Bacteria, a fact that made the Archaea with the exception of the methanogens initially indistinguishable from the Bacteria. This catabolic diversity allows the Archaea to thrive in even the most extreme environments. Many of the catabolic pathways can be readily identified through genome analyses exploiting the wealth of information on bacterial respiratory pathways. However, the many advances in exploring archaeal respiratory pathways have led to unexpected discoveries of unique archaeal respiratory signatures such as the formation of supercomplexes in oxygen respiration and methanogensis, the outward orientation of nitrate reductase and the simple menaquinol-linked electron transport chain in archaeal denitrification, unique archaeal enzymes in sulphur oxidation and reduction (Kletzin *et al.*, 2004) and haem modifications of respiratory enzymes. While we have significant insight into some of the respiratory pathways, archaeal metal ion reduction remains to be a challenge. In addition to enzyme purification a combination of tools that include comparative genomics and proteomics combined with a systems biology approach will help to decipher pathway components. Since most respiratory enzymes are membrane bound, contain unique cofactors, and are possibly post-translationally modified it will remain extremely difficult to elucidate their function by overexpression in heterologous bacterial hosts. Therefore, a future challenge will be to develop genetic systems for model Archaea to identify respiratory pathway components through gene replacement techniques. Yet another open question is how Archaea adapt to changing energy substrates in their environment. In some Archaea significant advances have been made using global expression profiling or proteomics approaches, however, little is known about the impact of oxygen, nitrate, sulphur, iron and other energy substrates on respiratory pathway gene expression. The understanding of respiratory pathways and how these are regulated is of prime importance as it is the activity of microbial respiration that shapes and balances our planet.

References

Abramson, J., Riistama, S., Larsson, G., Jasaitis, A., Svensson-Ek, M., Laakkonen, L., Puustinen, A., Iwata, S., and Wikström, M. (2000). The structure of the ubiquinol oxidase from *Escherichia coli* and its ubiquinone binding site. Nat. Struct. Biol. *7*, 910–917.

Afshar, S., Johnson, E., de Vries, S., and Schröder, I. (2001). Properties of a thermostable nitrate reductase from the hyperthermophilic archaeon *Pyrobaculum aerophilum*. J. Bacteriol. *183*, 5491–5495.

Afshar, S., Kim, C., Monbouquette, H.G., and Schröder, I. (1998). Effect of tungstate on nitrate reduction by the hyperthermophilic archaeon *Pyrobaculum aerophilum*. Appl. Environ. Microbiol. *64*, 3004–3008.

Anemüller, S., and Schäfer, G. (1990). Cytochrome aa_3 from *Sulfolobus acidocaldarius*. A single-subunit, quinol-oxidizing archaebacterial terminal oxidase. Eur. J. Biochem. *191*, 297–305.

Anemüller, S., Schmidt, C.L., Pacheco, I., Schäfer, G., and Teixeira, M. (1994). A cytochrome aa_3-type quinol oxidase from *Desulfurolobus ambivalens*, the most acidophilic archaeon. FEMS Microbiol. Lett. *117*, 275–280.

Babcock, G.T., and Wikström, M. (1992). Oxygen activation and the conservation of energy in cell respiration. Nature *356*, 301–309.

Bach, M., Reilander, H., Gärtner, P., Lottspeich, F., and Michel, H. (1993). Nucleotide sequence of a putative succinate dehydrogenase operon in *Thermoplasma acidophilum*. Biochim. Biophys. Acta *1174*, 103–107.

Balch, W.E., Fox, G.E., Magrum, L.J., Woese, C.R., and Wolfe, R.S. (1979). Methanogens: reevaluation of a unique biological group. Microbiol. Rev. *43*, 260–296.

Bandeiras, T.M., Salgueiro, C., Kletzin, A., Gomes, C.M., and Teixeira, M. (2002). *Acidianus ambivalens* type-II NADH dehydrogenase: genetic characterization and identification of the flavin moiety as FMN. FEBS Lett. *531*, 273–277.

Bandeiras, T.M., Salgueiro, C.A., Huber, H., Gomes, C.M., and Teixeira, M. (2003). The respiratory chain of the thermophilic archaeon *Sulfolobus metallicus*: studies on the type-II NADH dehydrogenase. Biochim. Biophys. Acta *1557*, 13–19.

Bäumer, S., Ide, T., Jacobi, C., Johann, A., Gottschalk, G., and Deppenmeier, U. (2000). The $F_{420}H_2$ dehydrogenase from *Methanosarcina mazei* is a redox-driven proton pump closely related to NADH dehydrogenases. J. Biol. Chem. *275*, 17968–17973.

Bertero, M.G., Rothery, R.A., Palak, M., Hou, C., Lim, D., Blasco, F., Weiner, J.H., and Strynadka, N.C. (2003). Insights into the respiratory electron transfer pathway from the structure of nitrate reductase A. Nat. Struct Biol. *10*, 681–687.

Blackburn, N.J., Barr, M.E., Woodruff, W.H., van der Oost, J., and de Vries, S. (1994). Metal-metal bonding in biology: EXAFS evidence for a 2.5 Å copper-copper bond in the Cu_A center of cytochrome oxidase. Biochemistry *33*, 10401–10407.

Blaut, M., and Gottschalk, G. (1984). Coupling of ATP synthesis and methane formation from methanol and molecular hydrogen in *Methanosarcina barkeri*. Eur. J. Biochem. *141*, 217–222.

Blaut, M., Müller, V., and Gottschalk, G. (1992). Energetics of methanogenesis studied in vesicular systems. J. Bioenerg. Biomembr. *24*, 529–546.

Bloch, D., Belevich, I., Jasaitis, A., Ribacka, C., Puustinen, A., Verkhovsky, M.I., and Wikström, M. (2004). The catalytic cycle of cytochrome *c* oxidase is not the sum of its two halves. Proc. Natl. Acad. Sci. USA *101*, 529–533.

Blöchl, E., Rachel, R., Burggraf, S., Hafenbradl, D., Jannasch, H.W., and Stetter, K.O. (1997). *Pyrolobus fumarii*, gen. and sp. nov., represents a novel group of Archaea, extending the upper temperature limit for life to 113 degrees C. Extremophiles *1*, 14–21.

Bott, M., and Thauer, R.K. (1987). Proton-motive-force-driven formation of CO from CO_2 and H_2 in methanogenic Bacteria. Eur. J. Biochem. *168*, 407–412.

Boucher, Y., Douady, C.J., Papke, R.T., Walsh, D.A., Boudreau, M.E., Nesbo, C.L., Case, R.J., and Doolittle, W.F. (2003). Lateral gene transfer and the origins of prokaryotic groups. Annu. Rev. Genet. *37*, 283–328.

Brock, T.D., Brock, K.M., Belly, R.T., and Weiss, R.L. (1972). *Sulfolobus*: a new genus of sulfur-oxidizing Bacteria living at low pH and high temperature. Arch. MikroBiol. *84*, 54–68.

Brock, T.D., and Gustafson, J. (1976). Ferric iron reduction by sulfur- and iron-oxidizing Bacteria. Appl. Environ. Microbiol. *32*, 567–571.

Brüggemann, H., Falinski, F., and Deppenmeier, U. (2000). Structure of the $F_{420}H_2$:quinone oxidoreductase of *Archaeoglobus fulgidus* identification and overproduction of the $F_{420}H_2$-oxidizing subunit. Eur. J. Biochem. *267*, 5810–5814.

Brzezinski, P., and Adelroth, P. (1998). Pathways of proton transfer in cytochrome *c* oxidase. J. Bioenerg. Biomembr. *30*, 99–107.

Brzezinski, P., and Larsson, G. (2003). Redox-driven proton pumping by heme–copper oxidases. Biochim. Biophys. Acta *1605*, 1–13.

Butland, G., Spiro, S., Watmough, N.J., and Richardson, D.J. (2001). Two conserved glutamates in the bacterial nitric oxide reductase are essential for activity but not assembly of the enzyme. J. Bacteriol. *183*, 189–199.

Cabello, P., Roldan, M.D., and Moreno-Vivian, C. (2004). Nitrate reduction and the nitrogen cycle in Archaea. Microbiology *150*, 3527–3546.

Castresana, J. (2001). Comparative genomics and bioenergetics. Biochim. Biophys. Acta *1506*, 147–162.

Castresana, J., and Moreira, D. (1999). Respiratory chains in the last common ancestor of living organisms. J. Mol. Evol. *49*, 453–460.

Castresana, J., and Saraste, M. (1995). Evolution of energetic metabolism: the respiration-early hypothesis. Trends Biochem. Sci. *20*, 443–448.

Chaban, B., Ng, S.Y., and Jarrell, K.F. (2006). Archaeal habitats – from the extreme to the ordinary. Can. J. Microbiol. *52*, 73–116.

Chan, C.S., Howell, J.M., Workentine, M.L., and Turner, R.J. (2006). Twin-arginine translocase may have a role in the chaperone function of NarJ from *Escherichia coli*. Biochem. Biophys. Res. Commun. *343*, 244–251.

Charnock, J.M., Dreusch, A., Korner, H., Neese, F., Nelson, J., Kannt, A., Michel, H., Garner, C.D., Kroneck, P.M.H., and Zumft, W.G. (2000). Structural investigations of the Cu-A centre of nitrous oxide reductase from *Pseudomonas stutzeri* by site-directed mutagenesis and X-ray absorption spectroscopy. Eur. J. Biochem. *267*, 1368–1381.

Childers, S.E., and Lovley, D.R. (2001). Differences in Fe(III) reduction in the hyperthermophilic archaeon, *Pyrobaculum islandicum*, versus mesophilic Fe(III)-reducing Bacteria. FEMS Microbiol. Lett. *195*, 253–258.

Cramm, R., Pohlmann, A., and Friedrich, B. (1999). Purification and characterization of the single-component nitric oxide reductase from *Ralstonia eutropha* H16. FEBS Lett. *460*, 6-10.

Darland, G., Brock, T.D., Samsonoff, W., and Conti, S.F. (1970). A thermophilic, acidophilic mycoplasma isolated from a coal refuse pile. Science *170*, 1416-1418.

Das, T.K., Gomes, C.M., Bandeiras, T.M., Pereira, M.M., Teixeira, M., and Rousseau, D.L. (2004). Active site structure of the aa_3 quinol oxidase of *Acidianus ambivalens*. Biochim. Biophys. Acta *1655*, 306-320.

de Vries, S., Albracht, S.P., Berden, J.A., and Slater, E.C. (1981). A new species of bound ubisemiquinone anion in QH_2: cytochrome *c* oxidoreductase. J. Biol. Chem. *256*, 11996-11998.

de Vries, S., and Schröder, I. (2002). Comparison between the nitric oxide reductase family and its aerobic relatives, the cytochrome oxidases. Biochem. Soc. Trans *30*, 662–667.

de Vries, S., Strampraad, M.J., Lu, S., Moenne-Loccoz, P., and Schröder, I. (2003). Purification and characterization of the MQH2: NO oxidoreductase (qNOR) from the hyperthermophilic Archaeon *Pyrobaculum aerophilum*. J. Biol. Chem.

Deppenmeier, U. (2002). The unique biochemistry of methanogenesis. Prog. Nucleic Acid Res. Mol. Biol. *71*, 223–283.

Deppenmeier, U. (2004). The membrane-bound electron transport system of *Methanosarcina* species. J. Bioenerg. Biomembr. *36*, 55–64.

Deppenmeier, U., Blaut, M., Mahlmann, A., and Gottschalk, G. (1990). Reduced coenzyme F_{420}:heterodisulfide oxidoreductase, a proton-translocating redox system in methanogenic Bacteria. Proc. Natl. Acad. Sci. USA *87*, 9449–9453.

Dirmeier, R., Keller, M., Hafenbradl, D., Braun, F.J., Rachel, R., Burggraf, S., and Stetter, K.O. (1998). *Thermococcus acidaminovorans* sp. nov., a new hyperthermophilic alkalophilic archaeon growing on amino acids. Extremophiles *2*, 109–114.

Doolittle, W.F. (1999). Phylogenetic classification and the universal tree. Science *284*, 2124–2129.

Elshahed, M.S., Savage, K.N., Oren, A., Gutierrez, M.C., Ventosa, A., and Krumholz, L.R. (2004). *Haloferax sulfurifontis* sp. nov., a halophilic archaeon isolated from a sulfide- and sulfur-rich spring. Int. J. Syst Evol. Microbiol. *54*, 2275–2279.

Emmel, T., Sand, W., König, W.A., and Bock, E. (1986). Evidence for the existence of a sulphur oxygenase in *Sulfolobus brierleyi*. J. Gen. Microbiol. *132*, 3415–3420.

Falb, M., Pfeiffer, F., Palm, P., Rodewald, K., Hickmann, V., Tittor, J., and Oesterhelt, D. (2005). Living with two extremes: conclusions from the genome sequence of *Natronomonas pharaonis*. Genome Res. *15*, 1336-1343.

Faxen, K., Gilderson, G., Adelroth, P., and Brzezinski, P. (2005). A mechanistic principle for proton pumping by cytochrome *c* oxidase. Nature *437*, 286-289.

Feinberg, L.F., and Holden, J.F. (2006). Characterization of dissimilatory Fe(III) versus NO_3^- reduction in the hyperthermophilic archaeon *Pyrobaculum aerophilum*. J. Bacteriol. *188*, 525–531.

Ferguson-Miller, S., and Babcock, G.T. (1996). Heme/copper terminal oxidases. Chem. Rev. *96*, 2889–2908.

Ferry, J.G. (1999). Enzymology of one-carbon metabolism in methanogenic pathways. FEMS Microbiol. Rev. *23*, 13–38.

Fischer, F., Zillig, W., Stetter, K.O., and Schreiber, G. (1983). Chemolithoautotrophic metabolism of anaerobic extremely thermophilic archaebacteria. Nature *301*, 511–513.

Fox, G.E., Magrum, L.J., Balch, W.E., Wolfe, R.S., and Woese, C.R. (1977). Classification of methanogenic Bacteria by 16S ribosomal RNA characterization. Proc. Natl. Acad. Sci. USA *74*, 4537–4541.

Francis, C.A., Roberts, K.J., Beman, J.M., Santoro, A.E., and Oakley, B.B. (2005) Ubiquity and diversity of ammonia-oxidizing Archaea in water columns and sediments of the ocean. Proc. Natl. Acad. Sci. USA *102*, 14683–14688.

Friedrich, C.G., Bardischewsky, F., Rother, D., Quentmeier, A., and Fischer, J. (2005). Prokaryotic sulfur oxidation. Curr. Opin. Microbiol. *8*, 253–259.

Friedrich, C.G., Rother, D., Bardischewsky, F., Quentmeier, A., and Fischer, J. (2001). Oxidation of reduced inorganic sulfur compounds by Bacteria: emergence of a common mechanism? Appl. Environ. Microbiol. *67*, 2873–2882.

Friedrich, T. (2001). Complex I: a chimaera of a redox and conformation-driven proton pump? J. Bioenerg. Biomembr. *33*, 169–177.

Friedrich, T., and Scheide, D. (2000). The respiratory complex I of Bacteria, Archaea and Eukarya and its module common with membrane-bound multisubunit hydrogenases. FEBS Lett. *479*, 1–5.

Fujiwara, T., and Fukumori, Y. (1996). Cytochrome *cb*-type nitric oxide reductase with cytochrome *c* oxidase activity from *Paracoccus denitrificans* ATCC 35512. J. Bacteriol. *178*, 1866-1871.

Garcia-Horsman, J.A., Puustinen, A., Gennis, R.B., and Wikström, M. (1995). Proton transfer in cytochrome bo_3 ubiquinol oxidase of *Escherichia coli*: second-site mutations in subunit I that restore proton pumping in the mutant Asp135 →Asn. Biochemistry *34*, 4428–4433.

Gärtner, P. (1991). Characterization of a quinole-oxidase activity in crude extracts of *Thermoplasma acidophilum* and isolation of an 18-kDa cytochrome. Eur. J. Biochem. *200*, 215–222.

Genest, O., Seduk, F., Ilbert, M., Mejean, V., and Iobbi-Nivol, C. (2006a). Signal peptide protection by specific chaperone. Biochem. Biophys. Res. Commun. *339*, 991–995.

Genest, O., Seduk, F., Theraulaz, L., Mejean, V., and Iobbi-Nivol, C. (2006b). Chaperone protection of immature molybdoenzyme during molybdenum cofactor limitation. FEMS Microbiol. Lett. *265*, 51–55.

Gilderson, G., Aagaard, A., Gomes, C.M., Adelroth, P., Teixeira, M., and Brzezinski, P. (2001). Kinetics of electron and proton transfer during O_2 reduction in cytochrome aa_3 from *A. ambivalens*: an enzyme lacking Glu(I-286). Biochim. Biophys. Acta *1503*, 261–270.

Girsch, P., and deVries, S. (1997). Purification and initial kinetic and spectroscopic characterization of NO reductase from *Paracoccus denitrificans*. Biochim. Biophys. Acta *1318*, 202–216.

Giuffre, A., Gomes, C.M., Antonini, G., D'Itri, E., Teixeira, M., and Brunori, M. (1997). Functional properties of the quinol oxidase from *Acidianus ambivalens* and the possible catalytic role of its electron donor – studies on the membrane-integrated and purified enzyme. Eur. J. Biochem. *250*, 383–388.

Giuffre, A., Stubauer, G., Sarti, P., Brunori, M., Zumft, W.G., Buse, G., and Soulimane, T. (1999). The heme–copper oxidases of *Thermus thermophilus* catalyze the reduction of nitric oxide: evolutionary implications. Proc. Natl. Acad. Sci. USA *96*, 14718–14723.

Gleissner, M., Kaiser, U., Antonopoulos, E., and Schäfer, G. (1997). The archaeal SoxABCD complex is a proton pump in *Sulfolobus acidocaldarius*. J. Biol. Chem. *272*, 8417–8426.

Gomes, C.M., Backgren, C., Teixeira, M., Puustinen, A., Verkhovskaya, M.L., Wikström, M., and Verkhovsky, M.I. (2001). Heme–copper oxidases with modified D- and K-pathways are yet efficient proton pumps. FEBS Lett. *497*, 159–164.

Gribaldo, S., and Brochier-Armanet, C. (2006). The origin and evolution of Archaea: a state of the art. Phil. Trans. R. Soc. Lond. B Biol. Sci. *361*, 1007–1022.

Grimaldi, S., Ostermann, T., Weiden, N., Mogl, T., Miyoshi, H., Ludwig, B., Michel, H., Prisner, T.F., and MacMillan, F. (2003). Asymmetric binding of the high-affinity Q(H)(*)(-) ubisemiquinone in quinol oxidase (*bo*$_3$) from *Escherichia coli* studied by multifrequency electron paramagnetic resonance spectroscopy. Biochemistry *42*, 5632–5639.

Haase, P., Deppenmeier, U., Blaut, M., and Gottschalk, G. (1992). Purification and characterization of $F_{420}H_2$-dehydrogenase from *Methanolobus tindarius*. Eur. J. Biochem. *203*, 527–531.

Haltia, T., Brown, K., Tegoni, M., Cambillau, C., Saraste, M., Mattila, K., and Djinovic-Carugo, K. (2003). Crystal structure of nitrous oxide reductase from *Paracoccus denitrificans* at 1.6 A resolution. Biochem. J. *369*, 77–88.

Hatzixanthis, K., Clarke, T.A., Oubrie, A., Richardson, D.J., Turner, R.J., and Sargent, F. (2005). Signal peptide-chaperone interactions on the twin-arginine protein transport pathway. Proc. Natl. Acad. Sci. USA *102*, 8460–8465.

Hederstedt, L. (2002). Succinate:quinone oxidoreductase in the Bacteria *Paracoccus denitrificans* and *Bacillus subtilis*. Biochim. Biophys. Acta *1553*, 74–83.

Hendriks, J., Gohlke, U., and Saraste, M. (1998). From NO to OO: nitric oxide and dioxygen in bacterial respiration. J. Bioenerg. Biomembr. *30*, 15–24.

Hendriks, J., Oubrie, A., Castresana, J., Urbani, A., Gemeinhardt, S., and Saraste, M. (2000). Nitric oxide reductases in Bacteria. Biochim. Biophys. Acta *1459*, 266-273.

Hubmacher, D., Matzanke, B.F., and Anemüller, S. (2003). Effects of iron limitation on the respiratory chain and the membrane cytochrome pattern of the Euryarchaeon *Halobacterium salinarum*. Biol. Chem. *384*, 1565–1573.

Ichiki, H., Tanaka, Y., Mochizuki, K., Yoshimatsu, K., Sakurai, T., and Fujiwara, T. (2001). Purification, characterization, and genetic analysis of Cu-containing dissimilatory nitrite reductase from a denitrifying halophilic archaeon, *Haloarcula marismortui*. J. Bacteriol. *183*, 4149–4156.

Inatomi, K.L., and Hochstein, L.I. (1996). The purification and properties of a copper nitrite reductase from *Haloferax denitrificans*. Curr. Microbiol. 32, 72–76.

Ishikawa, R., Ishido, Y., Tachikawa, A., Kawasaki, H., Matsuzawa, H., and Wakagi, T. (2002). *Aeropyrum pernix* K1, a strictly aerobic and hyperthermophilic archaeon, has two terminal oxidases, cytochrome *ba*$_3$ and cytochrome *aa*$_3$. Arch. Microbiol. *179*, 42–49.

Iwasaki, T., Kounosu, A., Aoshima, M., Ohmori, D., Imai, T., Urushiyama, A., Cosper, N.J., and Scott, R.A. (2002). Novel [2Fe–2S]-type redox center C in SdhC of archaeal respiratory complex II from *Sulfolobus tokodaii* strain 7.J. Biol. Chem. *277*, 39642–39648.

Iwasaki, T., Wakagi, T., and Oshima, T. (1995). Resolution of the aerobic respiratory system of the thermoacidophilic archaeon, *Sulfolobus* sp. strain 7. III. The archaeal novel respiratory complex II (succinate:caldariellaquinone oxidoreductase complex) inherently lacks heme group. J. Biol. Chem. *270*, 30902–30908.

Iwata, S., Ostermeier, C., Ludwig, B., and Michel, H. (1995). Structure at 2.8 Å resolution of cytochrome *c* oxidase from *Paracoccus denitrificans*. Nature *376*, 660–669.

Janssen, S., Schäfer, G., Anemüller, S., and Moll, R. (1997). A succinate dehydrogenase with novel structure and properties from the hyperthermophilic archaeon *Sulfolobus acidocaldarius*: genetic and biophysical characterization. J. Bacteriol. *179*, 5560–5569.

Jormakka, M., Richardson, D.J., Byrne, B., and Iwata, S. (2004). Architecture of NarGH reveals a structural classification of Mo-bisMGD enzymes. Structure *12*, 95–104.

Junemann, S. (1997). Cytochrome *bd* terminal oxidase. Biochim. Biophys. Acta *1321*, 107–127.

Keller, M., Braun, F.J., Dirmeier, R., Hafenbradl, D., Burggraf, S., Rachel, R., and Stetter, K.O. (1995). *Thermococcus alcaliphilus* sp. nov., a new hyperthermophilic archaeum growing on polysulfide at alkaline pH. Arch. Microbiol. *164*, 390–395.

Kelly, M., Lappalainen, P., Talbo, G., Haltia, T., van der Oost, J., and Saraste, M. (1993). Two cysteines, two histidines, and one methionine are ligands of a binuclear purple copper center. J. Biol. Chem. *268*, 16781–16787.

Kerscher, S., Drose, S., Zickermann, V., and Brandt, U. (2007). The three families of respiratory NADH dehydrogenases. Results Probl. Cell Differ. *Epub ahead of print.*

Klein, C., Garcia-Rizo, C., Bisle, B., Scheffer, B., Zischka, H., Pfeiffer, F., Siedler, F., and Oesterhelt, D. (2005). The membrane proteome of *Halobacterium salinarum*. Proteomics 5, 180–197.

Kletzin, A., Urich, T., Müller, F., Bandeiras, T.M., and Gomes, C.M. (2004). Dissimilatory oxidation and reduction of elemental sulfur in thermophilic Archaea. J. Bioenerg. Biomembr. *36*, 77–91.

Komorowski, L., and Schäfer, G. (2001). Sulfocyanin and subunit II, two copper proteins with novel features, provide new insight into the archaeal SoxM oxidase supercomplex. FEBS Lett. *487*, 351–355.

Komorowski, L., Verheyen, W., and Schäfer, G. (2002). The archaeal respiratory supercomplex SoxM from

S. *acidocaldarius* combines features of quinole and cytochrome *c* oxidases. Biol. Chem. *383*, 1791–1799.

König, H., Rachel, R., and Claus, H. (2007). Proteinaceous surface layers of Archaea: ultrastructure and biochemistry, In Archaea. Molecular and Cellular Biology, R. Cavicchioli, ed. (Washington, DC: ASM Press), pp. 315–340.

Könneke, M., Bernhard, A.E., de la Torre, J.R., Walker, C.B., Waterbury, J.B., and Stahl, D.A. (2005) Isolation of an autotrophic ammonia-oxidizing marine archaeon. Nature *437*, 543–546.

Kunow, J., Linder, D., Stetter, K.O., and Thauer, R.K. (1994). $F_{420}H_2$: quinone oxidoreductase from *Archaeoglobus fulgidus*. Characterization of a membrane-bound multisubunit complex containing FAD and iron–sulfur clusters. Eur. J. Biochem. *223*, 503–511.

Lancaster, C.R., and Kröger, A. (2000). Succinate: quinone oxidoreductases: new insights from X-ray crystal structures. Biochim. Biophys. Acta *1459*, 422–431.

Lanciano, P., Vergnes, A., Grimaldi, S., Guigliarelli, B., and Magalon, A. (2007). Biogenesis of a respiratory complex is orchestrated by a single accessory protein. J. Biol. Chem.

Leininger, S., Urich, T., Schloter, M., Schwark, L., Qi, J., Nicol, G.W., Prosser, J.I., Schuster, S.C., and Schleper, C. (2006) Archaea predominate among ammonia-oxidizing prokaryotes in soils. Nature *442*, 806–809.

Lienard, T., Becher, B., Marschall, M., Bowien, S., and Gottschalk, G. (1996). Sodium ion translocation by N5-methyltetrahydromethanopterin: coenzyme M methyltransferase from *Methanosarcina mazei* Göl reconstituted in ether lipid liposomes. Eur. J. Biochem. *239*, 857–864.

Lledo, B., Martinez-Espinosa, R.M., Marhuenda-Egea, F.C., and Bonete, M.J. (2004). Respiratory nitrate reductase from haloarchaeon *Haloferax mediterranei*: biochemical and genetic analysis. Biochim. Biophys. Acta *1674*, 50–59.

Lovley, D.R., Holmes, D.E., and Nevin, K.P. (2004). Dissimilatory Fe(III) and Mn(IV) reduction. Adv. Microb Physiol *49*, 219–286.

Lübben, M., Kolmerer, B., and Saraste, M. (1992). An archaebacterial terminal oxidase combines core structures of two mitochondrial respiratory complexes. Embo J. *11*, 805–812.

Lübben, M., and Morand, K. (1994). Novel prenylated hemes as cofactors of cytochrome oxidases. Archaea have modified hemes A and O.J. Biol. Chem. *269*, 21473–21479.

Lübben, M., and Schäfer, G. (1989). Chemiosmotic energy conversion of the archaebacterial thermoacidophile *Sulfolobus acidocaldarius*: oxidative phosphorylation and the presence of an F_0-related N,N'-dicyclohexylcarbodiimide-binding proteolipid. J. Bacteriol. *171*, 6106–6116.

Martinez-Espinosa, R.M., Marhuenda-Egea, F.C., and Bonete, M.J. (2001). Purification and characterisation of a possible assimilatory nitrite reductase from the halophile archaeon *Haloferax mediterranei*. FEMS Microbiol. Lett. *196*, 113–118.

Mattar, S., and Engelhard, M. (1997). Cytochrome *ba*$_3$ from *Natronobacterium pharaonis* – an archaeal four-subunit cytochrome-*c*-type oxidase. Eur. J. Biochem. *250*, 332–341.

Mattar, S., Scharf, B., Kent, S.B., Rodewald, K., Oesterhelt, D., and Engelhard, M. (1994). The primary structure of halocyanin, an archaeal blue copper protein, predicts a lipid anchor for membrane fixation. J. Biol. Chem. *269*, 14939–14945.

Melo, A.M., Bandeiras, T.M., and Teixeira, M. (2004). New insights into type II NAD(P)H:quinone oxidoreductases. Microbiol. Mol. Biol. Rev. *68*, 603–616.

Miller, M.J., and Gennis, R.B. (1985). The cytochrome *d* complex is a coupling site in the aerobic respiratory chain of *Escherichia coli*. J. Biol. Chem. *260*, 14003–14008.

Mincer, T.J., Church, M.J., Taylor, L.T., Preston, C., Karl, D.M., and Delong, E.F. (2007) [3]Monterey Bay Aquarium Research Institute, Moss Landing, CA 95039, U Quantitative distribution of presumptive archaeal and bacterial nitrifiers in Monterey Bay and the North Pacific Subtropical Gyre. Environ. Microbiol. *9*, 1162–1175.

Mitchell, P. (1976). Possible molecular mechanisms of the proton motive function of cytochrome systems. J. Theor Biol. *62*, 327–367.

Moenne-Loccoz, P. (2007). Spectroscopic characterization of heme iron-nitrosyl species and their role in NO reductase mechanisms in diiron proteins. Nat. Prod. Rep. *24*, 610–620.

Moir, J.W., and Wood, N.J. (2001). Nitrate and nitrite transport in Bacteria. Cell Mol. Life Sci. *58*, 215–224.

Moll, R., and Schäfer, G. (1988). Chemiosmotic H^+ cycling across the plasma membrane of the thermoacidophilic archaebacterium *Sulfolobus acidocaldarius*. FEBS Lett. *232*.

Müller, F.H., Bandeiras, T.M., Urich, T., Teixeira, M., Gomes, C.M., and Kletzin, A. (2004). Coupling of the pathway of sulphur oxidation to dioxygen reduction: characterization of a novel membrane-bound thiosulphate:quinone oxidoreductase. Mol. Microbiol. *53*, 1147–1160.

Nunoura, T., Sako, Y., Wakagi, T., and Uchida, A. (2003). Regulation of the aerobic respiratory chain in the facultatively aerobic and hyperthermophilic archaeon *Pyrobaculum oguniense*. Microbiology *149*, 673–688.

Nunoura, T., Sako, Y., Wakagi, T., and Uchida, A. (2005). Cytochrome *aa*$_3$ in facultatively aerobic and hyperthermophilic archaeon *Pyrobaculum oguniense*. Can J. Microbiol. *51*, 621–627.

Ogura, T., and Kitagawa, T. (2004). Resonance Raman characterization of the P intermediate in the reaction of bovine cytochrome *c* oxidase. Biochim. Biophys. Acta *1655*, 290–297.

Oren, A., Ginzburg, M., Ginzburg, B.Z., Hochstein, L.I., and Volcani, B.E. (1990). *Haloarcula marismortui* (Volcani) sp. nov. nom. rev., an extremely halophilic bacterium from the Dead Sea. Int. J. Syst. Bacteriol. *40*, 209–210.

Osborne, J.P., and Gennis, R.B. (1999). Sequence analysis of cytochrome *bd* oxidase suggests a revised topology for subunit I. Biochim. Biophys. Acta *1410*, 32–50.

Ostermeier, C., Harrenga, A., Ermler, U., and Michel, H. (1997). Structure at 2.7 Å resolution of the *Paracoccus denitrificans* two-subunit cytochrome *c* oxidase complexed with an antibody FV fragment. Proc. Natl. Acad. Sci. USA *94*, 10547–10553.

Peinemann, S., Müller, V., Blaut, M., and Gottschalk, G. (1988). Bioenergetics of methanogenesis from acetate by *Methanosarcina barkeri*. J. Bacteriol. *170*, 1369–1372.

Pereira, M.M., Bandeiras, T.M., Fernandes, A.S., Lemos, R.S., Melo, A.M., and Teixeira, M. (2004). Respiratory chains from aerobic thermophilic prokaryotes. J. Bioenerg. Biomembr. *36*, 93–105.

Pereira, M.M., Santana, M., and Teixeira, M. (2001). A novel scenario for the evolution of haem-copper oxygen reductases. Biochim. Biophys. Acta *1505*, 185–208.

Pereira, M.M., and Teixeira, M. (2004). Proton pathways, ligand binding and dynamics of the catalytic site in haem-copper oxygen reductases: a comparison between the three families. Biochim. Biophys. Acta *1655*, 340–346.

Philippot, L. (2002). Denitrifying genes in bacterial and archaeal genomes. Biochim. Biophys. Acta *1577*, 355–376.

Pohlschröder, M., Dilks, K., Hand, N.J., and Wesley Rose, R. (2004). Translocation of proteins across archaeal cytoplasmic membranes. FEMS Microbiol. Rev. *28*, 3–24.

Purschke, W.G., Schmidt, C.L., Petersen, A., and Schäfer, G. (1997). The terminal quinol oxidase of the hyperthermophilic archaeon *Acidianus ambivalens* exhibits a novel subunit structure and gene organization. J. Bacteriol. *179*, 1344–1353.

Puustinen, A., Finel, M., Haltia, T., Gennis, R.B., and Wikström, M. (1991). Properties of the two terminal oxidases of *Escherichia coli*. Biochemistry *30*, 3936-3942.

Richardson, D.J. (2000). Bacterial respiration: a flexible process for a changing environment. Microbiol. *146*, 551–571.

Richardson, D.J., Berks, B.C., Russell, D.A., Spiro, S., and Taylor, C.J. (2001). Functional, biochemical and genetic diversity of prokaryotic nitrate reductases. Cell Mol. Life Sci. *58*, 165–178.

Richardson, D.J., and Watmough, N.J. (1999). Inorganic nitrogen metabolism in Bacteria. Curr. Opin. Chem. Biol. *3*, 207–219.

Sako, Y., Nunoura, T., and Uchida, A. (2001). *Pyrobaculum oguniense* sp. nov., a novel facultatively aerobic and hyperthermophilic archaeon growing at up to 97 degrees C. Int. J. Syst Evol. Microbiol. *51*, 303–309.

Saraste, M., and Castresana, J. (1994). Cytochrome oxidase evolved by tinkering with denitrification enzymes. FEBBS Lett. *341*, 1–4.

Sato-Watanabe, M., Mogi, T., Ogura, T., Kitagawa, T., Miyoshi, H., Iwamura, H., and Anraku, Y. (1994). Identification of a novel quinone-binding site in the cytochrome *bo* complex from *Escherichia coli*. J. Biol. Chem. *269*, 28908–28912.

Schäfer, G., Anemüller, S., and Moll, R. (2002). Archaeal complex II: 'classical' and 'non-classical' succinate:quinone reductases with unusual features. Biochim. Biophys. Acta *1553*, 57–73.

Schäfer, G., Engelhard, M., and Müller, V. (1999). Bioenergetics of the Archaea. Microbiol. Mol. Biol. Rev. *63*, 570–620.

Schäfer, G., Purschke, W., and Schmidt, C.L. (1996). On the origin of respiration: electron transport proteins from Archaea to man. FEMS Microbiol. Rev. *18*, 173–188.

Scharf, B., and Engelhard, M. (1993). Halocyanin, an archaebacterial blue copper protein (type I) from *Natronobacterium pharaonis*. Biochem. *32*, 12894–12900.

Scharf, B., Wittenberg, R., and Engelhard, M. (1997). Electron transfer proteins from the haloalkaliphilic archaeon *Natronobacterium pharaonis*: possible components of the respiratory chain include cytochrome *bc* and a terminal oxidase cytochrome *ba*$_3$. Biochemistry *36*, 4471–4479.

Schauder, R., and Müller, E. (1993). Polysulfide as a possible substrate for sulfur-reducing Bacteria. Arch. Microbiol. *169*, 377–382.

Segerer, A.H., Burggraf, S., Fiala, G., Huber, G., Huber, R., Pley, U., and Stetter, K.O. (1993). Life in hot springs and hydrothermal vents. Orig. Life Evol. Biosph. *23*, 77–90.

Shand, R.F., and Betlach, M.C. (1991). Expression of the *bop* gene cluster of
Halobacterium halobium is induced by low oxygen tension and by light. J. Bacteriol. *173*, 4692–4699.

Solioz, M., Carafoli, E., and Ludwig, B. (1982). The cytochrome *c* oxidase of *Paracoccus denitrificans* pumps protons in a reconstituted system. J. Biol. Chem. *257*, 1579–1582.

Soulimane, T., Buse, G., Bourenkov, G.P., Bartunik, H.D., Huber, R., and Than, M.E. (2000). Structure and mechanism of the aberrant *ba*$_3$-cytochrome *c* oxidase from *Thermus thermophilus*. Embo J. *19*, 1766-1776.

Sreeramulu, K., Schmidt, C.L., Schäfer, G., and Anemüller, S. (1998). Studies of the electron transport chain of the euryarcheon *Halobacterium salinarum*: indications for a type II NADH dehydrogenase and a complex III analog. J. Bioenerg. Biomembr. *30*, 443–453.

Stetter, K.O. (1988). *Archaeoglobus fulgidus* gen. nov., sp. nov.: a novel taxon of extremely thermophilic archaebacteria. Syst Appl. Microbiol. *10*, 172–173.

Stetter, K.O. (2006a). History of discovery of the first hyperthermophiles. Extremophiles *10*, 357–362.

Stetter, K.O. (2006b). Hyperthermophiles in the history of life. Phil. Trans. R. Soc. Lond. B Biol. Sci. *361*, 1837–1842; discussion 1842–1833.

Stetter, K.O., Lauerer, G., Thomm, M., and Neuner, A. (1987). Isolation of extremely thermophilic sulfate reducers: evidence of a novel branch of archaebacteria. Science *236*, 822–824.

Suharti, Strampraad, M.J.F., Schröder, I., and de Vries, S. (2001). A novel copper A containing menaquinol NO reductase from *Bacillus azotoformans*. Biochemistry *40*, 2632–2639.

Tanaka, M., Ogawa, N., Ihara, K., Sugiyama, Y., and Mukohata, Y. (2002). Cytochrome aa_3 in *Haloferax volcanii*. J. Bacteriol. *184*, 840–845.

Tindall, B.J. (1989). Fully saturated menaquinones in the archaebacterium *Pyrobaculum islandicum*. FEMS Microbiol. Lett. *60*, 251–254.

Toledo-Cuevas, M., Barquera, B., Gennis, R.B., Wikström, M., and Garcia-Horsman, J.A. (1998). The cbb_3-type cytochrome *c* oxidase from *Rhodobacter sphaeroides*, a proton-pumping heme–copper oxidase. Biochim. Biophys. Acta *1365*, 421–434.

Trumpower, B.L., and Gennis, R.B. (1994). Energy transduction by cytochrome complexes in mitochondrial and bacterial respiration: the enzymology of coupling electron transfer reactions to transmembrane proton translocation. Annu. Rev. Biochem. *63*, 675–716.

Tsukihara, T., Aoyama, H., Yamashita, E., Tomizaki, T., Yamaguchi, H., Shinzawa-Itoh, K., Nakashima, R., Yaono, R., and Yoshikawa, S. (1996). The whole structure of the 13-subunit oxidized cytochrome *c* oxidase at 2.8 Å. Science *272*, 1136-1144.

Turner, R.J., Papish, A.L., and Sargent, F. (2004). Sequence analysis of bacterial redox enzyme maturation proteins (REMPs). Can J. Microbiol. *50*, 225–238.

Vadas, A., Monbouquette, H.G., and I. Schröder (1999). Identification and characterization of a novel ferric reductase from the hyperthermophilic Archaeon *Archaeoglobus fulgidus*. J. Biol. Chem. *274*, 36715–36721.

Vargas, M., Kashefi, K., Blunt-Harris, E.L., and Lovley, D.R. (1998). Microbiological evidence for Fe(III) reduction on early earth. Nature *395*, 65–67.

Victor, B.L., Baptista, A.M., and Soares, C.M. (2004). Theoretical identification of proton channels in the quinol oxidase aa_3 from *Acidianus ambivalens*. Biophys. J. *87*, 4316–4325.

Völkl, P., Huber, R., Drobner, E., Rachel, R., Burggraf, S., Trincone, A., and Stetter, K.O. (1993). *Pyrobaculum aerophilum* sp. nov., a novel nitrate reducing hyperthermophilic Archaeum. Appl. Environment. Microbiol. *59*, 2918–2926.

Vorholt, J.A., Hafenbradl, D., Stetter, K.O., and Thauer, R.K. (1997). Pathways of autotrophilic CO_2 fixation and of dissimilatory nitrate reduction to N_2O in *Ferroglobus placidus*. Arch. Microbiol. *167*, 19–23.

Wasser, I.M., de Vries, S., Moenne-Loccoz, P., Schröder, I., and Karlin, K.D. (2002). Nitric oxide in biological denitrification: Fe/Cu metalloenzyme and metal complex NOx redox chemistry. Chem. Rev. *102*, 1201–1234.

Weber, K.A., Achenbach, L.A., and Coates, J.D. (2006). Microorganisms pumping iron: anaerobic microbial iron oxidation and reduction. Nat. Rev. Microbiol. *4*, 752–764.

Wikström, M., and Verkhovsky, M.I. (2002). Proton translocation by cytochrome *c* oxidase in different phases of the catalytic cycle. Biochim. Biophys. Acta *1555*, 128–132.

Wilmanns, M., Lappalainen, P., Kelly, M., Sauer-Eriksson, E., and Saraste, M. (1995). Crystal structure of the membrane-exposed domain from a respiratory quinol oxidase complex with an engineered dinuclear copper center. Proc. Natl. Acad. Sci. USA *92*, 11955–11959.

Winstone, T.L., Workentine, M.L., Sarfo, K.J., Binding, A.J., Haslam, B.D., and Turner, R.J. (2006). Physical nature of signal peptide binding to DmsD. Arch. Biochem. Biophys. *455*, 89–97.

Woese, C.R., Kandler, O., and Wheelis, M.L. (1990). Towards a natural systems of organisms: proposal for the domains Archaea, Bacteria, and eucarya. Proc. Natl. Acad. Sci. USA *87*, 4576–4579.

Yang, C.F., and DasSarma, S. (1990). Transcriptional induction of purple membrane and gas vesicle synthesis in the archaebacterium *Halobacterium halobium* is blocked by a DNA gyrase inhibitor. J. Bacteriol. *172*, 4118–4121.

Yang, K., Zhang, J., Vakkasoglu, A.S., Hielscher, R., Osborne, J.P., Hemp, J., Miyoshi, H., Hellwig, P., and Gennis, R.B. (2007). Glutamate 107 in subunit I of the cytochrome *bd* quinol oxidase from *Escherichia coli* is protonated and near the heme *d*/heme *b*(595) binuclear center. Biochemistry *46*, 3270–3278.

Yoshikawa, S., Muramoto, K., Shinzawa-Itoh, K., Aoyama, H., Tsukihara, T., Shimokata, K., Katayama, Y., and Shimada, H. (2006). Proton pumping mechanism of bovine heart cytochrome *c* oxidase. Biochim. Biophys. Acta *1757*, 1110–1116.

Yoshimatsu, K., Iwasaki, T., and Fujiwara, T. (2002). Sequence and electron paramagnetic resonance analyses of nitrate reductase NarGH from a denitrifying halophilic euryarchaeote *Haloarcula marismortui*. FEBS Lett. *516*, 145–150.

Yoshimatsu, K., Sakurai, T., and Fujiwara, T. (2000). Purification and characterization of dissimilatory nitrate reductase from a denitrifying halophilic archaeon, *Haloarcula marismortui*. FEBS Lett. *470*, 216-220.

Zumft, W.G. (1997). Cell biology and molecular basis of denitrification. Microbiol. Mol. Biol. Rev. *61*, 533–616.

Zumft, W.G., and Kroneck, P.M. (2007). Respiratory transformation of nitrous oxide (N_2O) to dinitrogen by Bacteria and Archaea. Adv. Microb Physiol 52, 107–227.

Archaea–Metal Interactions: Metabolism and Strategies of Resistance

2

Elisabetta Bini

Abstract

Transition metals are required by all living organisms. Because of their chemical properties, they are important cofactors of numerous enzymes and are essential to carry out the most diverse functions, from basic metabolic processes to highly specialized ones. Essential metals are required only in trace levels, becoming toxic when their intracellular concentration exceeds the physiological level. Other metals, not required by biological processes, can be harmful at very low concentrations. This chapter attempts to provide an overview of the ways developed by Archaea for utilizing metals and their strategies to maintain metals within physiological ranges, and a review of the genetic determinants and regulators controlling the response to metals. There is a growing awareness that all microorganisms, including Archaea, have an important impact on the biogeochemical cycling of elements. It is evident that in-depth investigations of the mechanisms underlying metal homeostasis and resistance in Archaea are required. Fortunately, new and improved techniques of analysis, combined with an increasing number of genome sequences, are rapidly advancing the field of metal metabolism in Archaea.

Introduction

Metals had a major role in the origin and evolution of life. The autotrophic theory assumes that life originated on earth in a hydrothermal environment, presumably volcanic, where the formation of basic organic compounds was favoured by the presence of transition metals acting as catalysts (Huber and Wachtershauser, 1998, 2006; Martin and Russell, 2003; Wachtershauser, 2006). Consistently with this theory, the oldest microbial fossils reported were discovered in a 3.2 billion years old deep-sea volcanogenic massive sulphide deposit (Rasmussen, 2000). The idea that thermal springs at oceanic ridges constituted the first inhabited environments is also supported by phylogenetic analyses placing thermophiles on deep branches of the tree of life (Forterre, 1996). And finally, a large fraction of these deep branching microorganisms are sulphur dependent and use reductive Krebs cycles, suggesting that primordial reactions were non-enzymatic but possibly catalysed by metals (Maden, 1995). Given the abundance of metals in deep-sea vents environments, primitive life was probably challenged with the toxicity of the metals driving these ancient chemical reactions. Thus it is reasonable to assume that metal resistance systems are the most ancient mechanisms of defence evolved by cells. They are found virtually in all prokaryotic species, and some of the components of resistance systems have protein sequences well conserved across domains. For example, the archaeal copper-transporting ATPases maintain high sequence similarity to bacterial sequences and even to the human homologues ATP7A and ATP7B, and perform the same function, although the purpose of this function has changed during the transition to multicellularity, being copper resistance in the Archaea and redistribution of copper to target tissues in humans.

Early studies of the interactions between Archaea and metals focused mostly on thermophilic species because of their potential application in bioleaching. More recently, metal resistance in Archaea has increasingly attracted interest due to its potential impact on bioremediation. Sites contaminated with high levels of metals are usually found in mines where the activity of microbial communities leads to the generation of acid drainage. These environments are often rich in sulphur compounds, that through oxidation release sulphuric acid and metal. As a consequence, the microorganisms are constantly exposed to high concentrations of metal ions. Studies of microbial communities in extreme environments, including environments naturally metal-rich, such as hot and acidic springs or heavily contaminated environments like mines have contributed to the discover of new metal resistant strains and increased interest in the strategies of metal resistance. Furthermore, analyses of metal-rich environments have led to the discovery of species that can exploit their interactions with a variety of metals to generate energy. In the past decade this has resulted in a growing number of publications concerning the effect of metals on biological processes, and the impact of microorganisms on geochemical cycles, as it becomes evident from the distribution of the scientific literature concerning this topic over the past years.

Akin to Bacteria, Archaea have a love–hate relationship with metals. Metals are used in fundamental biological processes, either in assimilatory reactions where metal ions are incorporated into a protein or enzyme and function as cofactors, or in dissimilatory processes where metals are utilized as terminal electron acceptors for energy production. However, excessive exposure to any metal can cause great damage to cellular components and impair cell functions. Metals in metabolic processes and mechanisms of resistance are discussed separately in this chapter, but there are instances where this division between functions is not so clear-cut, for example oxidation or reduction of certain metals may serve the twofold purpose of respiration and detoxification. In contrast with Bacteria, where mechanisms of detoxification and resistance have been extensively investigated, much work is still required to fully elucidate metal resistance in Archaea. However, in a few cases, the interplay between genetic determinants of metal resistance has been investigated in more depth, and sections are dedicated to their review. In all other cases, strategies of resistance to metals are reviewed according to mechanism of detoxification, mostly based on active efflux and transformation by specific enzymatic activities, but also on metal exclusion and sequestration.

Use of metals in metabolic processes

Metabolism of metals is linked to a wide range of basic metabolic processes such as nitrogen fixation, respiration, and other processes relying on oxido-reduction reactions or electron transfer for energy production, because the enzymes of all these pathways contain metal cofactors, or in case of anaerobic respiration, the metal itself is used as terminal electron acceptor. Similarly to Bacteria, many archaeal species have the ability to derive energy from the reduction of a variety of metals. *Archaeoglobus fulgidus*, *Pyrococcus furiosus* and *Pyrodictium abyssi* are capable of reducing Fe (III), and two *Pyrobaculum* species can effectively grow respiring Fe(III) (Vargas *et al.*, 1998). At least one archaeal species, *Pyrobaculum arsenaticum*, can use arsenate as terminal electron acceptor for growth (Oremland and Stolz, 2003).

The following sections focus on the biological role of trace metals; if a metal is involved in different processes, it will appear in more than one section. Iron, which is a relatively abundant and ubiquitous metal but well tolerated by microorganisms because of its low solubility as Fe(III), will not be discussed in this chapter.

Molybdenum and vanadium

Methanogenic Archaea and Bacteria share determinants of nitrogen fixation (Fani *et al.*, 2000; Leigh, 2000). Nitrogenases are highly conserved across these domains, and for this reason either an early origin or horizontal transfer of these sequences has been hypothesized. Nitrogenases are complex enzymes made of two components including multiple subunits; one component contains iron, the second carries molybdenum, and in some cases vanadium (Rubio and Ludden, 2005). Because of the importance of nitrogenases in the biogeochemical cycling of nitrogen, there are extensive studies of the activities of microbial communities based on nitrogenases (Zehr *et al.*,

2003). While the presence of nitrogenases in methanogens is also confirmed by genomic analysis, archaeal enzymes, their functions and metal requirements have yet to be characterized.

Tungsten and molybdenum

Tungsten and molybdenum have similar chemical properties. Molybdenum is a trace metal required by virtually any species, and tungsten can replace molybdenum in some instances (Kletzin and Adams, 1996). In most enzymes both metals are found associated with an organic component, called molybdopterin, often containing dinucleotides (Johnson *et al.*, 1993). Early studies of tungsten-requiring enzymes in hyperthermophilic Archaea have led to the discovery that tungsten has a biological role. In fact, tungsten is an essential trace metal for some Archaea.

The aldehyde ferredoxin oxidoreductase (AOR) of *Thermococcus* strain ES-1 was purified and shown to contain tungsten, in addition to Fe and Mg (Heider *et al.*, 1995). *Thermococcus litoralis* uses another tungsten-containing enzyme, formaldehyde ferredoxin oxidoreductase (FOR), whose properties were investigated (Dhawan *et al.*, 2000). The hyperthermophile *Pyrococcus furiosus*, possesses at least three enzymes, AOR, FOR, and glyceraldehyde-3-phosphate oxidoreductase (GAPOR), that specifically require tungsten for their activity as well as for the synthesis of the enzymes (Mukund and Adams, 1996). Two additional enzymes containing tungsten in *Pyrococcus* are the aldehyde oxidoreductases WOR4 and WOR5 were characterized more recently (Bevers *et al.*, 2005; Roy and Adams, 2002).

Two types of systems for tungsten and molybdenum uptake have been identified in Archaea; these are the *wtp* and *mod* systems, both belonging to the ABC family of transporters. *Pyrococcus furiosus* has been recently reported to use the tungsten-binding protein WtpA for the selective uptake of molybdenum and tungsten. WtpA belongs to the ABC family of transporters. The *wtp* system is thought to be widespread in Archaea due to the presence of WtpA homologues in the proteomes of the Archaea *Thermococcus kodakarensis*, *Pyrococcus abyssi*, *Pyrococcus horikoshii*, *Methanococcus jannaschii*, *Archaeoglobus fulgidus*, *Methanococcus maripaludis*, *Methanosarcina acetivorans*, *Haloarcula marismortui*, and only in three Bacteria. In all cases, adjacent to genes encoding WtpA were also identified sequences encoding the permease component WtpB and the ATPase component WtpC of the ABC transporter (Bevers *et al.*, 2006).

In addition to the WtpABC system for tungsten and molybdenum, the uptake of molybdate in some Archaea, similarly to Bacteria, is also accomplished via ABC-type uptake systems employing the metal-binding proteins ModA, a periplasmic protein, ModB, a membrane protein, and the ATPase ModC (Self *et al.*, 2001). This mechanism is similar to the one present in Bacteria, where the system *modABC* encode components for specific transport of molybdenum and tungsten, and the specific tungsten transporter TupA, also belonging to the ABC family of transporters. The structure of the *Archaeoglobus* molybdate importer ModBC with its binding protein ModA was solved this year by using X-ray crystallography (Hollenstein *et al.*, 2007).

Nickel and cobalt

One of the earliest reports concerned with the interaction of Archaea with metals looked into the nickel, cobalt and molybdenum requirements for growth of *Methanobacterium thermoautothrophicum* (Schonheit *et al.*, 1979). The extensive studies that followed showed that nickel was also a requirement for the methanogenic Archaea *Methanobrevibacter smithii* and *Methanosarcina barkeri*, for incorporation into the cofactor F430, a yellow chromophore found in the methylreductase of *Methanobacterium* (Diekert *et al.*, 1981; Ellefson *et al.*, 1982). Jarrell and Sprott observed active and specific uptake of nickel in *Methanobacterium bryantii*, not ATP driven but probably coupled to proton movement (Jarrell and Sprott, 1982). Therefore nickel is an important requirement for methanogenesis in *Methanobacterium* strains (Hartzell and Wolfe, 1986).

Nickel is an essential cofactor for NiFe-hydrogenases, carbon monoxide dehydrogenase, methyl CoM reductase, urease, among others. Cobalt is mostly found in coenzyme B12. At least six systems for Ni/Co uptake are known, two of which were recently discovered by in silico analyses (Rodionov *et al.*, 2006). Such systems are required for high-affinity uptake of nickel or cobalt, since usually their environmental concentration is below the required levels. The NikABCDE multicomponent permeases, a family of ABC transporters, were characterized in *E. coli* (Navarro *et al.*, 1993). The NikA component of the permease is also thought to be involved in Ni sequestration when Ni is in excess (Eitinger and Mandrand-Berthelot, 2000). Using sequence similarity and additional criteria, like proximity to nickel or cobalt-requiring enzymes and Ni regulatory elements, transporters belonging to the NikABCDE family of permeases were discovered in the genomes of *Methanosarcina acetivorans*, *M. barkeri* and *M. mazei* (Rodionov *et al.*, 2006). Similar criteria were applied to find homologues of the NiCoT, a family of metal permeases with preferential affinity for nickel or cobalt, in the two Archaea *S. solfataricus* and *Thermoplasma acidophilum* (Rodionov *et al.*, 2006), interestingly both species are acidophiles. Sequences similar to the bacterial secondary transporters for nickel and cobalt, HupE/UreJ and UreH/SodT, were not found in Archaea. Two novel transport systems for nickel and cobalt uptake were present in most prokaryotic species, including 11 archaeal species. These systems were named CbiMNQO and NikMNQO, and possess a protein belonging to the ABC family of transporter, CbiO or NikO. Metal accumulation assays showed that the activity of the CbiMN component confirmed its hypothesized function of cobalt transporter (Rodionov *et al.*, 2006).

Arsenic

Energy can be generated both by oxidation or reduction of specific arsenic oxyanions. Arsenite oxidases, enzymes that convert As(III) to As(V), have been characterized in Bacteria (Muller *et al.*, 2003) and the structure of the *Alcaligenes faecalis* enzyme has been determined (Ellis *et al.*, 2001). Several bacterial species, including strains of *Pseudomonas*, *Chloroflexus*, *Thiobacillus*, *Alcaligenes*, *Thiobacillus*, and *Thermus*, can oxidize arsenite for energy metabolism. Archaeal genomes contain operons encoding arsenite oxidase, and a phylogenetic analysis of these sequences suggest a common origin before the separation between Archaea and Bacteria (Lebrun *et al.*, 2003). Interestingly, most sequence retrieved were of thermophilic microorganisms from volcanic environments (Lebrun *et al.*, 2003). Two types of arsenate reductase are known, which have different structure and functions. These are the respiratory reductases, discussed in this section, and ArsC type reductases employed for detoxification that will be considered later. Respiratory reductases have been described in a limited number of arsenate respiring Bacteria. These enzymes are dissimilar and do not appear to belong to a single family, they may have one or two subunits and may or may not contain a metal cofactor. This fact suggests that determinants of arsenate reduction may have evolved independently. A conserved domain characteristic of *arr* respiratory arsenate reductases is present in the genomes of *Hyperthermus butylicus* (BisC), *Pyrobaculum islandicus*, *P. aerophilum*, *P. arsenaticum*, *Archaeoglobus fulgidus*. Since these are anaerobic species, the presence of Arr-type sequence may indicate the ability to grow using arsenate as alternate electron acceptor. This ability to reduce arsenate for energy production was confirmed in *P. aerophilum* and *P. arsenaticum* (Huber *et al.*, 2000).

Selenium

Selenium is a component of the amino acid selenocysteine, through which it is incorporated in enzymes requiring this metal as cofactor. Selenocysteine insertion (SECIS) elements have been detected in archaeal genomes and form secondary structures in the 3′ untranslated region of genes requiring the recoding of the UGA codon with selenocysteine. SECIS sequences are highly conserved among *Methanococcus* species. (Wilting *et al.*, 1997). A report by Rother and coworkers demonstrated that one of these genes, encoding a subunit of the coenzyme F_{420}-reducing hydrogenase of *Methanococcus jannaschii*, is expressed and indeed produced a protein containing the amino acid selenocysteine (Rother *et al.*, 2001). A selenium-binding protein with possible function in transport of selenium was isolated from *Methanococcus vannielii*. The recombinant protein forms tetramers upon binding of selenium (Patteson *et al.*, 2005; Self *et al.*,

2004). The Archaea *Pyrobaculum arsenaticum* and *P. aerophilum* can respire selenate, which is reduced to elemental selenium (Huber *et al.*, 2000). Recent developments in studies of arsenic and selenium metabolism in prokaryotes have been recently reviewed by Stoltz and coworkers (Stolz *et al.*, 2006).

Toxicity of metals

Most metals acting as co-factors are only required in trace levels, becoming toxic when their intracellular concentration exceeds the physiological level. Other metals, not required by biological processes, can be toxic at very low concentrations. Depending on the metal, toxicity can derive from direct or indirect interaction with chemical groups of cell components causing a variety of possible outcomes. Most metals cause the formation of reactive oxygen and nitrogen species that react with and damage DNA bases and phospholipids. The extent of toxicity depends also on the form and oxidation state of the metal, which affect its bioavailability and activity. Some metals, mercury, cadmium and nickel, and arsenic in the form of arsenite bind sulphydryl groups of cysteine residues, while copper and arsenic can bind thiol groups inactivating the proteins affected (Valko *et al.*, 2005). Arsenate, less toxic than arsenite, can replace phosphate interfering with phosphate functions. Chromate is similar in size and charge to sulphate and competes with it (Smith and Gadd, 2000). Cadmium interferes with calcium homeostasis by affecting enzymes involved in calcium metabolism.

Strategies of metal resistance

Strategies employed by archaeal microorganisms are essentially the same observed for Bacteria, and include mostly efflux by ATP driven transporters and enzymatic detoxification and, to a lesser extent, exclusion and metal sequestration. Toxic metals obviously have no uptake systems and enter the cells using channels designed for useful ions and organic molecules. Known systems of defence rely on use of more specific uptake systems, so that the toxic metal can be excluded (Cervantes *et al.*, 1994). For example, arsenic enters the cell using transport systems, Pit or Pst, designed for phosphate uptake. In the case of nickel and cobalt, the expression of permeases for their specific uptake is instead repressed (Chivers and Sauer, 2000).

Metal sequestration

Sulfolobus metallicus, a species capable of growing in concentrations of copper up to 200 mM, uses a mechanism of resistance based on sequestration by organic phosphate (Remonsellez *et al.*, 2006), which is presumably followed by active efflux of the metal-phosphate complex, similarly to what is observed in *E. coli*, *Acinetobacter johnsonii*, and *Acidothiobacillus ferroxidans* (Alvarez and Jerez, 2004) that employ the inorganic phosphate transport system (Pit) also for transport of metal-phosphate complexes. *S. metallicus*, but not *S. acidocaldarius* and *S. solfataricus*, synthesizes and accumulate large amounts of organic polyphosphate (polyP), which is rapidly hydrolysed by an exopolyphosphatase in the presence of copper (Remonsellez *et al.*, 2006).

Metal-binding proteins

Sequestration of metal ions by metal-binding proteins is strategy to cope with excess of toxic metals. An important class of metal-binding protein is represented by metallothioneine. However, methallothioneins are not represented in the archaeal genomes. Members of the CutA family of metal-binding proteins are found in Archaea, Bacteria and eukaryotes. The crystal structure of the *Pyrococcus horikoshii* CutA has been solved (Tanaka *et al.*, 2004). The proposed role for CutA is the capture and precipitation of metal ions. Multimerization of CutA is induced by heavy metal binding. Interestingly, while the metal binding site of the *E. coli* homologue contains Cys and His residues, these amino acids are absent in the *Pyrococcus* protein. Therefore the archaeal protein might represent a new model of metal responsive multimerization.

Possibly one of the first studies of copper resistance in Archaea is represented by the work on *Methanobacterium bryantii*. This is a particularly copper-resistant methanogen species isolated from a copper mine in Michigan. Kim and coworkers discovered that in the presence of high external copper, *M. bryantii* excretes at least three copper response extracellular (CRX) proteins in the medium (Kim *et al.*, 1995), the transcripts encoding CRX proteins increase with copper exposure, indicating the existence of a copper responsive system. It was proposed that these proteins, by acting as copper chelators, would significantly decrease the copper concentration surrounding the cell.

Active efflux systems

P1B-type ATPases

The most widely used mechanism of metal resistance is based on active efflux. The simplest example is constituted by the type P1B subfamily of ATPases. These are single unit membrane proteins that transport ions of transition metals like zinc, lead, cadmium, cobalt, copper and silver, a recent review by (Arguello *et al.*, 2007a) includes the most recent advances on catalytic mechanisms and structural characteristics of the functional domains in this class of ATPases in prokaryotes and eukaryotes. Copper-transporting ATPases belong to this subfamily and are important components of systems controlling copper homeostasis in prokaryotes and eukaryotes. Although no structure has been solved for a P1B-type ATPase, recently structural and functional studies of individual domains of a copper-transporting ATPase were performed on the *Archaeoglobus* protein(Arguello *et al.*, 2007a, b; Sazinsky *et al.*, 2006a; Sazinsky *et al.*, 2006b). Copper transporters are characterized by the presence of one or more N-terminal heavy metal-associated (HMA) domains that are conserved domains of about 30 amino acids. These domains contain two cysteine residues, CxxC, responsible for binding and transfer of metal ions (Mana-Capelli *et al.*, 2003; Mandal and Arguello, 2003; Rice *et al.*, 2006). P1B-type ATPases also feature a variable number of transmembrane helices, usually eight, with a cytoplasmic domain for ATP hydrolysis containing the conserved motif DKTGT between the sixth and seventh transmembrane helices (Mandal *et al.*, 2004). Also, a distinctive CPX motif is present in one of the transmembrane segments (the sixth) and it is involved in the binding of the metal ion for its transport across the membrane. Residues located in segments H7 and H8 might also be associated with metal selectivity. A small cytoplasmic loop, the actuator domain A, causes the rearrangement of transmembrane domains following the hydrolysis of ATP (Sazinsky *et al.*, 2006a).

Distribution of metal-transporting ATPases in archaeal genomes

Metal-transporting ATPases have quite conserved sequences and are widespread throughout the three domains of life. A survey of the archaeal genomes has shown the presence of P-type ATPase sequences in Archaea, a superfamily of cation-transporting ATPase that includes the P1B-type subset of metal transporters. Three metal-binding motifs, also present in Bacteria and eukaryotes, were identified in the transporter sequences. These motifs characterize Cu(I)/Ag(I), Zn(II)/Cd(II)/Pb(II) and Cu(II)/Cu(I)/Ag(I) transport (De Hertogh *et al.*, 2004). Archaeal genomes completed and in progress were inspected for presence of P1B-type ATPases. For this analysis, sequences of characterized metal-transporting ATPases were used to search the National Centre for Biotechnology Information database (http://www.ncbi.nlm.nih.gov/genomes/lproks.cgi), the Comprehensive Microbial Resource at (http://cmr.tigr.org/tigr-scripts/CMR/CmrHomePage.cgi), the US Department of Energy Joint Genome Institute at http://wwwjgi.doe.gov/, and the archaeal proteomes at ExPASy (http://expasy.org/sprot/hamap/Archaea.html). Metal ATPases were present in all the genomes considered, with few exceptions (*Cenarchaeum symbiosum*, *Halorubrum lacusprofundi*, *Nanoarchaeum equitans*, *Methanopyrus kandleri*, *Pyrococcus horikoshii*). All sequences considered conserved the active site DKTGT, except for the *Thermoplasma volcanium* sequence, where the internal T was replaced by a S. A sequence annotated as copper-transporting ATPase in *Methanocaldococcus jannaschii*, and an ATPase of *Thermophilum pendens* lacked the CPx metal-binding motif which is a signature of metal-transporting ATPases. These two sequences were excluded from the alignment. The ATPases considered in the analysis possessed different number and types of metal-binding domain (MBD). Three MBDs were identified: a histidine-rich domain, always present in single copy in a subset of Cu(II) transporters, which were also characterized by a CPH motif in the sixth transmembrane helix. A second type of

MBD belongs to the YHS family, named after three of its most conserved amino acids. This domain also possesses two conserved cysteines and is frequently found in Cu ATPases. All sequences analysed had a single copy of the YHS domain. The third, and most frequent, domain is part the HMA family, and it could be present in up to three copies. HMA domains were present at the C-terminal end in some instances. Because of the dissimilarity between MBDs, they were removed prior to the analysis. A phylogenetic tree (Fig. 2.1) was generated using the sequences listed in Table 2.1. Sequences were aligned using ClustalX (Thompson *et al.*, 1997) and the align-

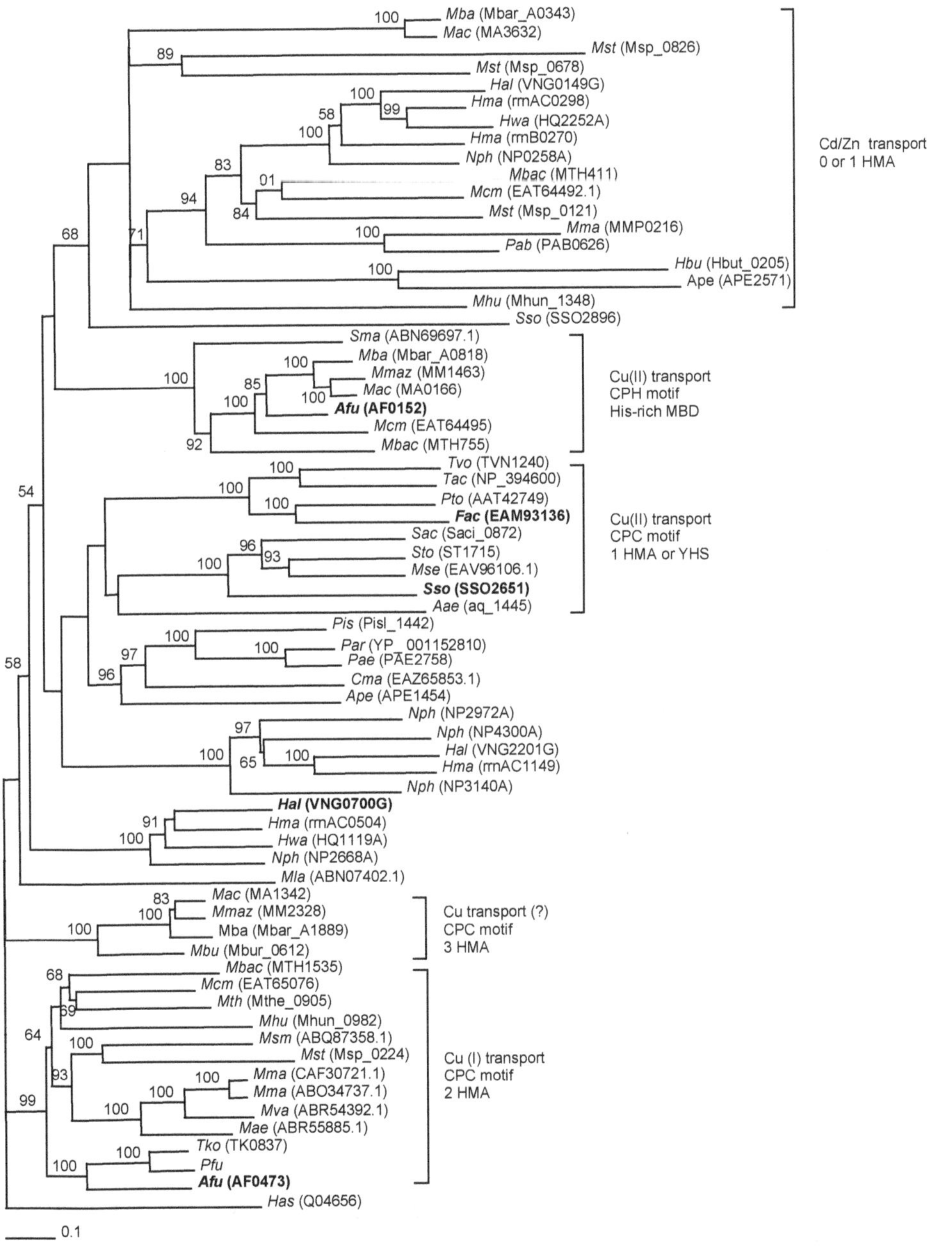

Figure 2.1 Phylogenetic tree of archaeal P1B-type ATPases. N-terminal and C-terminal binding domains were excluded from the analysis. Bootstrap values greater than 54 are indicated. The human copper-transporting ATPase ATP7A (Menkes ATPase) was used as outgroup. Abbreviations of species names and protein identifiers are listed in Table 2.2. Biochemical characterization, gene expression or physiology data were available for sequences in bold type: AF0152 (Mana-Capelli *et al.*, 2003), EAM93136 (Baker-Austin *et al.*, 2005), Sso2651 (Ettema *et al.*, 2006), VNG0700G (Kaur *et al.*, 2006), AF0473 (Yang *et al.*, 2007).

Table 2.1 Distribution of As-resistance genes based on biochemical analyses

Gene product	Name	Phyletic distribution
As(III)-transporter ATPase component	ArsA	Fac (?), Hal, Tac,
As(III)-transporter component	ArsB	Fac, Tac
Arsenate reductase	ArsC	Hal, Sac
Regulatory protein	ArsD	Hal
Arsenite oxidase		Sto, Ape, Sac
Methyl transferase	ArsM	Hal
As responsive regulator	ArsR	Fac, Hal, Tac

ment was edited manually before any further analysis, which was carried out using the programs Seqboot, Protdist, Neighbor, Consense and Fitch of the Phylip package for phylogenetic analysis (Felsenstein, 1989). Although MBDs were removed, the sequences still tended to cluster according the type/number of MBD possessed, indicating that characteristics of the core sequence co-occurred with specific MBD arrangements. The analysis showed a separation in two distinct groups representing Cu ATPases and Cd/Zn ATPases. Cu ATPases could be further divided in Cu(I) and Cu(II) transporters.

Systems for copper efflux

Copper is the cofactor of important enzymes responsible for redox reactions, including superoxide dismutases, cytochrome *c* oxidases and enzymes for iron metabolism, among many others. Since it is highly toxic in concentrations outside the physiological range, all microorganisms possess systems for controlling its homeostasis by regulating influx and efflux, and producing copper-binding proteins that readily complex the free copper.

In Archaea, similarly to Bacteria and yeast, genes for copper-transporting ATPases are often associated with genes encoding copper-binding proteins and a transcriptional regulator controlling the expression of the operons. *Sulfolobus solfataricus* possesses two P-type ATPases encoded by Sso2651 and Sso8962 (She *et al.*, 2001). Sso2651, encoding CopA, is specifically induced by exposure to copper and is under control of the copper responsive regulator CopT (Ettema *et al.*, 2006). The regulator CopT and the copper-binding protein CopM, whose coding gene partially overlaps *copA*, possess a TRASH domain, a sequence specifically involved in binding heavy metal ions and a component of metal responsive regulators and metal transporters, but that can also be found as a stand alone domain (Ettema *et al.*, 2003). The entire *cop* operon is constitutively expressed as a single transcript in the absence of copper, while copper causes a transient peak of *copA* transcription that decreases 10 min after induction (Villafane *et al.*, unpublished). Sso8967, encoding CopB, is transcribed in opposite direction with respect to Sso8968 and Sso11415, encoding a putative transcriptional regulator and copper-binding protein, respectively. The catalytic ATP-binding/phosphorylation domain of the *S. solfataricus* ATPase CopB was capable of hydrolysing ATP, and its activity was stimulated by Cu (II) (Deigweiher *et al.*, 2004). Two copper-transporting ATPases, CopA and CopB, were isolated and characterized in the hyperthermophilic *Archaeoglobus fulgidus* (Mana-Capelli *et al.*, 2003; Mandal and Arguello, 2003) and one atypical P-type ATPase had been identified in *Methanococcus jannaschii*. Later, however, this ATPase turned out to be unrelated with copper homeostasis (Bramkamp *et al.*, 2003). The extreme acidophile *Ferroplasma acidarmanus* possesses a *cop* operon for copper resistance, including genes encoding the transcriptional regulator CopY, the copper-binding protein CopZ and the putative copper-transporting ATPase CopB. Cotranscription of copZ and copB was shown to increase in response to copper, indicating a role in copper efflux (Baker-Austin *et al.*, 2005).

Systems for arsenic efflux

Genetic determinants of bacterial resistance to arsenic are reviewed elsewhere (Mukhopadhyay *et al.*, 2002). The operon *arsRBC* can be either

Table 2.2 Metal-transporting ATPases in archaea

Species	Species ID	Protein ID	Annotation	MBD
Aeropyrum pernix	Ape	APE1454	Hyp. cation-transporting ATPase	2 [a]
Aeropyrum pernix	Ape	APE2571	Hyp. heavy metal-transporting ATPase	0
Aquifex aeolicus	Aae	aq_1445	cation-transporter ATPase (ctrA2)	part. [b]
Archaeoglobus fulgidus	Afu	AF0152	Cu^{2+}-transporting ATPase (CopB)	H-rich N [c]
Archaeoglobus fulgidus	Afu	AF0473	Ag^{+}/Cu^{+}-transporting ATPase (CopA)	2, N/C [d]
Caldivirga maquilingensis	Cma	EAZ658853	heavy metal translocating P-type ATPase	1+ part.
Ferroplasma acidarmanus	Fac	EAM93136	Cu^{2+}-transporting ATPase (CopB)	YHS [e]
Haloarcula marismortui	Hma	rrnB0270	Zinc-transporting ATPase	1
Haloarcula marismortui	Hma	rrnAC0298	Zinc-transporting ATPase	2
Haloarcula marismortui	Hma	rrnAC0504	Copper-transporting ATPase CopA	2
Haloarcula marismortui	Hma	rrnAC1149	Cation-transporting ATPase	1
Halobacterium sp. NRC-1	Hal	VNG0149G	zntA zinc-transporting ATPase	1
Halobacterium sp. NRC-1	Hal	VNG0700G	Mo transport protein	2
Halobacterium sp. NRC-1	Hal	VNG2201G	Metal-transporting CPx-type ATPase	1
Haloquadratum walsbyi	Hwa	HQ1119A	Transport ATPase 1 (probably Cu/metal)	2
Haloquadratum walsbyi	Hwa	HQ2252A	Cd-transporting ATPase	2
Hyperthermus butylicus	Hbu	Hbut_0205	Cation-transporting ATPase	0
Metallosphaera sedula	Mse	EAV96106.1	Heavy metal-translocating P-type ATPase	1
Methanobrevibacter smithii	Msm	ABQ87358	Cation-transporting ATPase	2
Methanococcoides burtonii	Mbu	Mbur_0612	Copper-translocating P-type ATPase	3
Methanococcus aeolicus	Mae	ABR55885	Heavy metal translocating P-type ATPase	part.
Methanococcus maripaludis	Mma	MMP0216	Cation-transporting ATPase	0
Methanococcus maripaludis	Mma	CAF30721.1	Heavy metal transp/detox protein	1

Table 2.2 *continued*

Species	Species ID	Protein ID	Annotation	MBD
Methanococcus maripaludis	Mma	ABO34737	Heavy metal translocating P-type ATPase	1
Methanococcus vannielii	Mva	ABR54392	Heavy metal-translocating P-type ATPase	1
Methanocorpusculum labreanum	Mla	ABN07402.1	Heavy metal-translocating P-type ATPase	2, N/C
Methanoculleus marisnigris	Mcm	EAT65076	Copper-translocating P-type ATPase	2
Methanoculleus marisnigris	Mcm	EAT64495	Copper-translocating P-type ATPase	H-rich N
Methanoculleus marisnigris	Mcm	EAT64492.1	Cd/Co/Hg/Pb/Zn translocating P-type ATPase	0
Methanosaeta thermophila	Mth	Mthe_0905	Heavy metal-translocating P-type ATPase	2
Methanosarcina acetivorans	Mac	MA0166	Copper P-type ATPase	H-rich N
Methanosarcina acetivorans	Mac	MA1342	P-type copper-transporting ATPase	3
Methanosarcina acetivorans	Mac	MA3632	cadmium efflux ATPase	0
Methanosarcina barkeri	Mba	Mbar_A0343	Cadmium efflux ATPase	0
Methanosarcina barkeri	Mba	Mbar_A0818	Copper P-type ATPase	H-rich N
Methanosarcina barkeri	Mba	Mbar_A1889	P-type copper-transporting ATPase	3
Methanosarcina mazei	Mmaz	MM1463	Copper-exporting ATPase	H-rich N
Methanosarcina mazei	Mmaz	MM2328	Copper-exporting ATPase	3
Methanosphaera stadtmanae	Mst	Msp_0121	Predicted cation transport ATPase	H-rich N
Methanosphaera stadtmanae	Mst	Msp_0224	Predicted cation transport ATPase	2
Methanosphaera stadtmanae	Mst	Msp_0678	Predicted cation transport ATPase	0
Methanosphaera stadtmanae	Mst	Msp_0826	Predicted cation transport ATPase	0
Methanospirillum hungatei	Mhu	Mhun_0982	Copper-translocating P-type ATPase	2
Methanospirillum hungatei	Mhu	Mhun_1348	Heavy metal-translocating P-type ATPase	0
Methanothermobacter thermoautotroph	Mbac	MTH755	Heavy metal transporting CPx-type ATPase	H-rich N
Methanothermobacter thermoautotroph	Mbac	MTH1535	Heavy metal transp. CPx-type ATPase	2

Methanothermobacter thermoautotroph	Mbac	MTH411	Cd efflux ATPase	0
Natronomonas pharaonis	Nph	NP0258A	Transport ATPase (prob.substrates Zn/Cd)	1
Natronomonas pharaonis	Nph	NP2668A	transport ATPase	2
Natronomonas pharaonis	Nph	NP2972A	transport ATPase	1
Natronomonas pharaonis	Nph	NP3140A	transport ATPase	1
Natronomonas pharaonis	Nph	NP4300A	transport ATPase	1
Picrophilus torridus	Pto	AAT42749	Copper-transporting ATPase	YHS
Pyrobaculum aerophilum	Pae	PAE2758	Cation-transporting ATPase (P-type)	1
Pyrobaculum arsenaticum	Par	YP_001152810	Heavy metal translocating P-type ATPase	1
Pyrobaculum calidifontis	Pca	Pcal_0727	Heavy metal translocating P-type ATPase	1
Pyrobaculum islandicum	Pis	Pisl_1442	Heavy metal translocating P-type ATPase	1
Pyrococcus abyssi	Pab	PAB0626	Zinc-transporting ATPase	part.
Pyrococcus furiosus	Pfu	PF0740	Heavy metal-transporting cpx-type atpase	2
Sulfolobus acidocaldarius	Sac	Saci_0872	Truncated copper-transp P-type ATPase	1
Sulfolobus solfataricus	Sso	SSO2651	Cu^{2+}-transporting ATPase	1
Sulfolobus solfataricus	Sso	SSO2896	Copper-transporting ATPase	1
Sulfolobus tokodaii	Sto	ST1715	Hypothetical cation-transporter ATPase	1
Staphylothermus marinus	Sma	ABN69697	Copper-translocating P-type ATPase	0
Thermococcus kodakaraensis	Tko	TK0837	Heavy metal-transporting P-type ATPase	2
Thermoplasma acidophilum	Tac	NP_394600	Heavy metal-transporting P-type ATPase	YHS
Thermoplasma volcanium	Tvo	TVN1240	Cation transport ATPase	YHS
Homo sapiens	Has	Q04656	ATP7A Human Menkes Cu-ATPase	6

[a]Number of domains belonging to the HMA family.
[b]Partial metal-binding domain.
[c]Histidine-rich domain present.
[d]Two MBDs are present, one at each end of the protein.
[e]YHS metal-binding domain present.

chromosomally or plasmid encoded. Two additional genes encoding a second regulator *arsD* and an As(III) translocating ATPase *arsA* can be found on separate operons or plasmids both in Archaea and Bacteria (Lin *et al.*, 2006; Wu and Rosen, 1993; Xu *et al.*, 1998). The ATPase ArsA supplement the activity of the transporter ArsB, acting as a catalytic subunit (Dey and Rosen, 1995). Table 2.1 summarizes genes involved in arsenic resistance and metabolism, their functions and presence in archaeal species.

From the analysis of sequenced archaeal genomes emerged that *Thermoplasma acidophilum* possesses homologues of the *arsRB* operon and of the *arsA* gene in two distant loci, while the two species *S. solfataricus* and *S. todokaii* do not show any putative homologues of these genes (Dopson *et al.*, 2003). It is possible that some species use alternative strategies for arsenic resistance. For example, *Sulfolobus acidocaldarius* strain BC (renamed *Sulfolobus metallicus*), an archaeon of interest to bioleaching, oxidizes arsenite to the less toxic arsenate, using an As(III) oxidase located in the membrane (Sehlin and Lindstroem, 1992).

The acidophilic archaeon *Ferroplasma acidarmanus* can grow in high concentrations of arsenate (up to 133 mM) without effect on growth rate (Baker-Austin *et al.*, 2007). Interestingly, although *Ferroplasma* is capable of growth in such high levels of arsenate, its genome does not encode an arsenate reductase *arsC* or a phosphate transporter system Pst. The absence of arsenate reduction was confirmed by biochemical analysis. Hence, the authors speculate that additional genes for arsenate resistance should be present. The *Ferroplasma* chromosome possesses an *arsRB* homologue operon which is inducible by arsenite, but not by arsenate (Baker-Austin *et al.*, 2007; Gihring *et al.*, 2003). On a separate locus, a sequence similar to *arsA* was also identified in the *Ferroplasma* chromosome (Gihring *et al.*, 2003); however, this sequence might be a pseudogene in that it lacks an N-terminal domain and promoter (Baker-Austin *et al.*, 2007).

An extensive study of arsenic resistance in *Halobacterium* was conducted by the DasSharma and Rensing laboratories. Homologues of arsenic resistance genes were identified on the chromosome and on one of the two megaplasmids of *Halobacterium* sp. NRC-1. The megaplasmid pNRC100 encodes the gene cluster *arsADRC* and *arsR2M*. Deletion of *arsADRC* produced arsenite and antimonite sensitivity, but not to arsenate, indicating the presence of an export system specific for As(III) and Sb(III). Knockout of *arsB*, chromosomally encoded, did not affect the cell sensitivity to arsenic oxyanions, although *arsB* was shown to be inducible by arsenite and also antimonite (Wang *et al.*, 2004).

Enzymatic detoxification

Arsenate and chromium reductases

Arsenic resistance operons carry genes encoding an arsenic responsive regulator *arsR*, an As(III) transporter *arsB* and an arsenate reductase *arsC*. The arsenate reductase converts arsenate As(V) into arsenite As(III) that is actively transported outside the cell. Homologues of ArsC, the bacterial detoxifying arsenate reductase, are present in the annotation of several archaeal genomes (*Pyrococcus furiosus, Methanosarcina mazei, M. acetivorans, Methanothermobacter thermoautothrophicus, Archaeoglobus fulgidus, Candidatus nitrosopumilus, Haloquadratum walsbyi, Natromonas pharaonis, Thermococcus kodakarensis, Haloarcula marismortui*). Cr(IV) can be reduced to the less toxic form Cr(III) by Bacteria (Cervantes *et al.*, 2001). *Pyrobaculum islandicus* can reduce Cr(IV), in addition to several other metals (Kashefi and Lovley, 2000).

Methylation of arsenic oxyanions

Another mechanism of detoxification involves the methylation of inorganic forms of arsenic to volatile arsines and non-volatile species. Arsenic methylation has been studied in microbial communities (Islam *et al.*, 2005) and in pure cultures. The methanogenic Archaea *Methanobacterium bryantii, M. formicicum,* and *Methanosarcina barkeri* were reported to produce volatile forms of methylated arsenic (Michalke *et al.*, 2000). While methylation of arsenic has been described in eukaryotes, the mechanism is still unclear in microorganisms. ArsM is bacterial homologue of the rat methyltransferase and catalyses the formation of trimethylarsine from arsenite. A knockout of the gene encoding ArsM in *Halobacterium* resulted in sensitivity to arsenite, confirming its role as detoxifying enzyme (Wang *et al.*, 2004). ArsM has also been identified in 16 archaeal species including three *Methanosarcina* species and *Halobacterium salinarum* (Qin *et al.*, 2006).

Systems of mercury detoxification

Possibly, the mercury resistance system has been the first strategy of metal resistance to be described, with the isolation of an Hg resistance *Staphylococcus aureus* in 1960, and following studies of genetic determinants, regulation and mechanism of action of the mercury reductase MerA in *E. coli* and subsequently in other bacterial species (Barkay *et al.*, 2003). MerA catalyses the conversion of Hg(II) into the volatile Hg(0). In the crenarcheote *Sulfolobus* only recently a mercury detoxification system has been described. The *merRAHI* operon of *Sulfolobus* represents the most detailed study of mercury resistance in Archaea. Studies employing target gene disruption have shown that the *S. solfataricus* homologue of the bacterial mercury reductase MerA confers a low but measurable level of mercury resistance. A mutant carrying an inactive copy of *merR* expresses *merA* constitutively, indicating a negative regulatory role for the *Sulfolobus* MerR (Schelert *et al.*, 2004). Similarly to bacterial MerR factors, but presumably using different mechanisms, the archaeal MerR also induces transcription without leaving the *merA* promoter but producing conformational changes of the promoter itself in the presence of mercury (Schelert *et al.*, 2006). Two additional genes, named *merH* and *merI*, and located up- and downstream of *merA*, respectively, were identified in the *mer* operon. Both genes were cotranscribed with *merA* and induced by mercury. MerH contains the conserved metal-binding domain TRASH, for this reason it has been proposed to function as a chaperone for mercury mobilization (Schelert *et al.*, 2006). MerI is constitutively expressed from its own promoter, but in the presence of Hg its transcript increases due to co-transcription with MerHA from the *merH*p promoter. Disruption of *merI* did not affect the pattern of growth of the mutant strain, therefore the role of MerI is still unclear (Schelert *et al.*, 2006). Blast searches of archaeal genomes identified homologues of MerA also in *S. acidocaldarius*, *S. tokodaii*, *Aeropyrum pernix*, *Picrophilus torridus*, *Thermoplasma volcanium* and *Pyrobaculum aerophilum*.

Metal-responsive transcriptional regulators

Systems of metal resistance are all subject to regulation. The availability of genome sequences has allowed the discovery of metal-responsive regulators thanks to their proximity to detoxifying enzymes, permeases and transporters. Although the architecture of the archaeal transcriptional systems is characterized by eukaryotic-like traits, regulation of transcription is controlled by factors highly similar to bacterial transcriptional regulators. Major families of metal-responsive regulators are the MerR (Brown *et al.*, 2003), ArsR/SmtB (Busenlehner *et al.*, 2003), in Bacteria two-component systems such as CusRS and PcoRS (copper) (Munson *et al.*, 2000) and SilRS (silver) (Silver, 2003).

Some essential metals regulate expression of the metallo-enzymes, for which they act as cofactors. For example, molybdenum contributes to control the expression of the nitrate reductase operon in *E. coli* (Studholme and Pau, 2003). Microorganisms have evolved systems for the specific uptake of nickel and cobalt that are found in the environment only at very low concentrations. A strategy developed for resistance to higher levels of these metals is by repression of the uptake system, rather than active efflux. In *E. coli* the expression of nickel permease genes is repressed by the Ni-responsive regulator NikR that in the presence of high nickel concentrations acts in the form of a tetramer. The structure of the archaeal homologue, NikR, of *Pyrococcus horikoshii* was solved both in the apo form and with bound nickel. The structure can assume different conformations depending on the amount of nickel present, and binding of nickel allows the regulator to change its conformation and recognize the target promoter (Chivers and Tahirov, 2005).

Probably the first report concerning an archaeal metal responsive regulator is the one by Bell and coworkers (Bell *et al.*, 1999). The group investigated the binding properties of MDR1, the *Archaeoglobus fulgidus* homologue of the bacterial metal-responsive repressor DtxR. The recombinant MDR1 was shown to bind to specific locations of its own promoter in response to increased level of different metals in gel shift experiments, thereby repressing its own expression in response to metals.

MerR-like regulators have been assigned to the MerR family mainly because of the strong sequence similarity of the N-termini, the region encoding the DNA-binding domain. This family of transcription factors may have evolved from an ancient metal-responsive regulator that activated suboptimal promoters by reducing the space between specific recognition sequences. Regulators for specific metals would have maintained the mechanism of action while changing the C-terminal metal binding site to accommodate the new metal.

Future developments

Metal metabolism and resistance has been extensively studied in Bacteria, and these mechanisms have only recently started to be investigated in Archaea. Although studies have focused on a limited number of archaeal species to date, from the inspection of an increasing number of genome sequences it is becoming evident that genes for resistance and detoxification are widespread throughout the archaeal domain. Microbial genomes are an important source of clues on genetic determinants of metal resistance. Homology searches, using known genes, allow detection of genes related to metal metabolism in microbial genomes and enable studies of their distribution and evolution. Furthermore, application of genome sequences in global expression studies will help identifying novel systems for metal resistance, in Archaea and Bacteria. High throughput methods, such as microarray analysis and proteomics, generate large amount of data that can be analysed using systems-level approaches. Mathematical models are useful to describe biological processes. The quantitative modelling of a relatively simple system as the homeostasis of specific metal ions in Archaea can be accomplished. However, modelling requires availability of experimental data based on physiological responses, biochemical and structural information. A detailed mathematical model of copper efflux in *E. irae* has been proposed. The model describes how the *cop* operon responds to the presence of metal and can predict its behaviour following exposure to different copper concentrations (Pecou *et al.*, 2006). Attempts to describe metal homeostasis in *Halobacterium* using a systems approach have been made (Kaur *et al.*, 2006).

Another interesting, still unexplored, area concerns the presence of riboswitches in Archaea. RNA can perform a range of catalytic reactions and functions (Grundy and Henkin, 2006). Recent studies in Bacteria have shown evidence not only that metal-responsive proteins act as sensors and regulators of metal resistance genes, but also that riboswitches, special RNA secondary structures located at the 5′ end of mRNAs, can sense metal ions and interfere with the progress of transcription (Brantl, 2006). One example is the regulation by Mg(II) in the bacterial species *Salmonella*, *E. coli*, *Klebsiella* and *Citrobacter* (Cromie *et al.*, 2006). Although there are no reports of archaeal metal-sensing riboswitches, this field is just emerging and will be interesting to learn whether Archaea use this strategy to regulate their levels of metal ions.

References

Alvarez, S., and Jerez, C.A. (2004). Copper ions stimulate polyphosphate degradation and phosphate efflux in *Acidithiobacillus ferrooxidans*. Appl. Environ. Microbiol. *70*, 5177–5182.

Arguello, J.M., Eren, E., and Gonzalez-Guerrero, M. (2007a). The structure and function of heavy metal transport P1B-ATPases. Biometals *20*, 233–248.

Arguello, J.M., Eren, E., and Gonzalez-Guerrero, M. (2007b). The structure and function of heavy metal transport P(1B)-ATPases. Biometals.

Baker-Austin, C., Dopson, M., Wexler, M., Sawers, R.G., and Bond, P.L. (2005). Molecular insight into extreme copper resistance in the extremophilic archaeon '*Ferroplasma acidarmanus*' Fer1. Microbiology *151*, 2637–2646.

Baker-Austin, C., Dopson, M., Wexler, M., Sawers, R.G., Stemmler, A., Rosen, B.P., and Bond, P.L. (2007). Extreme arsenic resistance by the acidophilic archaeon '*Ferroplasma acidarmanus*' Fer1. Extremophiles *11*, 425–434.

Barkay, T., Miller, S.M., and Summers, A.O. (2003). Bacterial mercury resistance from atoms to ecosystems. FEMS Microbiol. Revs. *27*, 355–384.

Bell, S.D., Cairns, S.S., Robson, R.L., and Jackson, S.P. (1999). Transcriptional regulation of an archaeal operon in vivo and in vitro. Molecular cell *4*, 971–982.

Bevers, L.E., Bol, E., Hagedoorn, P.L., and Hagen, W.R. (2005). WOR5, a novel tungsten-containing aldehyde oxidoreductase from *Pyrococcus furiosus* with a broad substrate Specificity. J. Bacteriol. *187*, 7056–7061.

Bevers, L.E., Hagedoorn, P.L., Krijger, G.C., and Hagen, W.R. (2006). Tungsten transport protein A (WtpA) in *Pyrococcus furiosus*: the first member of a new class of tungstate and molybdate transporters. J. Bacteriol. *188*, 6498–6505.

Bramkamp, M., Gassel, M., Herkenhoff-Hesselmann, B., Bertrand, J., and Altendorf, K. (2003). The Methanocaldococcus jannaschii protein Mj0968 is not a P-type ATPase. FEBS Lett. *543*, 31–36.

Brantl, S. (2006). Bacterial gene regulation: metal ion sensing by proteins or RNA. Trends Biotechnol. *24*, 383–386.

Brown, N.L., Stoyanov, J.V., Kidd, S.P., and Hobman, J.L. (2003). The MerR family of transcriptional regulators. FEMS Microbiol. Revs. *27*, 145–163.

Busenlehner, L.S., Pennella, M.A., and Giedroc, D.P. (2003). The SmtB/ArsR family of metalloregulatory transcriptional repressors: Structural insights into prokaryotic metal resistance. FEMS Microbiol. Revs. *27*, 131–143.

Cervantes, C., Campos-Garcia, J., Devars, S., Gutierrez-Corona, F., Loza-Tavera, H., Torres-Guzman, J.C., and Moreno-Sanchez, R. (2001). Interactions of chromium with microorganisms and plants. FEMS Microbiol. Rev. *25*, 335–347.

Cervantes, C., Ji, G., Ramirez, J.L., and Silver, S. (1994). Resistance to arsenic compounds in microorganisms. FEMS Microbiol. Revs. *15*, 355–367.

Chivers, P.T., and Sauer, R.T. (2000). Regulation of high affinity nickel uptake in Bacteria. Ni^{2+}-dependent interaction of NikR with wild-type and mutant operator sites. J. Biol. Chem. *275*, 19735–19741.

Chivers, P.T., and Tahirov, T.H. (2005). Structure of *Pyrococcus horikoshii* NikR: nickel sensing and implications for the regulation of DNA recognition. J. Mol. Biol. *348*, 597–607.

Cromie, M.J., Shi, Y., Latifi, T., and Groisman, E.A. (2006). An RNA sensor for intracellular $Mg(^{2+})$. Cell *125*, 71–84.

De Hertogh, B., Lantin, A.C., Baret, P.V., and Goffeau, A. (2004). The archaeal P-type ATPases. J. Bioenerg. Biomembr. *36*, 135–142.

Deigweiher, K., Drell, T.L.t., Prutsch, A., Scheidig, A.J., and Lubben, M. (2004). Expression, isolation, and crystallization of the catalytic domain of CopB, a putative copper-transporting ATPase from the thermoacidophilic archaeon *Sulfolobus solfataricus*. J. Bioenerg. Biomembr. *36*, 151–159.

Dey, S., and Rosen, B.P. (1995). Dual mode of energy coupling by the oxyanion-translocating ArsB protein. J. Bacteriol. *177*, 385–389.

Dhawan, I.K., Roy, R., Koehler, B.P., Mukund, S., Adams, M.W., and Johnson, M.K. (2000). Spectroscopic studies of the tungsten-containing formaldehyde ferredoxin oxidoreductase from the hyperthermophilic archaeon Thermococcus litoralis. J. Biol. Inorg. Chem. *5*, 313–327.

Diekert, G., Konheiser, U., Piechulla, K., and Thauer, R.K. (1981). Nickel requirement and factor F430 content of methanogenic Bacteria. J. Bacteriol. *148*, 459–464.

Dopson, M., Baker-Austin, C., Koppineedi, P.R., and Bond, P.L. (2003). Growth in sulfidic mineral environments: metal resistance mechanisms in acidophilic micro-organisms. Microbiology *149*, 1959–1970.

Eitinger, T., and Mandrand-Berthelot, M.A. (2000). Nickel transport systems in microorganisms. Arch. Microbiol. *173*, 1–9.

Ellefson, W.L., Whitman, W.B., and Wolfe, R.S. (1982). Nickel-containing factor F430: chromophore of the methylreductase of *Methanobacterium*. Proc. Natl. Acad. Sci. USA *79*, 3707–3710.

Ellis, P.J., Conrads, T., Hille, R., and Kuhn, P. (2001). Crystal structure of the 100 kDa arsenite oxidase from Alcaligenes faecalis in two crystal forms at 1.64 A and 2.03 A. Structure *9*, 125–132.

Ettema, T.J., Brinkman, A.B., Lamers, P.P., Kornet, N.G., de Vos, W.M., and van der Oost, J. (2006). Molecular characterization of a conserved archaeal copper resistance (cop) gene cluster and its copper-responsive regulator in *Sulfolobus solfataricus* P2. Microbiology *152*, 1969–1979.

Ettema, T.J., Huynen, M.A., de Vos, W.M., and van der Oost, J. (2003). TRASH: a novel metal-binding domain predicted to be involved in heavy metal sensing, trafficking and resistance. Trends Biochem. Sci. *28*, 170–173.

Fani, R., Gallo, R., and Lio, P. (2000). Molecular evolution of nitrogen fixation: the evolutionary history of the nifD, nifK, nifE, and nifN genes. J. Mol. Evol. *51*, 1–11.

Felsenstein, J. (1989). PHYLIP – Phylogeny Inference Package (Version 3.2). Cladistics 5, 164–166.

Forterre, P. (1996). A hot topic: the origin of hyperthermophiles. Cell 85, 789–792.

Gihring, T.M., Bond, P.L., Peters, S.C., and Banfield, J.F. (2003). Arsenic resistance in the archaeon '*Ferroplasma acidarmanus*': new insights into the structure and evolution of the ars genes. Extremophiles 7, 123–130.

Grundy, F.J., and Henkin, T.M. (2006). From ribosome to riboswitch: control of gene expression in Bacteria by RNA structural rearrangements. Crits. Rev. Biochem. Mol. Biol. *41*, 329–338.

Hartzell, P.L., and Wolfe, R.S. (1986). Requirement of the nickel tetrapyrrole F430 for in vitro methanogenesis: reconstitution of methylreductase component C from its dissociated subunits. Proc. Natl. Acad. Sci. USA 83, 6726–6730.

Heider, J., Ma, K., and Adams, M.W. (1995). Purification, characterization, and metabolic function of tungsten-containing aldehyde ferredoxin oxidoreductase from the hyperthermophilic and proteolytic archaeon Thermococcus strain ES-1. J. Bacteriol. *177*, 4757–4764.

Hollenstein, K., Frei, D.C., and Locher, K.P. (2007). Structure of an ABC transporter in complex with its binding protein. Nature *446*, 213–216.

Huber, C., and Wachtershauser, G. (1998). Peptides by activation of amino acids with CO on (Ni,Fe)S surfaces: implications for the origin of life. Science *281*, 670–672.

Huber, C., and Wachtershauser, G. (2006). alpha-Hydroxy and alpha-amino acids under possible Hadean, volcanic origin-of-life conditions. Science *314*, 630–632.

Huber, R., Sacher, M., Vollmann, A., Huber, H., and Rose, D. (2000). Respiration of arsenate and selenate by hyperthermophilic Archaea. Syst. Appl. Microbiol. *23*, 305–314.

Islam, S.M., Fukushi, K., and Yamamoto, K. (2005). Development of an enumeration method for arsenic methylating Bacteria from mixed culture samples. Biotechnol. Lett. *27*, 1885–1890.

Jarrell, K.F., and Sprott, G.D. (1982). Nickel transport in *Methanobacterium bryantii*. J. Bacteriol. *151*, 1195–1203.

Johnson, J.L., Rajagopalan, K.V., Mukund, S., and Adams, M.W. (1993). Identification of molybdopterin as the organic component of the tungsten cofactor in four enzymes from hyperthermophilic Archaea. J. Biol. Chem. *268*, 4848–4852.

Kashefi, K., and Lovley, D.R. (2000). Reduction of Fe(III), Mn(IV), and toxic metals at 100 degrees C by *Pyrobaculum islandicum*. Appl. Environ. Microbiol. *66*, 1050–1056.

Kaur, A., Pan, M., Meislin, M., Facciotti, M.T., El-Gewely, R., and Baliga, N.S. (2006). A systems view of haloarchaeal strategies to withstand stress from transition metals. Genome Res. *16*, 841–854.

Kim, B.K., Pihl, T.D., Reeve, J.N., and Daniels, L. (1995). Purification of the copper response extracellular proteins secreted by the copper-resistant methanogen *Methanobacterium bryantii* BKYH and cloning, sequencing, and transcription of the gene encoding these proteins. J. Bacteriol. *177*, 7178–7185.

Kletzin, A., and Adams, M.W. (1996). Tungsten in biological systems. FEMS Microbiol. Rev. *18*, 5–63.

Lebrun, E., Brugna, M., Baymann, F., Muller, D., Lievremont, D., Lett, M.C., and Nitschke, W. (2003). Arsenite oxidase, an ancient bioenergetic enzyme. Mol. Biol. Evol. *20*, 686–693.

Leigh, J.A. (2000). Nitrogen fixation in methanogens: the archaeal perspective. Curr. Issues Mol. Biol. 2, 125–131.

Lin, Y.F., Walmsley, A.R., and Rosen, B.P. (2006). An arsenic metallochaperone for an arsenic detoxification pump. Proc. Natl. Acad. Sci. USA *103*, 15617–15622.

Maden, B.E. (1995). No soup for starters? Autotrophy and the origins of metabolism. Trends Biochem. Sci. *20*, 337–341.

Mana-Capelli, S., Mandal, A.K., and Arguello, J.M. (2003). *Archaeoglobus fulgidus* CopB is a thermophilic Cu^{2+}-ATPase: functional role of its histidine-rich-N-terminal metal binding domain. J. Biol. Chem. *278*, 40534–40541.

Mandal, A.K., and Arguello, J.M. (2003). Functional roles of metal binding domains of the *Archaeoglobus fulgidus* Cu(+)-ATPase CopA. Biochemistry *42*, 11040–11047.

Mandal, A.K., Yang, Y., Kertesz, T.M., and Arguello, J.M. (2004). Identification of the transmembrane metal binding site in Cu+-transporting PIB-type ATPases. J. Biol. Chem. *279*, 54802–54807.

Martin, W., and Russell, M.J. (2003). On the origins of cells: a hypothesis for the evolutionary transitions from abiotic geochemistry to chemoautotrophic prokaryotes, and from prokaryotes to nucleated cells. Philos. Trans. R. Soc. Lond B Biol. Sci. *358*, 59–83; discussion 83–55.

Michalke, K., Wickenheiser, E.B., Mehring, M., Hirner, A.V., and Hensel, R. (2000). Production of volatile derivatives of metal(loid)s by microflora involved in anaerobic digestion of sewage sludge. Appl. Environ. Microbiol. *66*, 2791–2796.

Mukhopadhyay, R., Rosen, B.P., Phung, L.T., and Silver, S. (2002). Microbial arsenic: from geocycles to genes and enzymes. FEMS Microbiol. Revs. *26*, 311–325.

Mukund, S., and Adams, M.W. (1996). Molybdenum and vanadium do not replace tungsten in the catalytically active forms of the three tungstoenzymes in the hyperthermophilic archaeon *Pyrococcus furiosus*. J. Bacteriol. *178*, 163–167.

Muller, D., Lievremont, D., Simeonova, D.D., Hubert, J.C., and Lett, M.C. (2003). Arsenite oxidase aox genes from a metal-resistant beta-proteobacterium. J. Bacteriol. *185*, 135–141.

Munson, G.P., Lam, D.L., Outten, F.W., and O'Halloran, T.V. (2000). Identification of a copper-responsive two-component system on the chromosome of Escherichia coli K-12. J. Bacteriol. *182*, 5864–5871.

Navarro, C., Wu, L.F., and Mandrand-Berthelot, M.A. (1993). The nik operon of Escherichia coli encodes a periplasmic binding-protein-dependent transport system for nickel. Mol. Microbiol. *9*, 1181–1191.

Oremland, R.S., and Stolz, J.F. (2003). The ecology of arsenic. Science *300*, 939–944.

Patteson, K.G., Trivedi, N., and Stadtman, T.C. (2005). *Methanococcus vannielii* selenium-binding protein (SeBP): chemical reactivity of recombinant SeBP produced in *Escherichia coli*. Proc. Natl. Acad. Sci. USA *102*, 12029–12034.

Pecou, E., Maass, A., Remenik, D., Briche, J., and Gonzalez, M. (2006). A mathematical model for copper homeostasis in *Enterococcus hirae*. Math. Biosci. *203*, 222–239.

Qin, J., Rosen, B.P., Zhang, Y., Wang, G., Franke, S., and Rensing, C. (2006). Arsenic detoxification and evolution of trimethylarsine gas by a microbial arsenite S-adenosylmethionine methyltransferase. Proc. Natl. Acad. Sci. USA *103*, 2075–2080.

Rasmussen, B. (2000). Filamentous microfossils in a 3,235-million-year-old volcanogenic massive sulphide deposit. Nature *405*, 676–679.

Remonsellez, F., Orell, A., and Jerez, C.A. (2006). Copper tolerance of the thermoacidophilic archaeon Sulfolobus metallicus: possible role of polyphosphate metabolism. Microbiology *152*, 59–66.

Rice, W.J., Kovalishin, A., and Stokes, D.L. (2006). Role of metal-binding domains of the copper pump from *Archaeoglobus fulgidus*. Biochem. Biophy. Res. Commun. *348*, 124–131.

Rodionov, D.A., Hebbeln, P., Gelfand, M.S., and Eitinger, T. (2006). Comparative and functional genomic analysis of prokaryotic nickel and cobalt uptake transporters: evidence for a novel group of ATP-binding cassette transporters. J. Bacteriol. *188*, 317–327.

Rother, M., Resch, A., Gardner, W.L., Whitman, W.B., and Bock, A. (2001). Heterologous expression of archaeal selenoprotein genes directed by the SECIS element located in the 3′ non-translated region. Mol. Microbiol. *40*, 900–908.

Roy, R., and Adams, M.W. (2002). Characterization of a fourth tungsten-containing enzyme from the hyperthermophilic archaeon *Pyrococcus furiosus*. J. Bacteriol. *184*, 6952–6956.

Rubio, L.M., and Ludden, P.W. (2005). Maturation of nitrogenase: a biochemical puzzle. J. Bacteriol. *187*, 405–414.

Sazinsky, M.H., Agarwal, S., Arguello, J.M., and Rosenzweig, A.C. (2006a). Structure of the actuator domain from the *Archaeoglobus fulgidus* Cu(+)-ATPase. Biochemistry *45*, 9949–9955.

Sazinsky, M.H., Mandal, A.K., Arguello, J.M., and Rosenzweig, A.C. (2006b). Structure of the ATP binding domain from the *Archaeoglobus fulgidus* Cu+-ATPase. J. Biol. Chem. *281*, 11161–11166.

Schelert, J., Dixit, V., Hoang, V., Simbahan, J., Drozda, M., and Blum, P. (2004). Occurrence and characterization of mercury resistance in the hyperthermophilic archaeon *Sulfolobus solfataricus* by use of gene disruption. J. Bacteriol. *186*, 427–437.

Schelert, J., Drozda, M., Dixit, V., Dillman, A., and Blum, P. (2006). Regulation of mercury resistance in the crenarchaeote *Sulfolobus solfataricus*. J. Bacteriol. *188*, 7141–7150.

Schonheit, P., Moll, J., and Thauer, R.K. (1979). Nickel, cobalt, and molybdenum requirement for growth of *Methanobacterium thermoautotrophicum*. Arch. Microbiol. *123*, 105–107.

Sehlin, H.M., and Lindstroem, E.B. (1992). Oxidation and reduction of arsenic by Sulfolobus acidocaldarius strain BC. FEMS Microbiology Letters *93*, 87–92.

Self, W.T., Grunden, A.M., Hasona, A., and Shanmugam, K.T. (2001). Molybdate transport. Res. Microbiol. *152*, 311–321.

Self, W.T., Pierce, R., and Stadtman, T.C. (2004). Cloning and heterologous expression of a Methanococcus vannielii gene encoding a selenium-binding protein. IUBMB Life 56, 501–507.

She, Q., Singh, R.K., Confalonieri, F., Zivanovic, Y., Allard, G., Awayez, M.J., Chan-Weiher, C.C., Clausen, I.G., Curtis, B.A., De Moors, A., *et al.* (2001). The complete genome of the crenarchaeon *Sulfolobus solfataricus* P2. Proc. Natl. Acad. Sci. USA *98*, 7835–7840.

Silver, S. (2003). Bacterial silver resistance: molecular biology and uses and misuses of silver compounds. FEMS Microbiol. Rev. *27*, 341–353.

Smith, W.L., and Gadd, G.M. (2000). Reduction and precipitation of chromate by mixed culture sulphate-reducing bacterial biofilms. J. Appl. Microbiol. *88*, 983–991.

Stolz, J.F., Basu, P., Santini, J.M., and Oremland, R.S. (2006). Arsenic and selenium in microbial metabolism. Annu. Rev. Microbiol. *60*, 107–130.

Studholme, D.J., and Pau, R.N. (2003). A DNA element recognised by the molybdenum-responsive transcription factor ModE is conserved in Proteobacteria, green sulphur Bacteria and Archaea. BMC Microbiol. *3*, 24.

Tanaka, Y., Tsumoto, K., Nakanishi, T., Yasutake, Y., Sakai, N., Yao, M., Tanaka, I., and Kumagai, I. (2004). Structural implications for heavy metal-induced reversible assembly and aggregation of a protein: the case of *Pyrococcus horikoshii* CutA. FEBS Lett. *556*, 167–174.

Thompson, J.D., Gibson, T.J., Plewniak, F., Jeanmougin, F., and Higgins, D.G. (1997). The CLUSTAL_X windows interface: flexible strategies for multiple sequence alignment aided by quality analysis tools. Nucleic Acids Res. *25*, 4876–4882.

Valko, M., Morris, H., and Cronin, M.T. (2005). Metals, toxicity and oxidative stress. Curr. Med. Chem. *12*, 1161–1208.

Vargas, M., Kashefi, K., Blunt-Harris, E.L., and Lovley, D.R. (1998). Microbiological evidence for Fe(III) reduction on early Earth. Nature *395*, 65–67.

Wachtershauser, G. (2006). From volcanic origins of chemoautotrophic life to Bacteria, Archaea and Eukarya. Phil. Trans. R. Soc. Lond. B Biol. Sci. *361*, 1787–1806; discussion 1806-1788.

Wang, G., Kennedy, S.P., Fasiludeen, S., Rensing, C., and DasSarma, S. (2004). Arsenic resistance in *Halobacterium* sp. strain NRC-1 examined by using an improved gene knockout system. J. Bacteriol. *186*, 3187–3194.

Wilting, R., Schorling, S., Persson, B.C., and Bock, A. (1997). Selenoprotein synthesis in Archaea: identification of an mRNA element of *Methanococcus jannaschii* probably directing selenocysteine insertion. J. Mol. Biol. *266*, 637–641.

Wu, J., and Rosen, B.P. (1993). The *arsD* gene encodes a second trans-acting regulatory protein of the plasmid-encoded arsenical resistance operon. Mol. Microbiol. *8*, 615–623.

Xu, C., Zhou, T., Kuroda, M., and Rosen, B.P. (1998). Metalloid resistance mechanisms in prokaryotes. J. Biochem. (Tokyo) *123*, 16-23.

Yang, Y., Mandal, A.K., Bredeston, L.M., Luis Gonzalez-Flecha, F., and Arguello, J.M. (2007). Activation of *Archaeoglobus fulgidus* Cu(+)-ATPase CopA by cysteine. Biochim. Biophys. Acta *1768*, 495–501.

Zehr, J.P., Jenkins, B.D., Short, S.M., and Steward, G.F. (2003). Nitrogenase gene diversity and microbial community structure: a cross-system comparison. Environ. Microbiol. *5*, 539–554.

Redox Enzymes in the Archaea

3

Edward J. Crane III, Charles S. Hummel and Evan T. Hall

Abstract

One of the hallmarks of living systems is their ability to use favourable redox reactions in the conversion of energy to forms that are useful to the cell. Microbes in the kingdom Archaea contain many unique redox enzymes, an observation which may result from the wide range of strategies they employ for energy conversion, the many extreme environments they inhabit, and the evolutionary separation of the Archaea from Bacteria that catalyse similar reactions. This chapter discusses archaeal redox enzymes, focusing mainly on enzymes that (1) are unique to this kingdom, (2) appear to provide a selective advantage under extreme conditions or (3) may be present outside the Archaea but have been best characterized from archaeal sources. Subjects covered include enzymes involved in sulphur metabolism, including both sulphur oxidation and reduction and the hydrogenases frequently associated with sulphur reduction, as well as several enzymes involved in electron transport. The central metabolic pathways of many archaeons utilize unique enzymes, and these enzymes are also discussed, with a focus on tungsten and ferredoxin-dependent enzymes and small proteins that participate in electron transfer. Enzymes of archaeal anti-oxidant systems, including the superoxide reductase system of *Pyrococcus* are also covered, as are some enzymes from the uniquely archaeal methanogenesis pathway.

Introduction

The kingdom Archaea contains a large range of organisms capable of supporting themselves by a wide range of energetic lifestyles, and as such they have an incredible array of redox enzymes capable of catalysing a variety of reactions. From the strictly anaerobic methanogens and sulphur reducers to the obligately aerobic *Sulfolobus* and phototrophic halophiles, with all of the facultative lifestyles in between, the range of redox environments occupied by and energy-conserving mechanisms utilized by the Archaea is immense and likely includes all available environments on earth (for an exhaustive review see (Amend and Shock, 2001)). The recent observation of archaeal hyperthermophiles capable of nitrogen fixation eliminated one of the few lifestyles believed to be absent in the kingdom Archaea (Mehta and Baross, 2006; Mehta *et al.*, 2003; Mehta *et al.*, 2005). Many of these pathways may be the surviving representatives of ancient metabolic strategies (such as the emphasis on the use of tungsten), although it should be kept in mind that these reactions are not present merely as vestigial remains; they have survived to this day because they are effective. As redox metabolism is the very energetic foundation on which life is based, this chapter is attempting to summarize a huge body of knowledge.

In order to keep this discussion specific to the microbes in the kingdom Archaea, it will focus on enzymes and systems (1) which appear to be unique to the Archaea and/or (2) which appear to confer an advantage to survival under extreme conditions of heat, salt, or acid, or (3) that may be of general importance, but have been best characterized in the Archaea. This review will also attempt to focus on systems whose presence (or absence) in the Archaea seems to be particularly revealing regarding the evolution of redox enzymes and pathways. Much of the work on the enzymes of hyperthermophiles has focused on the structural differences in proteins which confer thermostability, and excellent reviews are available on the considerable progress that has been made in this area (Razvi and Scholtz, 2006; Trivedi *et al.*, 2006), and for this reason this review will avoid that particular subject. It should be noted, however, that many of the enzymes discussed herein have provided much of what we know about the complex subject of enzyme thermostability.

Examples of enzymes of archaeal respiration and electron transport processes, with a focus on sulphur chemistry

Aerobic oxidation of S^0 via sulphur oxygenase and the TQO/'Dox' system

While the sulphur-dependent or sulphur-utilizing nature of many Archaea is well known, the precise mechanisms of many of these reactions have not been well characterized. An almost certain reason for this is a combination of their complex, membrane-bound nature and the many different systems evolution has designed to perform sulphur redox chemistry. Despite its important role in the geochemistry of many volcanic systems such as solfatars, continental hydrothermal springs and marine hydrothermal vents the chemolithoautotrophic dissimilatory oxidation of sulphur by Archaea has only begun to be studied. The system utilized by Archaea such as *Acidianus ambivalens* (optimal growth at 80°C, pH 2.5) is distinct from the Sox enzyme system that appears to be common among the sulphur oxidizing Bacteria (Friedrich *et al.*, 2005). The main enzyme of the archaeal system is a soluble sulphur oxygenase reductase (SOR), which catalyses the oxygen-dependent disproportional oxidation of sulphur to sulphite, thiosulphate and hydrogen sulphide. Homologues of the *A. ambivalens* SOR have been found in *A. tengchongensis, Sulfolobus tokodaii, Ferroplasma acidarmanus, Picrophilus torridus*, as well as in the hyperthermophilic Bacteria *Aquifex aeolicus* (Friedrich *et al.*, 2005; Urich *et al.*, 2006).

The structure for the SOR reveals that the enzyme has an unusual quaternary structure, consisting of a large spherical complex of 24 SOR monomers, with a large (~71 × 107 Å) positively charged compartment in the centre (Urich *et al.*, 2006). At the active site is a very low potential mononuclear centre (E_m = −268 mV, compared to values of +200–600 mV observed for similar centres) (Urich *et al.*, 2004). The iron shows a distorted octahedral coordination, with two histidines, a glutamate (coordinated through both of its carboxylate oxygens) and two waters serving as ligands. This iron centre is located in the interior of the monomer, and is accessible only via a narrow channel, providing evidence that the enzyme reacts with a linear form of sulphur such as polysulphide, rather than the more common S_8 ring form (Urich *et al.*, 2006). The iron seems to be the most likely site for oxygen activation, although there is no direct evidence to support its role in the reaction. There are also three cysteines present at the active site which are likely to react with the sulphur substrate. The exact mechanism of sulphur reactivity has not been determined, although it seems clear that formation of cysteine–cysteine disulphides at the active site is not possible, as the cysteines are too far apart (although significant conformational changes could allow for it). One of the active site cysteines is present as an unusual stable cysteine persulphide, a modification which was also seen on the SoxAX protein, suggesting that this modified amino acid may play a general role in catalysing reactions with elemental sulphur (Kappler *et al.*, 2005a).

In the further oxidation of the sulphur intermediates, a membrane bound thiosulfate-quinone oxidase (TQO) oxidizes thiosulfate to tetrathionate, using a quinone (decyl ubiquinone, DQ) or ferricyanide as an electron acceptor, while in membrane extracts the thiosulfate oxidation is linked to oxygen reduction (Kletzin *et al.*, 2004; Muller *et al.*, 2004). The enzyme appears to be a quinone-binding $\alpha_2\beta_2$ tetramer consisting of a glycosylated 28 kDa subunit and a smaller

16 kDa subunit, corresponding to DoxA and DoxD, respectively. The final oxidation of tetrathionate by oxygen is attributed to the DoxBC complex, a Cu_B and cytochrome aa_3-containing membrane complex. Homologues of the TQO/DoxDA of *A. ambivalens* have been found in *S. solfolobus and S. tokodaii* (Friedrich *et al.*, 2005).

S. acidocaldarius contains two terminal oxidases, SoxABCD and the SoxM 'supercomplex'

Sulfolobus represents one of the most well studied of the Archaea, and certainly of the aerobic respiratory Archaea. While it does contain an electron transport system that generates a proton gradient that is subsequently used to drive ATP synthesis, its terminal oxidases have several characteristics that make them unique among previously characterized terminal oxidases. Because they inhabit an acidic environment yet keep their internal pH near neutral, it is not surprising that *Sulfolobus* species would have unusual proton pumping mechanisms; *S. acidocaldarius* appears to have two terminal oxidases, as does the closely related species *Metallosphera sedula* (Kappler *et al.*, 2005b).

The first of these is the 'supercomplex' containing the product of the *SoxM* gene, which has not been isolated in a form with a functional proton pumping activity, probably because of its complex nature. The subunit encoded by the *SoxM* gene appears to be a fusion product of what would normally be subunit I and III of the terminal oxidase, and it co-purifies with *SoxC* and a Rieske protein (Lubben *et al.*, 1994). The complex is able to oxidize the nonphysiological substrate N,N,N′,N′-tetramethyl-*p*-phenylenediamine (TMPD), although it is expected to utilize caldariella quinone *in vivo* (Lubben *et al.*, 1994).

The *SoxABCD* terminal oxidase is made up of four subunits which contain a single copper site (present as a Cu_B-type site) and four haem *a* centres (Gleissner *et al.*, 1997). The product of the *SoxB* gene is a cytochrome aa_3 protein, with *SoxC* containing the other two a-type haems of the complex. The complex is able to efficiently oxidize TMPD and caldariella quinone, with reconstitution of the complex into liposomes allowing for the demonstration of proton pumping activity using a TMPD/ascorbate system (Gleissner *et al.*, 1997). It has been proposed that the *SoxABCD* complex utilizes a Q-cycle in its proton-translocating mechanism, an unusual characteristic for a terminal oxidase (Gleissner *et al.*, 1997), although it has also been argued that the translocation occurs via the more common pumping mechanism (Pereira *et al.*, 2001).

Evolution of archaeal terminal oxidases

Within the haem–copper superfamily of oxidases, it has been hypothesized that the archaeal terminal oxidases were obtained through horizontal gene transfer from the Gram-positive Bacteria, with no enzymes of that type present prior to the divergence of the Archaea and Bacteria from the last common (universal) ancestor (Gilderson *et al.*, 2001; Pereira *et al.*, 2001). The terminal oxygen-reductases in the Archaea belong to the type A, or mitochondrial-like oxidases, and more specifically to the type A2 subfamily, which contains a conserved -YSHPXV- motif; in this subfamily a conserved tyrosine appears to act in the place of the glutamate involved in the proton pumping mechanism in the other subfamilies (Gilderson *et al.*, 2001). Consistent with the bacterial origin of the archaeal terminal oxidases is the observation that these complexes tend to show higher sequence identity to the enzymes from Gram-positive Bacteria than they do to enzymes from each other.

Sulphur reduction and related hydrogenases

Elemental sulphur and sulphur-containing compounds (such as SO_2 and H_2S) are commonly found in volcanically influenced environments (Hedderich *et al.*, 1998; Kletzin *et al.*, 2004). Thus, it is not surprising that a variety of hyperthermophilic Archaea have found multiple ways to utilize them in their energy metabolism, in particular as a terminal electron acceptor for respiration. In many hyperthermophiles, the presence of sulphur is actually a necessity for growth (e.g. *Staphylothermus marinus*) (Hao and Ma, 2003; Schicho *et al.*, 1993). The cell size of some hyperthermophiles has also been shown to be proportional to the amount of sulphur in the growth media (Hao and Ma, 2003). In *S. marinus*, the minimal concentration to ensure optimal growth has been determined to be 0.05% S^0, and the organism is able to utilize sulphur in the polysulphide, sublimed or colloidal form. Cell-free extracts of *S. marinus* have been shown to contain sulphur reductase, hydrogenase, glutamate dehydrogenase and electron transfer activities, allowing for the transfer of electrons from peptide or amino acid oxidation to sulphur.

Sulphur reduction in *Pyrococcus*

P. furiosus is able to grow using protons or sulphur as terminal electron acceptors, resulting in the production of H_2 or H_2S, respectively. Growth in the presence of sulphur, however, is not only faster but also results in an increase in cell yield of about 50%, indicating that sulphur reduction is an energy-conserving process in *Pyrococcus* (Schicho *et al.*, 1993). None of the three currently determined pyrococcal genomes, however, contain homologues of the molybdopterin-dependent polysulphide reductase which functions in the sulphur respiration of the Bacteria *Wolinella succinogenes* and *Shewanella oneidensis* MR-1 (Hedderich *et al.*, 1998; Moser and Nealson, 1996). While *P. furiosus* contains two hydrogenases and a sulphide dehydrogenase that show sulphur-reducing activities *in vitro*, the expression of the hydrogenases decreases during growth on sulphur, and the so-called sulphide dehydrogenase is now believed to act as a ferredoxin:NADPH oxidoreductase (Schut *et al.*, 2007). A novel flavoprotein, which had previously been characterized as a coenzyme A disulphide reductase (Harris *et al.*, 2005), catalyses the NAD(P)H and coenzyme A-dependent reduction of sulphur, and transcription of this enzyme has been shown to increase during growth on sulphur (Schut *et al.*, 2007).

The role of heterodisulphide reductases in hydrogen oxidation and sulphate reduction in *Archaeaglobus profundis*

Sulphate-reducing Bacteria and Archaea play an important role in both the geochemical cycling of sulphur and in the creation and maintenance of anaerobic environments in sediments. Despite its important role in environmental microbiology, the precise mechanisms of sulphate reduction, and especially the pathways involved in electron transfer between the oxidative reactions and the sulphate reduction machinery, remain unknown. Because the sulphate-reducing archaeon *Archaeoglobus profundis* is an obligate hydrogenotroph, it should have a much simpler path of electron transfer between the oxidative and reductive pathways (Mander *et al.*, 2004). Two homologues of the heterodisulphide reductases involved in archaeal methanogenesis were found in *A. profundis*: a membrane-bound enzyme, HmeCD, which is proposed to serve in the transfer of electrons from the quinone pool to an unknown disulphide substrate, and a cytoplasmic complex, mVh:Hdl, which is proposed to transfer electrons directly from hydrogen to the unknown disulphide substrate pool (Mander *et al.*, 2004). The HmeCD consists of a haem-b-containing integral membrane protein and a peripheral protein containing two 4Fe–4S clusters, while the considerably more complicated mVh:Hdl complex contains a Ni–Fe containing subunit, several Fe–S containing subunits, and a 4Fe–4S and FAD-containing subunit. While based on their homology to the methanogenic enzymes it is assumed that both enzymes utilize disulphides as electron acceptors, the identity of these acceptors, and whether or not the two enzymes even use the same substrate as electron acceptors, remains to be determined. The mVh:Hdl complex appears to be unique to the sulphate reducing Archaea and Bacteria (Mander *et al.*, 2004).

Archaea-specific redox enzymes of central metabolism

GAPOR and GAPN

The modified Entner–Doudorf and Emden–Meyerhoff pathways utilized in the Archaea contain several enzymes which are common in this kingdom. In the modified Embden–Meyerhof pathway the oxidation of glyceraldehyde phosphate to 3-phosphoglycerate occurs via nonphosphorylative oxidation, which can be compared to the 1,3-bisphopsphoglycerate-producing reaction of glyceraldehyde 3-phosphate dehydrogenase (GAPDH) usually seen in the conventional Embden–Meyerhof pathway of Bacteria and eukaryotes (Fig. 3.1). GAPDH is found throughout the Archaea; however, it is used exclusively in the gluconeogenic direction in all cases except in

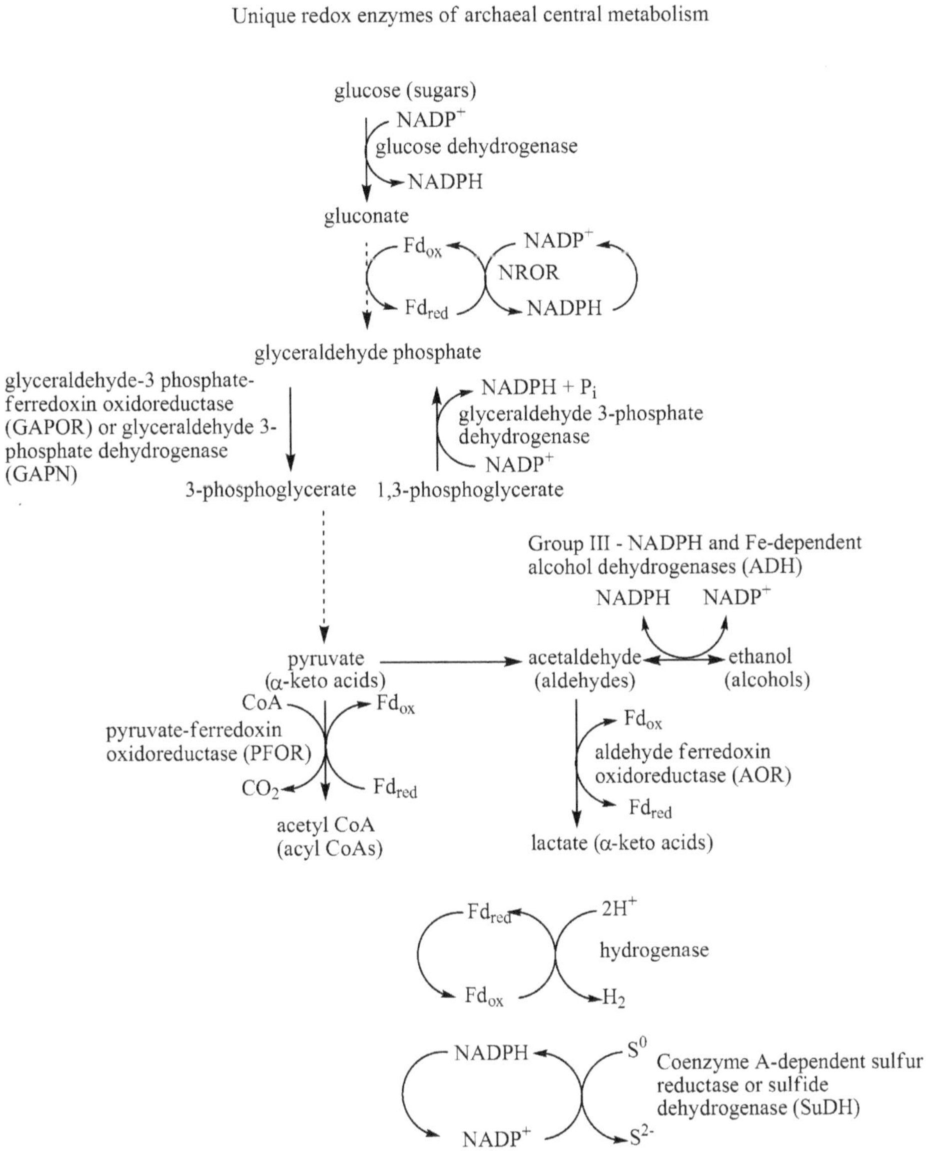

Figure 3.1 An overview of central metabolism of the Archaea showing the role of some of the enzymes which are specific to this kingdom.

that of the halophiles, and the enzymes show significant sequence and structural differences when compared with their non-archaeal counterparts (Littlechild *et al.*, 2004). In the catabolic direction the reaction is catalysed by either or both of the glyceraldehyde-3 phosphate–ferredoxin oxidoreductase (GAPOR), which is operative in *P. furiosus*, *Pyrobaculum aerophilum*, and *Archaeaglobus fulgidus* (Siebers and Schonheit, 2005) or the non-phosphorylative glyceraldehyde 3-phosphate dehydrogenase GAPN, operative in *Thermoproteus tenax* and *S. solfataricus* (Siebers and Schonheit, 2005). This step has been shown to be an important site of regulation in the Emden-Meyerhoff pathway and overall carbohydrate metabolism in *P. furiosus*, and under conditions where glycolysis is favoured the transcription of GAPOR is upregulated, while transcription of the opposing GAPDH is repressed (van der Oost *et al.*, 1998).

GAPOR was the third tungsten-dependent enzyme to be isolated from *P. furiosus* (Mukund and Adams, 1995). The enzyme is a monomer of 63 kDa which contains a cubane 4Fe4S centre and a tungsten atom liganded by pterin cofactors. GAPOR is a member of the general family of aldehyde:ferredoxin oxidoreductases that has been well characterized in *Pyrococcus*, and by comparison to the structures of the related AOR and FOR (discussed below) the tungsten centre is likely to be liganded to dithiolene sulphur atoms provided by two bound pterins present in the tricyclic form, with the FOR structure indicating the presence of an additional oxygen or sulphur ligand, probably in the form of W=O or W–OH (or W=S or W–SH) depending on the oxidation state (Fig. 3.2A). Based on EPR studies a second N or O non-enzyme ligand may be present in this class of enzymes, however, the general heterogeneity of the tungsten redox centre (which is present in three oxidation states) makes definitive identity of the W centre state difficult (Fig. 3.2A) (Hille, 2002; Hu *et al.*, 1999).

This heterogeneity represents an important characteristic of the group six tungsten and molybdenum cofactors, which are unique among biochemically relevant metals in that they have both one and two electron reduced/oxidized forms ($W^{IV}/W^{V}/W^{VI}$) accessible within a reasonably narrow range of potentials under physiological conditions (Bol *et al.*, 2006). The ability of these tungstopterin cofactors to perform single-step two electron redox reactions explains their importance in the Archaea in the catalysis of redox reactions with organic compounds such as aldehydes, and suggests that they may play a role similar to that of flavins, since both have in common the ability to perform one or two electron chemistry and to proceed either through direct electron transfer or through the formation of covalent intermediates.

The precise mechanism of the substrate reaction with the tungsten for the aldehyde oxidoreductases is not clear, however it has been proposed to be analogous to the reac-

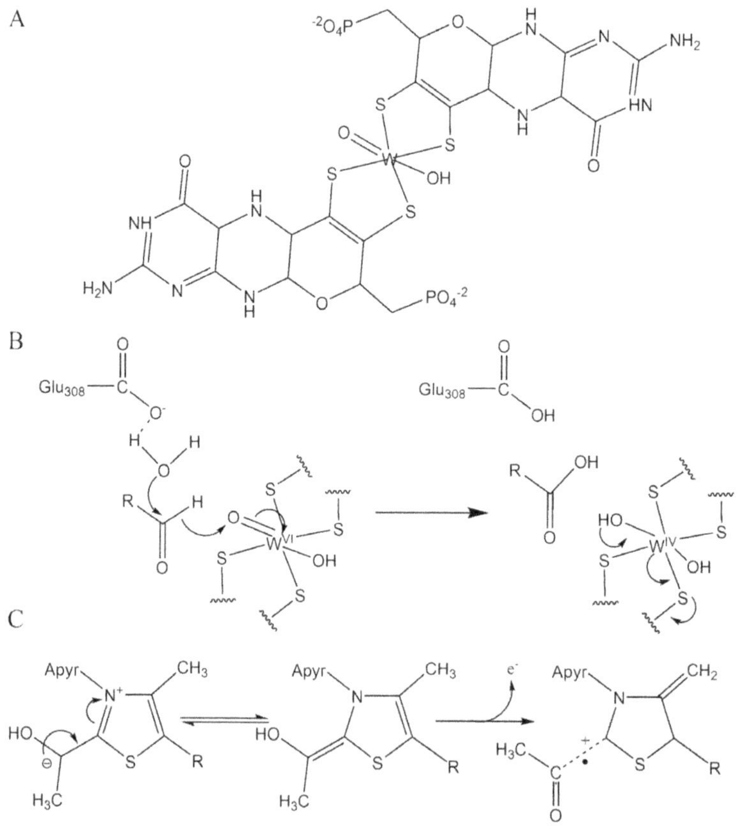

Figure 3.2 (A) The tungsto-bis-pterin centre of the aldehyde:ferredoxin family of enzymes. (B) Proposed mechanism for the tungsten centre of the aldehyde reductases of *Pyrococcus*, based on the proposed mechanisms for the molybdopterin oxidoreductase enzymes (Hu *et al.*, 1999; Huber *et al.*, 1996). (C)The proposed unique radical mechanism of the pyruvate–ferredoxin oxidoreductase of *Pyrococcus* (Chabriere *et al.*, 2001).

tion catalysed by the molybdopterin aldehyde oxidoreductase from the xanthine oxidase family of enzymes (Hu *et al.*, 1999; Huber *et al.*, 1996). This reaction proceeds through a coupled proton-electron transfer mechanism (Fig. 3.2B), with an active site base (Glu308 in FOR) activating a water for nucleophilic attack on the carbonyl position of the substrate, with the aldehyde group losing a hydride to the metal centre, although it is not clear which atoms of the tungsten complex are involved in this transfer (Hu *et al.*, 1999). Mechanistic studies with GAPOR indicate that following electron transfer from the substrate to the tungsten, transfer to the cubane centre occurs through one of the pterins, with the 4Fe–4S centre transferring electrons to the oxidizing ferredoxin molecule (Hagedoorn *et al.*, 1999). A likely path for electron transfer between the cofactors has been identified from the FOR structure with hydrogen bonds between the pterin and Cys491 and/or Arg180 or Lys175 involved in transfer to the FOR 4Fe–4S centre. The iron–sulphur centre of FOR and the electron-accepting ferredoxin iron sulphur centre appear to be linked by an interaction between Cys287 of FOR and Asp14 of the Fd (Hu *et al.*, 1999).

GAPN catalyses the nonphosphoylative irreversible oxidation of glyceraldehyde 3-phosphate via an oxidized pyridine nucleotide (glyceraldehyde-3-phosphate + $NAD(P)^+$ → 3-phosphoglycerate + NAD(P)H). *Thermoproteus* has been shown to contain two versions: 1) an $NADP^+$-dependent enzyme with a subunit m.w. of 39 kDa that showed the conventional phosphorylating activity and was believed to operate in the anabolic direction and 2) what was believed to be a 49 kDa NAD^+-specific non-phosphorylating version operating in the catabolic direction (Hensel *et al.*, 1987). The (supposedly) NAD^+-dependent GAPN from *Thermoproteus* is allosteric, with regulation occurring through control of NAD^+ binding. NADH and ATP decrease activity by lowering NAD^+ affinity, while AMP, ADP, glucose 1-phosphate, fructose 6-phosphate, fructose 1-phosphate, ribose 5-phosphate, phosphate and pyrophosphate act as positive effectors by increasing affinity for NAD^+ (Brunner *et al.*, 1998; Brunner and Hensel, 2001). This enzyme is the only aldehyde dehydrogenase which has been shown to be allosterically controlled.

In the initial studies of the *Thermoproteus* NAD^+-dependent enzyme, it was noted that $NADP^+$ acted as a tight binding competitive inhibitor of this enzyme, and when the crystal structure was determined it was found that the positions of NAD^+ and NADPH on the enzyme were almost identical, suggesting that $NADP^+$ should in fact act as a substrate. Further experiments in the presence of activators showed that $NADP^+$ is in fact a substrate, although its activity is highly dependent on the presence of activators, explaining the lack of activity observed in the earlier experiments (Lorentzen *et al.*, 2004). It is not known whether the less studied $NADP^+$-dependent GAPN found in *Thermoproteus* (Hensel *et al.*, 1987) might also show activity with either pyridine nucleotide substrate. $NADP^+$-dependent GAPNs have been found thoughout the eukaryotes and Bacteria, where they have been extensively studied and are generally thought to be important in anabolic rather than catabolic processes, as would be expected based on their substrate preference (Brunner *et al.*, 1998; Brunner and Hensel, 2001).

The determination of the *Thermoproteus* enzyme structure confirmed that the mechanism which had been proposed for class 2 aldehyde dehydrogenases (Wang and Weiner, 1995) was likely to be operational on this enzyme as well (Lorentzen *et al.*, 2004). In this mechanism the active site cysteine 297 is activated through thiolate formation, which may involve an essential glutamate (Glu268) acting as an active site base. This is followed by nucleophilic attack of the thiolate on the substrate and formation of the thiohemiacetal from which hydride transfer to the pyridine nucleotide occurs, resulting in the formation of a thioacyl intermediate. The essential glutamate then serves to activate the water that acts in the deacylation of the enzyme.

Aldehyde ferredoxin oxidoreductase (AOR), formaldehyde ferredoxin oxidoreductase (FOR) and broad-range aldehyde oxidoreductase (WOR5)

In addition to the GAPOR, four other tungsten-dependent enzymes have been characterized from *P. furiosus* (a fifth tungsten-containing enzyme from *P. furiosus* has been discovered, WOR4, but it is of unknown function (Roy and Adams, 2002b)). All three of these enzymes contain a tungstobispterin (Fig. 3.2A) and 4Fe–4S centre per subunit. The homodimeric AOR of *P. furiosus* was the first of these enzymes to be purified and characterized (Mukund and Adams, 1991), as well as being the first to have its structure determined, revealing the nature of the tungstobispterin centre (Chan *et al.*, 1995). A tungsten-dependent AOR was also recently isolated from the crenarchaeaon *Pyrobaculum aerophilum*, and that enzyme is distinguished from the pyrococcal enzyme by its use of a 7Fe ferrerdoxin which contains 3Fe–4S and 4Fe–4S iron centres (Hagedoorn *et al.*, 2005). The FOR from *P. furiosus* is a homotetrameric enzyme capable of oxidizing formaldehyde or larger aldehydes to their carboxylate derivatives (Bol *et al.*, 2006). FOR has been shown to react with the aldehyde rather than the hydrated form of the substrate (which in the case of formaldehyde would be methylene glycol) which predominates in aqueous solution. WOR5, as its name suggests, is capable of oxidizing a wide range of aldehyde substrates, including cinnamaldehyde and 3-phenylbutyraldehyde, with hexanal showing the highest enzymatic efficiency (Bevers *et al.*, 2005). WOR5 appears to be the only one of these enzymes which is specific to *P. furiosus*, as the genomes of *P. abyssi* and *P. horikoshii* do not appear to contain WOR5 homologues (Bevers *et al.*, 2005).

Glucose dehydrogenases from *S. solfataricus*, *Haloferax* and *Thermoplasma*

The first step in the non-phosphorylative or semi-phosphorylative Entner–Doudorf pathway operative in many Archaea is catalysed by glucose dehydrogenase (glucose/galactose + $NAD(P)^+$ → gluconate/galactonate + NADPH) (Ahmed *et al.*, 2005). This pathway is operative in a range of Archaea, including *S. solafataricus*, *Thermoplasma acidophilum*, *Thermoproteus tenax*, *Haloferax mediterranei*, *Picrophilus torridus* as well as the fungus *Aspergillus* (Lamble *et al.*, 2003; Siebers and Schonheit, 2005). In all cases except for *T. tenax* and *Aspergillus*, this enzyme appears to be promiscuous, allowing for the utilization of galactose and other carbohydrates by this pathway. It has been suggested that this promiscuity is an adaptation to extreme environments (Lamble *et al.*, 2003), and while a definite relationship has not been established, organisms such as *Sulfolobus* definitely inhabit extreme environments. The structure of the tetrameric enzyme from *T. acidophilum* has been determined, revealing similarities in the catalytic and nucleotide-binding domain between this enzyme and horse liver alcohol dehydrogenase which were not expected based on the low degree of sequence identity between these two enzymes (John *et al.*, 1994). Like alcohol dehydrogenase, the *Thermoplasma* glucose dehydrogenase structure contains two zincs. The catalytic zinc is liganded by a Cys, His and Asp, while the structural zinc, which is not essential for activity in alcohol dehydrogenase, is liganded to three Cys and an Asp residue.

Although the enzyme shows a double specificity for the reduced nucleotide substrate, the enzymes from *H. mediterranei* and *T. acidophilum* show a preference for $NADP^+$ (Esclapez *et al.*, 2007). In the crystal structure of the *H. mediterranei* enzyme an interesting 'cation cluster' of two potassium ions appears to be involved in the binding of the phosphate group of $NADP^+$, which the authors suggest may be a unique feature of halophilic enzymes (Britton *et al.*, 2006). In high-salt conditions this cation cluster may serve to facilitate phosphate binding to the enzyme by helping to overcome the 'competition' for the phosphate group presented by the high concentration of solvent cations.

Glutamate dehydrogenase (GDH)

This central enzyme in nitrogen and amino acid metabolism has been extensively studied in hyperthermophiles, and structures have been determined for GDH from *P. furiosus*, *T. littoralis*, and *Pyrobaculum islandicum* (Bhuiya *et al.*, 2005). While all of these examples are hexameric, the nucleotide specificity varies; GDHs are divided into NAD^+, $NADP^+$, or $NAD(P)^+$-specific classes. During the heterologous overexpression of the enzyme from *Thermococcus kodakaraensis* KOD1 in *E. coli* the enzyme was found to be present initially as a mixture of inactive dimers and tetramers and active hexamers, with heating of the enzyme preparation resulting in a 'maturation' of the enzyme mixture to a homogeneous solution of active hexameric enzyme (Imanaka and Atomi, 2002).

Although the well studied *P. furiosus* enzyme has been placed in the dual-specificity class, it has been suggested that due to its much lower activity with NAD^+ that it should be considered to be a $NADP^+$-specific enzyme (Bhuiya *et al.*, 2005). The enzyme from *Pyrococcus horikoshii* is also $NADP^+$ specific, showing no observable activity with NAD^+ in the oxidative deamination direction (Wang *et al.*, 2003); the enzymes from *Thermococcus waiotapuensis*, *Thermococcus profundus* and the hyperthermophilic archaeon ES4 are $NADP^+$ specific as well (DiRuggiero *et al.*, 1993; Lee *et al.*, 2002; Nakasako *et al.*, 2001). Given the reclassification of the *P. furiosus* enzyme, a careful review of the literature shows that archaeal enzymes appear to have a distinct preference for $NADP^+$, with the *P. islandicum* enzyme being the only hyperthermophilic enzyme found to be specific for NAD^{+-}, although both NAD^+ and $NADP^+$ dependent versions appear to be present in the archaeon *Halobacterium salinarum*, a mesophilic extreme halophile with four GDH genes (Ingoldsby *et al.*, 2005).

It is tempting to conclude from this observation that the true dual specificity enzymes evolved after the Archaea–Bacteria split, or that there is a preference for $NADP^+$ specific enzymes, or even for single specificity enzymes in general, in extreme environments. The switch in coenzyme specificities has occurred many times in evolution, so a simple inference regarding the split of the enzymes is not correct (Ingoldsby *et al.*, 2005). The preference for $NADP^+$ observed to this date in the Archaea and the lack of dual-specificity enzymes may simply be due to a coincidence due to the few GDHs characterized, and a greater number would provide more convincing evidence for a systematic preference for the $NADP^+$-specific enzymes in the hyperthermophiles. However, when considered in the context of what appears to be a repeating theme of unusual preferences for $NADP^+$/NADPH in archaeal systems that would normally prefer NAD^+/NADH in the Bacteria and eukaryotes, the case for a trend towards a $NADP^+$/NADPH preference in the Archaea and hyperthermophiles seems to become increasingly strong. Unfortunately, due to the frequent switches in specificity throughout evolution, it is not reliable to determine substrate specificity by looking at sequence, which means that the coenzyme specificity in the GDHs can only be determined by direct enzyme assay (Ingoldsby *et al.*, 2005). The relatively low values of K_m for the oxidized pyridine nucleotides in these enzymes suggests that the enzymes in this kingdom overall may prefer to operate in the catabolic direction.

Pyruvate–ferredoxin oxidoreductase (PFOR)

The pyruvate–ferredoxin oxidoreductase catalyses a reaction quite similar to the oxidative decarboxylation of pyruvate by the more widely known pyruvate dehydrogenase complex, except that in the case of PFOR ferredoxin rather than NAD^+ serves as the electron acceptor (pyruvate + CoA-SH + ferredoxin$_{ox}$ $\rightarrow$ Acetyl-CoA + ferredoxin$_{red}$ + CO_2) (Ma and Adams, 2001). In addition, unlike the pyruvate dehydrogenase complex the PFOR enzyme is capable of catalysing the reverse reaction, which is utilized in carbon fixation in methanogenesis and acetogenesis and by green photosynthetic Bacteria (Evans *et al.*, 1966; Tersteegen *et al.*, 1997). The enzyme is generally found in strict anaerobes, including Bacteria, however PFOR has been purified from a range of Archaea including *Pyrococcus furiosus* (Blamey, 1993), *Archaeaglobus fulgidus* (Kunow, 1995), *Methanosarcina barkeri* (Bock *et al.*, 1996), and *Halobacterium halobium* (Plaga *et al.*, 1992) as well as from the hyperthermophilic Bacteria *Thermotoga maritima (Blamey and Adams, 1994)*. The archaeal enzymes listed above, as well as the *Thermotoga* enzyme, all appear to be heterotetramers with m.w. 113–130 kDa, with the exception of the *Halobacterium* enzyme, which is a 256 kDa $\alpha_2\beta_2$ tetramer. The enzyme purified from a range of mesophilic anaerobic Bacteria is a homodimer of 240 kDa. All three types of enzymes, however, have in common the use of thiamin pyrophosphate (TDP) and 4Fe–4S cluster cofactors.

While based on their different subunit sizes and quaternary structure it might be expected that these three classes of PFORs are evolutionarily unrelated, phylogenetic analysis shows that the catalytic domains of all three are related, with the subunits of the homodimeric bacterial enzyme and the $\alpha_2\beta_2$ halobacterial enzyme containing domains which show significant identity to the four different subunits of the hyperthermophilic enzyme. These similarities have been proposed to have occurred through rearrangement of the four genes that are currently seen in the hyperthermophilic PFOR, with either gene duplications or recombinations resulting in the assembly of the separate domains into larger units (Kletzin and Adams, 1996).

PFOR is unique among TDP-dependent enzymes in that it is the only enzyme for which an acetyl thiamin dipohosphate-based free radical intermediate has been identified (Chabriere *et al.*, 2001). As with many TDP catalysed reactions, the PFOR catalytic cycle begins with formation of the reactive C2 carbanion on the thiazolium ring of TDP, followed by attack of the carbaninon on the carbonyl carbon of the pyruvate substrate, which is followed by substrate decarboxylation, resulting in the formation of the hydroxyethyl-TDP eneamine (Fig. 3.2C). It should be noted that the released CO_2 remains tightly bound and can be observed in the structure of the radical intermediate, an observation which is not surprising when the ability of this enzyme to catalyse the reverse carboxylation reaction is considered. At that point, a one-electron transfer occurs from the hydroxyethylthiamin to a 4Fe–4S centre on the enzyme (Fig. 3.2C), which is followed by the transfer of a second electron from the thiazolium to CoA, at which point condensation of the thiyl CoA radical with the acetyl radical results in the synthesis of acetyl CoA. In the absence of CoA, a relatively stable radical intermediate is formed by the one electron oxidation, with the surprising observation that the intermediate appears to be in the form of a dearomatized acetyl thiazolium radical, with the radical present as a long one-electron bond between the C2 position of the thiazole ring and the carbonyl carbon of the substrate (Fig. 3.2C) (Chabriere *et al.*, 2001).

Iron and NADPH-dependent group III alcohol dehydrogenases

Alcohol dehydrogenases (ADHs) can be divided into three broad groups: group I ADHs are large (350–900 amino acids) zinc-dependent enzymes, the group II ADHs are smaller (~250 amino acids) non-metal enzymes, while the much less well characterized group III enzymes are non-haem iron-dependent homodimeric, tetrameric and hexameric enzymes made up of subunits of ~400 amino acids. These group III enzymes, which are iron-*containing* and are so far found exclusively in hyperthermophiles, should not be confused with the iron-*activated* enzymes found in mesophiles and thermophiles, which are not purified as iron-containing enzymes and require that iron be present in the assay mixture (Radianingtyas and Wright, 2003). While hyperthermophiles contain enzymes of all three types (*P. furiosus* has been shown by genome analysis to contain 19 different alcohol dehydrogenases (Machielsen *et al.*, 2006)), the group III type appear to play an important role in these organisms. Examples of this enzyme have been found in *P. furiosus* (this enzyme also contains a bound zinc; Ma and Adams, 1999b), *Thermococcus litoralis (Ma et al., 1994)*, *T. hydrothermalis* (Antoine *et al.*, 1999), *Thermococcus* strain ES-1 (Ma *et al.*, 1995), and the hyperthermophilic bacterium *Thermotoga hypogea* (Radianingtyas and Wright, 2003; Ying *et al.*, 2007).

Because relatively few examples of this enzyme have been characterized, it is difficult to make definite conclusions regarding its evolution. It has been noted, however, that the iron-containing and iron-activated enzymes from hyperthermophilic Archaea show remarkable sequence identity in their nucleotide sequence as well as their amino acid sequence, consistent with the spread of these enzymes occurring via horizontal gene transfer (Jain *et al.*, 2002; Radianingtyas and Wright, 2003). The N-terminal sequence of the *Thermotoga* enzyme, however, showed no similarity to the archaeal Fe-containing ADHs, suggesting that the bacterial versions may have either diverged significantly from the archaeal enzymes, or developed in parallel.

Of the group III enzymes, all appear to be NADPH/$NADP^+$-dependent, an unusual substrate specificity for an enzyme believed to play a major role in catabolic pathways. These enzymes tend to have broad substrate specificities, with the *Thermotoga* enzyme showing activity towards 2–8 carbon primary alcohols, 3–5 carbon diols, and the aromatic substrate phenylethanol (Ying *et al.*, 2007). The enzymes are more efficient in the aldehyde to alcohol direction, with $k_{cat\ app}$/$K_{m\ app}$s that are 10- to 50-fold greater in that direction (Ying *et al.*, 2007). All of the organisms found to contain this enzyme so far are capable of producing ethanol as a fermentative end product, consistent with the observed preference for aldehydes and NADPH, suggesting that the enzyme plays a role in the consumption of excess reducing equivalents in these anaerobes (Ma and Adams, 1999b; Ying *et al.*, 2007). The ADH from the sulphur-reducing *Thermococcus* strain ES-1 is upregulated when sulphur concentrations are decreased, consistent with this role (Ma *et al.*, 1995): as the concentration of the electron acceptor sulphur drops, acetaldehyde's role as an electron acceptor becomes increasingly important, with a concomitant rise in the amount of the ADH.

Protein disulphide oxidoreductase (PDO) from hyperthermophiles

The correct formation of disulphides in proteins is of course essential for the activity of many proteins, and at least two families of proteins catalyse this reaction, including the disulphide bond forming proteins (Dsb) and their homologues in Bacteria, and the protein disulphide isomerase (PDI) and other PDI-homologues in eukaryotes (D'Ambrosio *et al.*, 2006; Pedone *et al.*, 2006b). Although early studies suggested that structural disulphides were rare in the Archaea, especially in intracellular proteins, newer work indicates that they may be more common than previously believed (Ladenstein and Ren, 2006; Mallick *et al.*, 2002). A third family of proteins showing little sequence identity to the PDIs or Dsbs but involved in forming and maintaining correct protein disulphide bonds in hyperthermophiles has been found and placed in a family called the protein disulphide oxidoreductases (PDO), with examples found in *P. furiosus* (Pedone *et al.*, 2004), *P. horikoshii* (Kuroita *et al.*, 2007), *Aeropyrum pernix* K1 (D'Ambrosio *et al.*, 2006), *Sulfolobus solfataricus* (Pedone *et al.*, 2006b), and *Aquifex aeolicus* (Pedone *et al.*, 2006a). The structures of the enzymes from *P. furiosus* and *A. aeolicus* have been determined, and both are made up of approximately 25 kDa monomers that contain two closely packed but independently operating thioredoxin domains (Pedone *et al.*, 2006a; Ren and Ladenstein, 2001).

PDOs appear to be limited to a subset of the hyperthermophilic Archaea and Bacteria that appear by genomic analysis to contain a larger than usual number of intracellular disulphide bonds, and rather than being present in a certain branch of the Archaea or Bacteria, the presence of a PDO seems to correlate only with the presence of those two criteria (Ladenstein and Ren, 2006): the protein does not appear to be present in mesophilic Archaea or in hyperthermophiles that contain low numbers of intracellular disulphides. In general, the PDOs seem to have much in common with the eukaryotic PDIs, as they show not only disulphide reductase and dithiol oxidase activities, but have documented protein disulphide isomerase activities as well (Ladenstein and Ren, 2006).

Proteins of this class operate via a mechanism in which an oxidized disulphide on the PDO (or PDI or Dsb) forms a mixed disulphide with a cysteine on the target enzyme, which then exchanges with a cysteine on the target enzyme, forming a disulphide on the target protein. This reaction can also proceed with the reduced PDO forming a mixed disulphide with another cysteine from a disulphide present on the target protein, with the mixed disulphide reacting to form a disulphide on the PDO, resulting in a reduced dithiol on the target enzyme. By using a combination of these two reactions the PDO can 'shuffle' protein disulphides on target proteins, resulting in the reduction of 'improper' disulphides and the formation of 'correct' disulphides (Ladenstein and Ren, 2006).

The two disulphide/dithiol pairs on PDO are present in CXXC motifs, and it is believed that the two motifs have different redox potentials and different functions, with one acting mostly as an oxidase, and the other acting mostly as a reductase/isomerase. On thioredoxin and the Dsbs the N-terminal cysteine in the motif generally has a low pK_a and is solvent exposed; it is this residue which reacts with target protein disulphides to form the mixed disulphide. The C-terminal cysteine in the pair is usually less solvent exposed, and appears to operate solely in the reduction of the mixed-disulphide, resulting in a reduced target protein and oxidized enzyme.

It should be noted that these observations have not been proven to hold for the PDOs, as at this time no cysteine pK_as or redox potentials for PDO dithiols/disulphides have been directly determined (Ladenstein and Ren, 2006). It has also not been determined which proteins act to control the redox state of the PDOs *in vivo*, although the archaeal thioredoxin reductase has been proposed to be one of the proteins which may fulfill this function (Ladenstein and Ren, 2006).

Archaeal small redox proteins

Ferredoxin and polyferredoxin

From the examination of metabolic redox enzymes presented above it becomes obvious that ferredoxins play an important role in the transfer of electrons in hyperthermophile metabolism. The structural and catalytic variety present in ferredoxins in general is also reflected in the ferredoxins from the Archaea, with examples including the Rieske-type 2Fe–2S proteins (discussed in the next section) and the 4Fe–4S protein that plays such an important role in the central metabolism of *Pyrococcus*. Several are unusual because of their size and the presence of a zinc centre, such as those from *Sulfolobus* (Imai, 2001; Iwasaki and Oshima, 2001; Kim *et al.*, 2001; Wang *et al.*, 1999). Determination of reduction potentials of the enzymes from *T. maritima* and *P. furiosus* across the temperature range of 20–95°C (and of the POR enzymes from these species as well) indicate that while the enzymes appear to have similar potentials under ambient conditions, their behaviour changes greatly with increased temperature, showing the necessity of determining reduction potentials for thermophilic proteins under physiological conditions (Smith *et al.*, 1995). While most ferredoxins fall under the category of small proteins, an unusual 'polyferredoxin' was discovered in a species of the methanogenic genus *Methanobacterium* that is made up of six ferrodoxin domains (Steigerwald *et al.*, 1992).

Sulredoxin (SDX) and oxygenase-associated Rieske-type ferredoxin (ARF)

SDX and ARF are relatively high potential Rieske-type 2Fe–2S proteins that differ from the more conventional ferredoxins (such as the plant ferredoxin utilized in electron transfer from photosystem I) by their asymmetric 2Fe–2S centre in which one iron is liganded to two cysteine thiolates while the other is liganded to the Nδ of two histidines, rather than the symmetrical 2Fe2S of conventional 2Fe2S ferredoxins that utilize four cysteine ligands (Dikanov *et al.*, 2004). The considerably high potentials of these proteins (E° for ARF = −62 mV, E° for SDX = 190 mV) has been attributed to the considerably different electronegativity of the histidine ligands, with the fine-tuning of the potential that results in the potential differences between SDX and ARF attributed to variations in the geometry of the histidine ligands as well (Dikanov *et al.*, 2004). SDX, which was isolated from *Sulfolobus tokodaii* strain 7, is proposed to be associated with cytochrome bc_1, and a similar protein was isolated from *S. acidocaldarius* (Brugna *et al.*, 1999). ARF was isolated from *Sulfolobus solfataricus* strain P-1 (Kounosu *et al.*, 2004).

Rubredoxin

Rubredoxin, the simplest of the iron–sulphur proteins, was discovered in 1965, and its structural simplicity and stability resulted in it becoming one of the most well-characterized protein structures (Jenney and Adams, 2001; Kurihara *et al.*, 2004). Ironically, knowledge regarding its actual physiological role has lagged far behind the structural studies. Rubredoxin appears to be isolated in the strict anaerobes, although rubredoxin domains occur in larger proteins throughout the prokaryotic world (Jenney and Adams, 2001). With the advent of genomics and the concomitant explosion of studies in hyperthermophiles, a role for rubredoxin consistent with its isolation in the strict anaerobes is beginning to be understood. The best understood example of this protein is the version from *P. furiosus*, an understanding of which has emerged (as so much else regarding *Pyrococcus* has) from the studies of pyrococal redox metabolism in M. Adams' laboratory.

Rubredoxins typically contain a single iron coordinated by four cysteines, and can be split into two main groups based on the spacing of the conserved cysteines: group 1, the most common type, have a CXXC...CXXC motif, while the type 2 enzymes have a CXXXXC...CXXC iron-binding motif, and the closely related but even smaller desulphoredoxin has a motif of CC... CXXXC, which results in a distorted tetrahedral binding site due to the close spacing of the CC residues (Rodrigues *et al.*, 2005). Rubredoxin's iron centre has a potential that is relatively high when compared to other small redox proteins, with the *P. furiousus* rubredoxin $E^{\circ\prime} = 0$ mV at 25°C (Smith *et al.*, 1995), and the two rubrerdoxins from *Archaeaglobus fulgidus* having E° of +40 and +75 mV (Rodrigues *et al.*, 2005). The *P. furiosus* enzyme is a 53 amino acid hyperstable (no loss of activity for days at 95°C) protein that can be non-enzymatically oxidized by superoxide, subsequently reducing superoxide to hydrogen peroxide ($rub_{red} + O_2^- + 2H^+ \rightarrow rub_{ox} + H_2O_2$). A large increase in rate is observed with the addition of the enzyme superoxide reductase, which is the basis for the antioxidant system shown in Fig. 3.3 and discussed below (Jenney *et al.*, 1999).

Sulphocyanin and halocyanin

Both sulphocayanin and halocyanin are archaeal examples of the type I copper proteins, a group of proteins which include azurin and rusticyanin (Giri *et al.*, 2004), and are similar to plastocyanin, the type I Cu-protein that transfers electrons from cytochrome b_6f to photosystem I in photosynthesis. These proteins have an active site made up of highly conserved ligand-ing amino acids that include two histidines, a cysteine and a methionine sulphur (Giri *et al.*, 2004). Sulphocyanin, coded for by the *SoxE* gene and associated with the *SoxM* supercomplex discussed above, has been isolated from *Sulofobus acidocaldarius* and shown to have an E° of 300 mV (Komorowski and Schafer, 2001). Halocyanin is a peripheral membrane protein with a potential of 183 mV which is associated with a cytochrome bc/nitrate reductase supercomplex in *Haloarcula marismortui* (Giri *et al.*, 2004; Yoshimatsu *et al.*, 2007).

Archaeal antioxidant systems and enzymes

The problem of protecting anaerobes from oxidative stress is an interesting one, as these organisms' lack of ability to react with oxygen, which forms the basis for their anaerobic nature, would appear to make it impossible for these organisms to mount an antioxidant defence. This does not appear to be the case, however. The superoxide reductase-based antioxidant defence of *P. furiosus* is the best defined of these systems in the hyperthermophilic Archaea, although a considerable amount of progress has been made in the study of the system in *A. fulgidus* as well. While this system is definitely not unique to the Archaea, as homologues of the SOR appear to be widespread among anaerobic and microaerophilic species (Adams *et al.*, 2002), because the initial characterization of the system was performed in hyperthermophiles, and because it

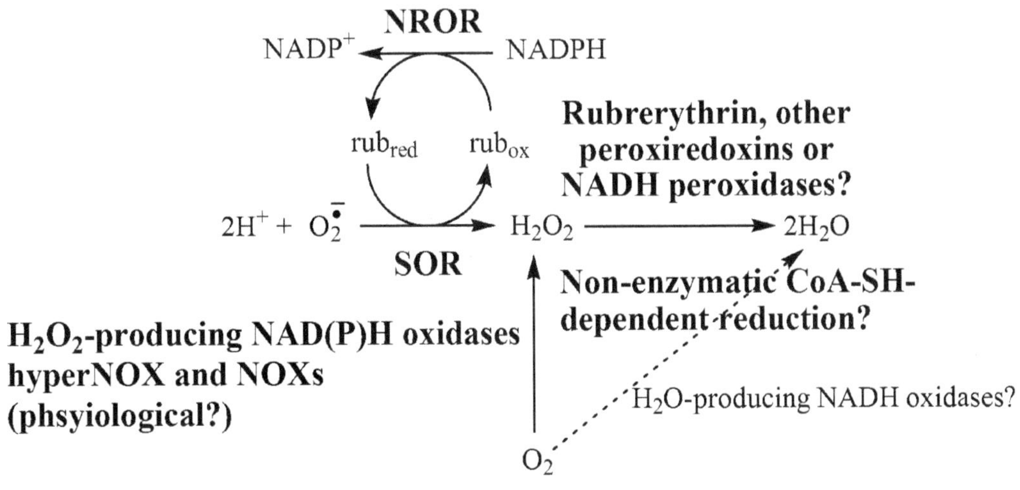

Figure 3.3 Summary of the antioxidant pathways operating in *Pyrococcus*.

appears to play an important role in the viability of hyperthermophiles we have chosen to discuss it here.

For strict anaerobes, the superoxide reductase reaction ($O_2^- + 2H^+ + \text{rubredoxin}_{red} \rightarrow H_2O_2 + \text{rubredoxin}_{ox}$) has the advantage of not generating oxygen as a product (Jenney *et al.*, 1999). The system does require, however, that a second protein come in and reduce the less toxic but still damaging peroxide. In the case of *P. furiosus* the SOR system has been shown to be reduced by the small protein rubredoxin, which is maintained in the reduced form via reduction by a NAD(P)H:rubredoxin oxidoreductase (NROR) as shown in Fig. 3.3 (Grunden *et al.*, 2005; Ma and Adams, 1999a). It is likely that other reductants are able to reduce SOR *in vivo*, since although all microaerophiles and anaerobes whose genomes have been sequenced to date contain SOR homologues, many, including *P. horikoshii*, are missing rubredoxin homologues (Grunden *et al.*, 2005).

DNA-binding SsDps and one-cys peroxidredoxin of *Sulfolobus* and their role in antioxidant reactions

SsDps-mediated reduction of hydrogen peroxide and possible DNA-binding capabilities of this archaeal Dps protein are mechanisms by which *S. solfataricus* mitigates oxidative damage (Wiedenheft *et al.*, 2005). This protein shows a diverse range of reactions, including catalase, iron mineralization, DNA binding, peroxidase, conjugation of N-acyl-amino acids, and engages in acetylation. A one-cys peroxidredoxin has also been characterized from *S. solfataricus*, in which its single essential cysteine is postulated to act as a scavenger of peroxides by cycling between sulphenic acid (Cys-SOH) and thiol redox states with the assistance of an as-yet undetermined reducing agent (Limauro *et al.*, 2006).

Superoxide reductase/neelaredoxin/desulphoferrodoxin

The superoxide reductases (to distinguish them from the superoxide dismutases) are members of a group of iron–sulphur proteins that catalyse the reduction of superoxide to peroxide, while using rubredoxin (and possibly other proteins or compounds) as a source of reducing equivalents. SORs are also referred to in the literature as neelaredoxins (based on the sanskrit word for blue, since the enzyme's non-haem iron imparts a blue colour), which, in spite of the nice visual imagery associated with the name 'neelaredoxin' results in unecessary confusion; for example, the *Archaeoglobus* SOR is referred to in the literature as a neelaredoxin (Abreu *et al.*, 2000; Abreu *et al.*, 2001; Abreu *et al.*, 2002; Rodrigues *et al.*, 2005) but in Genbank as an SOR. Some of this confusion may result from early studies that indicated that the *Archaeoglobus* SOR actually catalysed both the reduction and dismutation of superoxide (Abreu *et al.*, 2000; Abreu *et al.*, 2001), although later literature does not refer to the catalysis of a dismutation reaction by the *Archaeoglobus* SOR/neelaredoxin (Rodrigues *et al.*, 2006).

In addition to the one-iron SOR, some anaerobic hyperthermophiles, such as *Archaeoglobus*, contain a second larger homologue to the SORs, desulphoferredoxin (Dfx), which is named for its initial discovery in the sulphate-reducing Bacteria (Abreu *et al.*, 2000). Dfx contains an SOR domain that carries the iron site that catalyses superoxide reduction, and an additional domain carrying a more conventional second Fe–4Cys centre of unknown function, as removal of this second site does not appear to affect function either *in vivo* or *in vitro* (Abreu *et al.*, 2002; Emerson *et al.*, 2003). The SOR enzymes from *Pyrococcus furiosus* and *Archaeaglobus fulgidus* are probably the best studied, and their role in the antioxidant system of this and other organisms is indicated in Fig. 3.3 (Rodrigues *et al.*, 2006; Kurtz, 2004).

The pyrococcal enzyme is made up of a homotetramer of small (14.3 kDa) subunits, comprising seven β-strands with one turn of a 3_{10}-helix that occurs at the subunit interface and is involved in the tight intersubunit binding in this protein (Yeh *et al.*, 2000). The beta strands are divided into three- and four-strand antiparallel β-sheets. The iron centre is located in a relatively exposed position at the end of each subunit, and is coordinated to four histidines and a cysteine in the reduced form, while in the oxidized form two of the irons are pentacoordinate, while the other two become hexacoordinate with the addition of a glutamate ligand, as seen in Fig. 3.4. This unusual arrangement appears to be unique to the SOR family (Kurtz, 2004).

This change in coordination is associated with the movement of a loop (residues 9–15), which upon reduction pulls the glutamate ~5 Å away from the liganding position, with an additional, smaller movement occurring in the loop in the area of residues 36–40. It has been proposed that this glutamate could act to prevent binding of the superoxide to the oxidized enzyme, since access to the iron on two of the subunits in the oxidized enzyme would be fairly restricted by its hexacoodinate character, while upon reduction the pentacoordinate iron would have an area open for oxygen binding. This proposal is supported by the observation that the glutamate is surface exposed, so it is easy to see how its movement away from the iron would open the iron up for superoxide binding. Because superroxide can act as either a reductant or oxidant, this control may be responsible for the pyrococcal SOR's ablility to exclusively reduce superoxide while avoiding the O_2-generating disumtation pathway, despite the fact that the oxidation of superoxide by oxidized (ferric) SOR is thermodynamically favourable (Kurtz, 2004).

It has also been proposed that this glutamate is actually involved in assisting in peroxide product release, a hypothesis which is supported by the observation that when the coordinating glutamate in the SOR from *Desulfoarculus baarsii* is mutated to an alanine, it does not affect the kinetics of the formation of Fe^{3+}-peroxo species that are proposed to be intermediates in the reaction, and in one study it instead stabilizes the iron-peroxo intermediate (Mathe *et al.*, 2002). Because in later studies this mutation did not result in the reduction of any steps in the reaction of the intermediate, this role for Glu14 appears to have been discounted (Kurtz, 2004).

The mechanism by which SOR attracts the superoxide substrate to the catalytric iron is not readily apparent. The most visible option is that a lysine which is absolutely conserved among the SORs, corresponding to Lys15 in the pyrococcal enzyme, serves to attract the superoxide anion to the centre. Although the lysine is located 6–8 Å away from the iron in the reduced form of the enzyme, it is interesting to note that the conformational change that occurs upon oxidation results in a large swing of the side chain of this lysine away from the iron, so that it is pointing in the opposite direction (Fig. 3.4), consistent with the lysine moving away so that it will not attract the substrate into the oxidized active site, which could result in the enzyme following the dismutation rather than reduction path.

The reaction of the enzyme with superoxide is fast, with a nearly diffusion-controlled rate constant of ~10^9 M^{-1} s^{-1}. Although the release of the peroxide product at pH values >7 appears to occur at less than 5 s^{-1} from the enzyme from *Desulfoarculus baarsii* (Niviere *et al.*, 2004) and at ~50 s^{-1} from the *Desulfovibrio vulgaris*

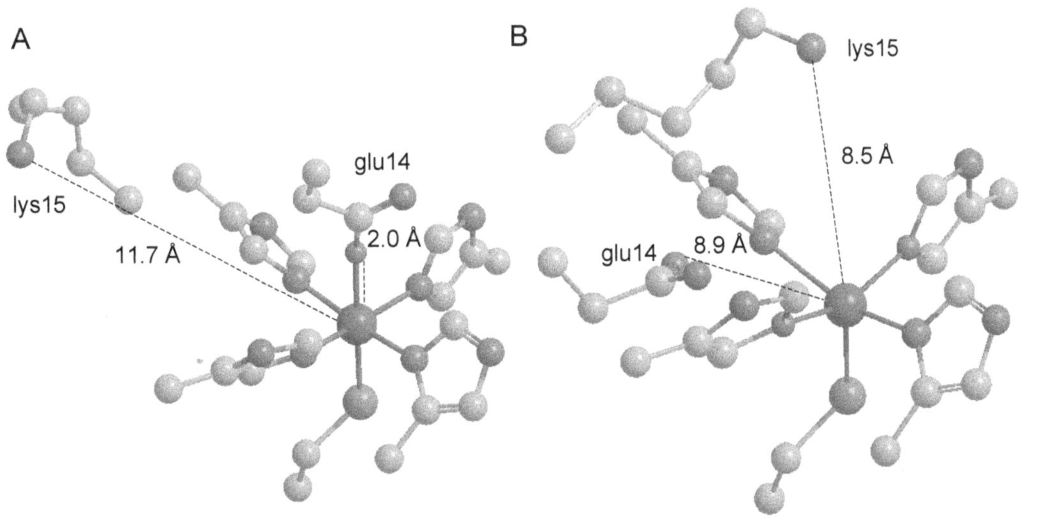

Figure 3.4 The active site of SOR, showing the iron active site with the liganding histidines, serine, and the glutamate and lysine that show significant movement. (A) Oxidized SOR, from the structure 1DO6 (http://www.rcsb.org). (B) Reduced SOR, from structure 1DQK.

Dfx (Kurtz, 2004), and the regeneration of reduced enzyme by rubredoxin or other reducing substrates appears to occur at a rate constant of ~ 10^6 M^{-1} s^{-1} (Kurtz, 2004; Niviere *et al.*, 2004), the slow step of the cycle is likely to be superoxide diffusion to the active site, since even *E. coli*, which is fairly aerotolerant, has an upper limit of superoxide tolerance of ~10^{-10} M. It has been pointed out that at that concentration there is an average of less than one superoxide anion per cell, making this an unusual case in which the enzyme is present at a concentration far in excess of its substrate (Kurtz, 2004).

It has been suggested that in addition to controlling access to the iron, the loop movements observed in the oxidized vs. reduced structures may play a role in controlling the binding of reduced rubredoxin (or other physiological donors) to the oxidized enzyme (Yeh *et al.*, 2000). Although proton donation appears to occur directly from solvent (Niviere *et al.*, 2004), other steps in the actual mechanism of superoxide reduction are controversial, despite being well studied with both the enzyme itself and non-enzymatic analogues of the SOR iron system (Theisen and Kovacs, 2005).

NAD(P)H:rubredoxin oxidoreductase (NROR)

Models of the superoxide detoxification pathway have proposed a role for an NAD(P)H-dependent reductase (NROR) which would allow NAD(P)H to serve as the ultimate electron donor in the detoxification pathway (Fig. 3.3). Confirming the role of this enzyme, however, has been hampered by difficulties in isolating sufficient quantities of enzyme to conduct these studies. Part of this difficulty has stemmed from the toxicity of the NROR gene product to *E. coli*, which appears to be due to the ability of this hyperthermophilic enzyme to operate at room temperature. While ambient conditions constitute a very low temperature for pyrococcal enzymes, this observation helps to explain some of the ability of *Pyrococcus* to tolerate oxygen at low temperatures (Grunden *et al.*, 2005). This difficulty was overcome by expressing the enzyme as an inactive fusion protein, which was then cleaved to yield active recombinant enzyme. The enzyme from *P. furiosus* is a 40 kDa FAD-dependent enzyme which is homologous to confirmed NRORs from mesophilic Bacteria. The rNROR is purified with only ~50% of the enzyme in the FAD bound form, although it is not clear whether attempts were made to fully reconstitute the enzyme.

The potential for rNROR was shown to be −173 mV at 23°C (−193 mV at 80°C), consistent with its proposed role in oxidizing NADPH ($E^{o\prime}$ = −320 mV) and reducing rubredoxin ($E^{o\prime}$ ~ 0 mV for the *P. furiosus* enzyme). The reaction shows a marked preference for NADPH over NADH, although the enzyme can catalyse the NADH-dependent reduction of benzyl viologen, albeit with a significantly higher K_m (Grunden *et al.*, 2005). The enzyme-bound FAD can be reduced by dithionite to what appears to be the $FADH^-$ or $FADH_2$ form, although it is not clear how many reducing equivalents are required for full reduction of the enzyme (Grunden *et al.*, 2005).

Using the rNROR, Grunden *et al.* were able to reconstitute for the first time both an NADPH-dependent SOR reductase system by combining rNROR, rubredoxin, and (oxidized) SOR in an anaerobic system, and an NADPH-dependent superoxide reductase system by combining all of the above in an aerobic system with superoxide generated by a xanthine/xanthine oxidase system, confirming the proposed role for all three components in the detoxification pathway shown in Fig. 3.3. Consistent with the ability of *Pyrococcus* to tolerate exposure to oxygen at low temperatures, the system was highly active at room temperature. Although the SOR potential (+31 mV) appears to be only slightly higher than the rubredoxin potential, a change in the potential of either electron donor or acceptor upon complex formation may drive the kinetics of the reaction, while the overall flux of electrons through the path may provide the overall thermodynamic driving force for the reaction.

Single-cysteine-containing NAD(P)H-dependent flavoprotein subclass of pyridine nucleotide oxidoreductases (PNDORs)

The single cysteine containing flavoproteins are a subgroup of the PNDOR enzymes represented by the true H_2O-producing NADH oxidase and NADH peroxidase of *Enterococcus faecalis* (Claiborne *et al.*, 1999). It initially appeared that these enzymes were not of especially great general importance, as they were found only sporadically in the Bacteria, mostly in facultative anaerobes. With the advent of genome sequencing, however, homologues of these enzymes were found throughout the Archaea and hyperthermophilic Bacteria, where they appear to be ubiquitous, with many hyperthermophiles having multiple versions; *P. furiosus* has at least three, and *A. fulgidus* has at least six. This observation suggested that these enzymes may confer some sort of advantage under extremeophilic conditions. These 'single cysteine' enzymes are distinguished by the presence of an absolutely conserved single redox-active cysteine at their active site that interacts with the C4a or N5 positions of the bound FAD. This cysteine is capable of cycling between either reduced thiol and oxidized sulfenic acid or thiol and oxidized mixed-disulphide forms. The other members of this group, frequently referred to as the flavoprotein disulphide reductases and represented by glutathione reductase and thioredoxin reductase, contain at least two active site cysteines that cycle between reduced dithiol and oxidized disulphide states.

In the Archaea, the single-cysteine containing enzymes can be divided into three groups, which are discussed in detail below: (1) The H_2O_2 or mixed H_2O/H_2O_2-producing NADH oxidases (NOXs), which tend to show sluggish oxidase activity, (2) coenzyme A disulphide reductases (CoADRs), which catalyse the reduction of CoA disulphide with NAD(P)H and have recently been shown to act in *P. furiosus* as a sulphur reductase (Schut *et al.*, 2007), and a new class, (3) the hyperactive NADH oxidases (hyperNOX), which are heterodimeric enzymes that catalyse a rapid peroxide-producing oxidase reaction. Archaeal or hyperthermophilic examples of the true H_2O-producing NADH oxidase ($O_2 + 2NADH + 2H^+ \rightarrow 2H_2O + 2NAD^+$) or NADH peroxidases ($H_2O_2 + NADH + H^+ \rightarrow 2H_2O + NAD^+$) have not been found to date despite considerable effort. It is somewhat of a mystery why the Archaea would not contain these two enzymes, since the three types discussed above show a high degree of sequence identity to them, suggesting that these enzymes evolved in the Bacteria after the Archaea/Bacteria divergence and they have not yet appeared in the Archaea, or that they lost the activity after the Archaea/Bacteria divergence. The lack of these enzymes is so striking that it makes this author wonder whether there may be something about extreme environments that selects against them. The robust nature of the NADH peroxidase from *E. faecalis* tends to argue against this; although the enzyme is not particularly thermostable, its long term-stability and generally durable nature, which make it an excellent subject for enzymological studies (Crane *et al.*, 1997), make it seem to be an easy protein for microbes to adapt to use under more extreme conditions.

H_2O_2 or mixed H_2O/H_2O_2-producing NADH oxidases (NOX)

An impressive number of these enzymes have been found and characterized from a range of hyperthermophilic Bacteria and Archaea, including examples from *P. furiosus*, *P. horikoshii*, *Archaeoglobus fulgidus* (Kengen *et al.*, 2003; Reed *et al.*, 2001; Ward *et al.*, 2001), and *Thermotoga hypogea* (Yang and Ma, 2005), with many uncharacterized homologues present in completed genomes. While it is clear that NADH (and sometimes NADPH) is the physiological reducing substrate for these enzymes, the identity of their physiological oxidizing substrate is not clear. Several lines of evidence argue against a role for oxygen reduction *in vivo*.

The recombinant enzyme from *P. furiosus* (*pfr*NOX) has been overexpressed in *E. coli* and characterized (Ward *et al.*, 2001). It is a thermostable and thermoactive homodimeric enzyme which, like many enzymes from hyperthermophiles, is relatively stable in organic solvents. This enzyme is an example of an H_2O and H_2O_2-producing enzyme, while others in the class produce only H_2O_2. The enzyme is fully reduced in two steps by two NADHs per flavin molecule, consistent with the reduction of a non-flavin redox centre which is most likely the absolutely conserved cysteine residue in a Cys-SOH or Cys-S-S-X form, followed by reduction of the bound FAD to $FADH_2$ or $FADH^-$. We have not been able to detect the release of a thiol from the enzyme upon reduction with NADH, arguing against the enzyme being in the form of a Cys-S-S-X mixed-disulphide (the enzyme active site Cys-SH is not detected either, most likely due to shielding of the active site from solvent, as seen in other cases in this family of enzymes (Claiborne *et al.*, 1999)). The four-electron reduction of the enzyme by dithionite occurs in the same sequence, providing additional evidence for the presence of a non-flavin redox centre.

While the easy four electron reduction of *pfr*NOX creates the oxygen reactive $FADH_2$ or $FADH^-$ intermediate, several lines of evidence suggest that oxygen is not the true oxidizing substrate for *pfr*NOX. The reaction is quite sluggish, with a $k_{cat\ app}$ of 4.8 s^{-1} at 80°C under ambient (20%) oxygen concentrations (Ward *et al.*, 2001), which can be compared with the rapid oxidase reaction of the hyperNOX discussed below (Yang and Ma, 2007). Two lines of evidence indicate that the oxidative half-reaction is responsible for the sluggishness of the reaction: 1) use of deuterated NADH results in an isotope effect of 1.0 on k_{cat}, and (2) the reductive half reaction, as measured by stopped-flow spectroscopy at 45°C, appears to be quite rapid (>1000 s^{-1}). The enzyme also cannot be saturated with oxygen under normal pressures, showing an almost linear increase in activity when the oxygen concentration is raised from 0.5% to 100%, an odd characteristic for an enzyme which would be expected to operate under relatively low concentrations of oxygen. If the physiological substrate for the enzyme is in fact not oxygen, the question remains as to what the relevant oxidizing substrate might be.

Hyperactive NADH oxidases (hyperNOX)

While the NOXs discussed above show only a sluggish reaction with O_2, an oxidase recently discovered in *Thermotoga maritima* shows a much more robust NADH oxidase activity (Yang and Ma, 2007). The enzyme, which is a heterodimer of 54 and 45 kDa subunits, is described here because each of the subunits has a homologue in the *P. furiosus* genome, making it likely that the enzyme is active in that and other hyperthermophilic Archaea. The 45 kDa subunit shows considerable identity to the single-cysteine flavoproteins and is homologous to the NOX3 of *P. furiosus*, while the 54 kDa subunit is homologous to another enzyme present in the *P. furiosus* genome. Because of its heterodimeric nature, it was necessary to discover and isolate this enzyme directly from *Thermotoga*. This result provides the lesson that not all enzymes can be discovered by searching genomes, at least with our current level of ability to predict protein-protein interactions. It also makes one wonder whether the NOXs discussed above might need another protein in order to become optimally active (although their generally homodimeric structure argues against this).

Unlike the NOXs discussed above, in addition to the hyperNOX's rapid reaction with oxygen (k_{cat}/K_m = ~1×10^7 M^{-1} s^{-1}) the enzyme is also able to utilize a range of oxidizing substrates such as ferricyanide and DTNB (98 and 7% relative activity, respectively). Both of hyperNOX's subunits contain FAD, with the large subunit containing iron in what appears to be a 2Fe–2S centre. Further characterization of this enzyme should shed light on its role in the ecophysiology of the hyperthermophiles.

Coenzyme A disulphide reductases and small thiols in the Archaea

CoADR was originally discovered in *Staphylococcus aureus*, and is the only known disulphide reductase that utilizes a single cysteine rather than a cysteine pair at its active site (delCardayre and Davies, 1998; delCardayre *et al.*, 1998). The discovery of the enzyme also led to the discovery that many microbes utilize coenzyme A in a manner analogous to the use of glutathione in other organisms. CoADRs have since been discovered in other organisms, including the pathogen *Borrelia burgdorferi*, which was also shown to use coenzyme A as a major intracellular thiol (Boylan *et al.*, 2005). An enzyme from *P. furiosus* and *P. horikoshii* previously believed to be a NOX was found to show a CoADR activity (*ph and pf*-CoADR), which was the first CoADR discovered in the hyperthermophiles or the Archaea (Harris *et al.*, 2005). This also led to the subsequent finding that *Pyrococcus* too appeared to utilize coenzyme A as its major intracellular thiol. The recent discovery of the sulphur reductase activity of the *P. furiosus* enzyme raises the possibility that the CoA-disulphide reductase activity discussed below is not physiologically relevant (Schut *et al.*, 2007). *Halobacterium halobium* has been shown to use a unique thiol/disulphide reductase system consisting of the γ-glutamylcysteine thiol and a bis-γ-glutamylcysteine disulphide reductase which is more similar to the glutathione/glutathione reductase system than to the CoA/CoADR system discussed above (Sundquist and Fahey, 1989).

The *ph*CoADR shows significant reactivity with NADH and NADPH, with the two giving roughly equal $k_{cat\ app}$ values, although the K_m for NADPH is considerably lower ($K_{m\ app}$ for NADPH > 9 μM, while the $K_{m\ app}$ for NADH = 73 μM) (Harris *et al.*, 2005). The enzyme was shown to be purified in the Cys-SOH oxidation state, and reduction of this enzyme with NAD(P)H or dithionite resulted in the reduction of the non-flavin redox centre, followed by little or no reduction of the FAD (Harris *et al.*, 2005). This type of reduction behaviour is typically seen in the disulphide reductases and NADH peroxidase, as these enzymes catalyse two-electron reductions, which means it makes sense for them to avoid a four-electron reduced state which would create an oxygen-sensitive reduced flavin intermediate which could then be oxidized in a peroxide-producing side reaction.

Reduced-nicotinamide-dependent 2,2′-dithiodiethanesulfonate reductase

Within the methanogens another unique disulphide redcutase was discovered in *Methanobacterium thermoautotrophicum* ΔH (Smith and Rouviere, 1990). The enzyme is able to use either NADH or NADPH, although the reaction is three times faster with NADPH. The enzyme is not believed to play a direct role in methanogenesis, as it is not able to reduce the heterodisulphide generated during that process. It is instead hypothesized to play a role similar to that of the disulphide reductases above. It may also be involved in reducing coenzyme M disulphide, since the oxidized form of CoM is not able to be utilized in methanogenesis.

Small thiols (CoenzymeA)

We have not been able to find, either by direct detection of the thiol or by the presence of homologues of its synthetic enzymes or its reductase in genomes, the presence of the common small thiol glutathione (which is found throughout the Bacteria and Eukaryotes) in the Archaea. We have suggested that the thiol coenzyme A plays a role similar to glutathione in the Archaea, especially in the hyperthermophiles, as it is present in high levels in hyperthermophiles and has been associated with a flavin-dependent disulphide reductase which is specific for coenzyme A (CoADR), and has recently been suggested to be involved in the reduction of sulphur in *Pyrococcus* (Harris *et al.*, 2005; Hummel *et al.*, 2005; Schut *et al.*, 2007). Coenzyme A has been implicated in similar roles in several Bacteria in which it has been detected at high levels along with the presence of the associated coenzyme A disulphide reductase (Boylan *et al.*, 2005; delCardayre and Davies, 1998; delCardayre *et al.*, 1998), an observation which suggests that the evolution of a coenzyme A-dependent antioxidant system evolved prior to the archaeal/bacterial split, with the glutathione-dependent system evolving after that split. The small thiols coenzyme M (the smallest known coneyzme) and coenzyme B (7-mercaptoheptanoylthreonine phosphate), which are essential in methanogenesis, are the only other small molecular weight thiol that we are aware of being detected in the Archaea (Hedderich *et al.*, 2005).

Enzymes of methanogenesis

F_{420}-dependent enzymes

Methanogenesis, and presumably anaerobic methane oxidation, are found to take place only in the Archaea (Hallam *et al.*, 2003). The central methanogenesis pathway contains a number of redox coenzymes and enzymes that appear to be unique to that pathway (Reeve *et al.*, 1997). These include the F_{420}-reducing dehydrogenase and F_{420}-dependent alcohol dehydrogenase, both of which utilize coenzyme F_{420}. F_{420} is a deazaflavin that was originally discovered in the methanogens and has also been found in sulphate-reducing and halophilic Archaea and in some Bacteria and eukaryotes, including mycobacteria and cyanobacteria (Deppenmeier *et al.*, 1990; Haase *et al.*, 1992; Warkentin *et al.*, 2001). This coenzyme is only capable of catalysing two-electron hydride transfers from the C5 position of the 5′-deazaflavin, making it more analogous to NAD((P)H in its reactivity than to flavins. A F_{420}–H_2 oxidase (FprA) has been characterized which is believed to act as an antioxidant enzyme in *Methanobrevibacter arboriphilus* (Seedorf *et al.*, 2004).

Methyl coenzyme M reductase and heterodisulphide reductase

Methyl CoM reductase has been found in all methanogens, and is considered to be a marker enzyme for methanogenesis, although recent results suggest that the enzyme is also likely to be found in the as-yet uncultivated anaerobic methane-oxidizing Archaea (Hallam *et al.*, 2003). This $(\alpha\beta\gamma)_2$ hexamer contains another unique coenzyme, the nickel porphinoid factor F_{430}, and catalyses the release of methane from methyl coenzyme M with the subsequent formation of a mixed disulphide between coenzyme M and coenzyme B in the terminal reaction of methanogenesis (Grabarse *et al.*, 2000). The CoM-CoB mixed disulphide is then reduced by a heterodisulphide reductase, which is either a iron–sulphur flavoprotein or a membrane-bound iron–sulphur and cytochrome b-containing enzyme, depending on the source (Mander *et al.*, 2004; Hedderich *et al.*, 2005).

CO dehydrogenases

CO dehydrogenases are found throughout the prokaryotic world, likely due to their importance as one of the three main methods for the fixation of carbon in autotrophic organisms (in addition to the Calvin and reductive TCA cycles) (Ferry, 1995). Because this enzyme has such an ancient origin, appearing well before the Archaea–Bacteria split, it has been proposed that it played an essential role in simple early chemotrophic lifestyles, and may have been the first pathway for carbon fixation. CO dehydrogenases catalyse three main reactions, all of which are parts of energy generating pathways. These reactions include oxidation of CO and the synthesis and cleavage of acetyl CoA (Ferry, 1995). CO dehydrogenases from aerobes contain Mo, Fe and a flavin cofactor, while the enzyme from anaerobes, which is discussed below, is a Ni and Fe-containing enzyme.

In the acetotrophic methanogenic Archaea, the enzyme is associated with an acetyl CoA synthase in the CO dehydrogenase/acetyl CoA sythase complex (CODH/ACS), which catalyses the reversible oxidation of CO to CO_2 ($CO + H_2O \leftrightarrow CO_2 + 2H^+ + 2e^-$), as well as the overall reaction of Acetyl CoA + H^+ + tetrahydromethanopterin $\rightarrow$ methyltetrahydromethanopterin + CO + CoA (Murakami and Ragsdale, 2000). This complex has been found in *Methanosarcina barkeri* and *M. thermophila* (Murakami and Ragsdale, 2000), as well as in the sulphate-reducing *Archaeaglobus fulgidus* (Dai *et al.*, 1998). The complex is composed of five subunits ($\alpha,\beta,\delta,\gamma,\varepsilon$), with the $\alpha\varepsilon$ dimer appearing to correspond to the CO dehydrogenase activity, and the other three domains constituting the acetyl CoA synthase domain. This dimer contains two redox centres which include one 4Fe–4S centre and a Ni–Ni–4Fe–4S centre.

The role of tungsten in anaerobic Archaea and the early anaerobic earth

The presence of tungsten in the biological systems of a variety of Archaea (as well as other prokaryotes) has been known for a relatively short time, and the story is constantly developing, with new discoveries appearing on a regular basis. Although an important, and often obligate, role for tungsten has been defined in certain cases (for example, in *P. furiosus* molybdenum cannot substitute for tungsten), a broad (and indeed systematic) understanding of tungsten's role and biological importance has yet to be established. It is certainly true that molybdenum-dependent enzymes are much more oxygen sensitive than their molybdenum-substituted counterparts, leading to the hypothesis that the appearance of an oxidizing atmosphere resulted in the isolation of tungsten-dependent enzymes in anerobic niches. Tungsten-dependent enzymes also have a lower potential than their molybdo-counterparts, a role indicative of their place in organisms found in reducing conditions.

A total of five tungsten-containing enzymes have been discovered in *Pyrococcus*: GAPOR, AOR, FOR, and an enzyme of unknown function, WOR4 (Roy and Adams, 2002a). The similarities between molybdenum and tungsten have been previously outlined (Kisker *et al.*, 1998), and owing to their relative positions on the periodic table of elements (both are in the same atomic group), the two metals have strikingly similar (and often identical) properties, particularly with respect to ionic and atomic radii, electronegativity, and double bond lengths with oxygen. Both of these elements are quite scarce; and in the case of W, the location of its abundance correlates significantly with the presence of organisms that rely on tungsten in their metabolism (Afshar *et al.*, 1998).

References

Abreu, I.A., Saraiva, L.M., Carita, J., Huber, H., Stetter, K.O., Cabelli, D., and Teixeira, M. (2000). Oxygen detoxification in the strict anaerobic archaeon *Archaeoglobus fulgidus*: superoxide scavenging by neelaredoxin. Mol. Microbiol. *38*, 322–334.

Abreu, I.A., Saraiva, L.M., Soares, C.M., Teixeira, M., and Cabelli, D.E. (2001). The mechanism of superoxide scavenging by *Archaeoglobus fulgidus* neelaredoxin. J. Biol. Chem. *276*, 38995–39001.

Abreu, I.A., Xavier, A.V., LeGall, J., Cabelli, D.E., and Teixeira, M. (2002). Superoxide scavenging by neelaredoxin: dismutation and reduction activities in anaerobes. J. Biol. Inorg Chem. *7*, 668–674.

Adams, M.W., Jenney, F.E., Jr., Clay, M.D., and Johnson, M.K. (2002). Superoxide reductase: fact or fiction? J. Biol. Inorg Chem. *7*, 647–652.

Afshar, S., Kim, C., Monbouquette, H.G., and Schroder, I. (1998). Effect of tungstate on nitrate reduction by the hyperthermophilic archaeon *Pyrobaculum aerophilum*. Appl. Environ. Microbiol. *64*, 3004–3008.

Ahmed, H., Ettema, T.J., Tjaden, B., Geerling, A.C., van der Oost, J., and Siebers, B. (2005). The semi-phosphorylative Entner–Doudoroff pathway in hyperthermophilic Archaea – a re-evaluation. Biochem. J. *390*, 529–540.

Amend, J.P., and Shock, E.L. (2001). Energetics of overall metabolic reactions of thermophilic and hyperthermophilic Archaea and Bacteria. FEMS Microbiol. Rev. *25*, 175–243.

Antoine, E., Rolland, J.L., Raffin, J.P., and Dietrich, J. (1999). Cloning and over-expression in *Escherichia coli* of the gene encoding NADPH group III alcohol dehydrogenase from *Thermococcus hydrothermalis*. Characterization and comparison of the native and the recombinant enzymes. Eur. J. Biochem. *264*, 880–889.

Bevers, L.E., Bol, E., Hagedoorn, P.L., and Hagen, W.R. (2005). WOR5, a novel tungsten-containing aldehyde oxidoreductase from *Pyrococcus furiosus* with a broad substrate specificity. J. Bacteriol. *187*, 7056–7061.

Bhuiya, M.W., Sakuraba, H., Ohshima, T., Imagawa, T., Katunuma, N., and Tsuge, H. (2005). The first crystal structure of hyperthermostable NAD-dependent glutamate dehydrogenase from *Pyrobaculum islandicum*. J. Mol. Biol. *345*, 325–337.

Blamey, J., and Adams, M. (1993). Purification and characterization of pyruvate ferredoxin oxidoreductase from the hyperthermophilic archaeon *Pyrococcus furiosus*. Biochim. Biophys. Acta *1161*, 19.

Blamey, J.M., and Adams, M.W. (1994). Characterization of an ancestral type of pyruvate ferredoxin oxidoreductase from the hyperthermophilic bacterium, *Thermotoga maritima*. Biochemistry *33*, 1000–1007.

Bock, A.K., Kunow, J., Glasemacher, J., and Schonheit, P. (1996). Catalytic properties, molecular composition and sequence alignments of pyruvate:ferredoxin oxidoreductase from the methanogenic archaeon *Methanosarcina barkeri* (strain Fusaro). Eur. J. Biochem. *237*, 35–44.

Bol, E., Bevers, L.E., Hagedoorn, P.L., and Hagen, W.R. (2006). Redox chemistry of tungsten and iron–sulfur prosthetic groups in *Pyrococcus furiosus* formaldehyde ferredoxin oxidoreductase. J. Biol. Inorg Chem. *11*, 999–1006.

Boylan, J.A., Hummel, C.S., Benoit, S., Garcia-Lara, J., Treglown-Downey, J., Crane, E.J., and Gherardini, F.C. (2005). *Borrelia burgdorferi* bb0728 encodes a coenzyme A disulphide reductase whose function suggests a role in intracellular redox and the oxidative stress response. Mol. Microbiol. *59*, 475–486.

Britton, K.L., Baker, P.J., Fisher, M., Ruzheinikov, S., Gilmour, D.J., Bonete, M.J., Ferrer, J., Pire, C., Esclapez, J., and Rice, D.W. (2006). Analysis of protein solvent interactions in glucose dehydrogenase from the extreme halophile *Haloferax mediterranei*. Proc. Natl. Acad. Sci. USA *103*, 4846–4851.

Brugna, M., Nitschke, W., Asso, M., Guigliarelli, B., Lemesle-Meunier, D., and Schmidt, C. (1999). Redox components of cytochrome bc-type enzymes in acidophilic prokaryotes. II. The Rieske protein of phylogenetically distant acidophilic organisms. J. Biol. Chem. *274*, 16766-16772.

Brunner, N.A., Brinkmann, H., Siebers, B., and Hensel, R. (1998). NAD^+-dependent glyceraldehyde-3-phosphate dehydrogenase from *Thermoproteus tenax*. The first identified archaeal member of the aldehyde dehydrogenase superfamily is a glycolytic enzyme with unusual regulatory properties. J. Biol. Chem. *273*, 6149–6156.

Brunner, N.A., and Hensel, R. (2001). Nonphosphorylating glyceraldehyde-3-phosphate dehydrogenase from *Thermoproteus tenax*. Methods Enzymol. *331*, 117–131.

Chabriere, E., Vernede, X., Guigliarelli, B., Charon, M.H., Hatchikian, E.C., and Fontecilla-Camps, J.C. (2001). Crystal structure of the free radical intermediate of pyruvate:ferredoxin oxidoreductase. Science *294*, 2559–2563.

Chan, M.K., Mukund, S., Kletzin, A., Adams, M.W., and Rees, D.C. (1995). Structure of a hyperthermophilic tungstopterin enzyme, aldehyde ferredoxin oxidoreductase. Science *267*, 1463–1469.

Claiborne, A., Yeh, J.I., Mallett, T.C., Luba, J., Crane, E.J., 3rd, Charrier, V., and Parsonage, D. (1999). Protein-sulfenic acids: diverse roles for an unlikely player in enzyme catalysis and redox regulation. Biochemistry *38*, 15407–15416.

Crane, E.J., 3rd, Vervoort, J., and Claiborne, A. (1997). 13C NMR analysis of the cysteine-sulfenic acid redox center of enterococcal NADH peroxidase. Biochemistry *36*, 8611–8618.

Dai, Y.R., Reed, D.W., Millstein, J.H., Hartzell, P.L., Grahame, D.A., and DeMoll, E. (1998). Acetyl-CoA decarbonylase/synthase complex from *Archaeoglobus fulgidus*. Arch. Microbiol. *169*, 525–529.

D'Ambrosio, K., Pedone, E., Langella, E., De Simone, G., Rossi, M., Pedone, C., and Bartolucci, S. (2006). A novel member of the protein disulfide oxidoreductase family from *Aeropyrum pernix K1*: structure, function and electrostatics. J. Mol. Biol. *362*, 743–752.

delCardayre, S.B., and Davies, J.E. (1998). *Staphylococcus aureus* Coenzyme a disulfide reductase, a new subfamily of pyridine nucleotide-disulfide oxidoreductase. Sequence, expression, and analysis of cdr. J. Biol. Chem. *273*, 5752–5757.

delCardayre, S.B., Stock, K.P., Newton, G.L., Fahey, R.C., and Davies, J.E. (1998). Coenzyme A disulfide reductase, the primary low molecular weight disulfide reductase from *Staphylococcus aureus*. Purification and characterization of the native enzyme. J. Biol. Chem. *273*, 5744–5751.

Deppenmeier, U., Blaut, M., Mahlmann, A., and Gottschalk, G. (1990). Reduced coenzyme F_{420}: heterodisulfide oxidoreductase, a proton- translocating redox system in methanogenic Bacteria. Proc. Natl. Acad. Sci. USA *87*, 9449–9453.

Dikanov, S.A., Shubin, A.A., Kounosu, A., Iwasaki, T., and Samoilova, R.I. (2004). A comparative, two-dimensional ^{14}N ESEEM characterization of reduced [2Fe–2S] clusters in hyperthermophilic archaeal high- and low-potential Rieske-type proteins. J. Biol. Inorg Chem. *9*, 753–767.

DiRuggiero, J., Robb, F.T., Jagus, R., Klump, H.H., Borges, K.M., Kessel, M., Mai, X., and Adams, M.W. (1993). Characterization, cloning, and in vitro expression of the extremely thermostable glutamate dehydrogenase from the hyperthermophilic Archaeon, ES4. J. Biol. Chem. *268*, 17767–17774.

Emerson, J.P., Cabelli, D.E., and Kurtz, D.M., Jr. (2003). An engineered two-iron superoxide reductase lacking the $[Fe(SCys)_4]$ site retains its catalytic properties *in vitro* and *in vivo*. Proc. Natl. Acad. Sci. USA *100*, 3802–3807.

Esclapez, J., Pire, C., Bautista, V., Martinez-Espinosa, R.M., Ferrer, J., and Bonete, M.J. (2007). Analysis of acidic surface of *Haloferax mediterranei* glucose dehydrogenase by site-directed mutagenesis. FEBS Lett. *581*, 837–842.

Evans, M.C., Buchanan, B.B., and Arnon, D.I. (1966). A new ferredoxin-dependent carbon reduction cycle in a photosynthetic bacterium. Proc. Natl. Acad. Sci. USA *55*, 928–934.

Ferry, J.G. (1995). CO dehydrogenase. Annu. Rev. Microbiol. *49*, 305–333.

Friedrich, C.G., Bardischewsky, F., Rother, D., Quentmeier, A., and Fischer, J. (2005). Prokaryotic sulfur oxidation. Curr. Opin. Microbiol. *8*, 253–259.

Gilderson, G., Aagaard, A., Gomes, C.M., Adelroth, P., Teixeira, M., and Brzezinski, P. (2001). Kinetics of electron and proton transfer during O_2 reduction in cytochrome aa_3 from *A. ambivalens*: an enzyme lacking Glu286. Biochim. Biophys. Acta *1503*, 261–270.

Giri, A.V., Anishetty, S., and Gautam, P. (2004). Functionally specified protein signatures distinctive for each of the different blue copper proteins. BMC Bioinformatics 5, 127.

Gleissner, M., Kaiser, U., Antonopoulos, E., and Schafer, G. (1997). The archaeal SoxABCD complex is a proton pump in *Sulfolobus acidocaldarius*. J. Biol. Chem. *272*, 8417–8426.

Grabarse, W., Mahlert, F., Shima, S., Thauer, R.K., and Ermler, U. (2000). Comparison of three methyl-coenzyme M reductases from phylogenetically distant organisms: unusual amino acid modification, conservation and adaptation. J. Mol. Biol. *303*, 329–344.

Grunden, A.M., Jenney, F.E., Jr., Ma, K., Ji, M., Weinberg, M.V., and Adams, M.W. W. (2005). In Vitro Reconstitution of an NADPH-dependent superoxide reduction pathway from *Pyrococcus furiosus*. Appl. Environ. Microbiol. *71*, 1522–1530.

Haase, P., Deppenmeier, U., Blaut, M., and Gottschalk, G. (1992). Purification and characterization of $F_{420}H_2$-dehydrogenase from *Methanolobus tindarius*. Eur. J. Biochem. *203*, 527–531.

Hagedoorn, P.L., Chen, T., Schroder, I., Piersma, S.R., de Vries, S., and Hagen, W.R. (2005). Purification and characterization of the tungsten enzyme aldehyde:ferredoxin oxidoreductase from the hyperthermophilic denitrifier *Pyrobaculum aerophilum*. J. Biol. Inorg. Chem. *10*, 259–269.

Hagedoorn, P.L., Freije, J.R., and Hagen, W.R. (1999). *Pyrococcus furiosus* glyceraldehyde 3-phosphate oxidoreductase has comparable W(6+/5+) and W(5+/4+) reduction potentials and unusual [4Fe–4S] EPR properties. FEBS Lett. *462*, 66–70.

Hallam, S.J., Girguis, P.R., Preston, C.M., Richardson, P.M., and DeLong, E.F. (2003). Identification of methyl coenzyme M reductase A (mcrA) genes associated with methane-oxidizing Archaea. Appl. Environ. Microbiol. *69*, 5483–5491.

Hao, X., and Ma, K. (2003). Minimal sulfur requirement for growth and sulfur-dependent metabolism of the hyperthermophilic archaeon *Staphylothermus marinus*. Archaea *1*, 191–197.

Harris, D.R., Ward, D.E., Feasel, J.M., Lancaster, K.M., Murphy, R.D., Mallet, T.C., and Crane, E.J., 3rd (2005). Discovery and characterization of a Coenzyme A disulfide reductase from *Pyrococcus horikoshii*. Implications for this disulfide metabolism of anaerobic hyperthermophiles. FEBS J. *272*, 1189–1200.

Hedderich, R., Hamann, N., and Bennati, M. (2005). Heterodisulfide reductase from methanogenic Archaea: a new catalytic role for an iron–sulfur cluster. Biol. Chem. *386*, 961–970.

Hedderich, R., Klimmek, O., Kroger, A., Dirmeier, R., Keller, M., and Stetter, K.O. (1998). Anaerobic respiration with elemental sulfur and with disulfides. FEMS Microbiol. Rev. *22*, 353–381.

Hensel, R., Laumann, S., Lang, J., Heumann, H., and Lottspeich, F. (1987). Characterization of two D-glyceraldehyde-3-phosphate dehydrogenases from the extremely thermophilic archaebacterium *Thermoproteus tenax*. Eur. J. Biochem. *170*, 325–333.

Hille, R. (2002). Molybdenum and tungsten in biology. Trends Biochem. Sci. *27*, 360–367.

Hu, Y., Faham, S., Roy, R., Adams, M.W., and Rees, D.C. (1999). Formaldehyde ferredoxin oxidoreductase from *Pyrococcus furiosus*: the 1.85 Å resolution crystal structure and its mechanistic implications. J. Mol. Biol. *286*, 899–914.

Huber, R., Hof, P., Duarte, R.O., Moura, J.J., Moura, I., Liu, M.Y., LeGall, J., Hille, R., Archer, M., and Romao, M.J. (1996). A structure-based catalytic mechanism for the xanthine oxidase family of molybdenum enzymes. Proc. Natl. Acad. Sci. USA *93*, 8846–8851.

Hummel, C.S., Lancaster, K.M., and Crane, E.J., 3rd (2005). Determination of coenzyme A levels in *Pyrococcus furiosus* and other Archaea: implications for a general role for coenzyme A in thermophiles. FEMS Microbiol. Lett. *252*, 229–234.

Imai, T., Taguchi K, Ogawara Y, Ohmori D, Yamakura F, Ikezawa H, Urushiyama A. (2001). Characterization and cloning of an extremely thermostable, *Pyrococcus furiosus*-type 4Fe ferredoxin from *Thermococcus profundus*. J. Biochem. (Tokyo) *130*, 649–655.

Imanaka, T., and Atomi, H. (2002). Catalyzing 'hot' reactions: enzymes from hyperthermophilic Archaea. Chem. Rec 2, 149–163.

Ingoldsby, L.M., Geoghegan, K.F., Hayden, B.M., and Engel, P.C. (2005). The discovery of four distinct glutamate dehydrogenase genes in a strain of *Halobacterium salinarum*. Gene *349*, 237–244.

Iwasaki, T., and Oshima, T. (2001). Ferredoxin and related enzymes from *Sulfolobus*. Methods Enzymol *334*, 3–22.

Jain, R., Rivera, M.C., Moore, J.E., and Lake, J.A. (2002). Horizontal gene transfer in microbial genome evolution. Theor. Popul. Biol. *61*, 489–495.

Jenney, F.E., Jr., and Adams, M.W. (2001). Rubredoxin from *Pyrococcus furiosus*. Methods Enzymol. *334*, 45–55.

Jenney, F.E., Jr., Verhagen, M.F., Cui, X., and Adams, M.W. (1999). Anaerobic microbes: oxygen detoxification without superoxide dismutase. Science *286*, 306-309.

John, J., Crennell, S.J., Hough, D.W., Danson, M.J., and Taylor, G.L. (1994). The crystal structure of glucose dehydrogenase from *Thermoplasma acidophilum*. Structure 2, 385–393.

Kappler, U., Hanson, G.R., Jones, A., and McEwan, A.G. (2005a). A recombinant diheme SoxAX cytochrome – implications for the relationship between EPR signals and modified heme-ligands. FEBS Lett. *579*, 2491–2498.

Kappler, U., Sly, L.I., and McEwan, A.G. (2005b). Respiratory gene clusters of *Metallosphaera sedula* – differential expression and transcriptional organization. Microbiology *151*, 35–43.

Kengen, S.W. M., van der Oost, J., and de Vos, W.M. (2003). Molecular characterization of H_2O_2-forming NADH oxidases from *Archaeoglobus fulgidus*. Eur. J. Biochem. *270*, 2885–2894.

Kim, C., Brereton, P.S., Verhagen, M.F., and Adams, M.W. (2001). Ferredoxin from *Pyrococcus furiosus*. Methods Enzymol. *334*, 30–40.

Kisker, C., Schindelin, H., Baas, D., Retey, J., Meckenstock, R.U., and Kroneck, P.M. (1998). A structural comparison of molybdenum cofactor-containing enzymes. FEMS Microbiol. Rev. *22*, 503–521.

Kletzin, A., and Adams, M.W. (1996). Molecular and phylogenetic characterization of pyruvate and 2-ketoisovalerate ferredoxin oxidoreductases from *Pyrococcus furiosus* and pyruvate ferredoxin oxidoreductase from *Thermotoga maritima*. J. Bacteriol. *178*, 248–257.

Kletzin, A., Urich, T., Muller, F., Bandeiras, T.M., and Gomes, C.M. (2004). Dissimilatory oxidation and reduction of elemental sulfur in thermophilic Archaea. J. Bioenerg. Biomembr. *36*, 77–91.

Komorowski, L., and Schafer, G. (2001). Sulfocyanin and subunit II, two copper proteins with novel features, provide new insight into the archaeal SoxM oxidase supercomplex. FEBS Lett. *487*, 351–355.

Kounosu, A., Li, Z., Cosper, N.J., Shokes, J.E., Scott, R.A., Imai, T., Urushiyama, A., and Iwasaki, T. (2004). Engineering a three-cysteine, one-histidine ligand environment into a new hyperthermophilic archaeal Rieske-type [2Fe–2S] ferredoxin from *Sulfolobus solfataricus*. J. Biol. Chem. *279*, 12519–12528.

Kunow, J., Linder, D., and Thauer, R.K. (1995). Pyruvate:ferredoxin oxidoreductase from the sulfate-reducing *Archeoglobus fulgidus*: molecular composition, catalytic properties, and sequence alignments. Archives of Microbiology *163*, 21.

Kurihara, K., Tanaka, I., Chatake, T., Adams, M.W., Jenney, F.E., Jr., Moiseeva, N., Bau, R., and Niimura, N. (2004). Neutron crystallographic study on rubredoxin from *Pyrococcus furiosus* by BIX-3, a single-crystal diffractometer for biomacromolecules. Proc. Natl. Acad. Sci. USA *101*, 11215–11220.

Kuroita, T., Kanno, T., Kawai, A., Kawakami, B., Oka, M., Endo, Y., and Tozawa, Y. (2007). Functional similarities of a thermostable protein-disulfide oxidoreductase identified in the archaeon *Pyrococcus horikoshii* to bacterial DsbA enzymes. Extremophiles *11*, 85–94.

Kurtz, D.M., Jr. (2004). Microbial detoxification of superoxide: the non-heme iron reductive paradigm for combating oxidative stress. Acc Chem. Res. *37*, 902–908.

Ladenstein, R., and Ren, B. (2006). Protein disulfides and protein disulfide oxidoreductases in hyperthermophiles. FEBS J. *273*, 4170–4185.

Lamble, H.J., Heyer, N.I., Bull, S.D., Hough, D.W., and Danson, M.J. (2003). Metabolic pathway promiscuity in the archaeon *Sulfolobus solfataricus* revealed by studies on glucose dehydrogenase and 2-keto-3-deoxygluconate aldolase. J. Biol. Chem. *278*, 34066-34072.

Lee, M.K., Gonzalez, J.M., and Robb, F.T. (2002). Extremely thermostable glutamate dehydrogenase (GDH) from the freshwater archaeon *Thermococcus waiotapuensis*: cloning and comparison with two marine hyperthermophilic GDHs. Extremophiles *6*, 151–159.

Limauro, D., Pedone, E., Pirone, L., and Bartolucci, S. (2006). Identification and characterization of 1-Cys peroxiredoxin from *Sulfolobus solfataricus* and its involvement in the response to oxidative stress. FEBS J. *273*, 721–731.

Littlechild, J.A., Guy, J.E., and Isupov, M.N. (2004). Hyperthermophilic dehydrogenase enzymes. Biochem. Soc. Trans *32*, 255–258.

Lorentzen, E., Hensel, R., Knura, T., Ahmed, H., and Pohl, E. (2004). Structural basis of allosteric regulation and substrate specificity of the non-phosphorylating glyceraldehyde 3-Phosphate dehydrogenase from *Thermoproteus tenax*. J. Mol. Biol. *341*, 815–828.

Lubben, M., Arnaud, S., Castresana, J., Warne, A., Albracht, S.P., and Saraste, M. (1994). A second terminal oxidase in *Sulfolobus acidocaldarius*. Eur. J. Biochem. *224*, 151–159.

Ma, K., and Adams, M.W. (2001). Ferredoxin:NADP oxidoreductase from *Pyrococcus furiosus*. Methods Enzymol *334*, 40–45.

Ma, K., and Adams, M.W. W. (1999a). A hyperactive NAD(P)H:rubredoxin oxidoreductase from the hyperthermophilic archaeon *Pyrococcus furiosus*. J. Bacteriol. *181*, 5530–5533.

Ma, K., and Adams, M.W. W. (1999b). An unusual oxygen-sensitive, iron- and zinc-containing alcohol dehydrogenase from the hyperthermophilic archaeon *Pyrococcus furiosus*. J. Bacteriol. *181*, 1163–1170.

Ma, K., Loessner, H., Heider, J., Johnson, M.K., and Adams, M.W. (1995). Effects of elemental sulfur on the metabolism of the deep-sea hyperthermophilic archaeon *Thermococcus* strain ES-1: characterization of a sulfur-regulated, non-heme iron alcohol dehydrogenase. J. Bacteriol. *177*, 4748–4756.

Ma, K., Robb, F.T., and Adams, M.W. (1994). Purification and characterization of NADP-specific alcohol dehydrogenase and glutamate dehydrogenase from the hyperthermophilic archaeon *Thermococcus litoralis*. Appl. Environ. Microbiol. *60*, 562–568.

Machielsen, R., Uria, A.R., Kengen, S.W., and van der Oost, J. (2006). Production and characterization of a thermostable alcohol dehydrogenase that belongs to the aldo-keto reductase superfamily. Appl. Environ. Microbiol. *72*, 233–238.

Mallick, P., Boutz, D.R., Eisenberg, D., and Yeates, T.O. (2002). Genomic evidence that the intracellular proteins of archaeal microbes contain disulfide bonds. Proc. Natl. Acad. USA *99*, 9679–9684.

Mander, G.J., Pierik, A.J., Huber, H., and Hedderich, R. (2004). Two distinct heterodisulfide reductase-like enzymes in the sulfate-reducing archaeon *Archaeoglobus profundus*. Eur. J. Biochem. *271*, 1106-1116.

Mathe, C., Mattioli, T.A., Horner, O., Lombard, M., Latour, J.M., Fontecave, M., and Niviere, V. (2002). Identification of iron(III) peroxo species in the active site of the superoxide reductase SOR from *Desulfoarculus baarsii*. J. Am. Chem. Soc. *124*, 4966–4967.

Mehta, M.P., and Baross, J.A. (2006). Nitrogen fixation at 92 degrees C by a hydrothermal vent archaeon. Science *314*, 1783–1786.

Mehta, M.P., Butterfield, D.A., and Baross, J.A. (2003). Phylogenetic diversity of nitrogenase (nifH) genes in deep-sea and hydrothermal vent environments of the Juan de Fuca Ridge. Appl. Environ. Microbiol. *69*, 960–970.

Mehta, M.P., Huber, J.A., and Baross, J.A. (2005). Incidence of novel and potentially archaeal nitrogenase genes in the deep Northeast Pacific Ocean. Environ. Microbiol. *7*, 1525–1534.

Moser, D.P., and Nealson, K.H. (1996). Growth of the facultative anaerobe *Shewanella putrefaciens* by elemental sulfur reduction. Appl. Environ. Microbiol. *62*, 2100–2105.

Mukund, S., and Adams, M.W. (1991). The novel tungsten-iron–sulfur protein of the hyperthermophilic archaebacterium, *Pyrococcus furiosus*, is an aldehyde ferredoxin oxidoreductase. Evidence for its participation in a unique glycolytic pathway. J. Biol. Chem. *266*, 14208–14216.

Mukund, S., and Adams, M.W. (1995). Glyceraldehyde-3-phosphate ferredoxin oxidoreductase, a novel tungsten-containing enzyme with a potential glycolytic role in the hyperthermophilic archaeon *Pyrococcus furiosus*. J. Biol. Chem. *270*, 8389–8392.

Muller, F.H., Bandeiras, T.M., Urich, T., Teixeira, M., Gomes, C.M., and Kletzin, A. (2004). Coupling of the pathway of sulphur oxidation to dioxygen reduction: characterization of a novel membrane-bound thiosulphate:quinone oxidoreductase. Mol. Microbiol. *53*, 1147–1160.

Murakami, E., and Ragsdale, S.W. (2000). Evidence for intersubunit communication during acetyl-CoA cleavage by the multienzyme CO dehydrogenase/acetyl-CoA synthase complex from *Methanosarcina thermophila*. Evidence that the beta subunit catalyzes C-C and C-S bond cleavage. J. Biol. Chem. *275*, 4699–4707.

Nakasako, M., Fujisawa, T., Adachi, S., Kudo, T., and Higuchi, S. (2001). Large-scale domain movements and hydration structure changes in the active-site cleft of unligated glutamate dehydrogenase from *Thermococcus profundus* studied by cryogenic X-ray crystal structure analysis and small-angle X-ray scattering. Biochemistry *40*, 3069–3079.

Niviere, V., Asso, M., Weill, C.O., Lombard, M., Guigliarelli, B., Favaudon, V., and Houee-Levin, C. (2004). Superoxide reductase from *Desulfoarculus baarsii*: identification of protonation steps in the enzymatic mechanism. Biochemistry *43*, 808–818.

Pedone, E., D'Ambrosio, K., De Simone, G., Rossi, M., Pedone, C., and Bartolucci, S. (2006a). Insights on a new PDI-like family: structural and functional analysis of a protein disulfide oxidoreductase from the bacterium *Aquifex aeolicus*. J. Mol. Biol. *356*, 155–164.

Pedone, E., Limauro, D., D'Alterio, R., Rossi, M., and Bartolucci, S. (2006b). Characterization of a multifunctional protein disulfide oxidoreductase from *Sulfolobus solfataricus*. FEBS J. *273*, 5407–5420.

Pedone, E., Ren, B., Ladenstein, R., Rossi, M., and Bartolucci, S. (2004). Functional properties of the protein disulfide oxidoreductase from the archaeon *Pyrococcus furiosus*. A member of a novel protein family related to protein disulfide-isomerase. Eur. J. Biochem. *271*, 3437–3448.

Pereira, M.M., Santana, M., and Teixeira, M. (2001). A novel scenario for the evolution of haem-copper oxygen reductases. Biochim. Biophys. Acta *1505*, 185–208.

Plaga, W., Lottspeich, F., and Oesterhelt, D. (1992). Improved purification, crystallization and primary structure of pyruvate:ferredoxin oxidoreductase from *Halobacterium halobium*. Eur. J. Biochem. *205*, 391–397.

Radianingtyas, H., and Wright, P.C. (2003). Alcohol dehydrogenases from thermophilic and hyperthermophilic Archaea and Bacteria. FEMS Microbiol. Rev. *27*, 593–616.

Razvi, A., and Scholtz, J.M. (2006). Lessons in stability from thermophilic proteins. Protein Sci. *15*, 1569–1578.

Reed, D.W., Millstein, J., and Hartzell, P.L. (2001). H_2O_2-Forming NADH oxidase with diaphorase (cytochrome) activity from *Archaeoglobus fulgidus*. J. Bacteriol. *183*, 7007–7016.

Reeve, J.N., Nolling, J., Morgan, R.M., and Smith, D.R. (1997). Methanogenesis: genes, genomes, and who's on first? J. Bacteriol. *179*, 5975–5986.

Ren, B., and Ladenstein, R. (2001). Protein disulfide oxidoreductase from *Pyrococcus furiosus*: structural properties. Methods Enzymol. *334*, 74–88.

Rodrigues, J.V., Abreu, I.A., Cabelli, D., and Teixeira, M. (2006). Superoxide reduction mechanism of *Archaeoglobus fulgidus* one-iron superoxide reductase. Biochemistry *45*, 9266–9278.

Rodrigues, J.V., Abreu, I.A., Saraiva, L.M., and Teixeira, M. (2005). Rubredoxin acts as an electron donor for neelaredoxin in *Archaeoglobus fulgidus*. Biochem. Biophys. Res. Commun. *329*, 1300–1305.

Roy, R., and Adams, M.W. (2002a). Characterization of a fourth tungsten-containing enzyme from the hyperthermophilic archaeon *Pyrococcus furiosus*. J. Bacteriol. *184*, 6952–6956.

Roy, R., and Adams, M.W. W. (2002b). Characterization of a fourth tungsten-containing enzyme from the hyperthermophilic archaeon *Pyrococcus furiosus*. J. Bacteriol. *184*, 6952–6956.

Schicho, R.N., Ma, K., Adams, M.W., and Kelly, R.M. (1993). Bioenergetics of sulfur reduction in the hyperthermophilic archaeon *Pyrococcus furiosus*. J. Bacteriol. *175*, 1823–1830.

Schut, G.J., Bridger, S.L., and Adams, M.W. (2007). Insights into the metabolism of elemental sulfur by the hyperthermophilic archaeon *Pyrococcus furiosus*: Characterization of a coenzyme A-dependent NAD(P)H sulfur oxidoreductase. J. Bacteriol. *189*, 4431–4441.

Seedorf, H., Dreisbach, A., Hedderich, R., Shima, S., and Thauer, R.K. (2004). F420H2 oxidase (FprA) from *Methanobrevibacter arboriphilus*, a coenzyme F_{420}-dependent enzyme involved in O_2 detoxification. Arch. Microbiol. *182*, 126-137.

Siebers, B., and Schonheit, P. (2005). Unusual pathways and enzymes of central carbohydrate metabolism in Archaea. Curr. Opin. Microbiol. *8*, 695–705.

Smith, E.T., Blamey, J.M., Zhou, Z.H., and Adams, M.W. (1995). A variable-temperature direct electrochemical study of metalloproteins from hyperthermophilic microorganisms involved in hydrogen production from pyruvate. Biochemistry *34*, 7161–7169.

Smith, S.G., and Rouviere, P.E. (1990). Purification and characterization of the reduced-nicotinamide-dependent 2,2'-dithiodiethanesulfonate reductase from *Methanobacterium thermoautotrophicum* delta H.J. Bacteriol. *172*, 6435–6441.

Steigerwald, V.J., Pihl, T.D., and Reeve, J.N. (1992). Identification and isolation of the polyferredoxin from *Methanobacterium thermoautotrophicum* strain delta H. Proc. Natl. Acad. Sci. USA *89*, 6929–6933.

Sundquist, A.R., and Fahey, R.C. (1989). The function of gamma-glutamylcysteine and bis-gamma-glutamylcystine reductase in *Halobacterium halobium*. J. Biol. Chem. *264*, 719–725.

Tersteegen, A., Linder, D., Thauer, R.K., and Hedderich, R. (1997). Structures and functions of four anabolic 2-oxoacid oxidoreductases in *Methanobacterium thermoautotrophicum*. Eur. J. Biochem. *244*, 862–868.

Theisen, R.M., and Kovacs, J.A. (2005). Role of protons in superoxide reduction by a superoxide reductase analogue. Inorg. Chem. *44*, 1169–1171.

Trivedi, S., Gehlot, H.S., and Rao, S.R. (2006). Protein thermostability in Archaea and eubacteria. Genet. Mol. Res. 5, 816–827.

Urich, T., Bandeiras, T.M., Leal, S.S., Rachel, R., Albrecht, T., Zimmermann, P., Scholz, C., Teixeira, M., Gomes, C.M., and Kletzin, A. (2004). The sulphur oxygenase reductase from *Acidianus ambivalens* is a multimeric protein containing a low-potential mononuclear non-haem iron centre. Biochem. J. *381*, 137–146.

Urich, T., Gomes, C.M., Kletzin, A., and Frazao, C. (2006). X-ray structure of a self-compartmentalizing sulfur cycle metalloenzyme. Science *311*, 996-1000.

van der Oost, J., Schut, G., Kengen, S.W., Hagen, W.R., Thomm, M., and de Vos, W.M. (1998). The ferredoxin-dependent conversion of glyceraldehyde-3-phosphate in the hyperthermophilic archaeon *Pyrococcus furiosus* represents a novel site of glycolytic regulation. J. Biol. Chem. *273*, 28149–28154.

Wang, P.L., Calzolai, L., Bren, K.L., Teng, Q., Jenney, F.E., Jr., Brereton, P.S., Howard, J.B., Adams, M.W., and La Mar, G.N. (1999). Secondary structure extensions in *Pyrococcus furiosus* ferredoxin destabilize the disulfide bond relative to that in other hyperthermostable ferredoxins. Global consequences for the disulfide orientational heterogeneity. Biochemistry *38*, 8167–8178.

Wang, S., Feng, Y., Zhang, Z., Zheng, B., Li, N., Cao, S., Matsui, I., and Kosugi, Y. (2003). Heat effect on the structure and activity of the recombinant glutamate dehydrogenase from a hyperthermophilic archaeon *Pyrococcus horikoshii*. Arch. Biochem. Biophys. *411*, 56–62.

Wang, X., and Weiner, H. (1995). Involvement of glutamate 268 in the active site of human liver mitochondrial (class 2) aldehyde dehydrogenase as probed by site-directed mutagenesis. Biochemistry 34, 237–243.

Ward, D.E., Donnelly, C.J., Mullendore, M.E., van der Oost, J., de Vos, W.M., and Crane, E.J., 3rd (2001). The NADH oxidase from *Pyrococcus furiosus*. Implications for the protection of anaerobic hyperthermophiles against oxidative stress. Eur. J. Biochem. *268*, 5816–5823.

Warkentin, E., Mamat, B., Sordel-Klippert, M., Wicke, M., Thauer, R.K., Iwata, M., Iwata, S., Ermler, U., and Shima, S. (2001). Structures of $F_{420}H_2$:$NADP^+$ oxidoreductase with and without its substrates bound. EMBO J. *20*, 6561–6569.

Wiedenheft, B., Mosolf, J., Willits, D., Yeager, M., Dryden, K.A., Young, M., and Douglas, T. (2005). From the cover: An archaeal antioxidant: Characterization of a Dps-like protein from *Sulfolobus solfataricus*. Proc. Natl. Acad. Sci. USA *102*, 10551–10556.

Yang, X., and Ma, K. (2005). Purification and characterization of an NADH oxidase from extremely thermophilic anaerobic bacterium *Thermotoga hypogea*. Arch. Microbiol. *183*, 331–337.

Yang, X., and Ma, K. (2007). Characterization of an exceedingly active NADH oxidase from anaerobic hyperthermophilic bacterium *Thermotoga maritima*. J. Bacteriol..

Yeh, A.P., Hu, Y., Jenney, F.E., Jr., Adams, M.W., and Rees, D.C. (2000). Structures of the superoxide reductase from *Pyrococcus furiosus* in the oxidized and reduced states. Biochemistry *39*, 2499–2508.

Ying, X., Wang, Y., Badiei, H.R., Karanassios, V., and Ma, K. (2007). Purification and characterization of an iron-containing alcohol dehydrogenase in extremely thermophilic bacterium *Thermotoga hypogea*. Arch. Microbiol. *187*, 499–510.

Yoshimatsu, K., Araya, O., and Fujiwara, T. (2007). *Haloarcula marismortui* cytochrome b_{561} is encoded by the narC gene in the dissimilatory nitrate reductase operon. Extremophiles *11*, 41–47.

Pentose Metabolism in Archaea

4

Harmen J.G. van de Werken, Stan J.J. Brouns
and John van der Oost

Abstract

Archaeal physiology has been studied extensively ever since the discovery that they constitute a distinct domain of life. The diversity of the archaeal metabolism is very high: they are able to grow fermentatively, but are also able to respire aerobically or anaerobically, and obtain their energy from light or (an)organic molecules. Initially, hexose conversions received most attention. Recently, however, significant insight has been gained in the archaeal pentose metabolism. Importantly, novel genomic, genetic and bioinformatical tools are applicable now to study the archaeal biology in more detail. We here compare the archaeal pentose pathways, the enzymes and their regulation to the bacterial and eukaryal counterparts, and also describe distinct archaeal metabolic features and their implications. The pentose metabolism in Archaea shows a mosaic of both bacterial and eukaryal anabolic and catabolic pathways, but has also unique archaeal conversions, novel enzymes and distinct regulatory features. This reflects the archaeal evolution of a variable metabolic shell that adjusts to the availability of the substrates and the extreme conditions under which many Archaea live.

Introduction

Since their discovery and classification Archaea have been recognized as a third domain of life, phylogenetically distinct from Bacteria and Eukarya (Woese and Fox, 1977; Woese *et al.*, 1990). Eukaryal organisms are clearly different from Bacteria and Archaea, because of their more complex structural organization: without exception they possess intracellular compartments and in addition they often have a multicellular composition. The morphologically of the prokaryotic Archaea and Bacteria is very similar, and by definition, these unicellular organisms do not have a nucleus, or have any other cytoplasmic compartments. Because of this similarity, the co-existence of two fundamentally different types of prokaryotes was not recognized before the introduction of molecular classification techniques in the 1970s. These molecular analyses strongly suggested that early in the cellular evolution two domains diverged within the prokaryotes: the Archaea (Archaebacteria) and the Bacteria (Eubacteria). The comparison of complete genomes that were released during the last decade, generally has confirmed the proposed division into three monophylic domains (Ciccarelli *et al.*, 2006; Snel *et al.*, 1999).

Although the majority of the Archaea was initially isolated from extreme environments (high temperature, high salt concentration, extreme pH), it has become clear that Archaea also thrive in non-extreme environments. Moreover, members of this domain of life are abundant in many ecosystems ranging from soil to marine environments (Pace, 1997; Schleper *et al.*, 2005; Sinninghe Damste *et al.*, 2002). Representatives of the archaeal classes discovered several decades ago (methanogen *Methanocaldococcus jannaschii*, halophile *Halobacterium salinarum* and (hyper)thermophiles, such as *Pyrococcus furiosus* and *Sulfolobus solfataricus*) have become model organisms for studying archaeal metabolism (physiology, biochemistry, genetics and genomics).

The metabolism of the Bacteria and Eukarya has been studied in great detail; especially the composition and capacity of the bacterial metabolic system is very versatile. Although the metabolism of Archaea has been investigated to a much lesser extent, it is clear that also their metabolism is very diverse. It ranges from fermentation to anaerobic and aerobic respiration and from photo- and chemolithotrophy to heterotrophy (Schonheit and Schafer, 1995). The archaeal hexose metabolism has been extensively studied in the last decades. Archaea degrade the hexose glucose via the modified Embden–Meyerhof–Parnas (EMP) or Entner–Doudoroff (ED) pathways (De Rosa *et al.*, 1984) and use distinct control mechanisms (for reviews see (Siebers and Schonheit, 2005; van der Oost and Siebers, 2007; Verhees *et al.*, 2003)). Pentose metabolism, however, has only recently been addressed in Archaea (Brouns *et al.*, 2006; Johnsen and Schonheit, 2004).

The aim of this chapter is to summarize some recent discoveries on the pentose metabolism of Archaea. Relevant pentose-converting routes, including novel enzymes and unique regulatory mechanisms, are compared to Bacteria and Eukarya.

Archaeal pentose anabolic metabolism

Pentose phosphate pathway

The pentose phosphate pathway (PPP) in Bacteria and Eukarya has a dual function. Firstly, it generates reducing power (NADPH) that serves as an electron donor in biosynthesis pathways. Secondly, it provides ribose-5-phosphate as building blocks for nucleotides and as precursor for histidine and coenzyme biosynthesis (riboflavin and NAD^+) and erythrose-4-phosphate for the synthesis of the aromatic amino acids (Fig. 4.1).

The PPP can be divided in an oxidative (OPPP) and a non-oxidative branch (NOPPP).

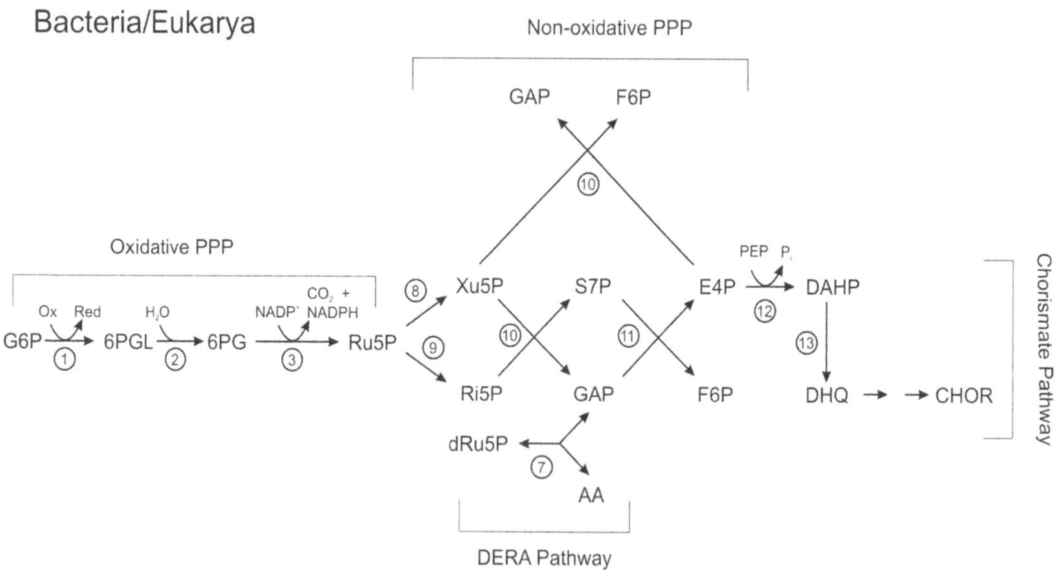

Figure 4.1 Pentose anabolic metabolism in Bacteria and Eukarya. Pathways that lead to the generation of pentoses and chorismate in Bacteria and Eukarya. Arrows represent enzymatic steps that are performed by known proteins described in Table 4.1. Abbreviated metabolites are: G6P, glucose-6-phosphate; 6PGL, 6-phosphoglucono-δ-lactone; 6PG, 6-phosphogluconate; Ru5P, ribulose-5-phosphate; GAP, glyceraldehyde-3-phosphate; AA, acetaldehyde; dRu5p, 2-deoxyribose-5-phosphate; Xu5p, xylulose-5-phosphate; Ri5P ribose-5-phosphate; S7P, sedoheptulose-7-phosphate; E4P, erythrose-4-phosphate; F6P; fructose-6-phosphate; PEP, phosphoenolpyruvate; Pi, inorganic phosphate; DAHP, 3-deoxy-d-arabino-2-heptulosonate-7-phosphate; DHQ, 3-dehydroquinate; CHOR chorismate. Pathway abbreviations are PPP, pentose phosphate pathway; DERA, 2-deoxyribose-5-phosphate aldolase. The figure was adapted from (Orita *et al.*, 2006).

The OPPP involves the step-wise oxidation of glucose-6-phosphate to ribulose-5-phosphate accompanied by the formation of two NADPH. The NOPPP links the glycolysis with the PPP by the transketolase and the transaldolase (Fig. 4.1). On the basis of comparative genomics, it has been proposed that ribose-5-phosphate can be converted to fructose-6-phosphate and glyceraldehyde-3-phosphate, by both enzymes, in Thermoplasmales, Methanococcales and *Cenarchaeum symbiosum* (for reviews see: (Soderberg, 2005; Verhees *et al.*, 2003)). Remarkably, no archaeal enzymes are known to catalyse the oxidative branch (see below), although candidates involved in 6-phosphogluconate oxidation have been proposed (Soderberg, 2005). In addition, it has been demonstrated that the methanogens *Methanococcus maripaludis* and *Methanococcus voltae* do not use the classical OPPP (Choquet *et al.*, 1994; Tumbula *et al.*, 1997; Yu *et al.*, 1994). Moreover, no OPPP enzyme activity has been measured in distinct methanogens, such as *Methanospirillum hungatei*, *Methanothermobacter thermautotrophicus* (formerly known as *Methanobacterium thermoautotrophicum*) Marburg and *Methanosarcina barkeri* Fusaro (Purwantini *et al.*, 1997) and in the crenarchaeon *Sulfolobus solfataricus* (De Rosa *et al.*, 1984).

A functional OPPP, however, can not be excluded for *M. hungatei*, *M. thermautotrophicus*, *Methanobrevibacter smithii*, *Methanosphaera stadtmanae*, *Methanosarcina barkeri* and *Methanobacterium bryantii*, since carbon labelling studies showed synthesis of labelled ribose-5-phosphate that was consistent with a functional OPPP (Choquet *et al.*, 1994; Eisenreich *et al.*, 1991; Ekiel *et al.*, 1983). Recently, White and co-workers showed that the PPP is not essential in *Methanocaldococcus jannaschii* (formerly known as *Methanococcus jannaschii*) for the production of ribose-5-phosphate and erythrose-4-phosphate (Grochowski *et al.*, 2005; White, 2004). This is in agreement with genome analysis that suggests either that the classical pathway is not present in methanogens, or that the enzymes involved are no orthologues of the bacterial and eukaryal counterparts.

The oxidative pentose phosphate pathway

Glucose-6-phosphate 1-dehydrogenase

This enzyme is responsible for the oxidation of glucose-6-phosphate to 6-phosphogluconolactone under the formation of one molecule of NADPH. This conversion is part of both the PPP and the ED pathway (Fig. 4.1/Table 4.1). NADP-dependent glucose-6-phosphate 1-dehydrogenase (G6PDH) activity has never been detected in cell-extracts, neither have orthologues of the bacterial G6PDH and eukaryal G6PDH (both belonging to the Clusters of Orthologous Groups of proteins COG0364 (Tatusov *et al.*, 2003)) been found in archaeal genomes sequenced to date.

Also the F_{420}-dependent glucose-6-phosphate dehydrogenase (FGD, COG2141) activity has not been determined in Archaea. However, close homologues of the *Mycobacterium smegmatis* FDG are identified in genome sequences in several methanogens and in *Archaeoglobus fulgidus* (Purwantini and Daniels, 1998). Unfortunately, the genomic neighbourhood of these archaeal genes is not conserved strong enough to predict a reliable function for the F_{420}-dependent enzymes.

6-Phosphogluconolactonase

The 6-phosphogluconolactonase hydrolyses the lactone ring, producing an aldonic acid. Until now 6-phosphogluconolactonase (COG0363) has only been found in Bacteria and Eukarya. Archaeal enzymes that hydrolyse this lactone might be related to COG2220 or COG3386 pentonolactonases (see: Clustering of genes involved in pentose oxidation). However, thermophilic Archaea may not require enzymatic lactonase activity, since these molecules are not very stable, and will hydrolyse spontaneously at non-limiting rates at high temperature (Brouns *et al.*, 2006).

6-Phosphogluconate dehydrogenase (decarboxylating)

The final step of the OPPP is the oxidation and decarboxylation of 6-phosphogluconate to ribulose-5-phosphate. At present, this step has

Table 4.1 Pentose anabolic enzymes that are involved in pathways that lead to the generation of pentoses and 3-dehydroquinate in Archaea and their bacterial and eukaryal counterparts

No.	Protein ID	Protein name	EC[a]	COG[b]	ABE[c]	Characterized in Archaea[d]	Reference
Oxidative pentose phosphate pathway							
1	G6PDH/ Zwf	Glucose-6-phosphate 1-dehydrogenase	1.1.1.49	COG0364	BE		
1	FGD	F_{420}-dependent glucose-6-phosphate dehydrogenase	1.1.1.-	COG2141	AB		
2	6PGL	6-Phosphogluconolactonase	3.1.1.31	COG0363	BE		
3	Gnd	6-Phosphogluconate dehydrogenase(decarboxylating)	1.1.1.44	COG0362	BE		
3		Predicted 6-phosphogluconate dehydrogenase	1.1.1.44	COG1023	AB		
3		Predicted 6-phosphogluconate dehydrogenase	1.1.1.44	COG2084	AB		
Non-oxidative pentose phosphate pathway							
8	Rpe	Ribulose-phosphate 3-epimerase	5.1.3.1	COG0036	ABE		
9	RpiA	Ribose-5-phosphate isomerase	5.3.1.6	COG0120	ABE	PH1375	(Ishikawa *et al.*, 2002)
9	RpiB	Ribose-5-phosphate isomerase	5.3.1.6	COG0698	B		
10	TktA	Transketolase	2.2.1.1	COG0021	BE		
10	Tkt1	N-terminal transketolase	2.2.1.1	COG3958	AB		
10	Tkt2	C-terminal transketolase	2.2.1.1	COG3959	AB		
11	TalB	Transaldolase	2.2.1.2	COG0176	ABE	MJ0960	(Soderberg and Alver, 2004)
Ribulose monophosphate pathway							
4	PGI	Glucose-6-phosphate isomerase	5.3.1.9	COG0166	ABE		
4	PGI	Glucose-6-phosphate isomerase	5.3.1.9	COG2140	A	PF0196	(Hansen *et al.*, 2001; Verhees *et al.*, 2001)
5	PHI	6-Phospho-3-hexuloisomerase	5.-.-.-	COG0794	AB	MJ1247	(Martinez-Cruz *et al.*, 2002)
6	HPS	3-Hexulose-6-phosphate synthase	4.1.2._	COG0269	AB		

No	Abbreviation	Enzyme	EC[a]	COG[b]	Domains[c]	Locus id[d]	Reference
2-Deoxyribose 5-phosphate aldolase (DERA) pathway							
–	PPM	Phosphopentomutase	5.4.2.7	COG1109	ABE	TK1777	(Rashid *et al.*, 2004)
–	PPM	Phosphopentomutase	5.4.2.7	COG1015	B		
7	DERA	2-Deoxyribose-5-phosphate aldolase	4.1.2.4	COG0274	AB	TK2104/ APE2437	(Rashid *et al.*, 2004)/(Sakuraba *et al.*, 2003)
AMP-metabolism pathway							
–	DeoA	Thymidine phosphorylase/AMP phosphorylase	2.4.2.4/2.4.2.–	COG0213	AB	TK0352	(Sato *et al.*, 2007)
–	E2b2	Translation initiation factor 2B subunit/Ribose-1,5-bisphosphate isomerase	-/5.3.1.-	COG1184	ABE	TK0185	(Sato *et al.*, 2007)
–	RbcL	Ribulose-bisphosphate carboxylase-oxygenase type III	4.1.1.39	COG1850	AB	TK2290	(Sato *et al.*, 2007)
Canonical chorismate pathway for 3-dehydroquinate pathway							
12	AroA	3-Deoxy-7-phosphoheptulonate synthase	2.5.1.54	COG2876	ABE		
12	AroG	3-Deoxy-7-phosphoheptulonate synthase	2.5.1.54	COG3200	B		
12	AroG	3-Deoxy-7-phosphoheptulonate synthase	2.5.1.54	COG0722	BE		
13	AroB	3-Dehydroquinate synthase	4.2.3.4	COG0337	ABE		
6-Deoxy-5-ketofructose-1-phosphate (DKFP) pathway							
14	AroA'	2-Amino-3,7-dideoxy-D-threo-hept-6-ulosonic acid synthase	-	COG1830	AB	MJ0400/ MMP0686	(White, 2004)/ (Porat *et al.*, 2006)
15	AroB'	3-Dehydroquinate synthase II	-	COG1465	A	MJ1249/ MMP0006	(White, 2004)/ (Porat *et al.*, 2006)
–	FBA	DhnA-type fructose-1,6-bisphosphate aldolase and related enzymes/DKFP transaldolase	-	COG1830	AB	MJ1585	(Siebers *et al.*, 2001)

No represents conversion numbers in Figs 4.1 or 4.2.

[a]EC, Enzyme Commission (Barthelmes *et al.*, 2007); [b]COG, Clusters of Orthologous Groups of Proteins (Tatusov *et al.*, 2003); [c]Domains are A: Archaea, B: Bacteria and E: Eukarya. [d]Locus id. The archaeal genome analyses is based on complete genome sequences of *Aeropyrum pernix*, *Hyperthermus butylicus*, *Sulfolobus acidocaldarius*, *Sulfolobus solfataricus*, *Sulfolobus tokodaii*, *Pyrobaculum aerophilum*, *Pyrobaculum islandicum*, *Thermofilum pendens*, *Archaeoglobus fulgidus*, *Haloarcula marismortui*, *Halobacterium* sp. NRC-1, *Haloquadratum walsbyi*, *Natronomonas pharaonis*, *Methanosphaera stadtmanae*, *Methanothermobacter thermautotrophicus*, *Methanocaldococcus jannaschii*, *Methanococcus maripaludis*, *Methanopyrus kandleri*, *Methanosaeta thermophila*, *Methanosarcina acetivorans*, *Methanosarcina barkeri*, *Methanosarcina mazei*, *Methanospirillum hungatei*, *Pyrococcus abyssi*, *Pyrococcus furiosus*, *Pyrococcus horikoshii*, *Thermococcus kodakarensis*, *Picrophilus torridus*, *Thermoplasma acidophilum*, *Thermoplasma volcanium* and *Nanoarchaeum equitans*.

only been demonstrated in Bacteria and Eukarya (COG0362, *gnd*), although it has been proposed that the genome of *Halobacterium salinarum* encodes a distantly related 6-phosphogluconate dehydrogenase (COG1023) (Soderberg, 2005). In addition, an even more distantly related orthologous group of proteins might be capable of catalysing the same conversion within the Archaea (COG2084). Despite the fact that the genome sequences of Archaea lack a classical OPPP, it cannot be ruled out that non-homologous proteins are capable of catalysing the similar conversion of 6-phosphogluconate. A second possibility is that some Archaea do possess a modified OPPP, which was not detected based on classical enzyme activity assays. In particular, Archaea that metabolize glucose via a modified ED pathway (*Sulfolobus* spp., *Thermoproteus* spp. and halophiles (Ahmed *et al.*, 2005)) might have the capacity to phosphorylate gluconate and subsequently catalyse the oxidative decarboxylation. Further research is needed to test these speculations.

The non-oxidative pentose phosphate pathway

The function of the non-oxidative pentose phosphate pathway is to convert pentoses to central metabolic intermediates such as glyceraldehyde-3-phosphate, fructose-6-phosphate, ribose-5-phosphate, and erythrose-4-phosphate, which serve as building blocks for the various biosynthesis pathways (Fig. 4.1). The interconversions of these compounds are accomplished by the classical enzymes ribose-5-phosphate isomerase, ribulose-phosphate 3-epimerase, transketolase and transaldolase (Table 4.1). Erythrose-4-phosphate is a precursor for 3-dehydroquinate (DHQ), which subsequently can be metabolized to chorismate (Fig. 4.1). DHQ can also be synthesized by a recently discovered 6-deoxy-5-ketofructose-1-phosphate pathway (see 6-deoxy-5-ketofructose-1-phosphate (DKFP) pathway, Fig. 4.2). The synthesis of DHQ in methanogens by the DKFP pathway is in agreement with ^{13}C-labelling experiments (Choquet *et al.*, 1994; Tumbula *et al.*, 1997; Yu *et al.*, 1994). Based on the genomic data, the NOPPP only seems to be complete in Thermoplasmales, Methanococcales and *Cenarchaeum symbiosum* (all species found in footnote of Table 4.1. were used for this analysis). Nevertheless, only two archaeal proteins, ribose-5-phosphate isomerase in *P. horikoshii* (Ishikawa *et al.*, 2002) and transaldolase in *Methanocaldococcus jannaschii* (Soderberg and Alver, 2004) have been characterized (Table 4.1).

Ribose-5-phosphate isomerase

The reversible isomerization reaction of ribulose-5-phosphate to ribose-5-phosphate can be catalysed by two distinct protein families: the non-inducible RpiA (COG0120) and the inducible bacterial RpiB (COG0698). RpiB is absent in Archaea, but RpiA seems to be ubiquitous in all Archaea. This contrasts to the fact that only a small portion of the Archaea possesses the complete NOPPP. Thus, the ribose-5-phosphate isomerase is probably involved in the NOPPP, as well as in alternative pathways that lead to the production of riboses. The crystal structure of the tetrameric ribose-5-phosphate isomerase of *Pyrococcus horikoshii*, which is homologous to RpiA, has been solved (Ishikawa *et al.*, 2002).

Ribulose-phosphate 3-epimerase

This enzyme catalyses the epimerization of ribulose-5-phosphate to xylulose-5-phosphate (X5P). Ribulose-phosphate 3-epimerase Rpe (COG0036) only appears to be encoded by the genomes of *Methanocaldococcus jannaschii*, *Methanococcus maripaludis*, *Thermoplasma acidophilum*, *Thermoplasma volcanium Picrophilus torridus* and *Cenarchaeum symbiosum*. In the latter case it is fused to the transketolase N-terminal subunit. Remarkably, these organisms encode also a transaldolase that is essential for the complete NOPPP.

Transketolase

To connect the pentoses R5P and X5P to the glycolysis the chemical compounds need to be converted to fructose-6-phosphate and glyceraldehyde-3-phosphate. These conversions need two transketolase and one transaldolase reactions that results in an overall formation of 2 fructose-6-phosphate molecules and one glyceraldehyde-3-phosphate out of three pentose molecules. The classical transketolase reactions in Bacteria and Eukarya are catalysed by a single enzyme TktA (COG0021) (Table 4.1). In Ar-

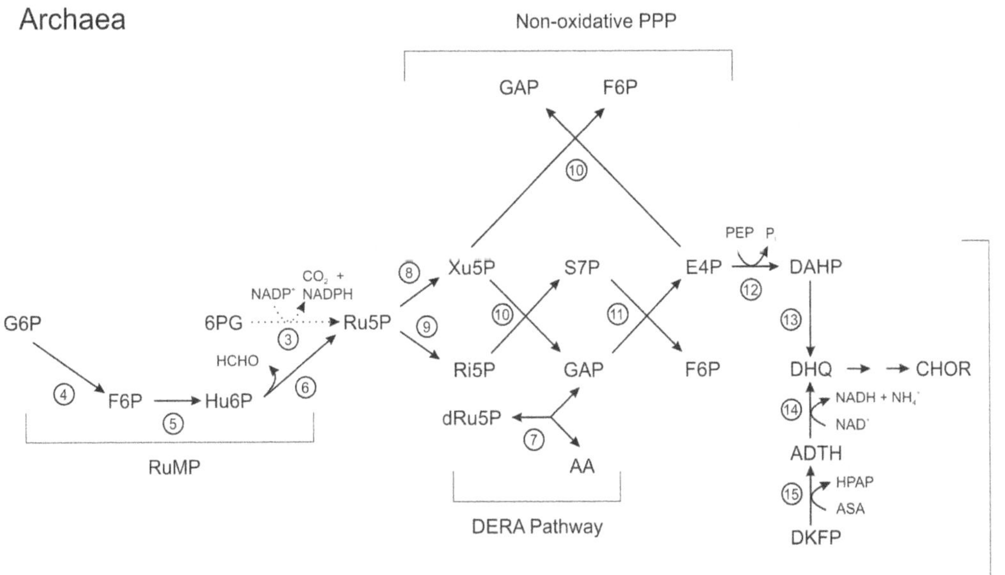

Figure 4.2 Pentose anabolic metabolism in Archaea. Pathways that lead to the generation of pentoses and chorismate in Archaea. Closed arrows represent enzymatic steps that are performed by known proteins, while dashed arrows are proteins with putative function; both are described in Table 4.1. Abbreviated metabolites are: G6P, glucose-6-phosphate; F6P, fructose-6-phosphate; Ru5P, ribulose-5-phosphate; GAP, glyceraldehyde-3-phosphate; AA, acetaldehyde; dRu5p, 2-deoxyribose-5-phosphate; Xu5p, xylulose-5-phosphate; Ri5P ribose-5-phosphate; S7P, sedoheptulose-7-phosphate; E4P, erythrose-4-phosphate; PEP, phosphoenolpyruvate; Pi, inorganic phosphate; DAHP, 3-deoxy-d-arabino-2-heptulosonate-7-phosphate; DHQ, 3-dehydroquinate; ADTH, 2-amino-3,7-dideoxy-d-threo-hept-6-ulosonic; DKFP, 6-deoxy-5-ketofructose-1-phosphate; ASA, l-aspartate semialdehyde; HPAP, hydroxypyruvaldehyde phosphate; CHOR chorismate. Pathway abbreviations are PPP, pentose phosphate pathway; RuMP ribulose monophosphate pathway; DERA, 2-deoxyribose-5-phosphate aldolase. The figure was adapted from Orita *et al.* (2006).

chaea this enzyme appears to be the result of a gene-fission, creating an N- and C-terminal part (COG3958/3959). These enzymes are present in most Archaea that can synthesis aromatic amino acids, such as *Sulfolobus solfataricus* and *Pyrococcus abyssi*; while they are lacking in almost all organisms that possess the novel 6-deoxy-5-ketofructose-1-phosphate pathway (see below). This indicates that the former organisms use the classical chorismate pathway, which requires erythrose-4-phosphate and phosphoenolpyruvate as starting compounds. Interestingly, these genes are frequently clustered with amino acid biosynthesis genes, supporting their involvement in erythrose-4-phosphate synthesis (Verhees *et al.*, 2003).

Transaldolase

The transaldolase MJ0960 (COG0176) from *Methanocaldococcus jannaschii* has been experimentally characterized. It catalyses the reaction of glyceraldehyde-3-phosphate and sedoheptulose-7-phosphate to form fructose-6-phosphate and erythrose-4-phosphate (Soderberg and Alver, 2004). The same reaction in Eukarya and Bacteria is involved in the recycling of the glycolytic compounds. In Archaea, the corresponding genes are only present in genomes that do encode the complete NOPPP.

Concluding remarks on the non-oxidative PP pathway in Archaea

The overall picture of the NOPPP in Archaea is that only a few (Thermoplasmales, Methanococcales and *Cenarchaeum symbiosum*) are capable of completely interconverting glycolytic intermediates and pentoses. However, several Archaea, such as *Sulfolobus* spp. and *Pyrobaculum* spp. can generate DHQ for aromatic amino acid biosynthesis, but have an incomplete NOPPP and are probably using the ribulose monophosphate pathway (see: below). Apart from the ribose-5-phosphate isomerase, the other Archaea do not possess obvious homologues of the proteins that catalyse these conversions in Bacteria and Eukarya.

Ribulose monophosphate pathway (RuMP)

Some Archaea do use the reverse ribulose monophosphate pathway to metabolize ribose-5-phosphate (Fig. 4.2). This pathway is used in methylotrophic Bacteria to fix formaldehyde with ribulose-5-phosphate to D-*arabino*-3-hexulose-6-phosphate, which is subsequently isomerized to fructose-6-phosphate. In *Thermococcus kodakaraensis*, and most likely in more Archaea, the fused-enzyme 3-hexulose-6-phosphate synthase and 6-phospho-3-hexuloisomerase have been demonstrated to be involved in the reverse reaction, i.e. in the conversion of fructose-6-phosphate to ribulose-5-phosphate (Orita *et al.*, 2006). This conversion has previously been postulated to occur in *M. jannaschii* (Grochowski *et al.*, 2005). A comparative analysis shows that in those archaeal organisms that do not possess the complete NOPPP, the RuMP fills the gaps. Surprisingly, halophiles are lacking both pathways and may have evolved a different, yet unknown solution (Soderberg, 2005; Verhees *et al.*, 2003). For organisms that depend on RuMP to synthesize nucleotides, the final step is synthesized by ribose-5-phosphate isomerase (see: NOPPP), which is encoded in all completely sequenced archaeal genomes. Finally, this pathway is in good agreement with results of the ^{13}C-labelling experiments in *Thermococcus zilligii* (Xavier *et al.*, 2000).

6-Phospho-3-hexuloisomerase

The first step in synthesizing ribulose-5-phosphate is the isomerization of fructose-6-phosphate to D-*arabino*-3-hexulose-6-phosphate by 6-phospho-3-hexuloisomerase (PHI). In Thermococcales, the enzyme PHI (COG0794) is fused to 3-hexulose-6-phosphate synthase (HPS) and it originally was suggested to be involved in formaldehyde fixation in *P. horikoshii* (Orita *et al.*, 2005). However, genetic analyses revealed that this gene is essential for the synthesis of nucleosides in *Thermococcus kodakaraensis* (Orita *et al.*, 2006). The fusion may be beneficial for these hyperthermophiles, because D-*arabino*-3-hexulose-6-phosphate is unstable at elevated temperatures (Kemp, 1974). Recently the crystal structure of the *M. jannaschii* PHI (MJ1247) was determined (Martinez-Cruz *et al.*, 2002).

3-Hexulose-6-phosphate synthase

The second step, cleaving of formaldehyde from D-*arabino*-3-hexulose-6-phosphate, results in ribulose-5-phosphate. Besides being fused to PHI, HPS (COG0269)-containing organisms also possess an additional domain corresponding to either a formaldehyde-activating enzyme (Vorholt *et al.*, 2000) (in methanogens and in *Archaeoglobus*; COG1795), or a tungsten-dependent formaldehyde ferredoxin oxidoreductase (Roy *et al.*, 1999)) (in *Pyrococcus furiosus*, COG2414) to detoxify formaldehyde. In addition, Soderberg (2005) concludes that only the *Sulfolobus solfataricus* genome does not contain an open reading frame of the latter two COGs. However, SSO0472 (COG1062), a gluthatione-independent formaldehyde dehydrogenase is a good candidate to fill the gap. The enzyme is homologous to the characterized enzyme in *Pseudomonas putida* (Ogushi *et al.*, 1984), but might have a different function in *E. coli*, because it is fused to a Ribonuclease E (RNase E) inhibitor. Nevertheless, in many Archaea PHI and HPI are essential in generating pentoses and COG1062 is good candidate, in some Archaea, to remove the toxic formaldehyde.

Metabolic link between pentoses, chorismate and central carbon metabolism

2-Deoxyribose 5-phosphate aldolase (DERA) pathway

In many Bacteria and Eukarya the link between the central carbohydrate metabolism and nucleosides is the conversion from acetaldehyde and glyceraldehyde-3-phosphate to pentoses. This reversible pathway is catalysed by a 2-deoxyribose-5-phosphate aldolase (DERA, Fig. 4.1) and a phosphopentomutase (PPM) to produce (deoxy)ribose 1-phosphate, which can be used as ribose moiety of nucleosides by nucleoside phosphorylases. A close homologue of the bacterial 2-deoxyribose-5-phosphate aldolase (DeoC/DERA; TK2104/APE2437, COG0274) was identified and characterized in *T. kodakaraensis* (Rashid *et al.*, 2004) and the crystal structure of the orthologue from *Archaeoglobus fulgidus* was solved (Sakuraba *et al.*, 2003). In addition, Rashid *et al.* characterized a novel phosphopentomutase

(TK1777, COG1109) which is involved in the isomerization of the pentose. As *T. kodakaraensis* is not able to grow on (deoxy)nucleosides, the 'DERA pathway' of some Archaea probably functions in the anabolic direction like in Eukarya, but in contrast to Bacteria. The phyletic distribution of DERA shows a huge variety in Archaea that do contain this enzyme: for example some methanogens do not have it, but it is present in all the halophiles. This observation suggests that the pathway in halophiles could have a role in the synthesis of pentoses.

Adenosine 5′- monophosphate (AMP) metabolism pathway

Recently, a novel adenosine 5′-monophosphate (AMP) degrading pathway was discovered (Sato *et al.*, 2007). In *Thermococcus kodakaraensis*, and several other anaerobic euryarchaea, the first step of this pathway is cleaving AMP molecule with phosphate into adenine and ribose-1,5-phosphate by an AMP phosphorylase (TK0352/COG0213, Table 4.1). The ribose-1,5-phosphate moiety can subsequently be isomerized by ribose-1,5-bisphosphate isomerase (TK0185/COG1184) to ribulose-1,5-bisphophate. Finally, the RuBisCO type III (TK2290/COG1850) converts, with CO_2 and H_2O, ribulose-1,5-bisphosphate to two molecules of the central carbon metabolite 3-phosphoglycerate. The product 3-phosphoglycerate can be further metabolized to produces ATP. Thus, Archaea could use the pathway when energy levels are low. A second function would be CO_2 fixation; however, *T. kodakaraensis* uses the RuMP to generate riboses and therefore would lose the fixated carbon through formaldehyde (Sato *et al.*, 2007).

6-Deoxy-5-ketofructose-1-phosphate (DKFP) pathway

A novel erythrose 4-phosphate-independent chorismate pathway has been identified in *Methanocaldococcus jannaschii* (Tumbula *et al.*, 1997; White, 2004), and recently in *Methanococcus maripulus* (Porat *et al.*, 2006). The classical chorismate synthesis pathway starts with the condensation of erythrose-4-phosphate and phosphoenolpyruvate and the subsequent conversion to 3-dehydroquinate (DHQ), a precursor of chorsimate (Fig. 4.1). The first step is catalysed by 3-deoxy-D-*arabino*-2-heptulosonate-7-phosphate (DAHP) synthase. Enzymes that are responsible for this conversion can be found in three different COGs (COG0722, COG3200, and COG2876, Table 4.1). Thus, proteins from different families convert the same reaction, which can be explained by the phenomenon that is called non-orthologous gene displacement (Koonin *et al.*, 1996). The enzymes catalysing the second step are, however, all members of one distinct COG: DHQ synthase (COG0337). Because of the deviating pentose metabolism of the Archaea, erythrose-4-phosphate is not generated; rather, an alternative 6-deoxy-5-ketofructose-1-phosphate (DKFP) pathway has evolved that leads to the synthesis of 3-hydroquinate (DHQ) (Fig. 4.2). Recently, the biosynthetic route of DKFP in *M. jannaschii* has been described (White and Xu, 2006). MJ1585 (COG1830), an archaeal Class I fructose-1,6-bisphosphate aldolase (Siebers *et al.*, 2001), also catalyses the transaldolase reaction of fructose-1-phosphate or fructose-1,6-bisphosphate with methylglyoxal to DKFP. Methylglyoxal can be generated chemically or enzymatically from the central metabolite glyceraldehyde-3-phosphate.

2-Amino-3,7-dideoxy-D -threo-hept-6-ulosonic (ADTH) synthase

The first step of the DKFP pathway is catalysed by ADTH synthase converting DKFP to 2-amino-3,7-dideoxy-D-threo-hept-6-ulosonate. This biochemical transaldolase activity was measured in *M. jannaschii*, and the enzyme was identified (MJ0400, a paralogue of MJ1585/COG1830) (White, 2004). In addition, a genetic study shows that the orthologous protein in *M. maripaludis* (MMP0686) is essential for growth without aromatic amino acids and the activity is inhibited during growth on aryl acids (Porat *et al.*, 2006).

3-Dehydroquinate synthase

The subsequent step is the synthesis of 3-dehydroquinate, which is catalysed by MJ1249 (COG1465) (White, 2004). The DHQ synthase activity could also be detected in *M. maripaludis*. Unexpectedly, a *M. maripaludis* mutant, in which the gene encoding the homologous Mmp0006 was disrupted, could still grow without aromatic amino acids. Therefore, it was concluded that methanococci have an alternative activity for this step (Porat *et al.*, 2006).

The occurrence of the 3-dehydroquinate synthase in archaeal genomes correlates, although not perfectly, with the absence of the transketolase genes (see: NOPPP) and the classical DHQ synthase gene in methanogens, halophiles, *Archaeoglobus fulgidus* and *Cenarchaeum symbiosum* (Soderberg, 2005; Verhees *et al.*, 2003). It also correlates with the organisms that have one or more paralogues of fructose-1,6-bisphosphate aldolase (Porat *et al.*, 2006; Soderberg, 2005) and, furthermore, both DKFP pathway genes are often in proximity to genes involved in aromatic amino acid biosynthesis (Porat *et al.*, 2006). Thus, genetic, biochemical and bioinformatic evidence has revealed the existence of two archaeal pathways to generate DHQ, the classical chorismate pathway and the novel DKFP pathway. The DFKP pathway is probably not unique to Archaea as homologous genes have been found in bacterial genomes.

Concluding remarks: pentose anabolism

The 'anabolic' pentose pathways in Archaea is a combination of unique conversions and novel proteins, but it also consists of general features and orthologous proteins that can be found in all three domains of life. The 'anabolic' metabolism of pentoses reflects the diversity of archaeal metabolism with a 'conserved archaeal core' and with a 'variable shell' (Makarova *et al.*, 1999) and confirms that metabolic pathways in Archaea are 'a playground of non-orthologous gene displacement' (Koonin and Galperin, 2003).

Archaeal pentose catabolic metabolism

Introduction

While the hexose metabolism in Archaea has been studied extensively for many years, studies on their pentose utilization appeared only recently (Brouns *et al.*, 2006; Johnsen and Schonheit, 2004). In contrast to growth on hexose compounds, only a few archaeal species can metabolize C_5 sugars, including some *Sulfolobus* spp. (Brouns *et al.*,; Grogan, 1989) and several halophiles (Tindall, 1992). The most abundant pentoses in nature (L-arabinose, D-arabinose, D-xylose and D-ribose) have been reported to serve as sole carbon and energy source for these Archaea.

In Bacteria, yeast and fungi, pentose utilization has been studied extensively. The pentoses D-arabinose, D-ribose, D-xylose and L-arabinose can be metabolized in three different pathways. First, in Bacteria an isomerase, kinase and, if necessary, an epimerase can convert D-/L-arabinose and D-xylose into D-xylulose-5-phosphate (Fig. 4.3A), which is an intermediate of the NOPPP and phosphoketolase pathways. The phosphoketolase pathway (PKP) is a catabolic pathway found in several Bacteria (Biesterveld *et al.*, 1994) and yeasts (Evans and Ratledge, 1984) and fungi. PKP converts hexoses, via the OPPP, or pentoses to xylulose-5-phosphate and subsequently split X5P with inorganic phosphate to glyceraldehyde-3-phosphate and acetyl-phosphate. The latter component can be further metabolized to acetate or ethanol. The enzyme xylulose-5-phosphate/D-fructose-6-phosphate phosphoketolase, which catalyses this unique step in PKP, was characterized in *Bifidobacterium lactis* (xfp/COG3957) (Meile *et al.*, 2001). Orthologues of Xfp can be found in Eukarya; however, in Archaea no orthologues are present, which is in agreement with the fact that PKP has never been detected in this domain of life. The pentose D-ribose is, in contrast to the other pentoses, directly phosphorylated by a ribokinase (RbsK, COG0524) and metabolized through the NOPPP, in Bacteria (Hope *et al.*, 1986; Woodson and Devine, 1994). Genes encoding the 'bacterial xylulose-5-phosphate pathway' are generally clustered in the bacterial genomes. In *Escherichia coli* the *araBAD* operon is involved in L-arabinose degradation (Lee *et al.*, 1986). In the Gram-positive bacterium *Bacillus subtilis* the three encoding genes are in the *araABDLMNPQ-abfA* operon together with the genes encoding the arabinose transporter and an α–arabinofuranosidase, which cleaves L-arabinose monomers from arabinose oligomers (Mota *et al.*, 1999). The *xylAB* genes in *E. coli* encode an isomerase and kinase involved in D-xylose degradation (Rosenfeld *et al.*, 1984). While the uncommon D-arabinose is degraded by proteins

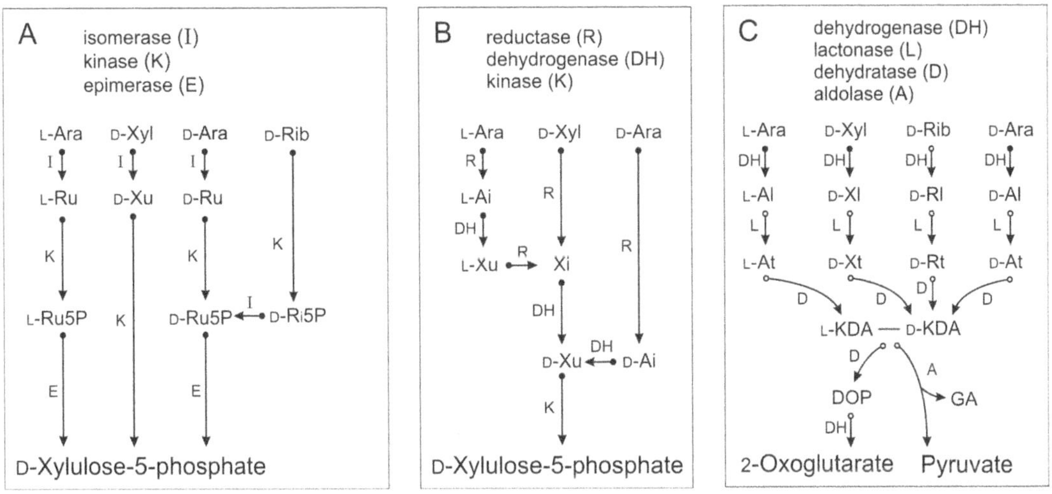

Figure 4.3 Pentose catabolism in Bacteria, Archaea and Eukarya. Schematic representation of three types of pentose degrading pathways (A, B and C). Arrows with an open or closed arrow tail represent enzymatic steps that are performed by unknown proteins or known proteins, respectively. Abbreviations: Ara, arabinose; Xyl, xylose; Ri(b), ribose; Ru, ribulose; Xu, xylulose; Ai, arabinitol; Xi, xylitol; Al, arabinonolactone; Xl, xylonolactone; Rl, ribonolactone; At, arabinonate; Xt, xylonate; Rt, ribonate; KDA, 2-keto-3-deoxy-arabinonate (also called 2-oxo-4,5-dihydroxypentanoate); DOP, 2,5-dioxopentanoate (also called 2-oxoglutarate semialdehyde); GA, glycolaldehyde. The figure was adapted from Brouns *et al.* (2006).

encoded by the *darK-fucPIK* gene cluster of *E. coli* (Elsinghorst and Mortlock, 1994).

Second, pentoses can be converted to D-xylulose-5-phosphate by reductases and dehydrogenases (Fig. 4.3B). These pathways are found in fungi, mammals and yeast but also in some Bacteria (Chiang and Knight, 1960; Fossitt *et al.*, 1964; Wojtkiewicz *et al.*, 1988).

Third, the pentoses L-/D-arabinose, D-xylose and D-ribose can be metabolized to pyruvate and glycolaldehyde, or to 2-oxoglutarate (a tricarboxylic acid cycle intermediate) (Fig. 4.3C). The first common steps of the pentose conversion proceeds via a pentose dehydrogenase, a pentonolactonase, and a pentonic acid dehydratase. Then, 2-keto-3-deoxypentonic acid can be cleaved by a 2-dehydro-3-deoxy-D-pentonate aldolase into pyruvate and glycolaldehyde. This variant, also called the Dahms-pathway, has been observed in *Pseudomonas* and *Bradyrhizobium* strains (Dahms and Anderson, 1969; Palleroni and Doudoroff, 1957; Pedrosa and Zancan, 1974). Alternatively, 2-keto-3-deoxypentonic acid can be converted by a 2-keto-3-deoxypentonic acid dehydratase and a 2,5-dioxopentanoate dehydrogenase to generate 2-oxoglutarate. This pathway occurs in several aerobic Bacteria of the genera *Pseudomonas* (Dagley and Trudgill, 1965; Dahms, 1974; Weimberg, 1961; Weimberg and Doudoroff, 1955) *Rhizobium* (Duncan, 1979; Duncan and Fraenkel, 1979) and *Azospirillum* (Watanabe *et al.*, 2006a) and has recently been demonstrated in Archaea as well. *Sulfolobus solfataricus* can degrade D-arabinose to 2-oxoglutarate in four consecutive steps, catalysed by the enzymes D-arabinose 1-dehydrogenase, D-arabinonate dehydratase, 2-dehydro-3-deoxy-D-arabinonate dehydratase and 2,5-dioxopentanoate (also called 2,5-dioxovalerate or α-ketoglutarate semialdehyde) dehydrogenase (Brouns *et al.*, 2006). Moreover, *Haloarcula marismortui* has been shown to utilize D-xylose as carbon and energy source; one of the enzymes involved D-xylose dehydrogenase, has been purified and characterized (Johnsen and Schonheit, 2004). The discussion below will focus on this class of prokaryotic pentose oxidation pathways (Fig. 4.3C) as these pathways are the only pentose degrading pathways in Archaea described to date.

Prokaryotic pentose oxidation pathways (PPOP)

The pentoses that can be metabolized via the third class of prokaryotic pentose oxidation pathways (Fig. 4.3C) are: D-arabinose in *Sulfolobus solfataricus* (Brouns *et al.*, 2006), L-arabinose in *Burkholderia thailandensis* (Moore *et al.*, 2004) and *Azospirillum brasilense* (Watanabe *et al.*, 2006a) and D-xylose in *Caulobacter crescentus* (Stephens *et al.*, 2007). In addition, this type of pathway has been predicted for D-xylose degradation in *Haloarcula marismortui* (Brouns *et al.*, 2006). Hence, this catabolic oxidation has only been described in *Sulfolobus* spp. and several halophiles in the domain Archaea, possibly reflecting the limited capacity of pentose utilization among Archaea. Some of the enzymes (Table 4.2) used in the pentose oxidation have been proposed to be similar to proteins that are possibly involved in hexaric acid and hydroxyproline catabolism in *Bacillus subtilis* and *Pseudomonas putida* (Brouns *et al.*, 2006).The first step is an oxidation of the pentose, after which, a possible pentonolactonase and dehydratase results in a 2-keto-3-deoxypentonic acid that can be dehydrated and oxidized to 2-oxoglutarate (Fig. 4.3C).

Pentose dehydrogenases

The oxidation of the pentose, D-/L-arabinose or D-xylose, is the first step in the prokaryotic pentose degradation. The dehydrogenation can be carried out by enzymes from different families. In *A. brasilense* the characterized L-arabinose 1-dehydrogenase (AraA) belongs to the COG0673 (Watanabe *et al.*, 2006a). AraA is 80 % identical to AraE of *B. thailandensis*, which is essential for growth on L-arabinose (Moore *et al.*, 2004). Moreover, the distantly related and characterized D-xylose dehydrogenase from the halophilic *Haloarcula marismortui* belongs to the same COG0673 (Johnsen and Schonheit, 2004). It has been shown that the recombinant version of this D-xylose 1-dehydrogenase (rrnAC3034) prefers D-xylose, but can also oxidizes D-ribose and to a lesser extent D-glucose. The homotetrameric protein complex has $NADP^+$ as preferred electron acceptor, which is not uncommon for Archaea (Snijders *et al.*, 2006). AraA, in contrast, is a monomeric enzyme which catalyses the oxidation of D-galactose as well (Watanabe *et al.*, 2006a). While it has been shown for *A. brasilense*, that it is capable of utilizing L-arabinose via the prokaryotic pentose oxidation pathway, a similar pathway has been proposed to be used by *H. marismortui* for converting D-xylose to 2-oxoglutarate, however, only a single enzymatic step has been characterized (Brouns *et al.*, 2006). The xylose 1-dehydrogenase activity has also been measured in the bacterium *Caulobacter crescentus* (Poindexter, 1964), and recently the genes of the complete D-xylose-degrading pathway were identified (Stephens *et al.*, 2007). The D-xylose dehydrogenase enzyme in *C. crescentus* XylB/CC0821 belongs to COG1028 and it uses NAD^+ as cofactor (Stephens *et al.*, 2007), while $NADP^+$ activity has been measured as well (Poindexter, 1964).

Finally, the archaeon *S. solfataricus* also degraded D-arabinose via the PPOP. The D-arabinose 1-dehydrogenase (AraDH; SSO1300) belongs to the COG1064 and forms a homotetrameric protein complex, with a clear cofactor preference for $NADP^+$ (Brouns *et al.*, 2006). Thus, the first oxidative step in the PPOP is used by various organisms to metabolize distinct pentoses that are catalysed by different protein families.

Pentonolactonases

The oxidation of a pentose yields a pentonolactone and can be hydrolysed by two different groups of proteins. First, COG3386, which includes the characterized D-gluconolactonase of *Zymomonas mobilis* (Kanagasundaram and Scopes, 1992), the recently identified L-arabinolactonase of *A. brasilense* (Watanabe *et al.*, 2006c) and the proposed D-xylonolactonase of *C. crescentus* (Stephens *et al.*, 2007). Second, the conversion can be catalysed by an alternative version of the enzyme that belongs to COG2220. In *H. marismortui*, the D-xylonolactonase is probably member of the latter group (rrnAC3033) (Brouns *et al.*, 2006). In *Sulfolobus solfataricus*, the enzyme responsible for the conversion of D-arabinolactone into D-arabinonic acid might be a member of the former group (COG3386; SSO2705/SSO3041). However, it has been suggested that *S. solfataricus* does not need an enzyme for this conversion at all, since the spontaneous lactone hydrolysis reaction, possibly, occurs at non-limiting rates at high temperature (Brouns *et al.*, 2006).

Pentonate dehydratase dehydratases

The third step of the PPOP is the dehydration of a pentonic acid, which yields a 2-oxo-4,5-

Table 4.2 Pentose catabolic enzymes in Archaea and their bacterial and eukaryal counterparts

Protein ID	Protein Name	EC[a]	COG[b]	ABE[c]	Characterized in Archaea[d]	Reference
Bacterial xylulose-5-phosphate pathway						
L-Arabinose						
AraA	L-Arabinose isomerase	5.3.1.4	COG2160	B		
AraB	L-Ribulokinase	2.7.1.16	COG1069	BE		
AraD	L-Ribulose-phosphate 4-epimerase	5.1.3.4	COG0235	ABE		
D-Xylose						
XylA	D-Xylose isomerase	5.3.1.5	COG2115	B		
XylA	D-Xylose isomerase	5.3.1.5	COG4952	B		
XylB	D-Xylulokinase	2.7.1.17	COG1070	ABE		
D-Arabinose						
FucI	L-Fucose isomerase/D-arabinose isomerase	5.3.1.25/5.3.1.3	COG2407	B		
DarK	D-Ribulokinase	2.7.1.47	COG1069	BE		
Rpe	Ribulose-phosphate 3-epimerase	5.1.3.1	COG0036	ABE		
D-Ribose						
RbsK	Ribokinase	2.7.1.15	COG0524	ABE		
RpiA	Ribose-5-phosphate isomerase	5.3.1.6	COG0120	ABE		
RpiB	Ribose-5-phosphate isomerase	‘	COG0698	B		
Phosphoketolase pathway (PKP)						
Xfp	D-Xylulose-5-phosphate/D-fructose-6-phosphate phosphoketolase	4.1.2.9/4.1.2.22.	COG3957	BE		

Table 4.2 *continued*

Protein ID	Protein Name	EC[a]	COG[b]	ABE[c]	Characterized in Archaea[d]	Reference
General xylulose-5-phosphate pathway						
AlrA	Aldose reductase	1.1.1.21	COG0656	ABE		
LadA	L-Arabinitol 4-dehydrogenase	1.1.1.12	COG1063	ABE		
DcxR	L-Xylulose reductase	1.1.1.10	COG1028	ABE		
Xdh	D-Xylulose reductase	1.1.1.9	COG1063	ABE		
XylB	D-Xylulokinase	2.7.1.17	COG1070	ABE		
Prokaryotic pentose oxidation pathways						
D-Arabinose						
AraDH	D-Arabinose 1-dehydrogenase (NAD(P)$^+$)	1.1.1.117	COG1064	ABE	SSO1300	(Brouns *et al.*, 2006)
–	D-Arabinose 1-dehydrogenase	1.1.1.116	–	–		
–	D-Arabinonolactonase	3.1.1.30	COG3386	ABE		
AraD	D-Arabinonate dehydratase	4.2.1.5	COG4948	AB	SSO3124	(Brouns *et al.*, 2006)
KdaD	2-Dehydro-3-deoxy-D-arabinonate dehydratase	4.2.1.–	COG3970	AB	SSO3118	(Brouns *et al.*, 2006)
DopDH	2,5-Dioxovalerate dehydrogenase	1.2.1.26	COG1012	ABE	SSO3117	(Brouns *et al.*, 2006)
–	2-Dehydro-3-deoxy-D-pentonate aldolase	4.1.2.28	–	–		
D-Xylose						
Xdh	D-Xylose 1-dehydrogenase (NADP$^+$)	1.1.1.179	COG0673	ABE	rrnAC3034	(Johnsen and Schonheit, 2004)
Xyl1	D-Xylose 1-dehydrogenase	1.1.1.175	COG0673	ABE		
XylB	D-Xylose 1-dehydrogenase	1.1.1.175	COG1028	ABE		
–	D-Xylonolactonase	3.1.1.68	COG2220	ABE		

XylC	D-Xylonolactonase	3.1.1.68	COG3386	ABE
–	D-Xylonate dehydratase	4.2.1.82	COG4948	AB
XylD	D-Xylonate dehydratase	4.2.1.82	COG0129	ABE
KdaD	2-Dehydro-3-deoxy-D-arabinonate dehydratase	4.2.1.-	COG3970	AB
DopDH	2,5-Dioxovalerate dehydrogenase	1.2.1.26	COG1012	ABE
L-Arabinose				
AraA/AraE	L-Arabinose 1-dehydrogenase	1.1.1.46	COG0673	ABE
AraB/AraI	L-Arabinonolactonase	3.1.1.15	COG3386	ABE
AraC/AraB	L-Arabinonate dehydratase	4.2.1.25	COG0129	ABE
AraD/AraD	2-Dehydro-3-deoxy-L-arabinonate dehydratase	4.2.1.43	COG0329	AB
AraE/AraC	2,5-Dioxovalerate dehydrogenase	1.2.1.26	COG1012	ABE
Dahms pathway				
–	2-Dehydro-3-deoxy-D-pentonate aldolase	4.1.2.28	–	B
Transcriptional regulators involved in pentose metabolism				
AraR	Transcriptional repressor of the L-arabinose operon	–	COG1609	B
AraC	Transcriptional activator of the L-arabinose operon	–	COG2207	B
XylR	D-Xylose transcriptional activator	–	COG1609/ COG2207	B
XylR	D-Xylose operon repressor		COG1940	AB
RbsR	Transcriptional repressor of D-ribose operon	–	COG1609	B
Xyr1/XlnR	Xylanase regulator 1/transcriptional activator XlnR	–	–	E

[a]EC, Enzyme Commission (Barthelmes *et al.*, 2007); [b]COG, Clusters of Orthologous Groups of Proteins (Tatusov *et al.*, 2003); [c]Domains are A: Archaea, B: Bacteria and E: Eukarya. [d]Locus id. The archaeal genome analysis is based on complete genome sequences of Table 4.1.

dihydroxypentanoate. In the bacterium *A. brasilense* the homodimer AraC is involved in the conversion of L-arabinonate into 2-keto-deoxy-L-arabinonate (Watanabe *et al.*, 2006c), and belongs to the dihydroxy-acid dehydratase (IlvD) and 6-phosphogluconate dehydratase (Edd) family (COG0129). IlvD/Edd is a huge protein family in which IlvD is involved in amino acid metabolism and Edd in the Entner–Doudoroff pathway. In *C. crescentus* four different proteins belong to this family and it has been proposed that CC0819/XylD is the D-xylonate dehydratase. Hence, the IlvD/Edd family has many different dehydratases that are not only involved in the amino acid and hexose catabolic pathways, but also in the distinct pentose catabolism.

In Archaea, unlike Bacteria, the D-arabinonate dehydratase from *S. solfataricus* (SSO3124) and the predicted D-xylonate dehydratase from *H. marismortui* (rrnAC3032) belong to the mandelate racemase/muconate lactonizing enzyme family and COG4948. SSO3124 forms a homooctameric complex, similarly as the homologous gluconate dehydratase (Kim and Lee, 2005). Interestingly, the enzyme that catalyses the first step of the modified Entner–Doudoroff pathway in *S. solfataricus*, the glucose dehydrogenase (SSO3198), exhibits a high catalytic efficiency for D-xylose (Milburn *et al.*, 2006). In addition, in this archaeon the dihydroxy-acid dehydratase enzyme (COG0129, SSO3107) shows D-xylonate dehydratase activity as well (Kim and Lee, 2006). This suggests that SSO3107 and SSO3198 can play a role in the D-xylose catabolism of *S. solfataricus*, because the oxidation of D-xylose and D-arabinose yield the same intermediate (2-oxo-4(S),5-dihydroxypentanoate) and can be further metabolized to 2-oxoglutarate by the enzymes mentioned below.

2-Keto-3-deoxy-pentanoate dehydratases

The two different products that are the result of the pentonic acid dehydratases (1) 2-keto-3-deoxy-L-arabinonate and (2) 2-keto-3-deoxy-D-arabinonate (also called 2-keto-3-deoxy-D-xylonate), in which the chiral differences between the two pentoses at the C-2 and C-3 atoms have been eliminated by the pentonate dehydratase, are dehydrated again. Although two distinct protein families are responsible for catalysing the dehydration of the two different compounds, the resulting product of this catalysis is the same: 2,5-dioxopentanoate (DOP, also called α-ketoglutaric semialdehyde). The first enzyme family, 2-keto-3-deoxy-L-arabinonate dehydratase (AraD), has first been identified in *A. brasilense* (COG0329) (Watanabe *et al.*, 2006a). The enzyme that generates α-ketoglutaric semialdehyde is homologous to the dihydrodipicolinate synthetase (DHDPS), but uses a unique mechanism. Whereas DHDSP catalyses a C-C bond formation, AraD drives a dehydration reaction.

A second 2-keto-3-deoxy-arabinonate (Kda) dehydratase type that can convert 2-keto-3-deoxy-D-arabinonate into 2,5-dioxypentanoic acid has been identified in *S. solfataricus* (KdaD, SSO3118). The latter enzyme and the predicted Kda dehydratase (XylX, CC0823) in *C. crescentus* belong to the fumarylacetoacetate hydrolase family (COG3970), consisting of enzymes with diverse functions, such as decarboxylases and hydratases. The homologous rrnAC1339 is probably responsible for de dehydration reaction in the D-xylose catabolic pathway in *H. marismortui*. Besides the dehydration of Kda by the enzyme, family members in mammals including humans are capable of hydrolysing fumarylacetoacetate, the final step of phenylalanine and tyrosine degradation (Bateman *et al.*, 2001). Moreover, the C-terminal decarboxylation domain of the HpcE from *E. coli* is also member of this family (Tame *et al.*, 2002).

2,5-Dioxopentanoate dehydrogenase

The final step of the pentose, hexaric acid and L-hydroxyproline oxidation in the prokaryotic pathway concerns the dehydrogenation of the aldehyde 2,5-dioxopentanoate (DOP). The 2,5-dioxopentanoate dehydrogenase, which is also called α-ketoglutaric semialdehyde dehydrogenase, has been characterized in *S. solfataricus* (SSO3117, COG1012) (Brouns *et al.*, 2006). The archaeal homotetrameric enzyme prefers $NADP^+$ over NAD^+ and is a close homologue of the putative aldehyde dehydrogenases that could catalyse the final step of the D-xylose catabolism in *H. marismortui* (rrnAC3036) and in *C. crescentus* (XylA, CC0822). Surprisingly, the bacterium *A. brasilense* possesses three homologous isozymes that oxidize the final step of the PPOP, but are induced separately, when grown on L-arabinose, D-hexaric acids or L-hydroxyproline (see clustering of genes involved in pentose oxidation). These three DOP dehydrogenases are together with the archaeal dehydrogenases members of COG1012. Also in archaeal genome sequences many paralogues of the DOP dehydrogenases can be identified and therefore more archaeal enzymes might be capable of producing the TCA-cycle intermediate 2-oxoglutarate from DOP.

2-Dehydro-3-deoxy-D-pentonate aldolase

L-/D-KDA can also be cleaved into pyruvate and glycolaldehyde by the 2-dehydro-3-deoxy-D-pentonate aldolase (Dahms, 1974) (Fig. 4.3C). Although the activity was detected in *Pseudomonas* sp. and in *Bradyrhizobium* sp., the gene has never been identified and it is therefore not possible to predict if the so-called Dahms pathway is active in more species.

Clustering of genes involved in pentose oxidation

Genomic context analysis is a very powerful tool to predict protein function (Koonin *et al.*, 2001; Osterman and Overbeek, 2003; von Mering *et al.*, 2003). In prokaryotes conserved gene neighbourhood or chromosomal gene clustering is a computational tool that can strongly suggest a certain function to a hypothetical protein, as turned out to be the case for the enzymes involved in the pentose, hexaric acids and L-hydroxyproline degradation pathways. In *S. solfataricus*, only the Kda dehydratase and DOP dehydrogenase genes of the identified enzymes of the D-arabinose pathway are located side by side. However, the comparative analysis of gene clusters in several aerobic proteobacteria that are member of the genera *Burkholderia, Rhizobium, Bradyrhizobium, Agrobacterium, Azospirillum,* and *Pseudomonas* correlates well with the ability of these organism to degrade pentoses, especially L-arabinose (Dagley and Trudgill, 1965; Duncan, 1979; Duncan and Fraenkel, 1979; Watanabe *et al.*, 2006a; Weimberg, 1961; Weimberg and Doudoroff, 1955). Based on genomic context analysis predictions, enzyme functions have been predicted for some missing links in the pathways (Brouns *et al.*, 2006). In the α-proteobacterium *C. crescentus* the D-xylose inducible promoter (Meisenzahl *et al.*, 1997) could be linked to D-xylose oxidation pathway, because the five downstream genes were assigned with enzymatical functions of the PPOP. Recently, almost all functional predictions have indeed been confirmed in this oligotrophic bacterium (Stephens *et al.*, 2007). The same D-xylose catabolic pathway was assigned to the archaeon *H. marismortui*. However, not all proteins are homologues to the enzymes of *C. crescentus* and the Kda dehydrates (rrnAC1339 and not rrnAC3039) is not in proximity to the other catabolic genes. Nevertheless, the enzymes can probably catalyse the same type of reactions, which has been shown for the D-xylose dehydrogenase enzyme (Johnsen and Schonheit, 2004).

The L-arabinose degradation gene cluster *araABCDEFGHI* in the pathogenic β-proteobacterium *Burkholderia thailandensis* consists of nine genes and was proposed to be responsible for the degradation to 2-oxoglutarate (Moore *et al.*, 2004). Disruption of the *araA, araC, araE,* and *araI* genes, which may encode a transcriptional regulator, DOP dehydrogenase, L-arabinose dehydrogenase and L-arabinolactonase, respectively, led to an L-arabinose negative phenotype. Interestingly, the *Azospirillum brasilense* has a gene cluster (*araBZYXADCR*) and a separately located DOP dehydrogenase (*araE*) for L-arabinose utilization. In this bacterium, these genes have been studied extensively and the L-arabinose degrading enzymes have been characterized (Watanabe *et al.*, 2006c). Strikingly, the gene clusters (*ycbC-ycbH*) with a homologue of the DOP dehydrogenase of *S. solfataricus* are most likely involved in hexaric acid degradation in *Bacillus* species (Sharma and Blumenthal, 1973). In addition, another gene cluster in *Pseudomonas putida* (PP1245–PP1249) is probably involved in the breakdown of L-hydroxyproline, which is a major constituent of collagen and plant cell wall proteins (Ramaswamy, 1984; Yoneya and Adams, 1961).

The degradation pathways of these diverse chemical compounds are, apparently, converging at the level of DOP. These mosaic pathways of orthologous and non-orthologous proteins involved in the catabolic reactions suggests that some of these enzymatic steps may have evolved by recruitment events (Jensen, 1976; Schmidt *et al.*, 2003), in which 2-oxoglutarate is the metabolic hub as the end product and may have been the driving force in the development of these pathways in aerobically respiring (TCA-cycle containing) Bacteria and Archaea.

Regulation of the pentose metabolism

In Bacteria and Eukarya, regulation of metabolism is executed at all levels. At the level of DNA by inducing or inhibiting transcription initiation, at the level of RNA by influencing transcription elongation and using attenuation mechanisms, and finally at the protein level through post-translational modification and allosteric regulation. Regulation at the protein level by allosteric regulation is certainly expected to be present in Archaea, because protein domains that are involved in allosteric regulation (e.g. ACT, RAM) are very common in Archaea (Aravind and Koonin, 1999; Ettema *et al.*, 2002). However, in the well-characterized archaeal glucose metabolic pathway, the Embden-Meyerhof-Parnas pathway (reviewed by (Siebers and Schonheit, 2005; van der Oost and Siebers, 2007; Verhees *et al.*, 2003)), the enzymes appear not to be regulated allosterically, but rather at the transcriptional level.

Regulation at the translational level (attenuation, anti-termination) has not been demonstrated to date in Archaea, but these mechanisms probably play an important role in archaeal biology as well. For instance, upstream of transport genes in *Thermoplasma* spp., thiamin riboswitches were predicted, which might be acquired by horizontal gene transfer (Rodionov *et al.*, 2002). In addition, an attenuation-like system has been proposed to regulate translation of the tryptophan regulator (TrpY) of *Methanobacterium thermoautotrophicus* (Xie and Reeve, 2005).

A relatively important site of regulating archaeal metabolism appears to be at the transcriptional level. The archaeal transcription initiation machinery is more similar to the eukaryal polymerase II system than to the bacterial machinery with its five subunit RNA polymerase (RNAP) and a series of sigma factors that interact with the Pribnow-box and the −35 region. The archaeal RNAP resembles the core of the eukaryal PolII system with (10–12 subunit RNAP, TATA-binding protein (TBP) and transcription factor B (TFB)) (Bell, 2005). Despite the similarity between the basal transcriptional machineries in Archaea and Eukarya, most archaeal transcriptional regulators more closely resemble their bacterial counterparts (Aravind and Koonin, 1999), but only a limited number of these regulators have been characterized (reviewed by (Bell, 2005; Geiduschek and Ouhammouch, 2005)). The archaeal transcriptional repressors often block the transcription initiation by competing with the transcription factors (TBP, TFB) or the RNAP for their respective binding site (e.g. Lrp-like regulators (Brinkman *et al.*, 2003)). Transcriptional activation has been shown in Archaea as well (Brinkman *et al.*, 2002; Ouhammouch *et al.*, 2003), little is known to date on specific interactions with the transcription initiation complex (TBP, TFB, RNAP).

The transcriptional regulator of the *mal* operon (TrmB) is the only transcriptional regulator involved in the archaeal carbohydrate metabolism that has been characterized. TrmB represses the transcription of the trehalose/maltose transport operon in *Thermococcus litoralis* (Lee *et al.*, 2003) and a separate maltodextrin ABC transporter in *P. furiosus* (Lee *et al.*, 2005). Recently, it was hypothesized that a TrmB homologue (Tgr) could be responsible for modulated gene expression of the archaeal glycolytic enzymes (van de Werken *et al.*, 2006); this has indeed been confirmed biochemically and genetically (Lee *et al.*, 2007; Kanai *et al.*, 2007). The transcriptional regulators of the archaeal pentose metabolism, however, are not known.

In Bacteria and Eukarya the oxidative branch of the pentose phosphate pathway is mainly regulated by the intracellular NADPH/NADP ratio. Glucose-6-phosphate dehydrogenase, for instance, is the rate-limiting step of the OPPP and this essentially irreversible conversion is allosterically regulated by the demand of the reducing power in the cell (Au *et al.*, 2000; Hansen *et al.*, 2002). The NOPPP is controlled, primarily, by the substrate availability (Berg *et al.*, 2002). As the OPPP is not likely to function in Archaea and several pathways are able to interconvert central metabolites with pentoses, different regulatory mechanisms are anticipated to occur in Archaea.

The bacterial pentose oxidation pathways are controlled mainly at the transcriptional level. In the Gram-positive bacterium *Bacillus subtilis* the AraR (COG1609, Table 4.2) negatively regulates the *araABDLMNPQ-abfA* operon and the *araE* and *araR* genes (Mota *et al.*, 1999) with L-arabinose as effector. In the Gram-negative organism *E. coli* the L-arabinose pathway is under control of AraC (COG2207), which activates in presence of L-arabinose the *araBAD* operon and the genes that transport L-arabinose (*araE* and *araF*) and its own product (Miyada *et al.*, 1984).

The XylR (COG1609/2207) of *E. coli* acts as an activator in the presence of D-xylose the transcription of the *xylAB* and the *xylFGHR* operons (Song and Park, 1997). In *B. subtilis* the XylR (COG1940) acts as a repressor of the *xylAB* operon by blocking its promoter (Dahl *et al.*, 1994). RbsR (COG1609) in both organisms represses the ribose operon, encoding the ribose kinase and the ribose transporter (Mauzy and Hermodson, 1992; Woodson and Devine, 1994).

The fungi *Hyprocrea jecorina* and *Aspergillus niger* are using the transcriptional activator Xyr1, XlnR respectively, to induce not only the transcription of the polysaccharide breakdown enzymes but also the D-xylose degradation genes (Hasper *et al.*, 2000; Stricker *et al.*, 2006). In Bacteria, the regulatory mechanism of the general xylulose-5-phosphate pathway is not fully known, but the enzymes involved in D-arabitol catabolism in *Aerobacter aerogenes* have been reported to be inducible (Wilson and Mortlock, 1973).

The prokaryotic pentose oxidation pathways (PPOP) in Bacteria seem to be modulated at the transcriptional level as well. Transcriptional profiling of *C. crescentus* showed up-regulation of the D-xylose gene cluster (2.8- to 11.6-fold) during growth on D-xylose vs. D-glucose (Hottes *et al.*, 2004), which confirms the D-xylose-dependent promoter (Meisenzahl *et al.*, 1997). In addition, reporter gene insertions showed that *araC* and *araE* gene expression of *Burkholderia thailandensis* was repressed during growth in D-glucose and was induced in L-arabinose media (Moore *et al.*, 2004). Finally, the three α-ketoglutarate semialdehyde dehydrogenase (KGSADH or DOP dehydrogenases) isozymes of *A. brasilense* are all inducible in different media: (KGSADH-I) is up-regulated grown on L-arabinose, (KGSADH-II) on D-glutarate/D-galactarate and (KGSADH-III) on hydroxy-L-proline (Watanabe *et al.*, 2006b; Watanabe *et al.*, 2007).

Although the PPOP gene clusters are regulated at the transcriptional level, no transcriptional regulators are characterized that are involved in the regulation of these genes. Most likely, however, the *araA* gene (COG0583) in *B. thailandensis* encodes a LysR-like positive regulator of the L-arabinose assimilation operon, since the *araA* knockout mutant was unable to grow on L-arabinose (Moore *et al.*, 2004).

Catabolite repression is the phenomenon that an organism will first fully consume the preferred carbohydrate (or the most efficient conversion, often glucose) and after a lag phase during which alternative enzymes are being produced, it will start degrading second-choice carbohydrates. This two-step utilization of carbohydrate mixes is known as 'catabolite repression'. In Bacteria and Eukarya catabolite repression is well studied, and important in pentose utilization. In Archaea, a catabolite repression-like system has been described in *Sulfolobus solfataricus* (Hoang *et al.*, 2004) (and reviewed by (Bini and Blum, 2001)). Recently, it has been shown that this catabolite repression-like system is also involved in regulation of the arabinose ABC-transporter (Lubelska *et al.*, 2006) and thus being an important mechanism in the pentose metabolism of this archaeon.

In *S. solfataricus* the complete D-arabinose pathway, including the arabinose transporter, was induced at transcriptome and proteome level grown on D-arabinose vs. D-glucose (Brouns *et al.*, 2006). This is similar with the results of the xylose dehydrogenase *H. marismortui*, which was induced during growth on xylose (Johnsen and Schonheit, 2004).

Recently, several functional genomic studies have been published (reviewed by van der Oost *et al.*, 2006). Moreover, integration of proteomics, transcriptomics and biochemistry has been successfully performed (Brouns *et al.*, 2006; Snijders *et al.*, 2006). The *Sulfolobus* D-arabinose pathway analysis showed the added value of a complete transcriptome and proteome study. Four enzymatical functions were elucidated, but despite the prediction of a binding motif upstream of the genes involved in D-arabinose assimilation, no transcriptional regulator could be identified. Other archaeal functional studies, such as, the microarray analysis in *Pyrococcus furiosus*: reveal co-expression of genes encoding the glycolytic EMP enzymes, and key enzymes of amino acid biosynthesis and transketolase (clustered with AA biosynthesis), during growth on maltose vs. peptides (Schut *et al.*, 2003). However, a combined microarray/proteomics study in *Sulfolobus solfataricus* revealed almost no fluctuation (Snijders *et al.*, 2006).

Concluding remarks on the PPOP in Archaea and regulation of pentose metabolism

Only a few Archaea (Sulfolobales and halophiles) are able to grow on pentoses as sole carbon and energy source. These Archaea are using a prokaryotic pentose oxidation pathway that yields 2-oxoglutaric acid. This TCA intermediate can be completely oxidized by aerobic species. The PPOP enzymatic steps have probably been evolved by recruitment events, yet another example of a 'conserved housekeeping core' and 'variable metabolic shell' that allows adjusting the metabolic infrastructure to available substrates (Makarova *et al.*, 1999). The regulatory mechanisms are still unknown and therefore interesting future research areas, especially with the new transcriptome and proteome tools.

Acknowledgements

This research is supported by the Earth and Life Sciences Foundation (ALW) and Chemische Wetenschappen (CW) project 050.50.206, which are subsidized by the Netherlands Organization for Scientific Research (NWO).

References

Ahmed, H., Ettema, T.J., Tjaden, B., Geerling, A.C., van der Oost, J., and Siebers, B. (2005). The semi-phosphorylative Entner–Doudoroff pathway in hyperthermophilic Archaea: a re-evaluation. Biochem. J. *390*, 529–540.

Aravind, L., and Koonin, E.V. (1999). DNA-binding proteins and evolution of transcription regulation in the Archaea. Nucleic Acids Res. *27*, 4658–4670.

Au, S.W., Gover, S., Lam, V.M., and Adams, M.J. (2000). Human glucose-6-phosphate dehydrogenase: the crystal structure reveals a structural $NADP^+$ molecule and provides insights into enzyme deficiency. Structure *8*, 293–303.

Barthelmes, J., Ebeling, C., Chang, A., Schomburg, I., and Schomburg, D. (2007). BRENDA, AMENDA and FRENDA: the enzyme information system in 2007. Nucleic Acids Res. *35*, D511–514.

Bateman, R.L., Bhanumoorthy, P., Witte, J.F., McClard, R.W., Grompe, M., and Timm, D.E. (2001). Mechanistic inferences from the crystal structure of fumarylacetoacetate hydrolase with a bound phosphorus-based inhibitor. J. Biol. Chem. *276*, 15284–15291.

Bell, S.D. (2005). Archaeal transcriptional regulation – variation on a bacterial theme? Trends Microbiol. *13*, 262–265.

Berg, J.M., Tymoczko, J.L., and Stryer, L. (2002). Biochemistry, 5th edition (New York, W.H. Freeman and company).

Biesterveld, S., Kok, M.D., Dijkema, C., Zehnder, A.J., and Stams, A.J. (1994). D-xylose catabolism in *Bacteroides xylanolyticus* X5–1. Arch. Microbiol. *161*, 521–527.

Bini, E., and Blum, P. (2001). Archaeal catabolite repression: a gene regulatory paradigm. Adv. Appl. Microbiol. *50*, 339–366.

Brinkman, A.B., Bell, S.D., Lebbink, R.J., de Vos, W.M., and van der Oost, J. (2002). The *Sulfolobus solfataricus* Lrp-like protein LysM regulates lysine biosynthesis in response to lysine availability. J. Biol. Chem. *277*, 29537–29549.

Brinkman, A.B., Ettema, T.J., de Vos, W.M., and van der Oost, J. (2003). The Lrp family of transcriptional regulators. Mol. Microbiol. *48*, 287–294.

Brouns, S.J., Walther, J., Snijders, A.P., van de Werken, H.J., Willemen, H.L., Worm, P., de Vos, M.G., Andersson, A., Lundgren, M., Mazon, H.F., *et al.* (2006). Identification of the missing links in prokaryotic pentose oxidation pathways: evidence for enzyme recruitment. J. Biol. Chem. *281*, 27378–27388.

Chiang, C., and Knight, S.G. (1960). Metabolism of D-xylose by moulds. Nature *188*, 79–81.

Choquet, C.G., Richards, J.C., Patel, G.B., and Sprott, G.D. (1994). Ribose biosynthesis in methanogenic cacteria. Arch. Microbiol. *161*, 481–488.

Ciccarelli, F.D., Doerks, T., von Mering, C., Creevey, C.J., Snel, B., and Bork, P. (2006). Toward automatic reconstruction of a highly resolved tree of life. Science *311*, 1283–1287.

Dagley, S., and Trudgill, P.W. (1965). The metabolism of galactarate, D-glucarate and various pentoses by species of *Pseudomonas*. Biochem. J. *95*, 48–58.

Dahl, M.K., Degenkolb, J., and Hillen, W. (1994). Transcription of the *xyl* operon is controlled in *Bacillus subtilis* by tandem overlapping operators spaced by four basepairs. J. Mol. Biol. *243*, 413–424.

Dahms, A.S. (1974). 3-Deoxy-D-pentulosonic acid aldolase and its role in a new pathway of D-xylose degradation. Biochem. Biophys. Res. Commun. *60*, 1433–1439.

Dahms, A.S., and Anderson, R.L. (1969). 2-keto-3-deoxyl-L-arabonate aldolase and its role in a new pathway of L-arabinose degradation. Biochem. Biophys. Res. Commun. *36*, 809–814.

De Rosa, M., Gambacorta, A., Nicolaus, B., Giardina, P., Poerio, E., and Buonocore, V. (1984). Glucose metabolism in the extreme thermoacidophilic archaebacterium *Sulfolobus solfataricus*. Biochem. J. *224*, 407–414.

Duncan, M.J. (1979). L-arabinose metabolism in rhizobia. J. Gen. Microbiol. *113*, 177–179.

Duncan, M.J., and Fraenkel, D.G. (1979). alpha-Ketoglutarate dehydrogenase mutant of *Rhizobium meliloti*. J. Bacteriol. *137*, 415–419.

Eisenreich, W., Schwarzkopf, B., and Bacher, A. (1991). Biosynthesis of nucleotides, flavins, and deazaflavins in *Methanobacterium thermoautotrophicum*. J. Biol. Chem. *266*, 9622–9631.

Ekiel, I., Smith, I.C., and Sprott, G.D. (1983). Biosynthetic pathways in *Methanospirillum hungatei* as determined by ^{13}C nuclear magnetic resonance. J. Bacteriol. *156*, 316-326.

Elsinghorst, E.A., and Mortlock, R.P. (1994). Molecular cloning of the *Escherichia coli* B L-fucose-D-arabinose gene cluster. J. Bacteriol. *176*, 7223–7232.

Ettema, T.J., Brinkman, A.B., Tani, T.H., Rafferty, J.B., and Van Der Oost, J. (2002). A novel ligand-binding domain involved in regulation of amino acid metabolism in prokaryotes. J. Biol. Chem. *277*, 37464–37468.

Evans, C.T., and Ratledge, C. (1984). Induction of xylulose-5-phosphate phosphoketolase in a variety of yeasts grown on D-xylose – the key to efficient xylose metabolism. Arch. Microbiol. *139*, 48–52.

Fossitt, D., Mortlock, R.P., Anderson, R.L., and Wood, W.A. (1964). Pathways of L-arabitol and xylitol metabolism in *Aerobacter Aerogenes*. J. Biol. Chem. *239*, 2110–2115.

Geiduschek, E.P., and Ouhammouch, M. (2005). Archaeal transcription and its regulators. Mol. Microbiol. *56*, 1397–1407.

Grochowski, L.L., Xu, H., and White, R.H. (2005). Ribose-5-phosphate biosynthesis in *Methanocaldococcus jannaschii* occurs in the absence of a pentose-phosphate pathway. J. Bacteriol. *187*, 7382–7389.

Grogan, D.W. (1989). Phenotypic characterization of the archaebacterial genus *Sulfolobus*: comparison of five wild-type strains. J. Bacteriol. *171*, 6710–6719.

Hansen, T., Oehlmann, M., and Schonheit, P. (2001). Novel type of glucose-6-phosphate isomerase in the hyperthermophilic archaeon *Pyrococcus furiosus*. J. Bacteriol. *183*, 3428–3435.

Hansen, T., Schlichting, B., and Schonheit, P. (2002). Glucose-6-phosphate dehydrogenase from the hyperthermophilic bacterium *Thermotoga maritima*: expression of the g6pd gene and characterization of an extremely thermophilic enzyme. FEMS Microbiol. Lett. *216*, 249–253.

Hasper, A.A., Visser, J., and de Graaff, L.H. (2000). The *Aspergillus niger* transcriptional activator XlnR, which is involved in the degradation of the polysaccharides xylan and cellulose, also regulates D-xylose reductase gene expression. Mol. Microbiol. *36*, 193–200.

Hoang, V., Bini, E., Dixit, V., Drozda, M., and Blum, P. (2004). The role of *cis*-acting sequences governing catabolite repression control of *lacS* expression in the archaeon *Sulfolobus solfataricus*. Genetics *167*, 1563–1572.

Hope, J.N., Bell, A.W., Hermodson, M.A., and Groarke, J.M. (1986). Ribokinase from *Escherichia coli* K12. Nucleotide sequence and overexpression of the *rbsK* gene and purification of ribokinase. J. Biol. Chem. *261*, 7663–7668.

Hottes, A.K., Meewan, M., Yang, D., Arana, N., Romero, P., McAdams, H.H., and Stephens, C. (2004). Transcriptional profiling of *Caulobacter crescentus* during growth on complex and minimal media. J. Bacteriol. *186*, 1448–1461.

Ishikawa, K., Matsui, I., Payan, F., Cambillau, C., Ishida, H., Kawarabayasi, Y., Kikuchi, H., and Roussel, A. (2002). A hyperthermostable D-ribose-5-phosphate isomerase from *Pyrococcus horikoshii* characterization and three-dimensional structure. Structure *10*, 877–886.

Jensen, R.A. (1976). Enzyme recruitment in evolution of new function. Annu. Rev. Microbiol. *30*, 409–425.

Johnsen, U., and Schonheit, P. (2004). Novel xylose dehydrogenase in the halophilic archaeon *Haloarcula marismortui*. J. Bacteriol. *186*, 6198–6207.

Kanagasundaram, V., and Scopes, R. (1992). Isolation and characterization of the gene encoding gluconolactonase from *Zymomonas mobilis*. Biochim. Biophys. Acta *1171*, 198–200.

Kanai, T., Akerboom, J., Takedomi, S., van de Werken, H.J.G., Blombach, F., van der Oost, J., Murakami, T., Atomi, H., and Imanaka, T. (2007). A global transcriptional regulator in *Thermococcus kodakaraensis* controls the expression levels of both glycolytic and gluconeogenic enzyme-encoding genes. J. Biol. Chem. (in press).

Kemp, M.B. (1974). Hexose phosphate synthase from *Methylcoccus capsulatus* makes D-*arabino*-3-hexulose phosphate. Biochem. J. *139*, 129–134.

Kim, S., and Lee, S.B. (2005). Identification and characterization of *Sulfolobus solfataricus* D-gluconate dehydratase: a key enzyme in the non-phosphorylated Entner–Doudoroff pathway. Biochem. J. *387*, 271–280.

Kim, S., and Lee, S.B. (2006). Catalytic promiscuity in dihydroxy-acid dehydratase from the thermoacidophilic archaeon *Sulfolobus solfataricus*. J. Biochem. (Tokyo) *139*, 591–596.

Koonin, E.V., and Galperin, M.Y. (2003). Sequence–Evolution–Function: Computational Approaches in Comparative Genomics Kluwer Academic Publishers).

Koonin, E.V., Mushegian, A.R., and Bork, P. (1996). Non-orthologous gene displacement. Trends Genet. *12*, 334–336.

Koonin, E.V., Wolf, Y.I., and Aravind, L. (2001). Prediction of the archaeal exosome and its connections with the proteasome and the translation and transcription machineries by a comparative-genomic approach. Genome Res. *11*, 240–252.

Lee, N., Gielow, W., Martin, R., Hamilton, E., and Fowler, A. (1986). The organization of the *araBAD* operon of Escherichia coli. Gene *47*, 231–244.

Lee, S.J., Engelmann, A., Horlacher, R., Qu, Q., Vierke, G., Hebbeln, C., Thomm, M., and Boos, W. (2003). TrmB, a sugar-specific transcriptional regulator of the trehalose/maltose ABC transporter from the hyperthermophilic archaeon *Thermococcus litoralis*. J. Biol. Chem. *278*, 983–990.

Lee, S.J., Moulakakis, C., Koning, S.M., Hausner, W., Thomm, M., and Boos, W. (2005). TrmB, a sugar sensing regulator of ABC transporter genes in *Pyrococcus furiosus* exhibits dual promoter specificity and is controlled by different inducers. Mol. Microbiol. *57*, 1797–1807.

Lee, S.J., Surma, M., Seitz, S., Hausner, W., Thomm, M., and Boos, W. (2007). Characterization of the TrmB-like protein, PF0124, a TGM-recognizing global transcriptional regulator of the hyperthermophilic archaeon *Pyrococcus furiosus*. Mol. Microbiol. *65*, 305–318.

Lubelska, J.M., Jonuscheit, M., Schleper, C., Albers, S.V., and Driessen, A.J. (2006). Regulation of expression of the arabinose and glucose transporter genes in the thermophilic archaeon *Sulfolobus solfataricus*. Extremophiles *10*, 383–391.

Makarova, K.S., Aravind, L., Galperin, M.Y., Grishin, N.V., Tatusov, R.L., Wolf, Y.I., and Koonin, E.V. (1999). Comparative genomics of the Archaea (Euryarchaeota): evolution of conserved protein families, the stable core, and the variable shell. Genome Res. *9*, 608–628.

Martinez-Cruz, L.A., Dreyer, M.K., Boisvert, D.C., Yokota, H., Martinez-Chantar, M.L., Kim, R., and Kim, S.H. (2002). Crystal structure of MJ1247 protein from *M. jannaschii* at 2.0 A resolution infers a molecular function of 3-hexulose-6-phosphate isomerase. Structure *10*, 195–204.

Mauzy, C.A., and Hermodson, M.A. (1992). Structural and functional analyses of the repressor, RbsR, of the ribose operon of *Escherichia coli*. Protein Sci. *1*, 831–842.

Meile, L., Rohr, L.M., Geissmann, T.A., Herensperger, M., and Teuber, M. (2001). Characterization of the D-xylulose 5-phosphate/D-fructose 6-phosphate phosphoketolase gene (*xfp*) from *Bifidobacterium lactis*. J. Bacteriol. *183*, 2929–2936.

Meisenzahl, A.C., Shapiro, L., and Jenal, U. (1997). Isolation and characterization of a xylose-dependent promoter from *Caulobacter crescentus*. J. Bacteriol. *179*, 592–600.

Milburn, C.C., Lamble, H.J., Theodossis, A., Bull, S.D., Hough, D.W., Danson, M.J., and Taylor, G.L. (2006). The structural basis of substrate promiscuity in glucose dehydrogenase from the hyperthermophilic archaeon *Sulfolobus solfataricus*. J. Biol. Chem. *281*, 14796-14804.

Miyada, C.G., Stoltzfus, L., and Wilcox, G. (1984). Regulation of the *araC* gene of *Escherichia coli*: catabolite repression, autoregulation, and effect on *araBAD* expression. Proc. Natl. Acad. Sci. USA *81*, 4120–4124.

Moore, R.A., Reckseidler-Zenteno, S., Kim, H., Nierman, W., Yu, Y., Tuanyok, A., Warawa, J., DeShazer, D., and Woods, D.E. (2004). Contribution of gene loss to the pathogenic evolution of *Burkholderia pseudomallei* and *Burkholderia mallei*. Infect. Immun. *72*, 4172–4187.

Mota, L.J., Tavares, P., and Sa-Nogueira, I. (1999). Mode of action of AraR, the key regulator of L-arabinose metabolism in *Bacillus subtilis*. Mol. Microbiol. *33*, 476–489.

Ogushi, S., Ando, M., and Tsuru, D. (1984). Formaldehyde dehydrogenase from *Pseudomonas putida*: a zinc metalloenzyme. J. Biochem. (Tokyo) *96*, 1587–1591.

Orita, I., Sato, T., Yurimoto, H., Kato, N., Atomi, H., Imanaka, T., and Sakai, Y. (2006). The ribulose monophosphate pathway substitutes for the missing pentose phosphate pathway in the archaeon *Thermococcus kodakaraensis*. J. Bacteriol. *188*, 4698–4704.

Orita, I., Yurimoto, H., Hirai, R., Kawarabayasi, Y., Sakai, Y., and Kato, N. (2005). The archaeon *Pyrococcus horikoshii* possesses a bifunctional enzyme for formaldehyde fixation via the ribulose monophosphate pathway. J. Bacteriol. *187*, 3636-3642.

Osterman, A., and Overbeek, R. (2003). Missing genes in metabolic pathways: a comparative genomics approach. Curr. Opin. Chem. Biol. *7*, 238–251.

Ouhammouch, M., Dewhurst, R.E., Hausner, W., Thomm, M., and Geiduschek, E.P. (2003). Activation of archaeal transcription by recruitment of the TATA-binding protein. Proc. Natl. Acad. Sci. USA *100*, 5097–5102.

Pace, N.R. (1997). A molecular view of microbial diversity and the biosphere. Science *276*, 734–740.

Palleroni, N.J., and Doudoroff, M. (1957). Metabolism of carbohydrates by *Pseudomonas saccharophila*. III. Oxidation of D-arabinose. J. Bacteriol. *74*, 180–185.

Pedrosa, F.O., and Zancan, G.T. (1974). L-Arabinose metabolism in *Rhizobium japonicum*. J. Bacteriol. *119*, 336-338.

Poindexter, J.S. (1964). Biological Properties and Classification of the Caulobacter Group. Bacteriol. Rev. *28*, 231–295.

Porat, I., Sieprawska-Lupa, M., Teng, Q., Bohanon, F.J., White, R.H., and Whitman, W.B. (2006). Biochemical and genetic characterization of an early step in a novel pathway for the biosynthesis of aromatic amino acids and p-aminobenzoic acid in the archaeon *Methanococcus maripaludis*. Mol. Microbiol. *62*, 1117–1131.

Purwantini, E., and Daniels, L. (1998). Molecular analysis of the gene encoding F_{420}-dependent glucose-6-phosphate dehydrogenase from *Mycobacterium smegmatis*. J. Bacteriol. *180*, 2212–2219.

Purwantini, E., Gillis, T.P., and Daniels, L. (1997). Presence of F_{420}-dependent glucose-6-phosphate dehydrogenase in *Mycobacterium* and *Nocardia* species, but absence from *Streptomyces* and *Corynebacterium* species and methanogenic Archaea. FEMS Microbiol. Lett. *146*, 129–134.

Ramaswamy, S.G. (1984). Hydroxyproline 2-epimerase of *Pseudomonas*. Subunit structure and active site studies. J. Biol. Chem. *259*, 249–254.

Rashid, N., Imanaka, H., Fukui, T., Atomi, H., and Imanaka, T. (2004). Presence of a novel phosphopentomutase and a 2-deoxyribose 5-phosphate aldolase reveals a metabolic link between pentoses and central carbon metabolism in the hyperthermophilic archaeon *Thermococcus kodakaraensis*. J. Bacteriol. *186*, 4185–4191.

Rodionov, D.A., Vitreschak, A.G., Mironov, A.A., and Gelfand, M.S. (2002). Comparative genomics of thiamin biosynthesis in procaryotes. New genes and regulatory mechanisms. J. Biol. Chem. *277*, 48949–48959.

Rosenfeld, S.A., Stevis, P.E., and Ho, N.W. (1984). Cloning and characterization of the *xyl* genes from *Escherichia coli*. Mol. Gen. Genet. *194*, 410–415.

Roy, R., Mukund, S., Schut, G.J., Dunn, D.M., Weiss, R., and Adams, M.W. (1999). Purification and molecular characterization of the tungsten containing formaldehyde ferredoxin oxidoreductase from the hyperthermophilic archaeon *Pyrococcus furiosus*: the third of a putative five-member tungstoenzyme family. J. Bacteriol. *181*, 1171–1180.

Sakuraba, H., Tsuge, H., Shimoya, I., Kawakami, R., Goda, S., Kawarabayasi, Y., Katunuma, N., Ago, H., Miyano, M., and Ohshima, T. (2003). The first crystal structure of archaeal aldolase. Unique tetrameric structure of 2-deoxy-D-ribose-5-phosphate aldolase from the hyperthermophilic Archaea Aeropyrum pernix. J. Biol. Chem. *278*, 10799–10806.

Sato, T., Atomi, H., and Imanaka, T. (2007). Archaeal type III RuBisCOs function in a pathway for AMP metabolism. Science *315*, 1003–1006.

Schleper, C., Jurgens, G., and Jonuscheit, M. (2005). Genomic studies of uncultivated Archaea. Nat. Rev. Microbiol. *3*, 479–488.

Schmidt, S., Sunyaev, S., Bork, P., and Dandekar, T. (2003). Metabolites: a helping hand for pathway evolution? Trends Biochem. Sci. *28*, 336-341.

Schonheit, P., and Schafer, T. (1995). Metabolism of Hyperthermophiles. World Journal of Microbiology & Biotechnology *11*, 26–57.

Schut, G.J., Brehm, S.D., Datta, S., and Adams, M.W. (2003). Whole-genome DNA microarray analysis of a hyperthermophile and an archaeon: *Pyrococcus furiosus* grown on carbohydrates or peptides. J. Bacteriol. *185*, 3935–3947.

Sharma, B.S., and Blumenthal, H.J. (1973). Catabolism of D-glucaric acid to alpha-ketoglutarate in *Bacillus megaterium*. J. Bacteriol. *116*, 1346-1354.

Siebers, B., Brinkmann, H., Dorr, C., Tjaden, B., Lilie, H., van der Oost, J., and Verhees, C.H. (2001). Archaeal fructose-1,6-bisphosphate aldolases constitute a new family of archaeal type class I aldolase. J. Biol. Chem. *276*, 28710–28718.

Siebers, B., and Schonheit, P. (2005). Unusual pathways and enzymes of central carbohydrate metabolism in Archaea. Curr. Opin. Microbiol. *8*, 695–705.

Sinninghe Damste, J.S., Rijpstra, W.I., Hopmans, E.C., Prahl, F.G., Wakeham, S.G., and Schouten, S. (2002). Distribution of membrane lipids of planktonic Crenarchaeota in the Arabian Sea. Appl. Environ. Microbiol. *68*, 2997–3002.

Snel, B., Bork, P., and Huynen, M.A. (1999). Genome phylogeny based on gene content. Nat. Genet. *21*, 108–110.

Snijders, A.P., Walther, J., Peter, S., Kinnman, I., de Vos, M.G., van de Werken, H.J., Brouns, S.J., van der Oost, J., and Wright, P.C. (2006). Reconstruction of central carbon metabolism in *Sulfolobus solfataricus* using a two-dimensional gel electrophoresis map, stable isotope labelling and DNA microarray analysis. Proteomics *6*, 1518–1529.

Soderberg, T. (2005). Biosynthesis of ribose-5-phosphate and erythrose-4-phosphate in Archaea: a phylogenetic analysis of archaeal genomes. Archaea *1*, 347–352.

Soderberg, T., and Alver, R.C. (2004). Transaldolase of *Methanocaldococcus jannaschii*. Archaea *1*, 255–262.

Song, S., and Park, C. (1997). Organization and regulation of the D-xylose operons in *Escherichia coli* K-12: XylR acts as a transcriptional activator. J. Bacteriol. *179*, 7025–7032.

Stephens, C., Christen, B., Fuchs, T., Sundaram, V., Watanabe, K., and Jenal, U. (2007). Genetic Analysis of a Novel Pathway for D-Xylose Metabolism in *Caulobacter crescentus*. J. Bacteriol. *189*, 2181–2185.

Stricker, A.R., Grosstessner-Hain, K., Wurleitner, E., and Mach, R.L. (2006). Xyr1 (xylanase regulator 1) regulates both the hydrolytic enzyme system and D-xylose metabolism in *Hypocrea jecorina*. Eukaryot Cell 5, 2128–2137.

Tame, J.R., Namba, K., Dodson, E.J., and Roper, D.I. (2002). The crystal structure of HpcE, a bifunctional decarboxylase/isomerase with a multifunctional fold. Biochemistry *41*, 2982–2989.

Tatusov, R.L., Fedorova, N.D., Jackson, J.D., Jacobs, A.R., Kiryutin, B., Koonin, E.V., Krylov, D.M., Mazumder, R., Mekhedov, S.L., Nikolskaya, A.N., *et al.* (2003). The COG database: an updated version includes eukaryotes. BMC Bioinformatics *4*, 41.

Tindall, B.J. (1992). The family *Halobacteriaceae*. In The prokaryotes. A handbook on the biology of Bacteria: ecophysiology, isolation, identification, applications (New York. N.Y., Springer-Verlag), pp. 768–808.

Tumbula, D.L., Teng, Q., Bartlett, M.G., and Whitman, W.B. (1997). Ribose biosynthesis and evidence for an alternative first step in the common aromatic amino acid pathway in *Methanococcus maripaludis*. J. Bacteriol. *179*, 6010–6013.

van de Werken, H.J., Verhees, C.H., Akerboom, J., de Vos, W.M., and van der Oost, J. (2006). Identification of a glycolytic regulon in the Archaea *Pyrococcus* and *Thermococcus*. FEMS Microbiol. Lett. *260*, 69–76.

van der Oost, J., and Siebers, B. (2007). The Glycolytic Pathways of Archaea – Evolution by Tinkering. In Archaea, R.A. Garret, and H.P. Klenk, eds. (Blackwell Publishing), pp. 247–259.

van der Oost, J., Walther, J., Brouns, S.J.J., van de Werken, H.J. G., Snijders, A.P. L., Wright, P.C., Andersson, A., Bernander, R., and de Vos, W.M. (2006). Functional Genomics of the Thermo-Acidophilic Archaeon *Sulfolobus Solfataricus*. In Extremophiles – Methods in Microbiology F.A. Rainey, and A. Oren, eds. (Amsterdam, Elsevier/Academic Press), pp. 201–231.

Verhees, C.H., Huynen, M.A., Ward, D.E., Schiltz, E., de Vos, W.M., and van der Oost, J. (2001). The phosphoglucose isomerase from the hyperthermophilic archaeon *Pyrococcus furiosus* is a unique glycolytic enzyme that belongs to the cupin superfamily. J. Biol. Chem. *276*, 40926–40932.

Verhees, C.H., Kengen, S.W., Tuininga, J.E., Schut, G.J., Adams, M.W., De Vos, W.M., and Van Der Oost, J. (2003). The unique features of glycolytic pathways in Archaea. Biochem. J. *375*, 231–246.

von Mering, C., Huynen, M., Jaeggi, D., Schmidt, S., Bork, P., and Snel, B. (2003). STRING: a database of predicted functional associations between proteins. Nucleic Acids Res. *31*, 258–261.

Vorholt, J.A., Marx, C.J., Lidstrom, M.E., and Thauer, R.K. (2000). Novel formaldehyde-activating enzyme in *Methylobacterium extorquens* AM1 required for growth on methanol. J. Bacteriol. *182*, 6645–6650.

Watanabe, S., Kodaki, T., and Makino, K. (2006a). Cloning, expression, and characterization of bacterial L-arabinose 1-dehydrogenase involved in an alternative pathway of L-arabinose metabolism. J. Biol. Chem. *281*, 2612–2623.

Watanabe, S., Kodaki, T., and Makino, K. (2006b). A novel alpha-ketoglutaric semialdehyde dehydrogenase: evolutionary insight into an alternative pathway of bacterial L-arabinose metabolism. J. Biol. Chem. *281*, 28876-28888.

Watanabe, S., Shimada, N., Tajima, K., Kodaki, T., and Makino, K. (2006c). Identification and characterization of L-arabonate dehydratase, L-2-keto-3-deoxyarabonate dehydratase, and L-arabinolactonase involved in an alternative pathway of L-arabinose metabolism. Novel evolutionary insight into sugar metabolism. J. Biol. Chem. *281*, 33521–33536.

Watanabe, S., Yamada, M., Ohtsu, I., and Makino, K. (2007). {alpha}-Ketoglutaric semialdehyde dehydrogenase Isozymes involved in metabolic pathways of D-glucarate, D-galactarate, and hydroxy-L-proline: molecular and metabolic convergent evolution. J. Biol. Chem. *282*, 6685–6695.

Weimberg, R. (1961). Pentose oxidation by *Pseudomonas fragi*. J. Biol. Chem. *236*, 629–635.

Weimberg, R., and Doudoroff, M. (1955). The oxidation of L-arabinose by *Pseudomonas saccharophila*. J. Biol. Chem. *217*, 607–624.

White, R.H. (2004). L-Aspartate semialdehyde and a 6-deoxy-5-ketohexose 1-phosphate are the precursors to the aromatic amino acids in *Methanocaldococcus jannaschii*. Biochemistry *43*, 7618–7627.

White, R.H., and Xu, H. (2006). Methylglyoxal is an intermediate in the biosynthesis of 6-deoxy-5-ketofructose-1-phosphate: a precursor for aromatic amino acid biosynthesis in *Methanocaldococcus jannaschii*. Biochemistry *45*, 12366-12379.

Wilson, B.L., and Mortlock, R.P. (1973). Regulation of D-xylose and D-arabitol catabolism by *Aerobacter aerogenes*. J. Bacteriol. *113*, 1404–1411.

Woese, C.R., and Fox, G.E. (1977). Phylogenetic structure of the prokaryotic domain: the primary kingdoms. Proc. Natl. Acad. Sci. USA *74*, 5088–5090.

Woese, C.R., Kandler, O., and Wheelis, M.L. (1990). Towards a natural system of organisms: proposal for the domains Archaea, Bacteria, and Eucarya. Proc. Natl. Acad. Sci. USA *87*, 4576–4579.

Wojtkiewicz, B., Szmidzinski, R., Jezierska, A., and Cocito, C. (1988). Identification of a salvage pathway for D-arabinose in *Mycobacterium smegmatis*. Eur. J. Biochem. *172*, 197–203.

Woodson, K., and Devine, K.M. (1994). Analysis of a ribose transport operon from *Bacillus subtilis*. Microbiology *140 (Pt 8)*, 1829–1838.

Xavier, K.B., da Costa, M.S., and Santos, H. (2000). Demonstration of a novel glycolytic pathway in the hyperthermophilic archaeon *Thermococcus zilligii* by ^{13}C-labeling experiments and nuclear magnetic resonance analysis. J. Bacteriol. *182*, 4632–4636.

Xie, Y., and Reeve, J.N. (2005). Regulation of tryptophan operon expression in the archaeon *Methanothermobacter thermautotrophicus*. J. Bacteriol. *187*, 6419–6429.

Yoneya, T., and Adams, E. (1961). Hydroxyproline metabolism. V. Inducible allohydroxy-D-proline oxidase of *Pseudomonas*. J. Biol. Chem. *236*, 3272–3279.

Yu, J.P., Ladapo, J., and Whitman, W.B. (1994). Pathway of glycogen metabolism in *Methanococcus maripaludis*. J. Bacteriol. *176*, 325–332.

DNA Replication in Archaea

5

Brian R. Berquist and Shiladitya DasSarma

Abstract

Much initial study of archaeal DNA replication came as a direct result of complete genome sequencing, as the difficulty of culturing most Archaea and their limited capability for genetic and biochemical manipulation in the laboratory had restricted the scope of studies. Here, we review bioinformatic, biochemical, and genetic work on identification of archaeal chromosomal DNA replication origins and the protein factors required for both DNA replication initiation and elongation. We highlight work done in both divisions of cultured Archaea: the euryarchaea and the crenarchaea and attempt to provide unifying themes from analyses in diverse archaeal organisms.

Introduction

DNA replication is a fundamental process common to all life. The structure of DNA, with two anti-parallel strands (Watson and Crick, 1953), and the semiconservative nature of DNA replication provides a mechanism by which the genetic information for life can be faithfully passed on from parent to progeny (Meselson and Stahl, 1958). In the three domains of life, Bacteria, Archaea, and Eukarya, there are generally conserved characteristics of DNA replication initiation and elongation, although the protein machinery used in the different domains may have originated twice independently, once in the bacterial lineage and once in the archaeal/eukaryotic lineage (Leipe *et al.*, 1999). This hypothesis is supported by the homology shared between archaeal and eukaryotic DNA replication proteins and the apparent lack of homology to factors used in Bacteria (Tables 5.1 and 5.2).

Generally, DNA replication initiates at one or more chromosomal DNA replication origin(s) by a protein factor or complex recognizing and binding to DNA elements in the replication origin, leading to local DNA helix unwinding through destabilization of AT-rich DNA unwinding elements (DUEs). Next, a replicative DNA helicase is recruited to the origin and aids in actively unwinding the two DNA strands. This leads to recruitment of additional protein factors and complexes leading to formation of a replisome complex comprising single-strand DNA (ssDNA) binding proteins, DNA directed RNA primase, DNA polymerases for leading and lagging strand synthesis, DNA polymerase processivity clamp, and proteins for Okazaki fragment maturation (RNA primer removal). Once the assembly of these factors at the origin has occurred DNA replication initiates from RNA primers synthesized at a replication origin.

The great advances in our understanding of DNA replication that have taken place over the past 50 years, especially in model organisms such as the bacterium *E. coli* and eukaryote *Saccharomyces cerevisiae* and related organisms, is a testament to the power of modern biology. The most productive systems have traditionally employed a combination of biochemistry and genetics and more recently genomics and structural biology. Many of these technologies are now being applied to a wide diversity of organisms, including a number of Archaea. In this article, we begin by briefly discussing some

Table 5.1 DNA replication components in the three domains of life

	Domain of life		
	Bacteria	Eukarya	Archaea
Protein function			
Origin recognition	DnaA	ORC 1–6 (heterohexamer)	Orc/Cdc6 (1–17 genes)
Helicase Recruiter	DnaC	Cdc6/Cdt1	Orc/Cdc6 (1–17 genes)
Replicative Helicase	DnaB (homohexamer)	MCM 2–7 (heterohexamer)	MCM (Homo- or potential hetero- hexamer)
ssDNA binding	Ssb (homotetramer)	RPA (heterotrimer)	RPA/SSB (heterotrimer/homotetamer)
Primase	DnaG	Primase – Pol-α	Eukaryotic-type primase (heterodimer)
Clamp loader	γ -Complex (5 subunit)	RFC (pentamer)	RFC (1 large + 1 or 2 small subunits, heteropentamer)
Processivity clamp	β -Clamp (dimer)	PCNA (homotrimer)	PCNA (homo-/hetero-trimer)
Replicative DNA polymerase(s)	Pol III (C Family)	Polδ & Polε (B Family)	Pol B (B family) + PolD (D family) or PolB alone
Removal of RNA primer	Pol I and RNaseH	FEN1 and RNaseH	Flap endonuclese and RNaseHII and/or RNase HI
Ligation of Okazaki fragments	DNA Ligase (NAD^+)	DNA Ligase (ATP)	DNA Ligase (ATP or NAD^+)

of the salient features of bacterial and eukaryotic replication systems for context and then provide a detailed review of the literature on archaeal DNA replication systems. Our focus is on the basic aspects of archaeal DNA replication rather than the biotechnological impact of studies of extremophilic DNA polymerases on the development and refinement of polymerase chain reaction technology (Majernik *et al.* 2004).

Bacterial replication

Typical bacterial DNA replication initiates at a specific site in the chromosome, *oriC*, and proceeds bidirectionally around the chromosome until replication forks reach the terminus, yielding two identically replicated chromosomes. DNA replication is initiated by binding of the bacterial origin recognition factor DnaA to specific DNA sequences (DnaA boxes), making up the bacterial chromosomal DNA replication origin, *oriC*. *oriC* is typically located proximal to the gene encoding DnaA and in some cases the *dnaN* gene, encoding the DNA polymerase processivity β clamp. In general, bacterial *oriC* comprise a number of DnaA boxes, from five to nineteen in different Bacteria, with AT-rich stretches, both proximal to the *dnaA* gene. In *Escherichia coli*, *oriC* is 260 bp with 5 DnaA boxes (5′-TTATNCACA-3′) and an AT-rich region at the 5′ end with three 13-bp repeats. For most completely sequenced Bacteria, the DNA replication origin can be identified via bioinformatics analysis utilizing nucleotide skews (Grigoriev, 1998; McLean *et al.*, 1998) or Z-curve analysis (Zhang *et al.*, 2003), which examine the differences between the number of certain nucleotides normalized to the total genomic content of these nucleotides.

For *E. coli* replisome formation, the DnaA protein binds to DnaA boxes at *oriC* and induces a kink at each site leading to helix destabilization and local unwinding of the AT-rich region. ssDNA-binding protein (SSB) coats the ssDNA to keep it unwound and the replicative helicase recruiter DnaC brings the hexameric replicative helicase DnaB to the origin. DnaB is loaded onto the DNA and further unwinds the duplex DNA in a 5′–3′ direction. DnaG, the bacterial-type primase, now associates and lays down RNA primers for nascent DNA synthesis. The DNA polymerase processivity β-clamp is loaded onto the primed template by the clamp loader, γ-complex, and association of the replicative DNA polymerase, DNA polymerase III, occurs, leading to primer extension, processive DNA polymerization, and duplication of genomic material (Messer, 2002).

Eukaryal replication

The mechanisms of DNA replication initiation in Eukarya are substantially more complex and varied between organisms. DNA replication initiates at multiple sites in multiple linear chromosomes, some at sequence specific sites depending on the organism examined, and proceeds bidirectionally. Specific DNA sequences acting as DNA replication origins have been identified in budding yeast, *Saccharomyces cerevisiae* (Brewer and Fangman, 1987), and in fission yeast, *Schizosaccharomyces pombe* (Clyne and Kelly, 1995). In *S. cerevisiae*, DNA replication origins comprise three to four 10–15 bp sequences which constitute the conserved and essential A element, also called the autonomously replicating sequence consensus (ACS) (5′-(A/T)TTTA(T/C)(A/G)TTT(A/T)-3′) (Brewer and Fangman, 1987), as well as additional B elements, which are less conserved between DNA replication origins. In *S. pombe*, DNA replication origins are larger, comprising 800–1000 bp of DNA with several 20–50 bp regions of AT-rich sequences, none of which shows sequence similarity to the ACS or B elements of *S. cerevisiae* (Clyne and Kelly, 1995). It appears that in *S. pombe*, the ability for intergenic sequences to function as a DNA replication origin is governed primarily by AT content and length (Dai *et al.*, 2005). Metazoan and mammalian DNA replication origins are even less well conserved, with some origins identified, but with no unifying sequence characteristics for either system.

For replisome formation in *S. cerevisiae*, there is initial formation of the pre-replicative complex (pre-RC), which involves binding of the hexameric origin recognition complex (ORC) to the ACS (A and B1 elements). ORC is made up of Orc1–6p, with Orc1p, Orc4p, and Orc5p having sequence similarity to each other as well as to Cdc6p (Giraldo and Diaz-Orejas, 2001; Neuwald *et al.*, 1999). ORC associates with the replicative helicase recruiter Cdc6p as well as Cdt1p to bring the hexameric replicative 3′–5′ helicase complex Mcm2–7p to the ORC-origin protein–DNA complex (pre-RC). Although Mcm2–7p are required for origin association, it appears as though the Mcm4/6/7p complex acts as the hexameric replicative helicase with Mcm2p, 3p, and 5p proteins acting as helicase inhibitors (Lee and Hurwitz, 2000), even though all Mcm proteins share significant sequence similarity with one another. Once the MCM complex is loaded onto origin DNA, ORC and Cdc6p can be removed without preventing DNA replication from proceeding. *In vivo* ORC, Cdc6p, and MCM are phosphorylated by cyclin-dependent kinases (CDKs) in order to prevent premature re-replication. A number of other protein factors are associated with DNA replication origins during the transition from initiation to elongation. Mcm10p is required for association of Cdc45p, which in turn is required for association of the heterotrimeric ssDNA-binding protein complex replication protein A (RPA), consistent with its presence at DNA replication origins prior to DNA unwinding. RPA is required for association of the polα-primase complex and DNA polymerase ε. Loading of the homotrimeric DNA polymerase sliding clamp proliferating cell nuclear antigen (PCNA) by replication factor C (RFC) requires the polα-primase complex to lay down RNA primers. Association of the DNA polymerase Pol δ occurs and subsequently processive DNA replication proceeds bidirectionally from each origin (Bell and Dutta, 2002).

Archaeal replication

DNA replication in Archaea appears to have characteristics of both of bacterial and eukaryal DNA replication systems, with genome structure being more similar to Bacteria and DNA replisome components more similar to Eukarya. Archaea possess circular chromosomes in which DNA replication initiates at one or more sequence-specific sites utilizing proteins homologous to many of those used in eukaryotes. Much initial study of archaeal DNA replication came as a direct result of complete genome sequencing, as the difficulty of culturing most Archaea and their limited capability for genetic and biochemi-

Table 5.2 Components of DNA replication in sequenced archaeal genomes

Organism[e or c]	OriC	Orc/ Cdc6	MCM	Primase: Pri small	Primase: Pri large	GINS: GINS 15	GINS: GINS 23	DNA polymerases: PolD small	DNA polymerases: PolD large	DNA polymerases: PolB
Aeropyrum pernix K1[c]	E*	2	1	1	1	1	1			3
Archaeoglobus fulgidus DSM 4304[e]	N	2	1	1	1	1		1	1	1
Haloarcula marismortui ATCC 43049[e]	C	17	3	1	1	1		1	1	2
Halobacterium sp. NRC-1[e]	E	10	1	1	1	1		1	1	2
Haloquadratum walsbyi DSM 16790[e]	C	7	1	1	1	1		1	1	1
Hyperthermus butylicus DSM 5456[e]	C*	2	1	1	1		1			2
Methanocaldococcus jannaschii DSM 2661[e]	N	0	4	1	1	1		1	1	1
Methanococcoides burtonii DSM 6242[e]	N	2	1	1	1	1		1	1	4
Methanococcus maripaludis S2[e]	N	0	4	1	1	1		1	1	1
Methanocorpusculum labreanum Z[e]	N	2	1	1	1	1		1	1	1
Methanopyrus kandleri AV19[e]	N	2	2	1	1	1		1	1	1
Methanosaeta thermophila Pt[e]	N	2	1	1	1	1	1	1	1	1
Methanosarcina acetivorans C2A[e]	C	2	2	1	1	1	1	1	1	1
Methanosphaera stadtmanae DSM 3091[e]	N	2	1	1	1	1		1	1	2
Methanospirillum hungatei JF-1[e]	N	2	1	1	1	1		1	1	2
Methanothermobacter thermautotrophicus str. Delta H[e]	C	2	1	1	1	1		1	1	2
Nanoarchaeum equitans Kin4-M[e]	N	1	1	1	1	1		1	1	2
Natronomonas pharaonis DSM 2160[e]	C	5	2	1	1	1		1	1	1
Picrophilus torridus DSM 9790[e]	N	1	1	1	1	1		1	1	2
Pyrobaculum aerophilum str. IM2[c]	N	1	1	1	1	1				3
Pyrococcus abyssi GE5[e]	E	1	2	1	1	1	1	1	1	1
Pyrococcus furiosus DSM 3638[e]	C	1	1	1	1	1	1	1	1	1
Pyrococcus horikoshii OT3[e]	C	1	1	1	1	1	1	1	1	1
Sulfolobus acidocaldarius DSM 639[c]	E	3	1	1	1	1	1			3
Sulfolobus solfataricus P2[c]	E	3	1	1	1	1	1			3

RFC		PCNA	RNaseH		Flap endonuclease	Single-strand DNA-binding proteins			
RFC large	RFC small		RNaseH2	RNaseH1		RPA 70	RPA 32	RPA 14	SSB
1	1	3	1		1				1
1	1	1	1		1	2	1		
1	2	1	1	1	1	5	3	1	
1	2	1	1	1	1	5	2	1	
1	2	1	1	2	1	5	2	1	
1		3	1		1				1
1	1	1	1		1	1			
1	1	1	1		1	4	2		
1	4	1	1		1	2	2		
1	2	1	1		1	3	2		
1	1	1	1		1	1			
1	1	1	1	1	1	3	1		
1	2	1	1		1	3	2		
1	1	1	1		1	2	2		
1	2	1	1	1	1	1	2		
1	1	1	1		1	1			
1	1	1	1		1				
1	2	1	1	1	1	5	2		
1	1	1	1		1	2	1		
1	2	2	1	1	1				1
1	1	1	1		1	1	1	1	
1	1	1	1		1	1	1	1	
1	1	1	1		1	1	1	1	
1	1	3	1	1	1				1
1	1	3	1		1				1

Table 5.2 *continued*

Organism[e or c]	OriC	Orc/Cdc6	MCM	Primase		GINS		DNA polymerases		
				Pri small	Pri large	GINS 15	GINS 23	PolD small	PolD large	PolB
Sulfolobus tokodaii str. 7[c]	C	3	1	1	1	1	1			3
Thermococcus kodakarensis KOD1[e]	C	1	3	1	1	1	1	1	1	1
Thermoplasma acidophilum DSM 1728[e]	N	2	1	1	1	1		1	1	1
Thermoplasma volcanium GSS1[e]	N	2	1	1	1	1		1	1	1

[e]Euryarchaea.
[c]Crenarchaea.
E, Origin experimentally identified; C – Candidate origin identified computationally; N, no origin locus identified.
*Recently identified in Robinson and Bell (2007).

cal manipulation in the laboratory had restricted the scope of studies. Here, we review work comprising archaeal chromosomal DNA replication origins and protein factors required for both DNA replication initiation and elongation. We highlight work done in both divisions of cultured Archaea: the euryarchaea and the crenarchaea and attempt to provide unifying themes from analyses in diverse archaeal organisms.

Archaeal chromosomal DNA replication origins

With the availability of three genome sequences of the hyperthermophilic euryarchaeal genus *Pyrococcus, in silico* analyses were used initially to predict the location of the chromosomal origin utilizing a methodology to identify overrepresented tetramer nucleotide sequences in *P. abyssi* and the closely related organisms *P. horikoshii* and *P. furiosus*. For the three *Pyrococcus* species, a single tetramer skew minimum and a single tetramer skew maximum was identified, which also corresponded to the only highly conserved large intergenic region in the three genome sequences, pointing to what could comprise an archaeal DNA replication origin. Experimentally, a strategy was implemented to identify an early replicating genomic fragment of *P. abyssi* to validate *in silico* predictions. DNA replication was first inhibited and subsequent newly synthesized DNA was labelled with ^{3}H-uracil. Subsequent examination of relative abundance of *P. abyssi* genomic fragments identified an 80 kb region that reproducibly had the highest specific activity of incorporated ^{3}H-uracil. This fragment correlated to the region encompassing the global minimum in tetramer skew analysis. Contained within this early replicating fragment was the only *orc/cdc6* gene (*orc1*) in the *P. abyssi* genome. 5′ to the *orc/cdc6* gene was a large intergenic region which was found to be conserved in other *Pyrococcus* spp. Additionally, located proximal to *orc1* were genes for a D family DNA polymerase subunit, a recombinase, and a DNA sliding clamp loader subunit (Myllykallio *et al.*, 2000). Subsequent analysis utilized neutral/neutral 2D (N/N 2D) agarose gel electrophoresis to more finely map the *P. abyssi* chromosomal DNA replication origin. Replication initiation was found to occur in a 1 kb window 5′ to the *P. abyssi orc/cdc6* gene. Within that 1 kb window a number of short repeats were identified having a consensus sequence of

5′-(C/T)TCCAG(A/T)GGAAA-3′

as well as two potential DNA unwinding elements. A larger 34 bp inverted repeat (IR) with the sequence

(5′-CCCCCAGAGTTTCATTTC-
CAGTGGAACTGGGTTG (240 bp)
CAACCCGGTTCCAGTGGAAAT-
GAAACCCTGGGGG-3′)

was also noted (Matsunaga *et al.*, 2001). Additional fine mapping of DNA replication initiation sites to the nucleotide level, using replication inition point (RIP) mapping, indicated that initiation in *P. abyssi* occurs immediately 5′ to the start of the 5′ half this large inverted repeat. Further *in silico* analysis showed the large

RFC			RNaseH			Single-strand DNA-binding proteins			
RFC large	RFC small	PCNA	RNaseH2	RNaseH1	Flap endonuclease	RPA 70	RPA 32	RPA 14	SSB
1	1	3	1	1	1				1
1	1	1	1		1	1	1	1	
1	1	1	1		1	2	1		
1	1	1	1		1	2	1		

inverted repeat structure to be highly conserved in sequence in the genome sequences of the two other *Pyrococcus* spp., thereby identifying a conserved sequence motif and precisely mapping the replication initation points within the single chromosomal replication origin of *Pyrococcus* spp. (Matsunaga *et al.*, 2003).

In silico GC skew analysis [(G+C)/(G-C) on a DNA strand] of the complete genome sequence of the mesophilic, extremely halophilic euryarchaeon, *Halobacterium* sp. NRC-1 indicated the potential for multiple replication origins in the Archaea, as the presence of four GC skew polarity switches for the *Halobacterium* sp. NRC-1 large chromosome were consistent with two origins of replication on the large chromosomal (but GC skew analysis was inconclusive for the two smaller pNRC minichromosomal replicons). Further analysis of the polarity switch regions of the large chromosome revealed the presence of two *orc/cdc6* homologues located proximal to these switch points, suggesting the association of these genes with replication origins (Kennedy *et al.*, 2001). A directed genetic approach was undertaken to attempt to identify autonomous replication ability of DNA fragments proximal to three chromosomally encoded *orc/cdc6* genes in this organism, based the previous studies in *Pyrococcus* spp. implicating a physical association between archaeal replication origins and *orc/cdc6* genes and GC skew data obtained for the *Halobacterium* sp. NRC-1 large chromosome. Selected regions of the genome proximal to the three chromosomal *Halobacterium* sp. NRC-1 *orc/cdc6* genes (named *orc6*, *orc7*, and *orc8*) were cloned into a plasmid vector which was not able to replicate in haloarchaea and assayed for endowment of autonomous replication ability. Regions proximal to the other chromosomally encoded *orc/cdc6* gene, *orc10*, were not included due to the presence of an IS element located close to the *orc10* translational start codon. Only plasmids containing the *orc7* gene plus 750 bp of 5′ DNA were able to replicate autonomously, all other plasmid constructs were observed to integrate into the *Halobacterium* sp. NRC-1 large chromosome and not replicate autonomously, strongly suggesting the use of a single chromosomal replication origin. Within the 750 bp 5′ to the *orc7* open reading frame (ORF) was a large 31 bp inverted repeat with the sequence

(5′-CAACCCCCCACCCCT-
TCGTTTCGGGTGGAAC (189 bp)
GTTCCACTCGAAACGAAGGGGT-
GGGGGGTTG-3′)

which flanked an extremely AT-rich stretch of 189 bp. A nearly perfect match of the 31 bp IR was found 5′ to *orc7* genes in the genomes of two distantly related haloarchaea, *Haloarcula marismortui* and *Haloferax volcanii*, with the former having a 28 bp IR containing two mismatches in sequence separated by an AT-rich region of 249 bp and the latter having a perfect 28 bp IR with a 2 bp insertion in the second half, flanking a 172 bp AT-rich region (Berquist and DasSarma, 2003).

Work done to map DNA replication origins in related crenarchaea *Sulfolobus solfataricus* and

S. acidocaldarius produced extremely interesting results. Examination of genomic loci of the three *S. solfataricus orc/cdc6* genes identified a single large intergenic region > 450 bp from one *orc/cdc6* gene, similar to the large intergeneic regions 5′ to *orc/cdc6* identified in the euryarchaea, *Pyrococcus* spp. and haloarchaea; *Halobacterium* sp. NRC-1, *H. marismortui*, and *H. volcanii*. Using 2D N/N agarose gel electrophoresis and probing regions proximal to two of three *orc/cdc6* homologues in the genome of *S. solfataricus*, surprisingly, indicated that two of these regions both acted as DNA replication origins, displaying for the first time that an organism outside of the domain Eukarya used multiple chromosomal DNA replication origins. Finer mapping of the replication initiation regions, via RIP mapping, revealed that replication initiation at *oriC1* occured 230 bp upstream of the *S. sofataricus orc/cdc6-1* gene, while replication initiation at *oriC2* was found to occur between 630 and 678 bp upstream of *orc/cdc6-3*. No active replication origin could be detected proximal to the other *orc/cdc6* gene in *S. sofataricus*, *orc/cdc6-2*. Located 5′ to the *orc/cdc6-1* open reading frame were three large 36 bp repeats deemed origin recognition boxes (ORBs)

(5′-ACCCATCTATTTCCTCT-GAAAATTACTTAATCTAGC (65 bp) ATCATATTAATAATTTTC-CAAACGAAACAATGGGGT (23 bp) TATTTACCTTAAGTTCTCCAGT-GGAAACAAAGGGGT-3′)

(Robinson *et al.*, 2004). Located 5′ to the *orc/cdc6-3* open reading frame were two smaller inverted sequence repeats named mini-ORB elements as well as direct sequence repeats further 5′ to the mORB elements (Robinson *et al.*, 2004).

In both *S. solfataricus* and *S. acidocaldarius*, marker frequency analysis using microarrays (MF) was performed to map DNA replication origins and identified an additional unique replication origin conserved in the genomes of both organisms, while identifying the two previously described DNA replication origins *oriC1* and *oriC2* in *S. sofataricus*. Intriguingly, *S. sofataricus oriC3* does not appear to be linked to an *orc/cdc6* gene, and ORB sequence elements could not be detected in the 40 kb window that the MF analysis had shown as an active replication origin. Using a procedure to partially synchronize *S. acidocaldarius* cells, it was found that the three DNA replication origins fire synchronously, but terminate asynchronously, with replication forks from *oriC2* and *oriC3* meeting first, followed by replication forks from *oriC1* and *oriC2*, and finally *oriC1* and *oriC3* (Lundgren *et al.*, 2004).

In order to more closely map the *S. sofataricus oriC3* 2D N/N agarose gel electrophoresis was performed and indicated that replication initiated from an intergenic region between open reading frames (ORFs) SSO0866 and SSO0867. Found within the intergenic region was a 12 bp inverted repeat

(5′-TTA(C/T)TACTACTA-3′)

that clearly differed from ORB elements identified at *oriC1* (Robinson *et al.*, 2007). Clearly this replication origin differs significantly for the other two *Sulfolobus* spp. chromosomal origins, having non-ORB sequence repeats and not being located proximal to an *orc/cdc6* gene.

Archaeal Orc/Cdc6

Encoded in archaeal genomes are 1–17 genes homologous to eukaryotic origin recognition complex (ORC) proteins 1, 4 and 5 as well as to the eukaryotic replicative helicase recruiter Cdc6 (Giraldo, 2003; Giraldo and Diaz-Orejas, 2001) (Fig. 5.1). Two *orc/cdc6* genes are most commonly found in archaeal genomes (Table 5.2), one usually linked to a chromosomal replication origin (experimentally determined or predicted) and one unlinked (Berquist and DasSarma, 2003). Halophilic Archaea are a group of archaeal organisms where a large expansion of the *orc/cdc6* gene family has occurred (Berquist *et al.*, 2005). Phylogenetic analysis of archaeal Orc/Cdc6 proteins indicated the presence of two major general archaeal clades. One clade shares the most homology to Orc7 from *Halobacterium* sp. NRC-1 and the other shares the most homology to Orc6 from *Halobacterium* sp. NRC-1 (Fig. 5.1). The general archaeal clades, Orc6 and Orc7, contain single members from each haloarchaeon in each clade. All other Archaea have at least one member in one of these clades, with most having one in each clade. Exceptions are *Sulfolobus* spp. which have two members in the Orc7 clade (Fig. 5.1) and three chromosomal replication

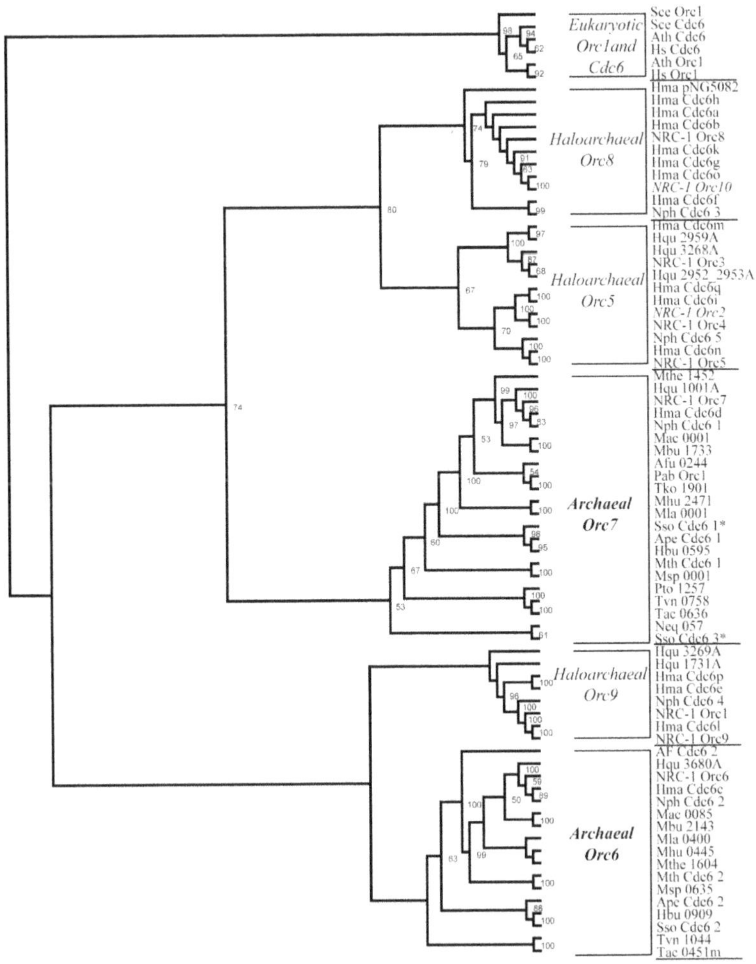

Figure 5.1 Neighbour-joining phylogenetic tree of archaeal Orc/Cdc6 proteins. Orc/Cdc6 protein sequences from *Aeropyrum pernix, Archaeoglobus fulgidis, Haloarcula marismortui, Halobacterium* sp. NRC-1, *Haloquadratum walsbyi, Hyperthermus butylicus, Methanococcoides burtonii, Methanocorpusculum labreanum, Methanosaeta thermophila, Methanosarcina acetivorans, Methanosphaera stadtmanae, Methanothermobacter* thermautotrophicus, *Nanoarchaeum equitans, Natronomonas pharaonis, Picrophilus torridus, Pyrobaculum aerophilum, Pyrococcus abyssi, Sulfolobus solfataricus, Thermococcus kodakarensis, Thermoplasma acidophilum*, and *Thermoplasma volcanium* were downloaded from NCBI and aligned using MCOFFEE. Three letter organism designations are used along with either published protein names or ORF numbers.

origins. In all sequenced haloarchaea five major clades of Orc/Cdc6 proteins have been identified (Berquist *et al.*, 2007). In addition to having a single member in both the Orc7 and Orc6 clades, each haloarchaeon has at least one member in the haloarchaeal specific Orc5, Orc8, and Orc9 clades, though typically multiple members from each organism are found within these clades. In addition to *in silico* analyses of archaeal Orc/Cdc6 proteins, biochemical and genetic experiments have been performed in order to attempt to elucidate the biological roles of these *orc/cdc6* gene products.

Genetic analysis of archaeal *orc/cdc6*

In *Halobacterium* sp. NRC-1 10 *orc/cdc6* genes are found, including *orc6, orc7, orc8,* and *orc10* genes located on the large chromosome, *orc1–5* located on large extrachromosomal replicon pNRC200, and *orc9* located on both the small and large extrachromosmal replicons, pNRC100 and pNRC200 respectively. The gene products are quite diverse in *Halobacterium* sp. NRC-1, ranging from 21% to 91% similarity. The halo-alkaliphilic archaeon, *N. pharaonis,* encodes the fewest number of *orc/cdc6* genes among haloarchaea (five), while *Haloarcula marismortui* encodes the most (seventeen). Each *orc/cdc6* gene in *Halobacterium* sp. NRC-1 was targeted for individual in-frame deletion. Based upon the requirement of *orc7* for minichromosome plasmid replicon autonomous replication in *Halobacterium* sp. NRC-1 (Berquist and DasSarma, 2003) and *orc6* conservation in the genome sequences of other Archaea, the expectation was that these two *orc* genes would have been essential. Neither *orc7* nor *orc6* were found to be essential, nor were *orc3*, *orc4*, *orc5*, *orc8*, or *orc9* since knockouts were readily obtained for those genes. No outward phenotypic consequences of deletion of any of these *orc/cdc6* genes individually were identified. Of ten *orc/cdc6* genes, only two were found to be essential for viability of *Halobacterium* sp. NRC-1; *orc10* encoded on the large chromosome and *orc2* encoded on pNRC200 (Berquist *et al.*, 2007).

Archaeal Orc/Cdc6 interaction with DNA, expression, and post-translational modification

In *P. abyssi*, chromatin immunoprecipitation (ChIP) studies displayed that *in vivo*, the *P. abyssi* Orc1 protein was able to stably associate with the 1 kb DNA replication origin region identified, in both replicating and non-replicating cells, suggesting that in *P. abyssi* origin binding by Orc1 is not the determinant for DNA replication initiation. In addition, immunoblot analysis estimated that ~5000–10,000 molecules of Orc1 were present in exponentially growing *P. abyssi* cells (Matsunaga *et al.*, 2001).

In vitro analysis of an Orc/Cdc6 homologue (AF0244) from the euryarchaeon *Archaeoglobus fulgidus* displayed that this protein, an Orc7 clade member, could bind to a DNA substrate containing a bubble, a forked DNA duplex, and a DNA duplex with a 5′ ssDNA tail. Although these substrates have potentially biologically relevant structures, they did not contain any specific sequence characteristics of typical archaeal DNA replication origins (Grainge *et al.*, 2003).

In the thermophilic methanogenic eurychaeon *Methanothermobacter thermoautotrophicus* more detailed biochemical and biophysical studies were performed on Orc/Cdc6:DNA association. *M. thermoautotrophicus* has two Orc/Cdc6 proteins. Upstream of one (mthOrc/Cdc6-1, Orc7 clade member) a 0.9 kb intergenic region was identified that had a series of symmetrically distributed repeats of 13 bp with the consensus

5′-TTACA(G/C)TTGAA(A/C)N-3′.

This region has been suggested as the likely DNA replication origin (Lopez *et al.*, 1999), although this has not yet been experimentally validated. Using an oligonucleotide substrate containing a single copy of the consensus repeat, *M. thermoautotrophicus* Orc/Cdc6-1 was shown to have a 40-fold preference of binding to the single repeat, while Orc/Cdc6-2 was shown to have some affinity for the single repeat sequence over random DNA. Using an oligonucleotide containing three repeats in direct orientation, Orc/Cdc6-1 was found to bind cooperatively with a 40-fold greater affinity than for the single site. Orc/Cdc6-2 showed a lower affinity for the triple repeat substrate than for the single repeat substrate and did not bind the triple repeat substrate preferentially over a substrate containing random sequence, indicating that Orc/Cdc6-2 from *M. thermoautotrophicus* does not preferentially bind to the putative origin sequence, but that Orc/Cdc6-1 does (Capaldi and Berger, 2004).

Archaeal Orc/Cdc6 proteins, from *M. thermoautotrophicus* (Orc/Cdc6-1 and Orc/Cdc6-2), also have been found to autophosphorylate *in vitro*, using either ATP or dATP. The autophosphorylation of these proteins is inhibited by either ssDNA or dsDNA, although the inhibitory effect of DNA on autophosphorylation is less for Orc/Cdc6-1 compared with OrcCdc6-2, which is completely inhibited (Grabowski and Kelman, 2001).

The situation is more complex in the crenarchaeon *S. solfataricus*, with three Orc/Cdc6 homologues and three DNA replication origins. DNase footprinting showed *S. solfataricus* Orc/Cdc6-1 (SSO0257, Orc7 clade member) to bind to *oriC1* with highest affinity for ORB1, then ORB2, and ORB3. *S. solfataricus* Orc/Cdc6-1 was also found to bind to *oriC2*. At *oriC2*, binding affinity was considerably lower and DNase footprinting revealed that 'mini-ORB' sequences

(5′-CGTTTCATCTGAAAATC (65 bp)
AATTTCATCTGAAACAT-3′)

in inverted orientation flanking an AT-rich stretch of 65 bp were protected. Orc/Cdc6-2 (SSO0771, Orc6 clade member) was also found to bind to both DNA replication origins. At *oriC1*, Orc/Cdc6-2 was found to bind three sites: a region overlapping ORB2, but not ORB1 or ORB3, a region downstream of ORB3 at the likely promoter site for *orc/cdc6-1*, and a region upstream of ORB3. Orc/Cdc6-1 binding was also found to produce DNase I hypersensitivity sites in ORB3. At *oriC2*, Orc/Cdc6-2 binds to two imperfect inverted repeats

(5′-TTTTACAGTAAATTTCT-TACGTTT-3′, 5′-TTATGCAGTAAATT-TCTTGTGATT-3′),

although these repeats were not the sequences recognized at *oriC1*. Orc/Cdc6-3 (SSO2184, Orc7 clade member) did not specifically bind to *oriC1*, but bound two direct repeats at *oriC2*

(5′-GTAAATTTCCTACGTTTC-3′),

one of which overlaps an Orc/Cdc6-2 binding site and one that overlaps an Orc/Cdc6-1 binding site. Using ChIP, *in vivo* association of the Orc/Cdc6 proteins was examined and corroborated the *in vitro* binding data, with all proteins associating with *oriC2* and Orc/Cdc6-1 and Orc/Cdc6-2 associating with *oriC1* (Robinson *et al.*, 2004). With recent identification of *oriC3* DNA binding of the three Orc/Cdc6 proteins at this locus were examined. All three *S. solfataricus* Orc/Cdc6 homologues bound to *oriC3*, though not centered at the repeat sequences identified. Orc/Cdc6-1 was found to bind at the end of the intergenic region proximal to SSO0867 ORF, while Orc/Cdc6-2 and Orc/Cdc6-3 were found to bind regions at both ends of the intergenic region into each flanking ORF. Orc/Cdc6-3 also was noted to induce DNaseI hypersensitivity sites, implying bending of the DNA (Robinson *et al.*, 2007). The three Orc/Cdc6 proteins also were shown to be present in exponential growth phase and levels of each significantly reduced in stationary phase cells, consistent with roles in actively replicating cells (Robinson *et al.*, 2004). In order to examine protein levels in partially synchronized cell cultures, *S. acidocaldarius* was used. *S. acidocaldarius* Orc/Cdc6-1 and Orc/Cdc6-3 showed similar protein profiles, with expression in G_1 and S phase cells, with Orc/Cdc6-2 being present in G_2 phase cells (Robinson *et al.*, 2004).

Archaeal MCM

The putative replicative helicase in Archaea is encoded by one or more genes homologous to the eukaryotic MCM2-7 (minichromosome maintenance) complex. All six subunits of the eukaryotic complex are homologous to one another as well as to MCM proteins in Archaea. Many archaeal genomes encode just a single *mcm* gene, although up to three *mcm* genes have been found encoded in a single archaeal genome (Table 5.2). Much effort has focused on biochemical and structural analyses of archaeal MCM proteins.

A number of biochemical and structural studies have focused on the archaeal MCM complex, which is found in ~200–400 molecules in exponentially growing *P. abyssi* cells (Matsunaga *et al.*, 2003). Initial purification of recombinant MCM from *M. thermoautotrophicus* displayed monomeric and oligomeric forms. The oligomeric MCM complex was at the appropriate weight for a dodecamer and a double hexamer structure could be observed in electron micrograph (EM) pictures. The oligomeric form was found to have ATP independent DNA-binding activity for both ssDNA and dsDNA, DNA dependent ATPase activity, and processive (~500 bp) ATP dependent 3′ to 5′ DNA helicase activity (Chong *et al.*, 2000; Kelman *et al.*, 1999; Shechter *et al.*, 2000). Subsequent analysis indicated that the C4 type zinc finger of the *M. thermoautotrophicus* MCM protein was required for helicase activity and mutating the zinc finger impaired both ssDNA binding and DNA dependent ATPase activity (Poplawski *et al.*, 2001). *M. thermoautotrophicus* MCM was found to be able to displace a strepdavidin:biotin complex from ssDNA and dsDNA, unlike the eukaryotic Mcm4/6/7 complex (Shin *et al.*, 2003b). MCM from *M. thermoautotrophicus* has also been found to be able to unwind DNA bound by archaeal DNA-binding proteins *in vitro*. Histone HMtA2 bound to a SELEX1 derived binding sequence posed no block to unwinding by *M. thermoautotrophicus* MCM, implying that MCM can release bound HMtA2 from DNA. When an archaeal repressor is bound, a reduced rate of

MCM-mediated unwinding initially was found, implying that an archaeal repressor protein may provide an initial impediment to helicase activity by MCM. When an archaeal transcriptional pre-initiation complex (PIC), containing *M. thermoautotrophicus* TBP, TFB, and RNAP was formed on a promoter, MCM was assayed for ability to unwind the bound DNA in both head-on and co-directional orientations of the PIC. As with the repressor bound to DNA, *M. thermoautotrophicus* MCM mediated unwinding appeared to be slightly reduced by the presence of the DNA bound PIC in both orientations (Shin *et al.*, 2007).

MCM from *S. solfataricus* was found, via gel filtration, to form hexamers, possesses DNA helicase activity preferentially on substrates with a 5′ tail, and an ATPase activity weakly stimulated by DNA (Carpentieri *et al.*, 2002). Fluorescence resonance energy transfer (FRET) analysis confirmed the hexameric structure of *S. solfataricus* MCM, as well as identifying the orientation of MCM on a Y-shaped substrate with a free 3′ end. MCM loading onto this DNA substrate occurs with the AAA^+ motor domain oriented closer to dsDNA and the N-terminal domain positioned near the 3′ end of the DNA substrate. Loading of MCM onto this substrate likely occurs via the AAA^+ domain binding the free 3′ ssDNA end and sliding along until dsDNA is reached (McGeoch *et al.*, 2005).

A number of structural studies on *M. thermoautotrophicus* MCM have been performed and yielded conflicting results regarding the nature of the oligomeric form. X-ray crystallization of the N-terminal 286 amino acids, out of 666, revealed a dodecameric architecture (Fletcher *et al.*, 2003); in contrast three separate EM reconstruction studies identified structures of MCM heptamers (Yu *et al.*, 2002), hexamers (Pape *et al.*, 2003), or a heterologous mix of heptamers, octomers, and helical filaments (Chen *et al.*, 2005b) depending on the study examined. However, gel filtration analyses and biochemical assays indicate that a dodecameric/double hexameric form of *M. thermoautotrophicus* is catalytically active.

Genetic analysis of *mcm* in Archaea has been conducted in *Halobacterium* sp. NRC-1. In the distantly related haloarchaeon, *H. marismortui*, three *mcm* genes are encoded, while in *Halobacterium* sp. NRC-1, only a single *mcm* gene is present. Phylogenetic reconstruction of archaeal MCM sequences indicates that the *H. marismortui* MCM protein are monophyletic, indicative of gene duplication events after the divergence of these haloarchaea (Berquist *et al.*, 2005). The single *mcm* gene of *Halobacterium* has been targeted for deletion and was found to be an essential gene, as no deletants could be observed except when a wild-type *mcm* gene was placed *in trans* on a replicating plasmid for *Halobacterium* sp. NRC-1 (Berquist *et al.*, 2007).

Archaeal Orc/Cdc6–MCM interactions

Archaeal Orc/Cdc6 proteins, from *M. thermoautotrophicus* (Orc/Cdc6-1 and Orc/Cdc6-2), have been found to interact with the putative archaeal replicative helicase MCM and modulate its helicase function. Interactions seem to occur via the N-terminal portion of MCM and the winged-helix domain of each Orc/Cdc6 protein. Both *M. thermoautotrophicus* Orc/Cdc6 proteins inhibit the helicase activity of MCM in a concentration dependent manner, although Orc/Cdc6-2 was found to be a better inhibitor of MCM helicase activity overall. Interactions with MCM in *M. thermoautotrophicus* also modulate autophosphorylation of Orc/Cdc6-1 and Orc/Cdc6-2, with Orc/Cdc6-1:MCM interaction stimulating Orc/Cdc6-1 autophosphorylation and Orc/Cdc6-2:MCM interaction inhibiting Orc/Cdc6-2 autophosphorylation (Kasiviswanathan *et al.*, 2005; Shin *et al.*, 2003a). Similar results were obtained for Orc/Cdc6:MCM interactions in the crenarchaeon *S. solfataricus*. All Orc/Cdc6 proteins from this organism were found to interact with MCM and autophosphorylate in the absence of DNA (De Felice *et al.*, 2006).

Work from the thermophilic, acidophilic euryarchaeon, *Thermoplasma acidophilum*, has indicated that an archaeal Orc/Cdc6 protein can in fact stimulate the helicase activity of an archaeal MCM complex. Two Orc/Cdc6 homologues are found in the *T. acidophilum* genome, one an Orc7 clade member and one an Orc6 clade member (Berquist and DasSarma, 2003) (Fig. 5.1). *In vitro* both proteins are found to interact MCM, although the *T. acidophilum* Orc/Cdc6-2 is the only one which has an effect on the helicase activity, stimulating MCM mediated unwinding. *T. acidophilum* Orc/Cdc6-1 was found to have neither inhibitory nor stimulatory activity on MCM (Haugland *et al.*, 2006).

Archaeal primase

For DNA-directed RNA priming in replication, Archaea possess homologues of eukaryotic-type primase small (p48 homologue, PriS) and large (p58 homologue, PriL) subunits (Table 5.2). Also encoded within most archaeal genomes is an ORF containing significant homology to the bacterial-type primase, DnaG, which currently has no known function in Archaea, although in *S. solfataricus* it has been reported to be associated with the archaeal exosome (Evguenieva-Hackenberg *et al.*, 2003).

The eukaryotic-type primase proteins have been investigated *in vitro* from the hyperthermophilic euryarchaea *Pyrococcus* spp. (Bocquier *et al.*, 2001; Liu *et al.*, 2001) and *Methanococcus jannaschii* (Desogus *et al.*, 1999), as well as in the thermophilic crenarchaeon *S. solfataricus*. The small subunit (p48) homologue alone from both euryarchaea has been found to synthesize oligoribonucleotides, but in *P. furiousus*, was found to also preferentially utilize deoxyribonucleotides to synthesize several kilobases of DNA, suggesting that not only primase but also primer extension activities were contained in this single polypeptide (Bocquier *et al.*, 2001).

Further careful bioinformatic analysis of the *P. furiousus* genome sequence identified an ORF immediately 5′ and overlapping the ORF of the PriS homologue which had homology to the eukaryotic p58 primase subunit. Analysis of additional archaeal complete genome sequences identified a single gene for this p58 homologue in *M. thermoautotrophicus*, *M. jannaschii*, *T. acidophilium*, *A. fulgidius*, and *Halobacterium* sp. NRC-1 (Liu *et al.*, 2001). *In vitro* analysis of the large subunit alone displayed no primase or polymerase activity, although DNA-binding activity could be detected. When both proteins for the large and small subunits were incubated together with DNA a single shifted species was observed, which differed from the bands seen with either subunit alone. This suggested that both of the proteins interact and form a complex *in vitro*. The large and small subunit complex together from *P. furiosus* indicated that the full primase complex can synthesize 12- to 40-base-long RNA primers and *de novo* DNA strands fewer than 700 bases in length, both of which could be extended by each replicative-type DNA polymerase from *P. furiosus*. The DNA strands produced by the complex were noticably shorter than those synthesized by the small subunit alone, suggesting that the large subunit can play a regulatory role on catalytic activities of the small subunit as well (Liu *et al.*, 2001). Further biochemical studies have shown that the N-terminus of the large subunit acts to regulate affinity for DNA until L_1S_1 complex has formed, as well as anchoring the large subunit onto the small subunit providing stable primase complex interactions (Matsui *et al.*, 2003).

Archaeal primase complex has been investigated biochemically and structurally from the crenarchaeon *S. solfataricus*. ORFs for the small and large subunits were identified in the complete genome sequence, with interesting gene loci. The gene encoding the small subunit was found to overlap one of three *pcna* genes (*pcnA2*) in the *S. solfataricus* genome and the gene for the large subunit was found to be encoded within 2 kb of a gene for a replicative-type DNA polymerase gene (*polB1*). Yeast 2-hybrid analysis indicated both subunits interact, and gel filtration of co-expressed PriL and PriS was consistant with formation of a heterodimer of 1:1 stoichiometry. The PriSL complex was found to be able to synthesize both RNA and DNA products (DNA up to 7 kb) and an additional 3′ terminal nucleotidyl transferase (TdT) activity was identified (Lao-Sirieix and Bell, 2004).

Although multiple biochemical and structural efforts have been exerted on archaeal primase, only a single genetic study has been conducted. Both the genes for the large and the small subunit have been examined *in vivo* in *Halobacterium* sp. NRC-1. Both genes were targeted for deletion and were both found to be essential for viability. Stastical analysis concluded that both genes had a >99.99% probability of being essential genes and complementation analysis displayed that the gene encoding the PriL subunit (*pri2* in *Halobacterium* sp. NRC-1) could only be deleted if a wild-type copy of the gene was placed *in trans* on a replicating plasmid vector, consistent with the predicted function of the archaeal primase complex at the DNA replication fork (Berquist *et al.*, 2007).

Archaeal GINS complex

Recently homologues of the eukaryotic GINS (go [SLD5], ichi [PSF1], nii [PSF2], san [PSF3]) have been detected and functionally analysed in Archaea. In eukaryotes, GINS is an important component of the replisome interacting with MCM and CDC45. In the crenarchaeon *S. solfataricus*, two genes have been identified, one that shares homology to PSF2 and PSF3 and another that shares homology with SLD5 and PSF1. A yeast two hybrid screen to identify interactors with MCM came up with a novel protein most homologous to eukarotic PSF2 and PSF3, although homology to SLD5 and PSF1 is also noted. The gene encoding this MCM interacting protein is encoded immediately 5′ and overlapping the *mcm* gene and due to homology was named *gins23*. Affinity purification of native Gins23 from *S. solfataricus* led to identification of two additional gene products which co-purified. One of these was found to be encoded immediately 3′ to the gene encoding the small subunit of primase (which is immediately 3′ to a gene for a PCNA homologue), was found to have homology to SLD5 and PMS1, and thus named Gins15. The other was a gene product having homology to a RecJ DNA-binding domain (DBD), thus named RecJdbh. Two-hybrid analysis using the three putative archaeal GINS components indicated that Gins15 interacted with Gins23, RecJdbh, and itself. Gins23 interacted with Gins15, MCM, and both large and small subunits of primase. RecJdbh could only interact with Gins15 (Marinsek *et al.*, 2006). Identification of this archaeal GINS complex provides an additional component likely at the archaeal replication fork linking primase to the replicative helicase. Interestingly, homologues of Gins23 appear only to be found in crenarchaea and *Pyrococcus* spp., while Gins15 homologues are found in both eury- and crenarchaea and are typically linked to the gene encoding the small (catalytic) subunit of primase (Makarova *et al.*, 2005) (Table 5.2), although only a single report of functional analysis of archaeal GINS has been reported to date.

Archaeal replicative-type DNA polymerases: PolB and PolD

In euryarchaea, two putative replicative DNA polymerase families are present, while in crenarchaea there is likely just a single replicative polymerase family, although multiple paragolous genes for this polymerase may be present. Early research uncovered that crenarchaea possess multiple B family DNA polymerase genes (Edgell *et al.*, 1997). It was thought unusual for euryarchaea to possess just a single B family DNA polymerase, as many euryarchaea do (Table 5.2), which led to the discovery of an entirely novel DNA polymerase family (D) in euryarchaea (Ishino and Cann, 1998; Uemori *et al.*, 1997). This novel D family DNA polymerase is a heterodimeric complex of PolD1 (~70 kDa) and PolD2 (~150 kDa) subunits (called PolD herein) (Cann *et al.*, 1998). The PolD2 (large) subunit contains the DNA polymerization active site and is regulated by the PolD1 (small) subunit, which also possesses 3′–5′ proofreading exonuclease activity of the Mre11/calcineurin-type phosphoesterase superfamily. The *in vitro* biochemical characteristics of both B and D family polymerases from *P. abyssi* were investigated. PolB was most active in slightly alkaline pH (8.5–9.0) buffer containing 3 mM $MgCl_2$ and 50–80 mM KCl. PolB also was inhibited by up to 1 mM aphidicolin, similar to other B family DNA polymerases. The PolD heterodimeric complex was most active in slightly acidic pH (6.5) conditions with 15–20 mM $MgCl_2$ and 40 mM KCl. PolD also was not inhibited significantly by 1 mM aphidicolin (Gueguen *et al.*, 2001). Further detailed *in vitro* analysis of the two DNA polymerases from *P. abyssi* indicated some significant differences between the two families. PolD prefers a primed template for DNA binding and uses that primer, regardless of the composition, for extension. PolD contains an individual strand displacement activity and requires PCNA for efficient DNA synthesis, with the interaction dependent on the presence of DNA. PolB is exclusively a DNA primer-directed DNA polymerase. It does not require PCNA for efficient DNA synthesis, but does require PCNA for efficient DNA strand displacement activity (Henneke *et al.*, 2005).

In the only genetic study, both components of the euryarchaeal specific D family DNA polymerase and multiple genes encoding multiple B family DNA polymerases were examined in *Halobacterium* sp. NRC-1. Four genes in *Halobacterium* sp. NRC-1 were targeted for deletion; chromosomally encoded *polA1* and *polA2* (D family small and large subunits respectively), chromosomally encoded *polB1* (B family), and extrachromosomally encoded *polB2* (B family). The extrachromosomal B family DNA polymerase gene was non-essential, while all chromosomally encoded genes were essential. This work was the first genetic examination of replicative-type DNA polymerases in Archaea and displayed that the presence of both B and D polymerases are essential for viability (Berquist *et al.*, 2007).

Archaeal DNA polymerase sliding clamp: PCNA

PCNA (proliferating cell nuclear antigen) in Archaea can be a homotrimeric complex, like found in eukaryotes, or, in some crenarchaea, a heterotrimer. Most Archaea possess a single gene for *pcn*, although the thermophilic crenarchaea, *S. solfataricus* and *A. pernix*, are exceptions with three *pcn* genes each, reminiscent of the eukaryotic 9-1-1 complex (Daimon *et al.*, 2002; Dionne *et al.*, 2003). PCNA acts to tether DNA polymerase to DNA and increase its procesivity. In addition to interaction with DNA polymerases, PCNA has been found to interact with a number of other proteins involved in DNA replication and repair, e.g. flap endonuclease, DNA ligase I, and uracil DNA glycosylase (Dionne *et al.*, 2003).

Archaeal PCNA was initially recognized for functional analysis in the genome sequence of *P. horikoshii*, whose sequence was used to clone the *P. furiosus pcn* gene. The *P. horikoshii* PCNA was expressed and purified and shown to be present in cell extract via immunoblot. Recombinant PCNA was demonstrated to enhance the primer extension activities of both replicative-type DNA polymerases in *P. furiosus*. PCNA could also be co-imunoprecipitated from cell extract with both DNA polymerases, displaying *in vivo* interactions. The *P. horikoshii* PCNA has also been shown to interact with the large subunit of RFC from this organism (Cann *et al.*, 2001). PCNA from another euryarchaeon, *A. fulgidus*, was shown to interact with RFC from this organism, both large and small subunits, in an ATP dependent manner, via pull down assays and gel filtration (Seybert *et al.*, 2002).

The gene for PCNA is known to be essential in eukaryotes (Bauer *et al.*, 1990), and the single gene from *Halobacterium* sp. NRC-1 was shown to be essential as well. Utilizing a targeted gene knockout system in *Halobacterium* sp. NRC-1 zero deletants of *pcn* could be identified, indicating with >99.999 % probability that this gene is essential, suggesting that in general the single *pcn* gene product found in all euryarchaea plays an essential role in DNA replication and likely DNA repair (Berquist *et al.*, 2007).

In some crenarchaea the situation is slightly different, as in *A. pernix* and *Sulfolobus* spp., three *pcn* genes are found. Initial work on PCNA from *A. pernix* indicated that all three gene products could be detected in cell extract, showing that all three genes were expressed, translated, and thus none were pseudogenes. Individually *A. pernix* PCNA1, PCNA2, and PCNA3 were found to stimulate the processivity of two B family DNA polymerases, with PCNA3 stimulation the most dramatic. All three PCNA proteins were found to be able to trimerize individually, although co-immunoprecipitation (co-IP) experiments using both purified recombinant proteins as well as cell extract indicated that PCNA1 and PCNA2 interacted, but that PCNA3 did not interact with either PCNA1 or PCNA2 alone. The implications are that in *A. pernix* PCNA3 likely acts at the replication fork for enhancing the processivity of replicative DNA polymerases, while PCNA1 and PCNA2 may be involved, either as homo- or heterotrimeric complexes, in different cellular processes, e.g. DNA repair (Daimon *et al.*, 2002).

In *S. solfataricus*, a different PCNA scenario was found. Pull down analysis of PCNA1, PCNA2, and PCNA3 indicated that homomeric complexes do not form, in contrast to what was found for *A. pernix* PCNA complexes. Using *in vitro* pull down, gel filtration, surface plasma resonance, and co-immunoprecipation from *S. solfataricus* cell extract, each *pcn* gene product was found to heterotrimerize. The PCNA1-PCNA2 heterodimer was found to be a prerequisite for inclusion of PCNA3 into a heterotrimer and stoichiometric measurements indicated a 1:1:1 ratio. Further examination of *S. solfataricus* PCNA1,2,3 interaction with various DNA replication proteins showed a marked preference for individual subunits for different DNA replication factors. DNA polymerase B1 from *S. solfataricus* interacted strongly with PCNA2 and had weak interaction with PCNA3, although the PCNA1,2,3 heterotrimer was required for stimulation of the polymerase activity. Similarly PCNA1,2,3 heterotrimer was required for stimulation of *S. solfataricus* DNA ligase I activity, but interaction was found to be between LigI and PCNA3. Interaction between *S. solfataricus* flap endonuclease and PCNA1 was identified, but with the heterotrimer required for stimulation of flap endonuclease cleavage. Interestingly, a stable complex of *S. solfataricus* flap endonuclease, PolB1, and LigI could be found when PCNA1,2,3 heterotrimer was present, indicating that PCNA complex can link these three proteins at the replication fork (Dionne *et al.*, 2003).

Archaeal sliding clamp loader: RFC

The structure of PCNA as a trimeric torsidal ring requires the structure to be opened to load onto DNA or for the structure to be closed once on DNA. The sliding clamp loader replication factor C (RFC) complex accomplishes this in Eukarya (5 subunits) and homologues to the eukaryotic p140 (large) and p40 (small) subunits of RFC are found in Archaea. Archaea contain two or three genes for RFC, with one encoding a large subunit homologous to eukaryotic RFC p140 and one or two encoding small subunits homologous to eukaryotic RFC p40 (Table 5.2). Functional analyses thus far has mostly focused on archaeal species containing just a single large RFC gene and a single small RFC gene, with a single study performed on an archaeon encoding three homologues of RFC.

Co-expression of the two *M. thermoautotrophicus* RFC homologues in *E. coli* yielded a likely hexameric complex of two large subunits and four small. The *M. thermoautotrophicus* RFC complex was found to be a DNA stimulated ATPase, as well as being able to hydrolyse dATP, could bind DNA, was efficient at loading PCNA onto DNA. The presence of *M. thermoautotrophicus* RFC and PCNA stimulated PolB DNA synthesis activity. PolB required both RFC and PCNA for DNA synthesis stimulation, as incubation of PolB with RFC caused an inhibition of PolB, likely due to RFC interaction with DNA (Kelman and Hurwitz, 2000).

RFC complex from *P. furiosus* appeared to have a stoichiometry that differered from what was identified in *M. thermoautotrophicus* with one large subunit and four small subunits, although the biochemistry determined was very similar. The gene products were shown to be present *in vivo* via immunoblot analysis, with an intein being exised out of *P. furiosus* RFCS protein. RFCL and RFCS complexed together, as displayed by gel filtration, and that the complex was an ATPase (also dATP) stimulated more by ssDNA or by a substrate that mimicked a primed DNA template. *P. furiosus* RFC also was shown to bind ssDNA and primed DNA template substrates with equal affinities, and RFC when included with PCNA and PolB or PolD, stimulation of DNA synthesis which appeared to be ATP independent (Cann *et al.*, 2001). RFC from the related thermophile *P. abyssi* was found to have two distinct protein complex stoichiometries: either two large four small and or one large two small. ATPase activity of *P. abyssi* RFC was found to be quite robust, DNA independent, and with a marked preference for ATP *vs.* dATP. When *P. abyssi* PCNA was included, a further stimulation in RFC ATPase activity was seen, which now became dependent on DNA. *P. abyssi* RFC also increased PCNA stimlated DNA synthesis by PolD, which was ATP independent (Henneke *et al.*, 2002).

RFC from *A. fulgidus* had a stoichiometry, shown by gel filtration and analytical ultracentrifugation, consistent with one large and four small subunits. *A. fulgidus* RFC had a stimulatory effect on PCNA aided DNA synthesis by PolB. Interestingly, the ATPase of *A. fulgidus* RFC differered from that of other euryarchaea, in that no major differences in DNA substrates, including ssDNA, dsDNA, 5′ and 3′ tailed DNA, primed template mimics, and DNA/RNA hybrids, had an effect on ATP hydrolysis. A general stimulation was also observed when *A. fulgidus* PCNA was included in the reactions, although only on substrates that had a free DNA end. The interaction of RFC and PCNA was found to be dependent on RFC ATP binding, but not hydrolysis (Seybert *et al.*, 2002; Seybert *et al.*, 2006).

Methanosarcina acetivorans, along with *Halobacterium* sp. NRC-1 and *H. marismortui*, possess three genes encoding RFC: one large and two small subunits. Co-expression of *M. acetivorans* RFC in *E. coli* indicated that the three gene products were required to be expressed together to purify a stable complex: RFCS1 and RFCL formed a large protein aggregate when expressed together and RFCS2 and RFCL had no detectable interaction. All three subunits were then examined individually as well as a complex in various biochemical assays. Gel filtration indicated that RFCS1 formed dimers, RFCS2 was monomeric, RFCL appeared to form aggregates or a large molecular weight complex, and the three together formed a complex of ~350 kDa, suggesting a multimeric form whose composition could not be determined due to the increased complexity of two small subunit gene products. RFCS1, RFCL, and the RFC complex could bind ssDNA, although the RFCS2 subunit alone could only bind weakly. Neither small subunits could bind to substrate designed to mimic a primer template complex, although the large subunit and the entire complex could bind. Both small subunits exhibited a DNA (ss or ds) dependent stimulation of ATPase activity. PCNA could also stimulate this activity. RFCL ATPase stimulation could only be accomplished by addition of ssDNA and PCNA. ATPase activity of the RFC complex was highest when substrate mimicking a primer template was included, although ssDNA, dsDNA, and PCNA alone could stimulate ATPase activity to lower levels respectively. Individually all subunits inhibited DNA synthesis by *M. acetivorans* PolB1 with PCNA, but the full complex plus PCNA stimulated DNA synthesis of up to 7 kb. Mutations in ATPase domains of RFCL and RFCS2 did not inhibit ability of a full complex composed of mutant subunits to stimulate PCNA aided DNA synthesis by PolB1, though if the ATPase mutation was in RFCS2 this activity was abolished (Chen *et al.*, 2005a).

RFC has only been investigated in one crenarchaeon, *S. solfataricus*. Two genes were found encoding archaeal RFCL and RFCS and these genes were found to be co-transcribed. Stoichiometry determined by gel filtration was consistant with a single large subunit and four small subunits. The complex was shown to have DNA stimulated ATPase activity, although no effect on ATP hydrolysis was seen with the small subunit alone. *S. solfataricus* RFC could also preferentially bind dsDNA substrates. When individual gene products of PCNA were included with RFC and PolB1 from *S. solfataricus*, DNA synthesis was found to be stimulated; no stimulation could be identified when PCNA gene products were omitted (Pisani *et al.*, 2000).

Archaeal Okazaki fragments and Okazaki fragment maturation proteins: RNaseH and Flap endonuclease

For removal of RNA primers in DNA replication and Okazaki fragment maturation, Archaea typically possess a single type 2 RNaseH gene, although *Halobacterium* sp. NRC-1, *S. tokodaii*, and *Pyrobaculum aerophilum* encode both a type 2 RNaseH as well as a type 1 RNaseH (Ohtani *et al.*, 2004b) (Table 5.2).

Archaeal RNaseH2

The gene for *Thermococcus kodakaraensis* RNaseH2 was initially cloned via complementation of an *E. coli* RNaseH, RecB mutant and the sequence obtained compared with complete genome sequences available at the time (Haruki *et al.*, 1998). Initial biochemical characterization of recombinant archaeal type 2 RNaseH from *T. kodakaraensis*, *A. fulgidus*, and *M. jannaschii* indicated a variety of metal ions (Co^{2+}, Mg^{2+}, Ni^{2+}, and Mn^{2+}) could be used for enzymatic activity (Chai *et al.*, 2001; Haruki *et al.*, 1998; Lai *et al.*, 2003). The *M. jannaschii* RNaseH2 was found to bind Ca^{2+} and Ba^{2+} but could not use these for enzymatic cleavage (Lai *et al.*, 2003). *T. kodakaraensis* RNaseH2 was found to be able to cleave a 12 bp RNA–DNA substrate at multiple sites, but could cleave a 29 bp DNA–RNA–DNA/DNA double-stranded substrate specifically. An RNA substrate and a DNA-RNA-DNA single-strand substrate were not cleaved, consistant with RNaseH not cleaving single-strand substrates. The enzyme preferentially was found to produce a DNA product containing a 5′ monoribonucleotide (Haruki *et al.*, 1998). *A. fulgidus* RNaseH2 was examined using similar substrates and was found to have similar activity to the *T. kokadaraensis* RNaseH2, although in addition it was found to be able to cleave at the junction between RNA–DNA/DNA double-stranded substrate (Chai *et al.*, 2001).

The gene for *P. furiosus* RNaseH2 was identified from the complete genome sequence and biochemical characteristics of recombinant *P. furiosus* RNaseH2 indicated that cleavage was similar to that from *T. kodakaraensis* and *A. fulgidus*, leaving a DNA containing monoribonucleotide from a DNA–RNA–DNA/DNA substrate. In order to assess the fate of the remaining monoribonucleotide, co-incubation of the substrate with RNaseH2 and recombinant flap endonuclease from *P. furiosus* led to release of the monoribonucleotide, displaying that RNaseH2 enzymatic cleavage provides a substrate for flap endonuclease (Sato *et al.*, 2003). The impliations are that RNaseH2 and flap endonuclease can act cooperatively to process Okazaki fragments on the lagging strand during DNA replication.

Archaeal RNaseH1

Interestingly, some Archaea also possess a type 1 RNase H gene. The type 1 RNaseH gene product from *Halobacterium* sp. NRC-1 (VNG0255) was initially used to complement a temperature sensitive *E. coli* RNaseH mutant. Subsequently, the gene product was overexpressed in *E. coli*, purified, and characteristics examined *in vitro*. RNaseH1 showed a divalent metal preference for Mn^{2+} and Mg^{2+}, but could also use Ni^{2+} or Co^{2+} in enzymatic cleavage of DNA–RNA. This enzyme was able to specifically cleave the RNA strand of a DNA/RNA duplex. It also was shown to cleave an Okazaki fragment substrate at the junction between the 3′ side of a ribonucleotide and a 5′ side of a deoxyribonucleotide, suggesting that *in vivo* RNaseH1 could remove the last ribonucleotide of the RNA primer of the Okazaki fragment during DNA replication (Ohtani *et al.*, 2004b).

A type1 RNaseH has also been investigated in the thermoacidophilic crenarchaeon *Sulfolobus tokadii*. Recombinant RNaseH1 from *S. tokadii* could cleave a 12 bp RNA/DNA substrate in the presence of a variety of divalent cations (Mn^{2+}, Mg^{2+}, Co^{2+}, Ni^{2+}) with a preference for Mn^{2+} and Mg^{2+}. RNaseH1 was shown to have similar activities as the *Halobacterium* RNaseH1 on RNA/DNA substrates. In addition, RNaseH1 also was found to be able to cleave dsRNA unlike other RNaseH enzymes examined (Ohtani *et al.*, 2004a).

Archaeal Okazaki fragments

In *P. abyssi* and *S. acidocaldarius*, the characterisics of archaeal Okazaki fragments were examined. RNA containing DNA (Okazaki) fragments were extracted from exponential phase cultures and found to be between 10 and 120 nucleotides in length, similar to what is found in Eukarya, but signifigantly shorter than the sizes found in Bacteria (Bielinsky and Gerbi, 1999). Taking into account replication speed in these two Archaea, 2.2 and 1.6 Okazaki fragments per second are initiated during DNA replication, faster than what has been found in the other two domains of life (Matsunaga *et al.*, 2003). Both the size of RNA containing DNA segments and the rate of Okazaki fragments formation indicate that primer synthesis and Okazaki fragments from Archaea have characteristics more similar to those from Eukarya than to those from Bacteria.

Archaeal single-strand DNA-binding proteins: SSB/RPA

Single-strand DNA-binding proteins of Archaea comprise a group of proteins with wide variety of functional domain composition and characteristics. The characteristics of single-strand DNA-binding proteins are features that divides Archaea into cren- and euryarchaeal lineages. Typically, crenarchaea have a single ORF with a single oligonucleotide-binding (OB) fold, while euryarchaea have varied numbers of ORFs (Table 5.2) with varied numbers of OB folds and additional protein domains (Robbins *et al.*, 2005). In Bacteria only one single-strand DNA-binding protein (SSB) is typically found as a homotetramer, similar to crenarchaea, while in eukaryotes single-strand DNA-binding proteins form a complex termed replication factor/protein A (RFA/RPA) composed of three protein components, RPA70, RPA32, and RPA14, which form a functional ssDNA-binding heterotrimer, similar to euryarchaea.

Euryarchaeal RPA

Initially, homology to eukaryotic RPA70 was identified in the genome of *M. jannaschii*. This gene product was predicted to have four repeats of the ssDNA-binding OB fold, while an ORF in *M. thermoautotrophicum* was found to contain five repeats, and 2 ORFs in *A. fulgidus* were identified which contained two OB folds each. Recombinant RPA70 *M. jannaschii* was examined biochemically and found to purify as a monomer, bind ssDNA with ~100 fold greater affinity than for dsDNA with ~15–20 nt necessary for RPA70 binding (Kelly *et al.*, 1998).

A study of RPA in *P. furiosus* identified two additional genes forming a heterotrimeric RPA complex. Using the OB sequence of the *P. furiosus* RPA70 homologue an additional ORF was identified that shared homology of the OB fold. Interestingly this gene was located directly downstream of the RPA70 gene, with a small intervening ORF between the two. The three genes formed an operon-like structure and were designated RPA70, RPA32, and RPA14. Searches of available archaeal genomes identified a number of additional RPA homologues. The three *P. furiosus* RPA genes were found in the genome of two other *Pyrococcus* spp. and shared synteny. In both *A. fulgidus* and *T. acidophilium*, two RPA 70 homologues were found, one containing a zinc finger and one without, as well as single RPA32 homologues. *Halobacterium* sp. NRC-1 was found to have four RPA70 homologues, two with zinc finger motifs and two without. In addition, two RPA32 homologues were identified, which were linked to the *Halobacterium* sp. NRC-1 genes encoding the zinc finger containing RPA70 homologues. A RPA32 homologue was also identified in *M. jannaschii*, but only RPA70 homologues were found in the crenarchaea, *S. solfataricus* and *A. pernix* (Komori and Ishino, 2001).

The three gene products from *P. furiosus* were recombinately expressed, purified and used in co-IP studies, which demonstrated that the three gene products did indeed form a complex. Gel filtration and co-IP analyses indicated that *P. furiosus* RPA32 interacted with both RPA70 and RPA14 from the same organism to form a heterotrimeric complex; interestingly, RPA70 and RPA14 could not interact directly. Unlike RPA70 from *M. jannaschii*, the *P. furiosus* RPA70 could not bind ssDNA on its own, although the RPA70,32 complex and the RPA70,32,14 full complex could bind ssDNA, with RPA32,14 binding weakly and only at high protein concentrations. The *P. furiosus* RPA was found to stimulate strand exchange with RadA ~3 fold. Co-IP experiments using various antisera against *P. furiosus* DNA replication, recombination, and repair proteins indicated that RPA70 co-precipitated with RadA, a Holliday junction nuclease (Hjc), B and D family DNA polymerases, PCNA, the small subunit of RFC, and the large primase subunit. RadB, an archaeal paralogue of RadA did not co-IP (Komori and Ishino, 2001).

Examination of another euryarchaeon, *M. acetivorans*, identified three RPA70 homologues, two containing a zinc finger domain and two OB fold domains and one without a zinc finger and containing four OB fold domains. Analysis of the recombinant proteins implied that RPA1 from *M. acetivorans* may form multimers in solution and each of the three RPA70 proteins was able to bind ssDNA. The RPA70 homologues containing zinc fingers were shown to bind zinc. In comparison with *M. thermoautotrophicus* RPA, all three *M. acetivorans* RPA70 proteins stimulated DNA synthesis by PolB1 (Robbins *et al.*, 2004).

When RPA was examined in the genome sequences of diverse Archaea, the most commonly found composition was an RPA containing two OB folds and a zinc finger. This type of RPA70 was identified in the genomes of *Halobacterium* sp. NRC-1, *Ferroplasma acidarmanus* (2 homologues), *Thermoplasma* sp., *Methanopyrus kandleri*, and *M. acetivorans*. Recombinant proteins corresponding to this RPA were produced from four euryarchaeal genes from *M. acetivorans*, *M. kandleri*, and *F. acidarmanus* and their characteristics examined. The subunit composition of purified RPA70 from different organisms differed, with a likely dimer composition from *M. acetivorans*, a likely dimer composition for one of the two *F. acidaramanus*

RPA homologues and a monomer subunit composition for the other, and an apparent trimeric composition for RPA from *M. kandleri*. Although RPA from these three organisms all bound ssDNA preferentially, FRET experiments indicated very different binding characteristics of these homologous RPA proteins. *M. acetivorans* and *M. kandleri* RPA homologues seemed to compact ssDNA substrate at low concentrations, but stretch DNA as concentrations increased. Interestingly, *F. acidarmanus* RPA homologues showed two distinct protein:DNA interactions, one compacting ssDNA and the other stretching ssDNA (Robbins *et al.*, 2005).

Archaeal RPA also appears to be involved in the response to DNA damage as well. *Halobacterium* sp. NRC-1 encodes six RPA homologues, four RPA70 and two RPA32. The two RPA32 homologues form operon-like gene structures with two RPA70 genes containing zinc finger motifs, while the other two RPA70 genes appear to be stand alone genes which do not contain zinc finger motifs. A similar arrangement is found in *H. marismortui*, with five RPA70 and three RPA32 homologues, including operon-like gene organization for the three RPA32 genes and the three RPA70 genes containing zinc finger motifs (Berquist *et al.*, 2005). One of the most highly upregulated groups of genes in response to UV irridation in *Halobacterium* sp. NRC-1 was found to be one operon of RPA. In addition a small ORF immediately 3′ to the RPA operon was co-upregulated with the preceeding RPA genes. A homologous ORF was then identified in *H. marismortui* linked to one of the RPA operons, displaying evolutionary conservation of this three gene RPA region. It appears that in haloarchaea, at least one RPA operon produces three gene products, which likely form a functional RPA complex that is involved in the response to DNA damage (UV) (McCready *et al.*, 2005).

Crenarchaeal SSB

Screening for ssDNA-binding activity in *S. solfataricus* identified a single protein composed of ~100 aa and a single OB fold. From the protein sequence an unannotated ORF was identified in the only cenarchaeal complete genome available in 2000, *A. pernix*, which shared extensive sequence homology to the ssDNA-binding protein of *S. solfataricus*. The *S. solfataricus* SSB appeared to act as a monomer in solution, but multiple molecules could bind to ssDNA with ~4–5 nt bound per monomer or ~20–25 nt per tetramer, indicating that ssDNA was required for formation of larger molecular weight SSB complexes. It was also estimated that *S. solfataricus* SSB made up ~2–5% of total cellular protein of *S. solfataricus* (Wadsworth and White, 2001). Another study found that *S. solfataricus* SSB primarily forms tetramers in solution, although dimeric and monomeric forms were also detected. Non-cooperative ssDNA binding was observed with *S. solfataricus* SSB, binding 5 nt per monomer. This SSB was also able to substitute for an *E. coli* temperature sensitive SSB mutant, indicating that it can act to fulfil the requirements for *E. coli* SSB. In addition, the crenarchaeal SSB could promote strand exchange activity of *E. coli* RecA (Haseltine and Kowalczykowski, 2002).

The *S. solfataricus* SSB has been found to be able to melt dsDNA, likely through successive trapping of transient ssDNA regions within dsDNA due to dsDNA helix breathing. A very interesting finding was the SSB could preferentially destabilize dsDNA substrates containing a wide variety of DNA mismatches, modifications, and lesions. SSB effectively melted DNA containing a fluorescein modification, a G/A mismatch, a C/T mismatch, and a cyclobutane pyrimidine dimer, but was not able to destabilize a dsDNA substrate containing no damage or modifications (Cubeddu and White, 2005).

S. solfataricus SSB has been found to interact and stimulate helicase activity of MCM from the same organism (Carpentieri *et al.*, 2002). SSB has also been found to functionally interact with RNAP, via the C-terminal tail of SSB, implying not only a role in DNA replication, but also a role in transcription. Significantly, SSB was found to be able to substitute at promoters for the basal transcription factor TBP, relieve transcriptional inhibition by *Sulfolobus* chromatin protein Alba, and help basal transcription factor (TBP/TFB) assembly on promoters (Richard *et al.*, 2004). *S. solfataricus* SSB has also been found to interact with and stimulate the positive supercoiling activity of reverse gyrase, an enzyme unique to thermophilic Archaea which is responsible for inducing positive supercoiling of DNA.

In addition to stimulation of reverse gyrase activity, SSB was found to co-immuoprecipitate with reverse gyrase in the presence of DNA (Napoli *et al.*, 2005). SSB interacting proteins were also identified via ssDNA:SSB pull down followed by MS analysis of specific protein bands. Identified in this analysis were SSB, RNAP, reverse gyrase, chromatin proteins Alba and Sul7d, Rad50, TopoI, XPB, and four proteins whose functions are not yet known (Cubeddu and White, 2005).

Archaeal DNA ligases

All Archaea examined to date possess at least one DNA ligase homologous to ATP-dependent eukaryotic DNA ligase I, though some halophilic Archaea encode a bacterial type NAD^+- dependent DNA ligase and homologues of eukaryotic DNA ligase III can also be found in select archaeal genomes. DNA ligase III homologues have not been examined to date, although it is presumed they will act in DNA repair like their known counterparts in eukaryotes. DNA ligase I homologues are the primary candidates for the main replicative DNA ligase, although in some organisms they appear to be the only ligase present and presumably work in DNA repair as well. NAD^+-dependent DNA ligases are the primary ligase involved in DNA replication in Bacteria but their exact function and importance in select haloarchaeal genomes is still unclear.

Archaeal ATP-dependent DNA Ligase: LigI

The first characterization of an archaeal DNA ligase was from *M. thermoautotrophicus*. Recombinant LigI from this species was found to be active in sealing nicked DNA as a monomer. This activity required divalent cations, preferentially Mg^{2+}, Mn^{2+}, or Co^{2+}, and ATP or dATP. This LigI was active at slightly alkaline pH and had optimal activity at a temperature of 60°C (Sriskanda *et al.*, 2000).

Subsequently an archaeal DNA ligase from *T. kodakaraensis* was cloned, and phylogenetically and biochemically characterized. Phylogenetic reconstruction displayed that *T. kodakaraensis* LigI grouped with other ATP dependent DNA ligases from Archaea, Eukarya, and phage, while being phylogenetically distinct from NAD^+-dependent DNA ligases found in Bacteria. Production and analysis of this archaeal recombinant LigI indicated a monomeric form, with optimal ligation activity at slightly alkaline pH, in the presence of ATP, Mg^{2+} or K^+. At low enzyme:substrate ratios, the temperature optimum found was 65°C, but with an increase in the ratio the temperature optimum shifted to 70–80°C, with residual activity observed at tempertures upwards of 100°C. An additional surprise was the fact that NAD^+ could substitute for ATP in ligation reactions, although the enzymatic activity with NAD^+ was found to be lower than that for ATP (Nakatani *et al.*, 2000). A similar result was found with a ligase from a related *Thermococcus* sp. (Rolland *et al.*, 2004). Examination of the substrate specificity of *T. kodakaraensis* LigI indicated that proper base-pairing at the 3′ side of a nick was required for efficient ligation, with mismatches at the 5′ side tolerated (Nakatani *et al.*, 2002). DNA ligase from the highly related organism *P. horikoshii* was found to use only ATP, not ADP or NAD^+ and a variety of divalent cations for enzymatic activity (Keppetipola and Shuman, 2005).

ATP dependent DNA ligases from crenarchaea have been examined from *Sulfolobus shibatae* and *A. pernix*. Ligase I from *S. shibatae* shared many biochemical properties with proteins from both *M. thermoautotrophicus* and *Thermococcus* sp., although NAD^+ could not be used as a cofactor, only ATP, similar to DNA ligase I from *P. horikoshii* (Lai *et al.*, 2002). DNA ligase I from *A. pernix* was found to be Mg^{2+} or Mn^{2+} dependent and could use ATP as well as ADP as cofactors for ligation.

Archaeal NAD^+-dependent DNA Ligase: LigN

Intriguingly, the first bacterial-type NAD^+-dependent DNA ligase found outside of the bacterial domain of life in a free living organism was identified in the genome sequence of *H. marismortui* in 2005 (Berquist *et al.*, 2005). Maximum-likelihood phylogenetic analysis of ATP-dependent DNA ligase sequences from Archaea and NAD^+-dependent DNA ligases from Bacteria and *H. marismortui* clearly displayed the NAD^+-dependent DNA ligase sequence from *H. marismortui* to be distinct from archaeal DNA ligases and quite similar to the NAD^+-dependent ligases found in the bacterial domain of life. In addition to this DNA ligase, *H. marismortui* was found to encode a typical archaeal-type ATP dependent DNA ligase (LigI), as well as a homologue of eukaryotic DNA ligase III. In 2006, *Haloferax volcanii*, a closely related relative of *H. marismortui*, was found to have both ATP and NAD^+ dependent DNA ligases. The genes encoding these two DNA ligases were examined by gene disruption analysis. Interestingly, individually each DNA ligase gene was found not to be essential for *H. volcanii*, but no double gene disruption strains were viable, indicating that an essential function, likely in the ligation of Okazaki fragments during DNA replication, was shared by the gene products of the archaeal-type ATP and bacterial-type NAD^+-dependent DNA ligase genes (Zhao *et al.*, 2006). Biochemical characterization of the recombinant NAD^+-dependent DNA ligase from *H. volcanii* (LigN) verified that this gene product is a NAD^+-dependent DNA ligase capable of adenlyation using NADP but not ATP. Differing from the other archaeal DNA ligases examined this ligase appears to be of bacterial origin, as for the putative *H. marismortui* NAD^+-dependent DNA ligase sequence, and has a high salt optimum (3.2 M KCl), owing to the hypersaline environment and cytoplasm of the host organism, *H. volcanii*. As with the ATP/ADP dependent DNA ligases, a variety of divalent cations could be used for ligation reactions, with the highest activity seen with Mg^{2+} and Mn^{2+} (Poidevin and MacNeill, 2006).

Conclusions and future prospects

Many advances in the field of archaeal DNA replication have been made since the first archaeal complete genome sequence was published (Bult *et al.*, 1996; Fraser *et al.* 2004). The genome sequencing and genomics era have provided a wealth of information and unparalled opportunities to gain insights into functional processes in the third domain of life. Coupled to the vast amount of sequence information, biochemical, structural, and genetic studies have begun to define essential elements of a fundamental process: DNA replication. Like vast differences found in different groups of bacterial and eukaryotic organisms, differences have been found in specific mechanisms and components of DNA replication initiation and elongation in archaeal organisms. Perhaps most interestingly, elements common to all Archaea have begun to be defined.

In the emerging model, archaeal typical chromosomal DNA replication origins are composed of fairly large DNA sequence repeats, typically inverted, in an intergenic region proximal to an *orc/cdc6* gene which falls into the Orc7 clade of archaeal Orc/Cdc6 homologues. These large repeats likely serve as binding sequences for the origin-binding protein(s), Orc7 clade members. Binding of origin recognition protein(s) would lead to local DNA helix destabilization of AT-rich sequences proximal to the origin repeats, allowing for recruitment of the MCM helicase complex, potentially by an Orc6 clade Orc/Cdc6 protein. Local DNA helix opening and MCM recruitment then allows association of other replisome components. MCM has been found to interact with GINS which in turn interacts with primase, linking it to the replication fork. Once primase lays down initial RNA primers at the origin, a substrate for PCNA binding is formed. PCNA is then able to load onto the primed template, via action of RFC. PCNA in turn tethers the replicative DNA polymerase(s) to the DNA, PolD and PolB in euryarchaea and PolB only in crenarchaea. With DNA synthesis leading and lagging strands of DNA replication are formed as are Okazaki fragments, substrates for RNase H2, and RNaseH1 in some Archaea, and flap endonuclease. RNA primer removal produces substrates for DNA ligase which acts to seal nicks in DNA. For most Archaea, the ATP dependent DNA ligase will act at the replication fork, although in some haloarchaea a NAD^+-dependent DNA ligase may also serve this function. During DNA replication elongation, PolB may act as the leading strand DNA polymerase and PolD may act as the lagging strand DNA polymerase for processive and faithful duplication of the genome in euryarchaea. In crenarchaea, it is possible the multiple B family DNA polymerases may be present at the replication fork and may have specialized roles in leading and/or lagging strand synthesis.

There is significant work remaining to be continued in the DNA replication field in Archaea. Many unanswered questions remain relating to organismic differences in the DNA replication process, novel components of the DNA replication machinery, post-translational modifications and their effects, and elements of the archaeal cell cycle. As more sophisticated genetic tools become available (Berquist *et al.* 2006; Sowers *et al.* 2007) more *in vivo* studies will provide complementary information to the pleothora of *in vitro* biochemical studies performed and ongoing, leading to further refined models of DNA replication in Archaea.

References

Bauer G.A., and Burgers, P.M. (1990). Molecular cloning, structure and expression of the yeast proliferating cell nuclear antigen gene. Nucleic Acids Res. 18:261–265.

Bell, S.P., and Dutta, A. (2002). DNA replication in eukaryotic cells. Annu Rev. Biochem. *71*, 333–374.

Berquist, B.R., DasSarma, P., and DasSarma, S. (2007). Essential and non-essential DNA replication genes in the model halophilic archaeon, Halobacterium sp. NRC-1.

Berquist, B.R. J. Soneja, and DasSarma, S. (2005). Comparative genomic survey of information transfer systems in two diverse extremely halophilic Archaea, Halobacterium sp. strain NRC-1 and Haloarcula marismortui., In Adapatation to life at high salt concentrations in Archaea, Bacteria, and Eukarya, N. Gunde-Cimerman, A. Oren, and A. Plemenitas, eds. (Dordrecht, Netherlands: Springer), pp. 148–182.

Berquist, B.R., J.A. Müller, and S. DasSarma. 2005. Chapter 27. Genetic Systems for Halophilic Archaea. In Oren, A. and Rainey, F. (eds.), *Methods in Microbiology, Vol. 35*, Elsevier/Academic Press, pp. 649–680.

Berquist, B.R., and DasSarma, S. (2003). An archaeal chromosomal autonomously replicating sequence element from an extreme halophile, Halobacterium sp. strain NRC-1. J. Bacteriol. *185*, 5959–5966.

Bielinsky, A.K., and Gerbi, S.A. (1999). Chromosomal ARS1 has a single leading strand start site. Mol. Cell *3*, 477–486.

Bocquier, A.A., Liu, L., Cann, I.K., Komori, K., Kohda, D., and Ishino, Y. (2001). Archaeal primase: bridging the gap between RNA and DNA polymerases. Curr. Biol. *11*, 452–456.

Brewer, B.J., and Fangman, W.L. (1987). The localization of replication origins on ARS plasmids in S. cerevisiae. Cell *51*, 463–471.

Bult, C.J., White, O., Olsen, G.J., Zhou, L., Fleischmann, R.D., Sutton, G.G., Blake, J.A., FitzGerald, L.M., Clayton, R.A., Gocayne, J.D., *et al.* (1996). Complete genome sequence of the methanogenic archaeon, *Methanococcus jannaschii*. Science *273*, 1058–1073.

Cann, I.K., Ishino, S., Yuasa, M., Daiyasu, H., Toh, H., and Ishino, Y. (2001). Biochemical analysis of replication factor C from the hyperthermophilic archaeon *Pyrococcus furiosus*. J. Bacteriol. *183*, 2614–2623.

Cann, I.K., Komori, K., Toh, H., Kanai, S., and Ishino, Y. (1998). A heterodimeric DNA polymerase: evidence that members of Euryarchaeota possess a distinct DNA polymerase. Proc. Natl. Acad. Sci. USA *95*, 14250–14255.

Capaldi, S.A., and Berger, J.M. (2004). Biochemical characterization of Cdc6/Orc1 binding to the replication origin of the euryarchaeon Methanothermobacter thermoautotrophicus. Nucleic Acids Res. *32*, 4821–4832.

Carpentieri, F., De Felice, M., De Falco, M., Rossi, M., and Pisani, F.M. (2002). Physical and functional interaction between the mini-chromosome maintenance-like DNA helicase and the single-stranded DNA binding protein from the crenarchaeon *Sulfolobus solfataricus*. J. Biol. Chem. *277*, 12118–12127.

Chai, Q., Qiu, J., Chapados, B.R., and Shen, B. (2001). *Archaeoglobus fulgidus* RNase HII in DNA replication: enzymological functions and activity regulation via metal cofactors. Biochem. Biophys. Res. Commun. *286*, 1073–1081.

Chen, Y.H., Kocherginskaya, S.A., Lin, Y., Sriratana, B., Lagunas, A.M., Robbins, J.B., Mackie, R.I., and Cann, I.K. (2005a). Biochemical and mutational analyses of a unique clamp loader complex in the archaeon *Methanosarcina acetivorans*. J. Biol. Chem. *280*, 41852–41863.

Chen, Y.J., Yu, X., Kasiviswanathan, R., Shin, J.H., Kelman, Z., and Egelman, E.H. (2005b). Structural polymorphism of Methanothermobacter thermautotrophicus MCM. J. Mol. Biol. *346*, 389–394.

Chong, J.P., Hayashi, M.K., Simon, M.N., Xu, R.M., and Stillman, B. (2000). A double-hexamer archaeal minichromosome maintenance protein is an ATP-dependent DNA helicase. Proc. Natl. Acad. Sci. USA *97*, 1530–1535.

Clyne, R.K., and Kelly, T.J. (1995). Genetic analysis of an ARS element from the fission yeast Schizosaccharomyces pombe. Embo J. *14*, 6348–6357.

Cubeddu, L., and White, M.F. (2005). DNA damage detection by an archaeal single-stranded DNA-binding protein. J. Mol. Biol. *353*, 507–516.

Dai, J., Chuang, R.Y., and Kelly, T.J. (2005). DNA replication origins in the Schizosaccharomyces pombe genome. Proc. Natl. Acad. Sci. USA *102*, 337–342.

Daimon, K., Kawarabayasi, Y., Kikuchi, H., Sako, Y., and Ishino, Y. (2002). Three proliferating cell nuclear antigen-like proteins found in the hyperthermophilic archaeon Aeropyrum pernix: interactions with the two DNA polymerases. J. Bacteriol. *184*, 687–694.

De Felice, M., Esposito, L., Rossi, M., and Pisani, F.M. (2006). Biochemical characterization of two Cdc6/ORC1-like proteins from the crenarchaeon *Sulfolobus solfataricus*. Extremophiles *10*, 61–70.

Desogus, G., Onesti, S., Brick, P., Rossi, M., and Pisani, F.M. (1999). Identification and characterization of a DNA primase from the hyperthermophilic archaeon *Methanococcus jannaschii*. Nucleic Acids Res. *27*, 4444–4450.

Dionne, I., Nookala, R.K., Jackson, S.P., Doherty, A.J., and Bell, S.D. (2003). A heterotrimeric PCNA in the hyperthermophilic archaeon *Sulfolobus solfataricus*. Mol. Cell *11*, 275–282.

Edgell, D.R., Klenk, H.P., and Doolittle, W.F. (1997). Gene duplications in evolution of archaeal family B DNA polymerases. J. Bacteriol. *179*, 2632–2640.

Evguenieva-Hackenberg, E., Walter, P., Hochleitner, E., Lottspeich, F., and Klug, G. (2003). An exosome-like complex in *Sulfolobus solfataricus*. EMBO Rep. *4*, 889–893.

Fletcher, R.J., Bishop, B.E., Leon, R.P., Sclafani, R.A., Ogata, C.M., and Chen, X.S. (2003). The structure and function of MCM from archaeal *M. Thermoautotrophicum*. Nat. Struct Biol. *10*, 160–167.

Giraldo, R. (2003). Common domains in the initiators of DNA replication in Bacteria, Archaea and Eukarya: combined structural, functional and phylogenetic perspectives. FEMS Microbiol. Rev. *26*, 533–554.

Giraldo, R., and Diaz-Orejas, R. (2001). Similarities between the DNA replication initiators of Gram-negative Bacteria plasmids (RepA) and eukaryotes (Orc4p)/Archaea (Cdc6p). Proc. Natl. Acad. Sci. USA *98*, 4938–4943.

Grabowski, B., and Kelman, Z. (2001). Autophosphorylation of archaeal Cdc6 homologues is regulated by DNA. J. Bacteriol. *183*, 5459–5464.

Grainge, I., Scaife, S., and Wigley, D.B. (2003). Biochemical analysis of components of the pre-replication complex of *Archaeoglobus fulgidus*. Nucleic Acids Res. *31*, 4888–4898.

Grigoriev, A. (1998). Analyzing genomes with cumulative skew diagrams. Nucleic Acids Res. *26*, 2286-2290.

Gueguen, Y., Rolland, J.L., Lecompte, O., Azam, P., Le Romancer, G., Flament, D., Raffin, J.P., and Dietrich, J. (2001). Characterization of two DNA polymerases from the hyperthermophilic euryarchaeon *Pyrococcus abyssi*. Eur. J. Biochem. *268*, 5961–5969.

Haruki, M., Hayashi, K., Kochi, T., Muroya, A., Koga, Y., Morikawa, M., Imanaka, T., and Kanaya, S. (1998). Gene cloning and characterization of recombinant RNase HII from a hyperthermophilic archaeon. J. Bacteriol. *180*, 6207–6214.

Haseltine, C.A., and Kowalczykowski, S.C. (2002). A distinctive single-strand DNA-binding protein from the Archaeon *Sulfolobus solfataricus*. Mol. Microbiol. *43*, 1505–1515.

Haugland, G.T., Shin, J.H., Birkeland, N.K., and Kelman, Z. (2006). Stimulation of MCM helicase activity by a Cdc6 protein in the archaeon *Thermoplasma acidophilum*. Nucleic Acids Res. *34*, 6337–6344.

Henneke, G., Flament, D., Hubscher, U., Querellou, J., and Raffin, J.P. (2005). The hyperthermophilic euryarchaeota *Pyrococcus abyssi* likely requires the two DNA polymerases D and B for DNA replication. J. Mol. Biol. *350*, 53–64.

Henneke, G., Gueguen, Y., Flament, D., Azam, P., Querellou, J., Dietrich, J., Hubscher, U., and Raffin, J.P. (2002). Replication factor C from the hyperthermophilic archaeon *Pyrococcus abyssi* does not need ATP hydrolysis for clamp-loading and contains a functionally conserved RFC PCNA-binding domain. J. Mol. Biol. *323*, 795–810.

Ishino, Y., and Cann, I.K. (1998). The euryarchaeotes, a subdomain of Archaea, survive on a single DNA polymerase: fact or farce? Genes Genet. Syst *73*, 323–336.

Kasiviswanathan, R., Shin, J.H., and Kelman, Z. (2005). Interactions between the archaeal Cdc6 and MCM proteins modulate their biochemical properties. Nucleic Acids Res. *33*, 4940–4950.

Kelly, T.J., Simancek, P., and Brush, G.S. (1998). Identification and characterization of a single-stranded DNA-binding protein from the archaeon *Methanococcus jannaschii*. Proc. Natl. Acad. Sci. USA *95*, 14634–14639.

Kelman, Z., and Hurwitz, J. (2000). A unique organization of the protein subunits of the DNA polymerase clamp loader in the archaeon *Methanobacterium thermoautotrophicum* deltaH. J. Biol. Chem. *275*, 7327–7336.

Kelman, Z., Lee, J.K., and Hurwitz, J. (1999). The single minichromosome maintenance protein of *Methanobacterium thermoautotrophicum* DeltaH contains DNA helicase activity. Proc. Natl. Acad. Sci. USA *96*, 14783–14788.

Kennedy, S.P., Ng, W.V., Salzberg, S.L., Hood, L., and DasSarma, S. (2001). Understanding the adaptation of Halobacterium species NRC-1 to its extreme environment through computational analysis of its genome sequence. Genome Res. *11*, 1641–1650.

Keppetipola, N., and Shuman, S. (2005). Characterization of a thermophilic ATP-dependent DNA ligase from the euryarchaeon *Pyrococcus horikoshii*. J. Bacteriol. *187*, 6902–6908.

Komori, K., and Ishino, Y. (2001). Replication protein A in *Pyrococcus furiosus* is involved in homologous DNA recombination. J. Biol. Chem. *276*, 25654–25660.

Lai, B., Li, Y., Cao, A., and Lai, L. (2003). Metal ion binding and enzymatic mechanism of *Methanococcus jannaschii* RNase HII. Biochemistry *42*, 785–791.

Lai, X., Shao, H., Hao, F., and Huang, L. (2002). Biochemical characterization of an ATP-dependent DNA ligase from the hyperthermophilic crenarchaeon Sulfolobus shibatae. Extremophiles *6*, 469–477.

Lao-Sirieix, S.H., and Bell, S.D. (2004). The heterodimeric primase of the hyperthermophilic archaeon *Sulfolobus solfataricus* possesses DNA and RNA primase, polymerase and 3′-terminal nucleotidyl transferase activities. J. Mol. Biol. *344*, 1251–1263.

Lee, J.K., and Hurwitz, J. (2000). Isolation and characterization of various complexes of the minichromosome maintenance proteins of *Schizosaccharomyces pombe*. J. Biol. Chem. *275*, 18871–18878.

Leipe, D.D., Aravind, L., and Koonin, E.V. (1999). Did DNA replication evolve twice independently? Nucleic Acids Res. *27*, 3389–3401.

Liu, L., Komori, K., Ishino, S., Bocquier, A.A., Cann, I.K., Kohda, D., and Ishino, Y. (2001). The archaeal DNA primase: biochemical characterization of the p41-p46 complex from *Pyrococcus furiosus*. J. Biol. Chem. *276*, 45484–45490.

Lopez, P., Philippe, H., Myllykallio, H., and Forterre, P. (1999). Identification of putative chromosomal origins of replication in Archaea. Mol. Microbiol. 32, 883–886.

Lundgren, M., Andersson, A., Chen, L., Nilsson, P., and Bernander, R. (2004). Three replication origins in Sulfolobus species: synchronous initiation of chromosome replication and asynchronous termination. Proc. Natl. Acad. Sci. USA *101*, 7046–7051.

Majernik AI, Jenkinson ER, Chong JP. DNA replication in thermophiles. Biochem. Soc. Trans. 2004 Apr;32:236-239.

Makarova, K.S., Wolf, Y.I., Mekhedov, S.L., Mirkin, B.G., and Koonin, E.V. (2005). Ancestral paralogs and pseudoparalogs and their role in the emergence of the eukaryotic cell. Nucleic Acids Res. *33*, 4626–4638.

Marinsek, N., Barry, E.R., Makarova, K.S., Dionne, I., Koonin, E.V., and Bell, S.D. (2006). GINS, a central nexus in the archaeal DNA replication fork. EMBO Rep. *7*, 539–545.

Matsui, E., Nishio, M., Yokoyama, H., Harata, K., Darnis, S., and Matsui, I. (2003). Distinct domain functions regulating de novo DNA synthesis of thermostable DNA primase from hyperthermophile *Pyrococcus horikoshii*. Biochemistry *42*, 14968–14976.

Matsunaga, F., Forterre, P., Ishino, Y., and Myllykallio, H. (2001). In vivo interactions of archaeal Cdc6/Orc1 and minichromosome maintenance proteins with the replication origin. Proc. Natl. Acad. Sci. USA *98*, 11152–11157.

Matsunaga, F., Norais, C., Forterre, P., and Myllykallio, H. (2003). Identification of short 'eukaryotic' Okazaki fragments synthesized from a prokaryotic replication origin. EMBO Rep. *4*, 154–158.

McCready, S., Muller, J.A., Boubriak, I., Berquist, B.R., Ng, W.L., and Dassarma, S. (2005). UV irradiation induces homologous recombination genes in the model archaeon, Halobacterium sp. NRC-1. Saline Systems *1*, 3.

McGeoch, A.T., Trakselis, M.A., Laskey, R.A., and Bell, S.D. (2005). Organization of the archaeal MCM complex on DNA and implications for the helicase mechanism. Nat. Struct Mol. Biol. *12*, 756–762.

McLean, M.J., Wolfe, K.H., and Devine, K.M. (1998). Base composition skews, replication orientation, and gene orientation in 12 prokaryote genomes. J. Mol. Evol. *47*, 691–696.

Meselson, M., and Stahl, F.W. (1958). The replication of DNA. Cold Spring Harb. Symp. Quant. Biol. *23*, 9–12.

Messer, W. (2002). The bacterial replication initiator DnaA. DnaA and oriC, the bacterial mode to initiate DNA replication. FEMS Microbiol. Rev. *26*, 355–374.

Myllykallio, H., Lopez, P., Lopez-Garcia, P., Heilig, R., Saurin, W., Zivanovic, Y., Philippe, H., and Forterre, P. (2000). Bacterial mode of replication with eukaryotic-like machinery in a hyperthermophilic archaeon. Science *288*, 2212–2215.

Nakatani, M., Ezaki, S., Atomi, H., and Imanaka, T. (2000). A DNA ligase from a hyperthermophilic archaeon with unique cofactor specificity. J. Bacteriol. *182*, 6424–6433.

Nakatani, M., Ezaki, S., Atomi, H., and Imanaka, T. (2002). Substrate recognition and fidelity of strand joining by an archaeal DNA ligase. Eur. J. Biochem. *269*, 650–656.

Napoli, A., Valenti, A., Salerno, V., Nadal, M., Garnier, F., Rossi, M., and Ciaramella, M. (2005). Functional interaction of reverse gyrase with single-strand binding protein of the archaeon *Sulfolobus*. Nucleic Acids Res. *33*, 564–576.

Neuwald, A.F., Aravind, L., Spouge, J.L., and Koonin, E.V. (1999). AAA+: A class of chaperone-like ATPases associated with the assembly, operation, and disassembly of protein complexes. Genome Res. *9*, 27–43.

Ohtani, N., Yanagawa, H., Tomita, M., and Itaya, M. (2004a). Cleavage of double-stranded RNA by RNase HI from a thermoacidophilic archaeon, *Sulfolobus tokodaii* 7. Nucleic Acids Res. *32*, 5809–5819.

Ohtani, N., Yanagawa, H., Tomita, M., and Itaya, M. (2004b). Identification of the first archaeal Type 1 RNase H gene from *Halobacterium* sp. NRC-1: archaeal RNase HI can cleave an RNA-DNA junction. Biochem. J. *381*, 795–802.

Pape, T., Meka, H., Chen, S., Vicentini, G., van Heel, M., and Onesti, S. (2003). Hexameric ring structure of the full-length archaeal MCM protein complex. EMBO Rep. *4*, 1079–1083.

Pisani, F.M., De Felice, M., Carpentieri, F., and Rossi, M. (2000). Biochemical characterization of a clamp-loader complex homologous to eukaryotic replication factor C from the hyperthermophilic archaeon *Sulfolobus solfataricus*. J. Mol. Biol. *301*, 61–73.

Poidevin, L., and MacNeill, S.A. (2006). Biochemical characterisation of LigN, an NAD+-dependent DNA ligase from the halophilic euryarchaeon *Haloferax volcanii* that displays maximal in vitro activity at high salt concentrations. BMC Mol. Biol. *7*, 44.

Poplawski, A., Grabowski, B., Long, S.E., and Kelman, Z. (2001). The zinc finger domain of the archaeal minichromosome maintenance protein is required for helicase activity. J. Biol. Chem. *276*, 49371–49377.

Richard, D.J., Bell, S.D., and White, M.F. (2004). Physical and functional interaction of the archaeal single-stranded DNA-binding protein SSB with RNA polymerase. Nucleic Acids Res. *32*, 1065–1074.

Robbins, J.B., McKinney, M.C., Guzman, C.E., Sriratana, B., Fitz-Gibbon, S., Ha, T., and Cann, I.K. (2005). The euryarchaeota, nature's medium for engineering of single-stranded DNA-binding proteins. J. Biol. Chem. *280*, 15325–15339.

Robbins, J.B., Murphy, M.C., White, B.A., Mackie, R.I., Ha, T., and Cann, I.K. (2004). Functional analysis of multiple single-stranded DNA-binding proteins from *Methanosarcina acetivorans* and their effects on DNA synthesis by DNA polymerase BI. J. Biol. Chem. *279*, 6315–6326.

Robinson, N.P. and Bell, S.D. (2007). Extrachromosomal element capture and the evolution of multiple replication origins in archaeal chromosomes. PNAS *104*, 5806–5811

Robinson, N.P., Blood, K.A., McCallum, S.A., Edwards, P.A., and Bell, S.D. (2007). Sister chromatid junctions in the hyperthermophilic archaeon *Sulfolobus solfataricus*. Embo J. *26*, 816–824.

Robinson, N.P., Dionne, I., Lundgren, M., Marsh, V.L., Bernander, R., and Bell, S.D. (2004). Identification of two origins of replication in the single chromosome of the archaeon *Sulfolobus solfataricus*. Cell *116*, 25–38.

Rolland, J.L., Gueguen, Y., Persillon, C., Masson, J.M., and Dietrich, J. (2004). Characterization of a thermophilic DNA ligase from the archaeon *Thermococcus fumicolans*. FEMS Microbiol. Lett. *236*, 267–273.

Sato, A., Kanai, A., Itaya, M., and Tomita, M. (2003). Cooperative regulation for Okazaki fragment processing by RNase HII and FEN-1 purified from a hyperthermophilic archaeon, *Pyrococcus furiosus*. Biochem. Biophys. Res. Commun. *309*, 247–252.

Seybert, A., Scott, D.J., Scaife, S., Singleton, M.R., and Wigley, D.B. (2002). Biochemical characterisation of the clamp/clamp loader proteins from the euryarchaeon *Archaeoglobus fulgidus*. Nucleic Acids Res. *30*, 4329–4338.

Seybert, A., Singleton, M.R., Cook, N., Hall, D.R., and Wigley, D.B. (2006). Communication between subunits within an archaeal clamp-loader complex. Embo J. *25*, 2209–2218.

Shechter, D.F., Ying, C.Y., and Gautier, J. (2000). The intrinsic DNA helicase activity of *Methanobacterium thermoautotrophicum* delta H minichromosome maintenance protein. J. Biol. Chem. *275*, 15049–15059.

Shin, J.H., Grabowski, B., Kasiviswanathan, R., Bell, S.D., and Kelman, Z. (2003a). Regulation of minichromosome maintenance helicase activity by Cdc6. J. Biol. Chem. *278*, 38059–38067.

Shin, J.H., Jiang, Y., Grabowski, B., Hurwitz, J., and Kelman, Z. (2003b). Substrate requirements for duplex DNA translocation by the eukaryal and archaeal minichromosome maintenance helicases. J. Biol. Chem. *278*, 49053–49062.

Shin, J.H., Santangelo, T.J., Xie, Y., Reeve, J.N., and Kelman, Z. (2007). Archaeal minichromosome maintenance (MCM) helicase can unwind DNA bound by archaeal histones and transcription factors. J. Biol. Chem. *282*, 4908–4915.

Sowers, K.R., P. Blum, and S. DasSarma. 2007. Genetics of Archaea. *In* C.A. Reddy, T.J. Beveridge, J.A. Breznak, G.A. Marzluf, and T.M. Schmidt (ed.), *Methods for General and Molecular Microbiology*. American Society for Microbiology, Washington, D.C. (in press).

Sriskanda, V., Kelman, Z., Hurwitz, J., and Shuman, S. (2000). Characterization of an ATP-dependent DNA ligase from the thermophilic archaeon *Methanobacterium thermoautotrophicum*. Nucleic Acids Res. *28*, 2221–2228.

Uemori, T., Sato, Y., Kato, I., Doi, H., and Ishino, Y. (1997). A novel DNA polymerase in the hyperthermophilic archaeon, *Pyrococcus furiosus*: gene cloning, expression, and characterization. Genes Cells *2*, 499–512.

Wadsworth, R.I., and White, M.F. (2001). Identification and properties of the crenarchaeal single-stranded DNA binding protein from *Sulfolobus solfataricus*. Nucleic Acids Res. *29*, 914–920.

Watson, J.D., and Crick, F.H. (1953). Molecular structure of nucleic acids; a structure for deoxyribose nucleic acid. Nature *171*, 737–738.

Yu, X., VanLoock, M.S., Poplawski, A., Kelman, Z., Xiang, T., Tye, B.K., and Egelman, E.H. (2002). The *Methanobacterium thermoautotrophicum* MCM protein can form heptameric rings. EMBO Rep. *3*, 792–797.

Zhang, C.T., Zhang, R., and Ou, H.Y. (2003). The Z curve database: a graphic representation of genome sequences. Bioinformatics *19*, 593–599.

Zhao, A., Gray, F.C., and MacNeill, S.A. (2006). ATP- and NAD+-dependent DNA ligases share an essential function in the halophilic archaeon *Haloferax volcanii*. Mol. Microbiol. *59*, 743–752.

Recombination Processes and Proteins in the Archaea

6

Michael L. Rolfsmeier and Cynthia A. Haseltine

Abstract

Since the identification of Archaea as a distinct branch of the tree of life, general interest in these curious microbes has been steadily increasing. Archaea bear a strong physical resemblance to Bacteria and generally approach central metabolic activities in a comparable manner. Their methods for DNA information processing, however, are strikingly similar to mechanisms found in eukaryotes. In recent years, significant advances have been made in understanding recombination processes in members of the archaeal domain. This review highlights both *in vivo* genetic and *in vitro* biochemical approaches that have helped to reveal mechanistic details of recombination and takes a look forward to the future of DNA recombination studies in Archaea.

Introduction

In 1990, Carl Woese and his colleagues suggested a radical change in the accepted phylogenetic paradigm by introducing the addition of a third domain of life (Woese *et al.*, 1990). Until then, all cells lacking a nucleus had been grouped together as bacterial organisms. Cleverly disguised as simple Bacteria with rather strange environmental preferences, analysis of mutational variation in rRNA genes of these 'archaeabacteria' demonstrated that, despite their superficial resemblance to Bacteria, they deserved a separate classification to distinguish them from other prokaryotes. Six years later, Archaea went mainstream via news reports from major media outlets touting the discovery of a 'new form of life'. While *Methanococcus jannaschii* was hardly a newly discovered life form, the completion of its entire genome sequence (Bult *et al.*, 1996) effectively vindicated Woese's original theory and the archaeal branch of life was clearly delineated from the bacterial and eukaryotic branches. Ever since, members of this domain have been steadily recovering from this taxonomically imposed identity crisis.

Archaea are characterized by a set of unique features including ether-linked membrane lipids, unusual cell wall components, and distinctive rRNA motifs. The recent explosion of genome sequence availability has revealed that along with exclusively archaeal attributes, they share a mixture of other features with both the bacterial and eukaryal domains. Metabolically related activities appear to show more similarities to bacterial counterparts, while genes related to transcription, translation, replication and recombination are most similar to those of eukaryotes. Cultivated Archaea fall into two phylogenetic groups: euryarchaea and crenarchaea. Cultured euryarchaea are represented by three broad phenotypic classes: sulphur-metabolizing thermophiles, extreme halophiles, and methanogens. Many of the crenarchaea in culture are acidophiles, but all are thermophilic sulphur-metabolizers.

This chapter is devoted to recombination processes employed by members of the archaeal domain. Recombination activity is necessary for accurate reproduction, damage repair, resistance to deleterious evolutionary change, and preservation of evolutionary innovation; it has been observed in all three branches of the phylogenetic tree of life. Historically, there has been considerable interest in studying recombination and repair systems in members of the archaeal branch because they often have to contend with harsh environmental conditions that presumably threaten genomic stability. Several general observations have hinted at the presence of recombination mechanisms in archaeal species. Perhaps one of the earliest activities noticed was simple conjugative pyrimidine marker exchange in *Haloferax volcanii* in the 1980s (Mevarech and Werczberger, 1985; Rosenshine *et al.*, 1989). Nearly concurrently, a mutation was isolated in *Halobacterium halobium* that correlated with sequence homologies between insertion elements and a region upstream of the bacterio-opsin gene, implying a recombination event (Pfeifer *et al.*, 1984). Plasmid transfer between various *Sulfolobus* isolates was noted early on by Wolfram Zillig and coworkers (Schleper *et al.*, 1995). At about the same time, repair of double-strand breaks resulting from heat and gamma irradiation in *Pyrococcus furiosus* was documented (DiRuggiero *et al.*, 1997; Peak *et al.*, 1995). As the presence of multiple chromosome equivalents in quickly growing cells of many Bacteria is related to speedy repair of DNA lesions (Krasin and Hutchinson, 1977) and resistance to ionizing radiation (Harsojo *et al.*, 1981; Minton and Daly, 1995), multiple genome equivalents in archaeal species could serve the same purpose. Rolf Bernander and coworkers observed multiple genome equivalents in some archaeal species, strongly suggesting recombination as a probable mechanism for repair of double-strand DNA breaks (Bernander and Poplawski, 1997; Malandrin *et al.*, 1999). And, indeed, publication of the first complete archaeal genome sequence from *M. jannaschii* revealed numerous sequence homologues for genes responsible for recombination processes in eukaryotes (Bult *et al.*, 1996). Taken together, these early observations strongly hinted at interesting recombination mechanisms in these curious organisms. In the following sections we will examine recombination processes that have been studied both *in vivo* and *in vitro* in the Archaea, with the aim of better understanding how these organisms accomplish this essential function.

in vivo recombination

Conjugation

In Bacteria, conjugative mechanisms allow for the unidirectional transfer of genetic material from a donor cell to a recipient cell through direct cellular contact (Fig. 6.1A). Genes successfully transferred into the recipient can replace their equivalents in the recipient genome through recombination events. Similar transfer of genetic material between cells through conjugation also occurs in the Archaea (Hansen *et al.*, 2005; Mevarech and Werczberger, 1985). Marker exchange is different from that observed in bacterial Hfr conjugation, however. All cells are equally capable of acting as donor or recipient and the presence of a self-transmissible plasmid in a donor strain and absence of plasmid in a recipient strain is not required.

The first observations of conjugational activities were in *Haloferax volcanii* (Mevarech and Werczberger, 1985). In this very simple study, various auxotrophic mutations were used to test for conjugation activity. Genetic transfer was not unidirectional and required cell-cell contact. Similar early observations with *Sulfolobus acidocaldarius* also indicated direct cell contact without directionality (Grogan, 1996). Crosses between spontaneous *S. acidocaldarius* pyrimidine auxotrophs have yielded significant genetic data for this organism. This system has allowed resolution of closely spaced chromosomal mutations, identification of rearrangements and deletions, and mapping of defined deletion mutations (Reilly and Grogan, 2001). Recombination events between mutations separated by as few as 28 basepairs have been detected. Comparatively, approximately 25 kilobasepairs between markers is required for observed recombination in bacterial Hfr conjugation (Low, 1991; Reilly and Grogan, 2001). This short physical separation between recombined mutations indicates intragenic recombination through marker exchange is dominated by nonreciprocal events involving a small average patch size (Schmidt *et al.*, 1999).

Subsequently, a number of conjugative plasmids have been identified in *Sulfolobus* species (Brugger *et al.*, 2002; Prangishvili *et al.*, 1998; Zillig *et al.*, 1998; Rice *et al.*, 2001;

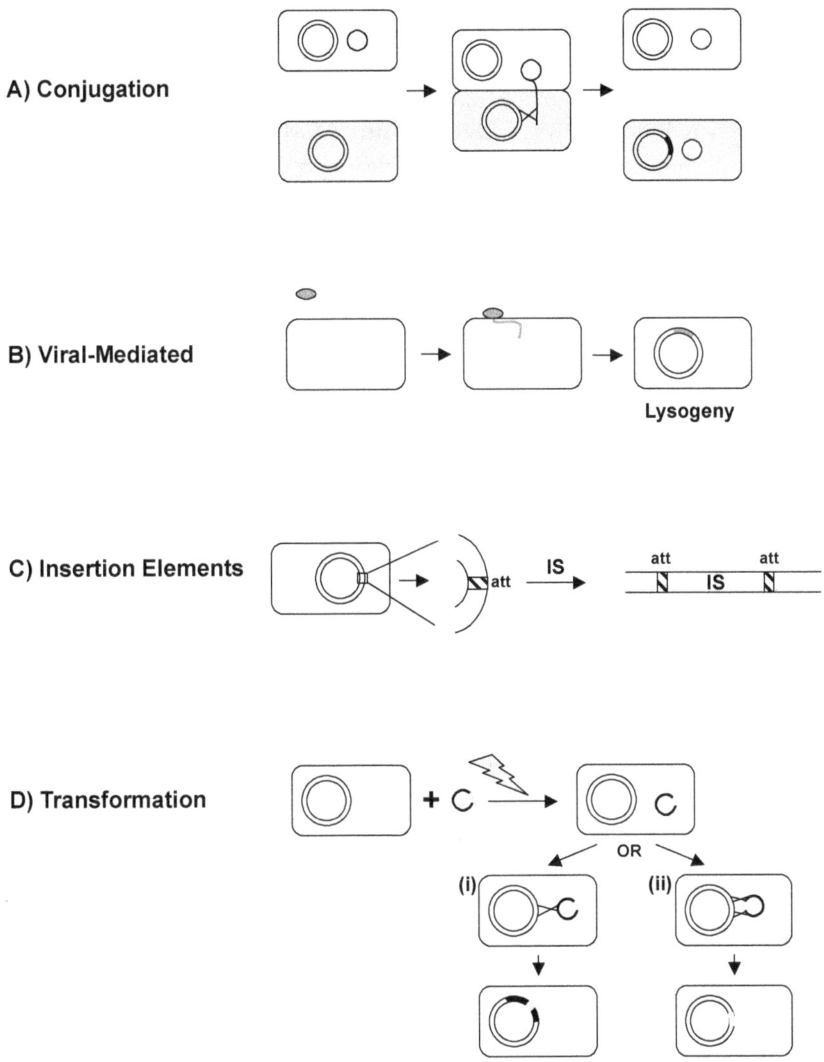

Figure 6.1 *in vivo* recombination activities. (A) Conjugation requires direct cell to cell contact. DNA passed to the recipient cell (shaded) can be integrated into the host genome or maintained as an extrachromosomal element. (B) Viral particles transfer their DNA to the cell where it integrates into the genome, resulting in infection. This integration is generally site-specific. (C) Insertion elements (IS) mobilize across the genome through action of an integrase and insert at specific attachment sites (att). (D) Exogenous DNA can be transformed into the target cell. The foreign DNA then recombines into the genome by i) a single crossover event or ii) directly replacing a gene through a double crossover event.

Brugger *et al.*, 2002; Erauso *et al.*, 2006). The first conjugative plasmid identified was pNOB8, a large 45 kilobasepair plasmid from *Sulfolobus* strain NOB8H2. Following cell mixing, cell to cell contact seen between plasmid donor and recipients hinted at conjugative transfer activity. Sequence analysis of this plasmid exposed an integrase gene, which is probably responsible for site-specific integration of the plasmid in the *Sulfolobus* chromosome (She *et al.*, 2004). Several other plasmids have been isolated from *S. islandicus*, including multiple variants of the pING family (Stedman *et al.*, 2000). Transpositions and site-specific recombination events are common within this plasmid family, with an increase in the variety of recombinants occurring after cellular stress. Most recently, two smaller plasmids that probably recombine site-specifically into the *Sulfolobus* chromosome have been reported (Erauso *et al.*, 2006). pSOG1 and pSOG2 were isolated from a single *S. islandicus* strain, SOG2/4. Both plasmids have an integrase gene sequence that may facilitate reversible integration via site-specific recombination between a pair of 43 basepair attachment sites, the plasmid attachment site *attP* and its corresponding chromosomal *attB* site.

Virus mediated

Site-specific recombination is a mechanism commonly used by temperate bacteriophage and other autonomous genetic elements for insertion into the bacterial chromosome (Fig. 6.1B). Generally, the recombination event is localized to a short region of perfect homology between the phage DNA and the host chromosome, with sequence-specific recombinases required for the integration process. During the study of a virus, it was first noticed that DNA can enter and leave archaeal chromosomes through integrase activity. SSV1 (*Sulfolobus shibatae virus 1*) is found in two intracellular states: as a covalently closed circular form and integrated site-specifically in an arginine tRNA gene (Yeats *et al.*, 1982; Reiter *et al.*, 1989). One of the open reading frames encoded by the viral DNA has significant homology to the consensus sequence of integrase family site-specific recombinases (Argos *et al.*, 1986; Palm *et al.*, 1991). This integrase promotes insertion and excision from the downstream half of a $tRNA^{Arg}$ gene in *S. shibatae* and is co-transcribed with an upstream open reading frame that encodes a protein similar to the Xis proteins of temperate phages (Muskhelishvili *et al.*, 1993; Reiter *et al.*, 1989). SSV1 integration is through an integrase-promoted recombination event between identical 44 basepair sequences present in both the arginine tRNA gene and the viral genome. Following insertion, the integrase gene is partitioned into two fragments, which border the linearized virus in the chromosome and are flanked by the 44 basepair direct repeats. A single copy of the attachment core sequence direct repeat is also present in the viral DNA (Reiter *et al.*, 1989). The 44 basepair common core contains sufficient information for synapsis, strand transfer, and resolution, while flanking sequences enhance the efficiency of intermolecular recombination. Recombination at these direct repeats results in excision and produces the circular virus and an intact integrase gene. After exiting the chromosome, the SSV1 provirus leaves behind an 'empty' attachment site (Muskhelishvili *et al.*, 1993). There is some flexibility in the core sequence requirements, since a nearly identical tRNA gene in *Sulfolobus solfataricus* is also an integration site for SSV1. SSV1 integration into the *S. solfataricus* genome occurs at a sequence with one mismatch relative to the *attA* of the natural host (Schleper *et al.*, 1992). The SSV1 lysogenic system is different from the bacterial lambda system in that there is an *in vitro* preference for linear substrates and no apparent requirement for accessory proteins for recombination (Muskhelishvili *et al.*, 1993).

A second virus, SSV2, has also been found to integrate into the *Sulfolobus* genome with site specificity (Contursi *et al.*, 2006). This virus is naturally found in *Sulfolobus islandicus* REY15/4 but also infects *S. solfataricus*. SSV2 virus encodes a tyrosine integrase that permits specific integration at a single tRNA site in the genome (Esposito and Scocca, 1997; Nunes-Duby *et al.*, 1998). SSV2 host-chromosome integration may be sequence-specific, since the virus only integrates at a sequence-identical *attA* site in both *S. islandicus* REY15/4 and *S. solfataricus* P2. However, there is no sequence identical to the SSV1 *attA* site present in the complete genome of *S. solfataricus* P2 (She *et al.*, 2001b). SSV2 host chromosome integration may therefore be sequence-specific or the virus could integrate into a site containing an SSV1 *attA* sequence (a near match) in the absence of a perfectly matched foreign host target.

Recently, an unusual two-tailed virus that infects *Acidianus convivator* has been described (Prangishvili *et al.*, 2006). Like SSV1, ATV (*Acidianus two-tailed virus*) cell infection results in lysogenization. This lysogeny can be interrupted by addition of mitomycin C or exposure to ultraviolet radiation, hinting at a recombination based mechanism. The ATV genome encodes an integrase that is maintained as an intact gene following integration, although the exact site for integration has not been described.

Integration of extrachromosomal elements like plasmids and viruses is a known mechanism for gene capture and acquisition of additional gene function (She *et al.*, 2001a; Greve *et al.*,

2004). Partitioned integrase genes have been noted in several archaeal chromosomes, implying that integrase activity facilitates insertion and deletion events as a general mechanism in these organisms. In *S. solfataricus*, one such partitioned sequence is on the border of the inserted plasmid pXQ1 from the *Sulfolobus* pRN family (Peng *et al.*, 2000) and another is on the edge of a large integron-type segment (She *et al.*, 2001a). Integrase-based mechanisms are also likely at work in *A. pernix* and *P. horikoshii*, since partitioned integrase sequences are also apparent in these genomes (She *et al.*, 2001a).

Insertion element mediated

Genomes are not static, but are dynamic and constantly evolving. Many different events influence genome evolution and direct genomic changes, most commonly point mutations, short insertions, and deletions. Generally, these alterations affect only a single gene or operon. Short DNA sequences known as 'hot spots' can also be present and are regions where intramolecular recombination occurs. In fact, dramatic changes can arise through homologous recombination at repeat sequences, causing significant genomic rearrangements (Arber, 2000). Insertion sequence (IS) elements and other mobile genomic elements can provide a basis for genome rearrangements through homologous recombination (Fig. 6.1C). Their movements influence the loss or acquisition of genes and can ultimately provoke horizontal gene transfer (Syvanen, 1994).

IS elements are common in the archaeal domain, but the number varies greatly between different genomes (Brugger *et al.*, 2002). They range in size from 0.5 to 2 kilobasepairs and encode transposases and enzymes like DNA resolvases. Terminal inverted repeats are also common in genomes, ranging in size up to 68 basepairs, but sometimes imperfect in sequence. These inverted repeats provide binding sites for the transposases that insert into target sites of varying size and degree of sequence conservation (Mizuuchi, 1992; Mahillon and Chandler, 1998). DNA-based mechanisms mobilize archaeal IS elements through either copy or excision methods. RNA intermediates, common in eukaryotic transposition events (Okada *et al.*, 1997), have not been observed for the Archaea. Additionally, there is a vast diversity of IS element types throughout the archaeal domain.

The current *Sulfolobus* gene pool is relatively highly conserved, with about 75% of the genes found in *S. acidocaldarius* also represented in either *S. solfataricus* or *Sulfolobus tokodaii*. Many of these genes are arranged in operons, but the overall gene order is not well conserved (Brugger *et al.*, 2004). These three *Sulfolobus* genomes have a wide range of putatively mobile IS elements. *S. solfataricus* contains the highest number of putatively mobile elements, comprising about 10% of the genome. There are 201 IS elements and at least 48 partial copies (Brugger *et al.*, 2002; She *et al.*, 2002; 2001b). While few archaeal IS elements have been demonstrated to have activity, some have been shown to transpose at a high frequency (Martusewitsch *et al.*, 2000; Schleper *et al.*, 1994). Several mutational events identified in *S. solfataricus* are attributed to IS element 'hops' (for examples see: (Schleper *et al.*, 1994; Martusewitsch *et al.*, 2000; Worthington *et al.*, 2003). IS element transposition was even noted during sequencing of the *S. solfataricus* genome, when an IS element located in one clone was often absent in overlapping clones (She *et al.*, 2001b). *S. tokodaii* has 34 IS elements, some of which are also present in *S. solfataricus* (Brugger *et al.*, 2002; Kawarabayasi *et al.*, 2001; She *et al.*, 2002). Many of the partial IS elements present in both genomes keep terminal repeat sequences at one end that may allow them to contribute to genome rearrangements through a transposase binding site. The *S. acidocaldarius* genome, however, has no apparent intact IS elements; instead, only four partial elements are present (Chen *et al.*, 2005). This apparent absence may explain mutational rate measurements that failed to detect any active IS elements (Grogan and Hansen, 2003).

Halobacterium NRC-1 maintains 82 IS elements, 24 of which are in the chromosome while the remaining 58 are present in two megaplasmids (Ng *et al.*, 2000).

The genomes of *Thermoplasma volcanium* and *Archaeoglobus fulgidus* have fewer than 30 IS elements (Kawashima *et al.*, 2000; Klenk *et al.*, 1997), while *M. jannaschii* has a single IS element copy (Bult *et al.*, 1996). Of the *Pyrococcus* strains, at least 23 full length IS elements are present in the genome of *P. furiosus* (Kanoksilapatham *et al.*, 2004) but IS elements seem to be missing from the genomes of *P. horikoshii* and *P. abyssii* (Diruggiero *et al.*, 2000; Lecompte *et al.*, 2001).

IS elements have been found in both chromosomal and plasmid copies within a single organism. The conjugative plasmid pNOB8 from *Sulfolobus* has two IS elements (She *et al.*, 1998) and pING from *Sulfolobus* carries one (Stedman *et al.*, 2000), both of which correlate to IS elements in the chromosome. Of the IS elements located on *Halobacterium* NRC-1 megaplasmids, 17 are also represented in the chromosome (Ng *et al.*, 2000). This arrangement supports potential IS element-mediated sequence transfer between plasmids and genomes as well as horizontal transfer between species. Indeed, horizontal transfer of a 16 kb IS element flanked fragment has been proposed to have occurred between *P. furiosus* and *Thermococcus litoralis* (Diruggiero *et al.*, 2000).

Archaea also contain non-autonomous mobile elements consisting of miniature inverted repeat transposable elements (MITEs). MITEs are common in eukaryotic genomes and appear to result from a deletion within an IS element or accumulation of terminal inverted repeats (Oosumi *et al.*, 1996). MITEs are present in *Sulfolobus* but their mobility has not yet been experimentally demonstrated (Brugger *et al.*, 2002; Redder *et al.*, 2001; She *et al.*, 2002). These elements are small, ranging in size from 79 to 355 basepairs, and have inverted terminal repeats but lack open reading frames. Mobility is believed to be mediated through transposases encoded by intact IS elements that recognize the inverted terminal repeat sequences (Redder *et al.*, 2001). There are two varieties of MITEs. In type I MITEs, the sequence between the terminal repeats resembles a partial IS element missing all or part of its transposase gene. Type II MITEs share the inverted terminal repeat sequences with IS elements, but not the intervening sequences. 143 MITEs have been identified in the *S. solfataricus* genome and all are type II. These elements range in size from 79 to 184 basepairs and appear to have corresponding IS elements present (Redder *et al.*, 2001). The *S. tokodaii* genome harbours 61 MITEs, most of which are type II. Nine copies, however, are of type I. No MITE-like sequences have been identified in the *S. acidocaldarius* genome, in agreement with the absence of IS elements and their mobilizing transposases (Brugger *et al.*, 2004; Chen *et al.*, 2005).

Examination of fairly closely related microorganisms reveals that the degree of genome rearrangement increases with an increase in the number of sequence repeats in the form of prophages, gene duplications and IS elements (Mira *et al.*, 2002; Zivanovic *et al.*, 2002). Mechanisms of genome maintenance required by organisms containing large numbers of potentially mobile elements are not well understood. In *S. solfataricus*, some regions of the genome are packed with mobile elements, several of which are even inserted into other elements making complex compound elements (Brugger *et al.*, 2002). The potential for genomic instability through uncontrolled homologous recombination in these organisms is enormous, so it is likely that significant regulation of transposase expression and homologous recombination events is necessary to limit deleterious genomic rearrangements.

Transformation

Bacterial transformation is a process in which a recipient cell acquires DNA sequences from the surrounding medium. While transformation occurs in nature, it is a powerful tool in the laboratory and serves as the foundation for microbial and molecular genetics. Transformation begins with DNA uptake by the recipient cell; the end result is either retention of an extrachromosomal element (i.e. a plasmid) or recombination based exchange between the donor DNA and homologous regions in the recipient chromosome (Fig. 6.1D). To better understand archaeal organisms, considerable effort has been invested in the development of reliable genetic transformation systems (Allers and Mevarech, 2005). Archaea can be encouraged to take up foreign DNA using a variety of methods. Both *T. kodakaraensis* and *M. voltae* appear to be naturally competent (Bertani and Baresi, 1987; Sato *et al.*, 2003). *Halobacterium halobium, H. volcanii, Methanococcus maripauldis*, and *P. abyssii* can be transformed following spheroplasting to remove the cell wall (Charlebois *et al.*, 1987; Cline and

Doolittle, 1987; Lucas *et al.*, 2002; Whitman *et al.*, 1997). *P. furiosus* and *S. acidocaldarius* have been transformed by use of more traditional bacterial $CaCl_2$ methods, but most success with *Sulfolobus* strains has been through the use of electroporation approaches (Aagaard *et al.*, 1996; Aravalli and Garrett, 1997).

Aside from finding a way to get DNA into archaeal species, there is the additional concern of stability *in vivo*. Exogenous DNA must be maintained in the cell long enough to allow for gene expression or recombination activity. Histidine auxotrophs in *H. volcanii* can be efficiently transformed using unmodified cosmid clones propagated in methylase deficient *E. coli* cells (Conover and Doolittle, 1990; Holmes *et al.*, 1991). In at least some cases, pre-treatment of DNA with a *SuaI* pattern of methylation (using *HaeIII* methylase) is effective in reducing degradation, but may result in reduced recovery of recombinants (Kurosawa and Grogan, 2005).

Efficiency of recombination of transformed DNA in archaeal species varies with the form of exogenous DNA that is introduced and likely reflects differences in homologous recombination systems in these organisms. In *Methanococcus voltae*, transformation efficiency of integrating plasmids increases nearly 400-fold if the plasmid is linearized prior to transformation (Patel *et al.*, 1994). Heterozygotic intermediates in this organism form from intact integrative plasmids through a single reciprocal crossover with exceptional stability (Gernhardt *et al.*, 1990). Circular constructs in *T. kodakaraenis* donate selectable sequences to the genome without self-integration, indicating a lack of single-crossover recombination (Sato *et al.*, 2003). Similarly, in plasmid-based transformations of *S. acidocaldarius*, no vector sequences have been observed to integrate into the genome, also implying an absence of single-crossover events (Kurosawa and Grogan, 2005).

Targeted gene disruption by homologous recombination has been reported in the Archaea. Both *Thermococcus kodakaraensis* KOD1 and *S. solfataricus* strain 98/2 permit double-crossover recombination between a chromosomal mutant gene and an intact plasmid version (Sato *et al.*, 2003; Worthington *et al.*, 2003). In *T. kodakaraensis* KOD1, linear DNA-based gene disruption is observed when the homologous regions are relatively long; regions of 1,020 and 753 basepairs are needed for efficient double-crossover recombination at the target gene. Much shorter homologous regions might also contribute to gene disruptions, but these events may well be masked by the activity of intracellular exonucleases.

Allele replacement has been used along with counter selection to produce archaeal genomic mutations (Peck *et al.*, 2000; Pritchett *et al.*, 2004). Silent mutations in *S. solfataricus* have been fashioned using selection methods followed by loss of the selectable marker, vector, and target allele via spontaneous segregation (Worthington *et al.*, 2003). By coupling site-specific mutagenesis with allele replacement, specific genomic mutations have been introduced in *S. solfataricus* strain 98/2 (Schelert *et al.*, 2006). In this case, recombinants produced through homologous recombination between a plasmid and a cloned copy of a modified DNA sequence were seen to segregate at a frequency of about 50%. Moreover, the use of linearized DNA for transformations in *Sulfolobus* gives greater control over the crossover event by preventing unwanted recombination events and precluding the need for two successive single-crossover events (Schelert *et al.*, 2006).

It is interesting to note that *S. acidocaldarius* is capable of oligonucleotide transformation (Kurosawa and Grogan, 2005). The phenomenon observed in this archaeal species resembles activities in yeast and *E. coli*, where oligonucleotides can replace mutations in the chromosome and transformations of point mutations exhibit clear strand bias. Double-stranded oligonucleotides with a maximum of 18 nucleotide overlaps on either side of a chromosomal deletion were incorporated into the genome of *S. acidocaldarius*. The small size of this overlap is significantly below the threshold required for efficient reciprocal recombination in Bacteria and fungi (Shen and Huang, 1986; Jinks-Robertson *et al.*, 1993). Incorporation of oligonucleotide sequences is not easily explained through reciprocal crossover events, however, and may involve oligonucleotide annealing to transient replication-associated single-strand gaps on the chromosome. This possibility is consistent with the observation that one strand is often more effective than its complement in oligonucleotide mediated recombination. Intriguingly, even some single- stranded oligonucleotides were found to be fully effective in *S. acidocaldarius* transformations.

Within populations

Recombination is an important force in the generation of diversity and adaptive novelty in prokaryotes. Recombination activity at the population level has, historically, been difficult to ascertain due to experimental restraints that required propagation of individual species in a laboratory setting. PCR methods have expanded opportunities for exploring population recombination events within species in their natural environmental habitats. Indeed, recombination events within populations have been investigated in a number of archaeal species including *Ferroplasma ssp., Halorubrum,* and *Sulfolobus islandicus.*

Through a random shotgun sequencing approach, *Ferroplasma* type II strains in a biofilm were found to have undergone homologous recombination within the population (Tyson *et al.*, 2004). Genome sequences reconstructed from three *Ferroplasma* spp. revealed mosaic genomes resulting from recombination between the genome types. Recombination break points were located about every 5 kilobasepairs when nucleotide sequences had greater than 98% similarity. Relatively little recombination was seen between more diverged sequences. It appears that most recent evolution of the population has been dominated by homologous recombination events between the strains.

Examination of a *Halorubrum* spp. population revealed that the majority of recent recombination events, as determined through single-locus variations, were between individuals with little sequence divergence (Papke *et al.*, 2004). Allelic profiles among isolates were highly mosaic and indicated random mating and recombination occur within and between *Halorubrum* spp. populations.

The most thorough investigation to date was made with a population *S. islandicus*. This study assessed recombination events in six single copy loci that were well distributed across the genome (Whitaker *et al.*, 2005). Recombinant genotypes in the population were determined through examination of the genealogical relationships at each locus. The effect of recombination relative to mutation was established by determining the number of nucleotide differences between alleles in individuals with single-locus variation. Single-locus variations occur in a short evolutionary time-frame, meaning that the probability of more than one mutation occurring in the same locus is small. The genetic structure of the *S. islandicus* population could not be explained through periodic selection. Instead, recombination was found to occur at rates higher than mutation, with most diversity in the form of mosaic combinations of variant alleles. Therefore, recombination is not only generating diversity in this population but also serves to maintain recombinant genotypes by preventing selective repair of single adaptive genotypes.

The mechanism of recombination in natural archaeal populations is not understood. In some cases, cell to cell contact may be required to transfer DNA. For example, prototrophic recombinants of *Haloferax* can be recovered following autotrophic strain mixing and plating on solid medium; the prototrophy may result from the formation of intercellular bridges and subsequent transfer of DNA sequences (Rosenshine *et al.*, 1989; Dyall-Smith *et al.*, 2003). Genetic exchange in *S. acidocaldarius* requires cell-cell contact and potentially the formation of such bridges (Grogan, 1996; Reilly and Grogan, 2001). Recombination also occurs readily between chromosomes and introduced homologous DNA fragments in several species (Cline and Doolittle, 1992; Wang *et al.*, 2004). Genetic exchange likely serves as a cellular mechanism for maintaining chromosomal integrity in response to DNA damage in Archaea. The apparent high rates of recombination in several archaeal species point to recombination as a fundamental force for maintaining diversity in natural populations.

in vitro recombination activities

DNA access

Based on *in vivo* and *in vitro* studies, a mechanistic model of homologous recombination in bacterial and eukaryotic organisms has been proposed (Fig. 6.2). Most of the biochemical recombination experiments that are the founda-

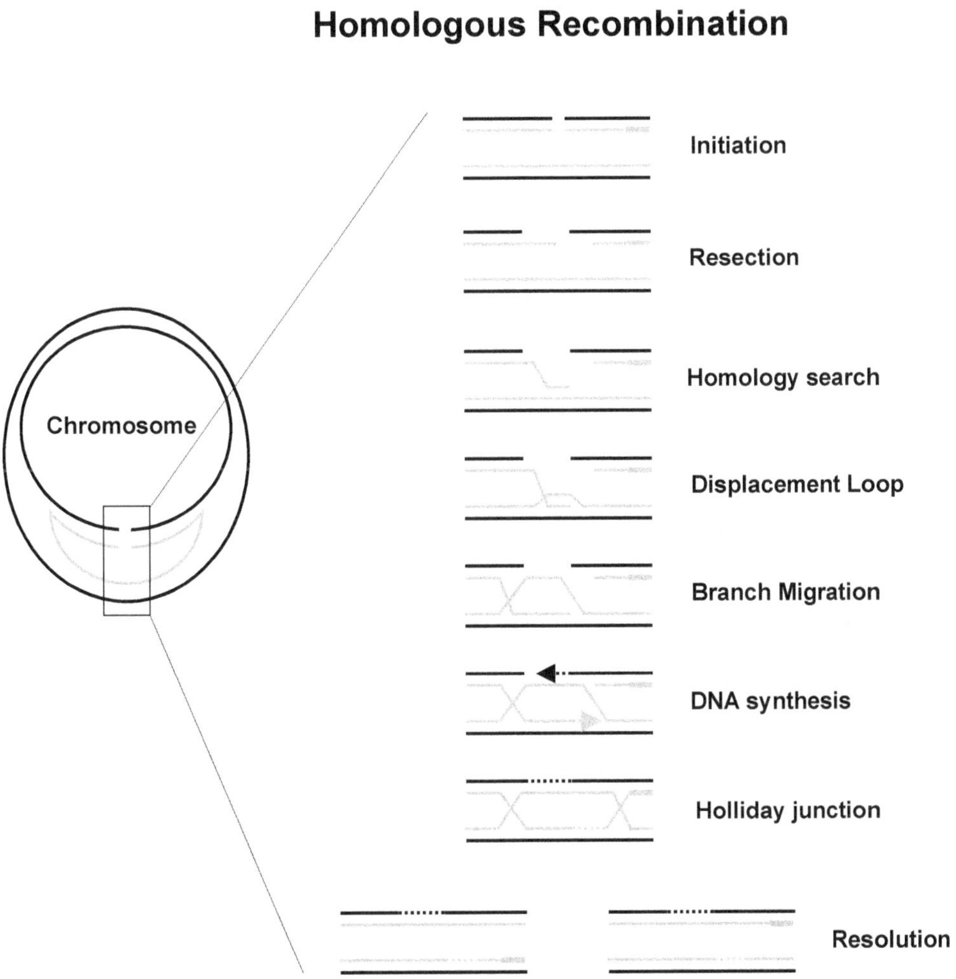

Figure 6.2 Formation of a double-strand break (DSB) and repair through homologous recombination. (A) Initiation: a DSB is formed in a DNA strand. (B) Resection: the DSB ends are processed, leaving 3′ overhangs; the strand exchange protein can now form the nucleoprotein filament. (C) Homology search: the fully formed nucleoprotein filament samples a homologous template for its specific matching sequence. (D) Displacement loop: the complementary strand is displaced to permit annealing of the strand exchange protein coated single-strand DNA to pair with its complementary sequence. (E) Branch migration: as the nucleoprotein filament anneals to its complementary sequence, the displaced strand further deanneals and the DNA junction migrates. (F) DNA synthesis: synthesis occurs to replace the information lost at the break site. At this time, a second (non-invading) end can anneal to its complement (the displaced loop) and prime DNA synthesis. G) Holliday junction: following ligation of the newly synthesized DNA, two junctions are formed. (H) Resolution: the resulting molecules will vary depending on the orientation with which the resolvase cuts the double-Holliday junction. The final event yields either a crossover or non-crossover product (orientation indicated by the shaded box).

tion of this model required the use of unadorned DNA substrates. This is not, however, the natural state of DNA. Cells must compact DNA through the association of DNA-binding proteins. These chromatin proteins are used by all organisms to permit reversible compaction of their genomes, thereby allowing gene expression, DNA replication, repair, and recombination. Archaeal homologues for eukaryotic chromatin proteins were first identified in *Methanothermus fervidus* (Sandman *et al.*, 1990) and most of the euryarchaea have similar histone homologues. These archaeal histones are generally shorter than their eukaryotic cousins and made up of the histone-core fold without tails. They can compact DNA, alter the mobility of DNA in gel-shift assays, and form structures that are analogous to histones. Little is known about the role of archaeal histones *in vivo*, although alteration in levels of compaction or distribution of histone subtypes could be used to modulate access to DNA (Sandman *et al.*, 1994).

It seems that histones are largely absent from the crenarchaeal domain (Cubonova *et al.*, 2005). Instead of classical histones, most crenarchaea appear to use highly abundant DNA-binding proteins to compact DNA. The most well studied crenarchaeal chromatin proteins are those from the *Sulfolobus* genus. The Sul7d protein is a 7-kDa monomeric DNA-binding protein that is highly expressed and accounts for about 5% of total soluble cellular protein (Edmondson and Shriver, 2001). Sul7d binds DNA non-cooperatively with a footprint of about 4 basepairs and may compact relaxed or positively supercoiled DNA (McAfee *et al.*, 1996; Gao *et al.*, 1998; Napoli *et al.*, 2002).

Recently, a novel protein called CC1 (crenarchaeal chromatin protein 1) has been reported (Luo *et al.*, 2006). Homologues of CC1 are found only in *Thermoproteus tenax*, *Pyrobaculum aerophilum*, and *Aeropyrum pernix* and have no clear similarity to any known protein at the amino acid level. CC1 protein is physically quite similar to the *Sulfolobus* Sul7 protein and likely represents a typical archaeal chromatin protein. The CC1 protein from *T. tenax* binds both single-strand and double-strand DNA (ssDNA and dsDNA) with strong cooperativity and is highly expressed *in vivo* (Luo *et al.*, 2006).

First identified as a highly abundant *Sulfolobus* protein, the Alba protein has non-sequence specific DNA-binding capacity. Homologues to Alba protein are present in the Archaea as well as some eukaryotes (Forterre *et al.*, 1999; Bell *et al.*, 2002). In Archaea, Alba seems specific for thermophiles and hyperthermophiles. A homodimer in solution, it can induce negative supercoiling of DNA and is associated with a wide range of genomic loci, suggesting a chromatin protein function (Xue *et al.*, 2000; Marsh *et al.*, 2005). Although Alba is not believed to compact DNA in any significant manner, it may be involved in organizing higher-order folding of DNA-histone or DNA-Sul7d nucleoprotein assemblies (Lurz *et al.*, 1986; Bell *et al.*, 2002). Another version of Alba has recently been identified in the Archaea (Jelinska *et al.*, 2005). Alba2 forms obligate heterodimers with Alba and seems to alter the higher-order DNA packing mediated by Alba. Differential expression of both Alba and Alba2 may well modulate the nucleoid structure in *Sulfolobus*.

While considerable progress has been made in understanding archaeal chromatin proteins, much remains to be discovered about how these proteins interact in the context of macromolecular assemblies. Protein access to packaged DNA in eukaryotes is controlled through mediation of internucleosome interactions via covalent modification of histone tails and the activity of ATP-utilizing complexes that reposition nucleosomes. A number of these complexes have been described, each of which is characterized by an ATPase subunit from the SNF-2 like family of DNA-stimulated ATPases (Fyodorov and Kadonaga, 2001; Aalfs and Kingston, 2000; Langst and Becker, 2001). The complexes function through increasing nucleosome mobility (sliding) without displacement from DNA and allowing access to DNA-binding factors (Hamiche *et al.*, 1999; Langst *et al.*, 1999; Whitehouse *et al.*, 1999). Homologous recombination is sensitive to local chromatin organization, suggesting that chromatin remodelling accompanies the process of DNA strand exchange (Sugawara *et al.*, 1995).

One eukaryotic protein that has been found to be very involved in recombination processes is the Rad54 protein. This protein is essential for recombination repair and appears to have important functions at multiple recombination steps. Rad54 has been dubbed a 'Jack of all trades' and the 'Swiss army knife' of recombination (for reviews, see: (Tan *et al.*, 2003; Heyer *et al.*, 2006). Specifically, it is a dsDNA-dependent ATPase that can stabilize the Rad51 filament, stimulate Rad51-dependent joint molecule formation, and influence the DNA heteroduplex extension step. Most importantly for the discussion here, Rad54 functions in nucleosome remodelling activities (Alexeev *et al.*, 2008).

Examination of all of the currently available archaeal genomic sequences reveals that only one archaeon has an open reading frame that could produce a protein with obvious homology to eukaryotic Rad54 (Durr *et al.*, 2005; Seitz *et al.*, 2001). The Rad54 homologue in *S. solfataricus* retains seven helicase motifs and a leucine zipper motif that are found in the yeast Rad54 protein. A portion of this protein, consisting of the seven helicase domains (the c-terminal half), has been crystallized (Durr *et al.*, 2005). This dsDNA-translocating SWI2/SNF2 ATPase core structure binds duplex DNA through the minor groove along a central protein cleft. The protein likely travels along the dsDNA minor grove without strand separation, generating torque for remodelling processes. Biochemical characterization of purified full-length Rad54 protein reveals that the *S. solfataricus* protein is a dsDNA-dependent ATPase that can stimulate RadA mediated strand exchange and remodel DNA structures (Haseltine and Kowalczykowski, in preparation). Identification of this protein in just a single archaeon suggests not that it is unique to *Sulfolobus*, but instead that sequence divergence has masked the identity of analogous proteins in the rest of the domain. In fact, there are a number of sequences present in other archaeal genome sequences that contain similar arrangements of helicase domains that may be the functional equivalent of the *S. solfataricus* Rad54 protein (Haseltine and Kowalczykowski, in preparation).

Processing at the break

Repair of DNA double-strand breaks through homologous recombination requires DNA end-processing to yield 3′ ssDNA tails. In *S. cerevisiae*, this is accomplished by three proteins: Rad50, Mre11, and Xrs2. They form a complex that is involved in homologous recombination, non-homologous end-joining, telomere maintenance, and double-strand break formation during meiosis (Pâques and Haber, 1999; Sung *et al.*, 2000). Rad50 is an ATP-dependent DNA-binding protein with DNA unwinding activities (Raymond and Kleckner, 1993). The Mre11 protein is a ssDNA endonuclease with 3′-to-5′ exonuclease activity (Furuse *et al.*, 1998; Paull and Gellert, 1998; Usui *et al.*, 1998). Rad50 and Mre11 act together as an endonucleolytic complex. In humans, the Rad50/Mre11 complex interacts with a protein called NBS1 (Dolganov *et al.*, 1996). Nbs1 helps the Rad50/Mre11 complex destabilize DNA hairpins and unwind duplex DNA in an ATP-dependent manner (Paull and Gellert, 1999). The function of yeast Xrs2 remains unknown (Johzuka and Ogawa, 1995).

Archaeal sequences with homology to eukaryotic Mre11 and Rad50 are organized in an apparent operon consisting of four open reading frames. In *S. acidocaldarius*, all four of these open reading frames are co-transcribed, implying a functional interaction (Constantinesco *et al.*, 2004). The Mre11 and Rad50 proteins from *P. furiosus* have been overexpressed and purified (Hopfner *et al.*, 2000a). Mre11 has manganese dependent ssDNA specific endonuclease activity that forms a stable complex with Rad50. It has been speculated that Rad50 can regulate DNA binding and release through its association with Mre11. The complex itself has 3′-to-5′ ssDNA exonuclease activity and can digest linear dsDNA, activities that are both ATP-dependent (Hopfner *et al.*, 2000a). Crystal structures of the ATP-bound and unbound forms of the Rad50 catalytic domain show that in the presence of ATP, the two catalytic domains associate to form a putative DNA binding groove (Hopfner *et al.*, 2000b).

The other two open reading frames in the archaeal Rad50/Mre11 operon have been cloned and their proteins overexpressed and purified. The NurA (for Nuclease repair of Archaea) protein from *S. acidocaldarius* has endonuclease activity with circular ssDNA substrates as well as 5′-to-3′ exonuclease activity with ss or dsDNA. This exonuclease has opposite directionality from the *P. furiosus* Mre11 (Constantinesco *et al.*, 2002). The other open reading frame has been designated HerA (for helicase repair of Archaea, identified in *S. acidocaldarius*) or MlaA (for Mre11 complex-linked ATPase in Archaea, identified in *P. abyssii*) (Constantinesco *et al.*, 2004; Manzan *et al.*, 2004). Biochemical characterization of the HerA protein shows it to function as a DNA helicase that can use either 3′ or 5′ ssDNA extensions for loading and DNA duplex unwinding coupled with multidirectional DNA translocation activity (Constantinesco *et al.*, 2004). The MlaA protein is an ATPase that preferentially binds to dsDNA containing secondary structure (Manzan *et al.*, 2004). Electron microscopy indicates that the protein forms a hexameric ring, meaning that the protein is likely to be a member of the hexameric AAA family of ATPases. This protein may be involved with Mre11 during processing of stalled replication/ recombination intermediates. The clustering of these genes in archaeal genomes implies that the Rad50/Mre11 complex could act with the NurA 5′–3′ exonuclease and the HerA bipolar DNA helicase during the initiation step of homologous recombination.

Apparent sequence homologues of Xrs2 and NBS1 are notably absent in the genome sequences of Archaea. In eukaryotes, Xrs2/ Nbs1 appears to regulate the endonuclease activity of the Rad50/Mre11 complex, ultimately resulting in the cleavage of 3′ DNA tails (Paull and Gellert, 1999). It is currently unknown if this activity is indeed unnecessary in Archaea or if instead the protein(s) involved are yet to be discovered.

Single-strand DNA-binding proteins

All organisms encode some sort of single-stranded DNA-binding protein or protein complex that is indispensable for recombination. Early in the process, these proteins help to remove ssDNA secondary structure and prevent reinvasion of displaced ssDNA after complementary strand annealing (Kowalczykowski *et al.*, 1994). In Bacteria, this protein is SSB (single-strand binding) and functions as a homotetramer. In eukaryotes, this function is carried out by a heterotrimeric complex comprising RPA70, RPA32, and RPA14 (Gomes and Wold, 1995; Gomes and Wold, 1996). Numerous ssDNA-binding proteins have been identified in the Archaea and multiple protein 'architectures' are observed.

Archaeal ssDNA-binding proteins were first identified in *M. jannaschii*, *M. thermoautotrophicum*, and *A. fulgidus* through their sequence homology to the RPA 70 subunit of eukaryotic ssDNA-binding proteins (Chédin *et al.*, 1998; Kelly *et al.*, 1998). The methanogen proteins are large single polypeptide chains that contain four ssDNA-binding domains arranged in tandem, each of which is homologous to the other. Additionally, these proteins have a strongly conserved zinc-finger domain. The ssDNA-binding protein from *M. jannaschii* has been purified and characterized (Kelly *et al.*, 1998; Seitz *et al.*, 2001). This protein efficiently binds ssDNA and functions as a monomer. The ssDNA-binding proteins from *A. fulgidus*, *P. abysii*, *P. horikoshii*, *P. furiosus*, and *Halobacterium* sp. NRC-1, however, comprise two subunits, each with two DNA-binding domains (Seitz *et al.*, 2001). In *P. furiosus*, the RPA protein more closely resembles eukaryotic RPA and is a functional heterotrimer. This protein complex efficiently binds ssDNA and can enhance RadA mediated strand exchange reactions (Komori and Ishino, 2001). Immunoprecipitation experiments with the heterotrimer showed interactions with numerous proteins, including replication proteins, indicating involvement in replication processes. In *Methanosarcina acetivorans*, a similar set of three ssDNA-binding proteins have been found (Robbins *et al.*, 2004). These proteins, rpa1, rpa2, and rpa3 appear to function individually as ssDNA-binding proteins instead of acting as a heterotrimeric complex. Each protein may play a different role in the replication and repair activities of this methanogen.

Crenarchaeal ssDNA-binding proteins from *Aeropyrum pernix* and *S. solfataricus* have been also been described (Haseltine and Kowalczykowski, 2002; Seitz *et al.*, 2001). These proteins consist of a single subunit with one ssDNA-binding domain and an acidic c-terminal region, resembling the *E. coli* SSB structure. The protein from *Sulfolobus* has been purified and extensively characterized (Haseltine and Kowalczykowski, 2002; Seitz *et al.*, 2001; Wadsworth and White, 2001). It efficiently binds ssDNA, can stimulate the ATPase and strand exchange activities of RadA (Haseltine and Kowalczykowski, unpublished observations), interacts with multiple *Sulfolobus* proteins including the MCM1-like helicase (Carpentieri *et al.*, 2002), RNA polymerase (Richard *et al.*, 2004), reverse gyrase (Napoli *et al.*, 2005), and has been crystallized (Kerr *et al.*, 2003). The structure reveals an OB-fold domain with much more similarity to eukaryotic RPA than to bacterial SSB (Kerr *et al.*, 2003). The protein appears to have functional activity as both a monomer and a tetramer (Wadsworth and White, 2001; Haseltine and Kowalczykowski, 2002; Richard *et al.*, 2004; Cubeddu and White, 2005; Haseltine and Kowalczykowski, unpublished observations). The need for this protein to participate in the presynaptic step of thermophilic archaeal recombination is unclear, since secondary structure is unstable in ssDNA at elevated temperatures. Consequently, ssDNA-binding proteins in thermophiles may instead have a primary role in postsynaptic steps.

RadA and RadA paralogues

Strand exchange is the central core of homologous recombination. It is during this step that identification and synapsis of a DNA strand with a homologous template is accomplished through the activity of a protein from the RecA family (Kowalczykowski and Eggleston, 1994; Kuzminov, 1999; Bianco and Kowalczykowski, 2005). First described in *E. coli* as RecA, a strand exchange protein homologue is found in all Bacteria as well as in eukaryotes (called Rad51). The archaeal version has been named RadA and was identified through its similarity to both bacterial RecA and eukaryotic Rad51 (Sandler *et al.*, 1996). Homologues of this protein have been identified in every sequenced archaeal genome and all are more closely related to eukaryotic Rad51/DMC1 proteins (~40% identity at the amino acid level) than they are to their bacterial counterparts. RecA protein family members share such a striking resemblance at the structural level that functional interactions of eukaryotic Rad51 protein with the damage repair protein Brca2 can be extrapolated from the archaeal RadA crystal structure (Shin *et al.*, 2003). The role of RadA in homologous recombination was first established by the deletion of the *radA* gene in *Haloferax volcanii* (Woods and Dyall-Smith, 1997). The resulting null mutation shows increased sensitivity to DNA damage caused by UV irradiation and ethylmethane sulfonate (EMS) as well as defective recombination activity.

The protein has subsequently been purified from a number of different archaeal sources including *T. kodakaraensis* KOD1, *S. solfataricus*, *Desulfurococcus amylolyticus*, *Pyrobaculum islandicum*, *P. furiosus*, *Archaeoglobus fulgidus*, and *M. voltae* (Rashid *et al.*, 1996; Seitz *et al.*, 1998; Kil *et al.*, 2000; Spies *et al.*, 2000; Komori *et al.*, 2000b; McIlwraith *et al.*, 2001; Wu *et al.*, 2005). RadA shares many of the same biochemical characteristics of RecA and Rad51 proteins (for review, see Seitz *et al.*, 2001). It is a DNA-dependent ATPase that forms nucleoprotein filaments, assists in generation of joint molecules, and is crucial for DNA pairing and strand exchange. Production of the protein appears to be moderately inducible by UV treatment (Reich *et al.*, 2001; Williams *et al.*, 2007). Additionally, RadA may have some preference for DNA sequences that are biased toward G and T residues (Seitz and Kowalczykowski, 2000). The C-terminal portion of the protein is required for DNA-dependent ATPase activity and DNA strand exchange activity, but the presence of the N-terminus is necessary for optimal activity (Komori *et al.*, 2000a). Extensive crystal structure analysis of the RadA protein from *M. voltae* has revealed that it can be encouraged to assume an active form and increase DNA affinity through the addition of either monovalent or divalent cations, providing a mechanistic trigger for assembly of functional nucleoprotein filaments (Qian *et al.*, 2006).

During strand exchange, the main strand exchange protein (RecA, Rad51, or RadA) can encounter obstacles that interfere with ssDNA binding or efficient DNA strand exchange and heteroduplex extension. Competition with ssDNA-binding proteins for DNA-binding sites can occur. The battle for DNA-binding sites is mitigated by the activity of mediator proteins that help DNA strand exchange proteins bind ssDNA. Bacterial RecF, RecO, and RecR proteins assist RecA protein in binding to SSB coated ssDNA (Umezu *et al.*, 1993; Webb *et al.*, 1997; Kuzminov, 1999). In yeast, two proteins with limited homology to Rad51 have been identified that may serve a similar purpose (Sung *et al.*, 2000). Yeast Rad55 and Rad57 proteins act as a complex that has a pre-synaptic accessory role in recombination where it assists the Rad51 protein in forming more continuous filaments through lessening RPA binding to ssDNA (Pâques and Haber, 1999; Sung, 1997; Pâques and Haber, 1999; Sung *et al.*, 2000).

In humans, there are also paralogues to the Rad51 protein. Rad51B, Rad51C, Rad51D, Xrcc2, and Xrcc3 all share 20–30% amino acid homology with Rad51 and each other; these proteins act in multi-protein complexes to aid Rad51 in the recombination process (Albala *et al.*, 1997; Rice *et al.*, 1997; Cartwright *et al.*, 1998; Dosanjh *et al.*, 1998; Liu *et al.*, 1998). These five Rad51 paralogues have been shown to interact with each other (Schild *et al.*, 2000), but their function in recombination is poorly understood. What is known is that Xrcc2 and Xrcc3 have been implicated in efficient repair of double-strand break pathways involving recombination (Cui *et al.*, 1999; Pierce *et al.*, 1999), while Rad51C has been associated with Holliday-junction processing (Liu *et al.*, 2004).

RadA protein paralogues are readily apparent in archaeal genomes and may serve the same purpose in presynaptic recombination steps that is accomplished by Rad51 paralogues in eukaryotic systems. Called RadB, sequence comparisons of these proteins from euryarchaeal species show more similarity to the *S. cerevisiae* Rad51 protein than to *E. coli* RecA (Bult *et al.*, 1996; Klenk *et al.*, 1997; Smith *et al.*, 1997; Kawarabayasi *et al.*, 1998; Komori *et al.*, 2000c). The first biochemically characterized archaeal RadB protein was from *P. furiosus* (Komori *et al.*, 2000c). RadB displays weak DNA-independent ATPase activity. It has a higher affinity for ssDNA binding than RadA protein and hinders RadA mediated D-loop formation. Strand exchange reactions are inhibited by RadB as well, unless RadA is allowed to form a nucleoprotein filament on ssDNA first. As yet, no interaction between RadB and the homologous RadA protein has been observed. This is quite different from the direct contact seen between Rad51 and the Rad55/57 heterodimer in *S. cerevisiae* (Johnson and Symington, 1995; Sung, 1997). Instead, RadB has been shown to interact *in vitro* with both PolD and the Holliday junction cleavage protein Hjc (Komori *et al.*, 2000c). RadB also inhibits the junction cleavage activity of Hjc in the absence of ATP, hinting that RadB may act during replication fork stalling through modulation of Hjc activity (Komori *et al.*, 2000a).

RadB appears to be exclusive to the euryarchaeal branch of the archaeal domain and contains a diagnostic KHR motif as identified in *H. volcanii* (Guy *et al.*, 2006). This is not to say that RadA-like proteins are not present in the crenarchaeal branch of the archaeal domain. While not formally recognized as RadB proteins, proteins with homology to RadA and *E. coli* RecA protein have been identified in these organisms (Seitz *et al.*, 2001; Kawarabayasi *et al.*, 2001; Chen *et al.*, 2005; Haseltine and Kowalczykowski, unpublished observations). The crenarchaeal RadA-like proteins are smaller than the RecA protein, but have 25–27% amino acid similarity across the entire protein. The function of these crenarchaeal RadA paralogues has not yet been determined.

Holiday junction resolvases and branched intermediate nucleases

Found in all three domains of life, Holiday junction resolvases are proteins involved in the resolution of the junction formed following successful recombination between two DNA strands (Aravind *et al.*, 2000). These proteins are structure-specific, metal-dependent endonucleases that share little to no homology at the amino acid level but bind and cleave four-way DNA junctions (Fig. 6.3C). There are two well described types of Holliday junction resolvases. Type 1 enzymes cleave at precise dinucleotide sequences, with specificity likely limiting cleavage solely to Holliday junctions. Type 2 enzymes have little or no substrate specificity and can cleave a wide variety of structures including three-way junctions (Fig. 6.3C) and mismatches (White *et al.*, 1997). The first archaeal activity that could be attributed to a Holliday junction-type protein was found in *P. furiosus* (Komori *et al.*, 1999). Hjc (for Holliday junction cleavage) is a type 2 enzyme that can introduce symmetrical nicks into a pair of DNA strands with similar polarity. It resolves Holliday junctions through paired cuts located 3′ to the point of strand exchange and does not demonstrate any apparent sequence specificity. This protein is unique and shares no sequence similarity with known resolvases in the bacterial and eukaryotic domains. The protein does appear to be highly conserved among archaeal species, however, and similar sequences are apparent in a number of other available genomes (Komori *et al.*, 2000a).

In *S. solfataricus*, two Holiday junction resolving activities have been described. The *S. solfataricus* Hjc protein was found through homology with the *P. furiosus* protein and the proteins share 34% amino acid sequence identity.

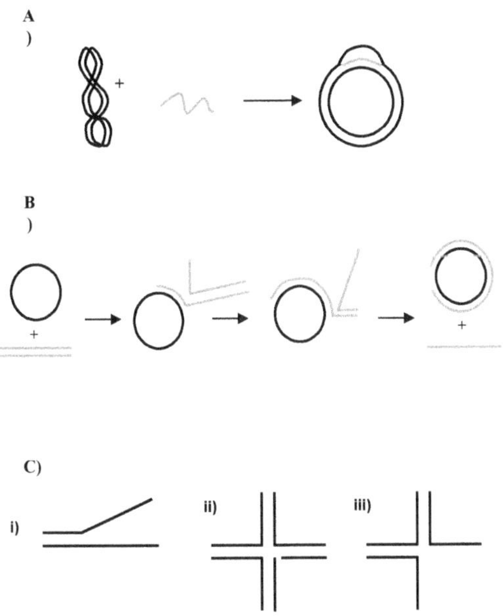

Figure 6.3 DNA-based assays and structures. (A) D-loop assay. Double-strand DNA (dsDNA) is combined with a complementary oligonucleotide. The reaction is catalysed by the strand exchange protein, which forms a nucleoprotein filament on the oligonucleotide and invades the DNA duplex. The oligonucleotide can be monitored by agarose gel migration, and the speed of migration is altered when the oligonucleotide invades dsDNA. (B) Strand exchange assay. Also called the 3-strand reaction, a single-stranded closed circular DNA is combined in a reaction with linear dsDNA. The activity of the strand exchange protein promotes the exchange of strands between linear dsDNA and ssDNA. One of the complementary strands from the dsDNA substrate anneals to the single-stranded circular DNA. Formation of the joint molecule intermediates and nicked circular product can be monitored by agarose gel electrophoresis. (C) Examples of Holliday junction endonuclease susceptible substrates including (i) a 3′ flap structure, (ii) a synthetic nicked Holliday junction, and (iii) a three-way junction.

The *Sulfolobus* protein can specifically bind four-way DNA junctions with magnesium dependence and cleave 3′ to the centre of the junction, perhaps with some cleavage sequence-specificity (Kvaratskhelia and White, 2000b). The structure of Hjc shows extensive homology with a superfamily of nucleases, including type II restriction enzymes (Daiyasu *et al.*, 2000). Additionally, a second Holliday-junction cleavage activity has been found in *Sulfolobus*. Dubbed Hje for Holiday junction endonuclease, this activity has only been partially purified, but resolves four way junctions by introducing paired nicks in opposing strands and producing nicked duplex DNA products (Kvaratskhelia and White, 2000a). Hje does not have apparent sequence specificity for cleavage at the junction, indicating that this protein is not a type 1 enzyme. The protein distinguishes continuous from exchanging strands of the four-way DNA junction more strongly than any other known Holliday-junction cleavage enzyme, implying cleavage specificity in the absence of specific local nucleotide sequence (Kvaratskhelia and White, 2000a).

Complex branched DNA structures often occur in cells as a direct consequence of DNA damage or as intermediates in recombination pathways (for examples, see Figs 6.3–6.5). Eukaryotes use the nucleases XPF-ERCC1 and Mus81-Eme1/Mms4 to process these substrates containing 3′ branches and duplex bubbles (Fricke *et al.*, 2005; Park *et al.*, 1995). Archaeal genomic sequence examination reveals clear homologues of XPF. In *P. furiosus*, the homologue is named Hef (Helicase-associated endonuclease for forked-structured DNA). The Hef protein cleaves numerous fork and flap structures that act as substrates for the eukaryotic Mus81 complex (Komori *et al.*, 2002). Structural analysis of Hef shows it has similarity to a superfamily of nucleases including restriction enzymes and the *S. solfataricus* Holliday junction resolving enzyme Hjc (Nishino *et al.*, 2003). Crenarchaeal XPF homologues appear to be smaller, simpler versions of the euryarchaeal protein, made up of the core nuclease and helix-hairpin-helix domains (Roberts *et al.*, 2003). These proteins cut splayed duplex structures and show a pronounced preference for Mus81-type substrates like D-loops, 3′ flaps, and nicked four-way junctions (Roberts *et al.*, 2003; Roberts and White, 2005a). *Sulfolobus* XPF can remove numerous types of DNA lesions and processively digest a nicked DNA strand in the 3′–5′ direction (Roberts and White, 2005b). Interestingly, the activity of the *Sulfolobus* XPF protein is entirely dependent upon association with the sliding clamp protein, PCNA. PCNA therefore serves as an essential cofactor for XPF and the target sites for XPF activity may, in part, be determined by PCNA loading (Roberts and White, 2005a).

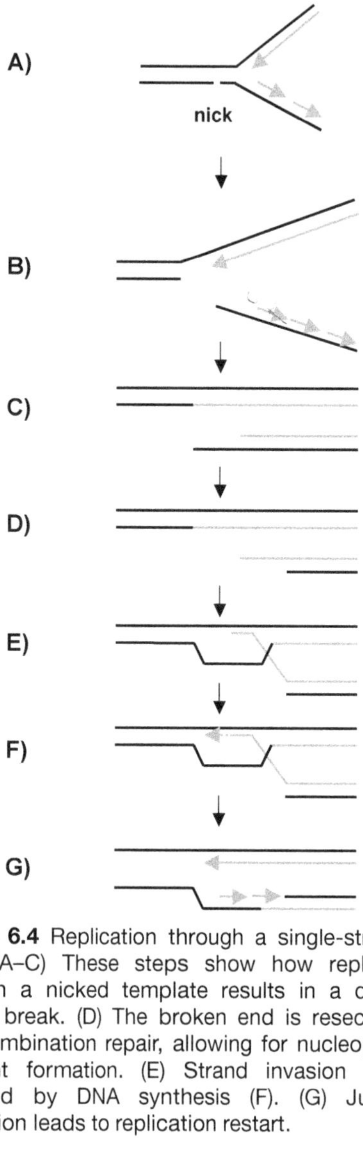

Figure 6.4 Replication through a single-stranded nick. (A–C) These steps show how replication through a nicked template results in a double-strand break. (D) The broken end is resected as in recombination repair, allowing for nucleoprotein filament formation. (E) Strand invasion occurs followed by DNA synthesis (F). (G) Junction resolution leads to replication restart.

Archaeal recombination as a model

Eukaryotic and bacterial systems

Archaeal recombination protein studies have significantly advanced our understanding of homologous systems in eukaryotes, perhaps most strongly through determination of protein structures. Crystallization of mesophilic proteins with large flexible regions or hinges is experimentally problematic, but structures of thermophilic protein versions of these proteins have been especially informative. Indeed, archaeal crystal structures have led to important advancements in understanding eukaryotic recombination mechanisms and bacterial nucleoprotein filament formation.

It was the topoisomerase II from *S. shibatae* that ultimately revealed the function of eukaryotic Spo11 protein. Spo11 protein was known to be required for initiation of meiotic recombination, but its mechanism of action was elusive. Similarity between the archaeal and eukaryotic proteins was coupled with the crystal structure of the *S. shibatae* protein. This comparison ultimately led to the demonstration that Spo11 is the protein responsible for initiating programmed double-strand breaks in meiotic recombination (Bergerat *et al.*, 1997; Nichols *et al.*, 1999).

Several crystal structures from *P. furiosus* have been pivotal in understanding the activities of eukaryotic recombination proteins. The structure of Mre11 has been determined and its interaction with Rad50 examined. Mapping homologues from human and yeast protein homologues onto the completed archaeal structure has uncovered macromolecular interaction sites in the eukaryotic proteins (Hopfner *et al.*, 2001). The *P. furiosus* RadA structure and its astonishing interaction with Brca2 led to the interface exchange hypothesis and a suggested role for Brca2 in formation of the Rad51 nucleoprotein filament (Shin *et al.*, 2003).

It is generally accepted that the extended form of RecA family nucleoprotein filaments form the active scaffold that facilitates strand exchange. Despite limited sequence homology, the physical structure of the *M. voltae* RadA shares strong similarity with *E. coli* RecA (Wu *et al.*, 2004). Crystallization of the *M. voltae* RadA has yielded the first high-resolution structure of the extended filament form. This remarkable structure will certainly contribute greatly to understanding the dynamic process of active filament formation.

Examination of archaeal recombination systems underscores the prevalence and persistence of specific recombination protein mechanisms across all of biology. It is not clear, though, whether recombination systems are the result of acquisition of complete pathways through lateral gene transfer or if instead these mechanisms evolved protein by protein in the individual domains of life.

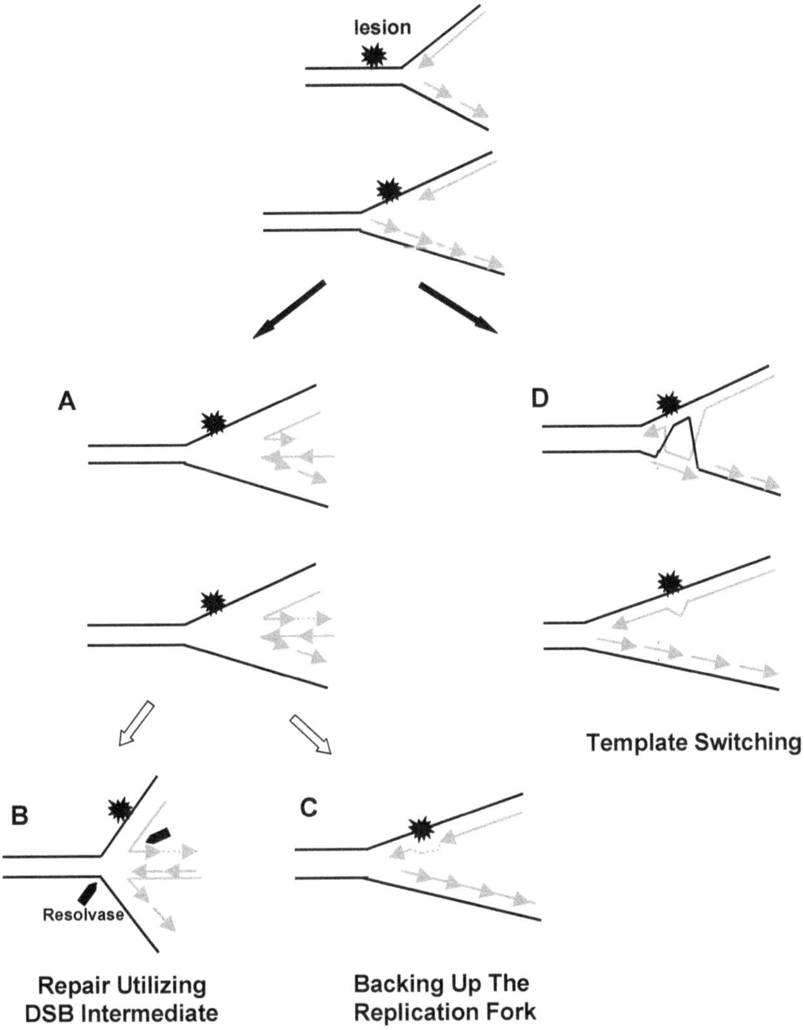

Figure 6.5 Active replication and repair of a DNA lesion through recombinational repair. The DNA lesion is represented by the starburst. (A) The replication fork can move backwards, allowing new DNA synthesis using the intact complementary strand as a template. After synthesis beyond the lesion, the resulting structure (resembling a 'chicken foot') can be processed as a double Holliday junction (B) or the leading strand can deanneal from its current template and reanneal to its original template at a location beyond the lesion (much like synthesis-dependent strand annealing) as in (C). (D) Template switching. The leading strand encounters the lesion and switches strands, via a strand invasion event, to utilize the unblocked strand as a template. After replication past the lesion, the strands deanneal and then reanneal with their original complementary strands before DNA synthesis continues.

Unique archaeal features

It was the core metabolic functions and processes involved with a prokaryotic lifestyle that resulted in the initial classification of Archaea as archaebacteria. Subsequently, genome sequencing projects have exploded and a flood of sequence data has shown that, surprisingly, archaeal genomes have a large number of open reading frames with no apparent homologues in either the bacterial or eukaryotic domains (Makarova and Koonin, 2003). As it had widely been assumed that Archaea were chimaeras of these two branches, this observation raises the question of the function of these mysterious proteins.

In accounting for the identity of these unidentified proteins, some simple observations of these organisms point to likely suspects. For example, generation and maintenance of unique archaeal membranes likely requires equally unique proteins. Also, certain metabolic pathways (like methanogenesis) are only present in Archaea. When it comes to recombination proteins, there have been some unique activities identified in the Archaea. Euryarchaeal histones, while similar in structure to their eukaryotic counterparts, lack their tails and commensurate modifications. The DNA-packaging proteins of the crenarchaea lack structural similarity to eukaryotic histones, but have demonstrable acetylation sites (for reviews see: Sandman and Reeve (2005) and Cubonova *et al.* (2005). The RadA protein from *T. kodakarensis* KOD1 has ATPase activity, as do all members of the RecA family, but also displays surprising DNase activity (Rashid *et al.*, 1999). The Hje protein establishes a new class of Holliday junction-resolving enzymes that is clearly different from previously identified junction-resolving proteins (Kvaratskhelia and White, 2000a).

The future of recombination studies in Archaea

It is truly an exciting time in archaeal science. With the availability of 19 complete archaeal genome sequences, the template for life in these unusual organisms is at long last being revealed (Makarova and Koonin, 2003). Thus far, 'sequence gazing' has been the primary method for identifying proteins with homology to those in eubacteria and eukaryotes (Table 6.1). Similarity does not, however, necessarily correlate with conservation of function. Many important genes will likely continue to escape identification as a result of insufficient sequence similarity to known genes. Functional overlaps may occur, and the most similar protein may not always also be the functional analogue. So while there is considerable enthusiasm over genome project sequence revelations, it must be remembered that dramatic divergence can effectively hide relevant genes. To discover archaeal recombination proteins without clear sequence homologues in other organisms, greater emphasis must be placed on the discovery of novel protein interactions. Two- and three-hybrid library screens have been successfully used to find replication related proteins (Barry and Bell, 2006) and are likely to reveal previously unknown recombination proteins. Ventures into archaeal microarrays promise to help elucidate functions of genes that are homologous only within the Archaea (Andersson *et al.*, 2005; Bonneau *et al.*, 2004; Hamilton-Brehm *et al.*, 2005; Williams *et al.*, 2007). The use of traditional reverse genetics including gene knockouts, combined with library approaches will elucidate the function of unknown genes in archaeal genomes. Blending these approaches will yield great insights into our understanding of archaeal recombination mechanisms.

There has been significant progress in development of genetic systems for Archaea since the revelation of the existence of their domain. In the last several years, there has been a virtual explosion of methods, especially for the less tractable thermophiles. Notably absent, however, is an *in vivo* recombination assay approach that bridges biochemical studies with genetic phenotypes.

Table 6.1 Proteins Involved in Homologous Recombination

Recombination Step	Crenarchaea	Euryarchaea	Expected Contributors
Initiation (DNA packaging)	Sul7d CC1 Alba1/Alba2	Histones	?
Resection	Rad50-Mre11	Rad50-Mre11	Xrs2 or Nbs1 homologue
Homology Search	RadA	RadA RadB	Rad52 homologue RadA paralogues
Displacement Loop	RadA	RadA	?
Branch Migration	RadA Rad54	RadA	Rad54 homologue
DNA Synthesis	?	?	DNA polymerase
Holliday Junction Resolution	Hjc Hje	Hjc	Mus81-Eme1/Mms4 Sgs1 Top3 homologue

More subtle recombination activities, as expected from the activity of accessory proteins, may not be detectable through standard genotoxic stress assays. Instead, methods like chromatin immunoprecipatation will be needed for detection of subtle kinetic changes in DNA repair.

Understanding archaeal recombination activities promises to yield new insights into similar eukaryotic mechanisms. For example, the discovery of the archaeal Holliday junction resolvase (Kvaratskhelia *et al.*, 2001) has bolstered efforts to find the elusive eukaryotic version. And while the close relationship between eukaryotic and archaeal DNA metabolism is clear, there still remains the possibility that Archaea can provide a link to activities thus far undetected in the bacterial branch. Finding the archaeal Rad54 protein, once thought to be an exclusively eukaryotic recombination component, is a tantalizing hint for the possible presence of an as yet unidentified functional homologue in eubacteria.

As the opportunity for homologous recombination in a haploid organism should be severely limited in order to maintain genomic integrity, recombination would likely be restricted to occurring during, or shortly after, replication. To successfully pass genetic information to the next generation, an efficient mechanism for recovery from this potentially damaging event is necessary. An unrepaired single-strand break or nick in the DNA template will become a double-strand break following replication fork passage (Fig. 6.4); just one lingering double-strand break can lead to cell death. There are several potential methods a cell might employ to cope with a collapsed or stalled replication fork (Fig. 6.5). Recombination during DNA replication is likely to be a preferred way for the cell to handle stalled or even collapsed replication forks. Recently, some halophilic Archaea have been shown to be polyploidy at all times in the life cycle (Breuert *et al.*, 2006). Some Archaea even have an extended G2 stage before cell division (Hjort and Bernander, 2001). In retrospect, then, the concept of haploid character is not completely accurate; meroploidy or polyploidy is the norm for extended periods in the life cycle, thus providing ample access to templates for homologous recombination based repair processes. Information about how Archaea effectively handle these life-threatening events is enticingly close.

Aside from recombination mechanism similarities between all three domains of life, it is quite likely that Archaea have unique mechanistic features all their own. Considering the extreme environmental niches inhabited by some archaeal species, it is not impractical to suggest that these organisms developed unusually effective strategies for surviving ssDNA breaks. Regions consisting of SSB or RPA protein-coated ssDNA or similarly coated single-strand DNA gaps are implicated in signalling the initiation of repair, and when not repaired efficiently, can lead to a double-strand break (Eder *et al.*, 2001). Archaeal mechanisms may involve novel biochemical features or may instead include many familiar biochemical mechanisms but utilize fewer or different proteins to accomplish the same tasks. Thus, archaeal pathways could be simplified versions of complex eukaryotic pathways. Investigation of recombination mechanisms from extremophilic Archaea may ultimately reveal microbial solutions for maintenance of genomic stability and help elucidate the mechanistic requirements necessary for survival in the most severe habitats, perhaps even those found elsewhere in the universe.

Acknowledgements

We would like to thank Becky Reeves for her helpful comments on the manuscript. We also wish to express our deep regret to our colleagues whose primary work could not be cited due to space limitations.

References

Aagaard, C., Leviev, I., Aravalli, R.N., Forterre, P., Prieur, D., and Garrett, R.A. (1996). General vectors for archaeal hyperthermophiles: strategies based on a mobile intron and a plasmid. FEMS Microbiol. Rev. *18*, 93–104.

Aalfs, J.D., and Kingston, R.E. (2000). What does 'chromatin remodeling' mean? Trends Biochem. Sci. *25*, 548–555.

Albala, J.S., Thelen, M.P., Prange, C., Fan, W., Christensen, M., Thompson, L.H., and Lennon, G.G. (1997). Identification of a novel human RAD51 homolog, RAD51B. Genomics *46*, 476–479.

Alexeev, A., Mazin, A., and Kowalczykowski, S.C. (2003). Rad54 protein possesses chromatin-remodeling activity stimulated by the Rad51-ssDNA nucleoprotein filament. Nat. Struct Biol. *10*, 182–186.

Allers, T., and Mevarech, M. (2005). Archaeal genetics – the third way. Nat. Rev. Genet. *6*, 58–73.

Andersson, A., Bernander, R., and Nilsson, P. (2005). Dual-genome primer design for construction of DNA microarrays. Bioinformatics *21*, 325–332.

Aravalli, R.N., and Garrett, R.A. (1997). Shuttle vectors for hyperthermophilic Archaea. Extremophiles *1*, 183–191.

Aravind, L., Makarova, K.S., and Koonin, E.V. (2000). Holliday junction resolvases and related nucleases: identification of new families, phyletic distribution and evolutionary trajectories. Nucleic Acids Res. *28*, 3417–3432.

Arber, W. (2000). Genetic variation: molecular mechanisms and impact on microbial evolution. FEMS Microbiol. Rev. *24*, 1–7.

Argos, P., Landy, A., Abremski, K., Egan, J.B., Haggard-Ljungquist, E., Hoess, R.H., Kahn, M.L., Kalionis, B., Narayana, S.V., Pierson, L.S., 3rd, *et al.* (1986). The integrase family of site-specific recombinases: regional similarities and global diversity. Embo J. *5*, 433–440.

Barry, E.R., and Bell, S.D. (2006). DNA replication in the Archaea. Microbiol. Mol. Biol. Rev. *70*, 876–887.

Bell, S.D., Botting, C.H., Wardleworth, B.N., Jackson, S.P., and White, M.F. (2002). The interaction of Alba, a conserved archaeal chromatin protein, with Sir2 and its regulation by acetylation. Science *296*, 148–151.

Bergerat, A., de Massy, B., Gadelle, D., Varoutas, P.C., Nicolas, A., and Forterre, P. (1997). An atypical topoisomerase II from Archaea with implications for meiotic recombination. Nature *386*, 414–417.

Bernander, R., and Poplawski, A. (1997). Cell cycle characteristics of thermophilic Archaea. J. Bacteriol. *179*, 4963–4969.

Bertani, G., and Baresi, L. (1987). Genetic transformation in the methanogen Methanococcus voltae PS. J. Bacteriol. *169*, 2730–2738.

Bianco, P.R., and Kowalczykowski, S.C. (2005). RecA protein. In Encyclopedia of Life Sciences (Chichester, John Wiley & Sons, Ltd).

Bonneau, R., Baliga, N.S., Deutsch, E.W., Shannon, P., and Hood, L. (2004). Comprehensive de novo structure prediction in a systems-biology context for the Archaea *Halobacterium* sp. NRC-1. Genome Biol. 5, R52.

Breuert, S., Allers, T., Spohn, G., and Soppa, J. (2006). Regulated polyploidy in halophilic Archaea. PLoS ONE *1*, e92.

Brugger, K., Redder, P., She, Q., Confalonieri, F., Zivanovic, Y., and Garrett, R.A. (2002). Mobile elements in archaeal genomes. FEMS Microbiol. Lett. *206*, 131–141.

Brugger, K., Torarinsson, E., Redder, P., Chen, L., and Garrett, R.A. (2004). Shuffling of *Sulfolobus* genomes by autonomous and non-autonomous mobile elements. Biochem. Soc. Trans *32*, 179–183.

Bult, C.J., White, O., Olsen, G.J., Zhou, L., Fleischmann, R.D., Sutton, G.G., Blake, J.A., Fitzgerald, L.M., Clayton, R.A., and Gocayne, J.D. (1996). Complete Genome Sequence of the Methanogenic Archaeon, *Methanococcus jannaschii*. Science *273*, 1058–1073.

Carpentieri, F., De Felice, M., De Falco, M., Rossi, M., and Pisani, F.M. (2002). Physical and functional interaction between the mini-chromosome maintenance-like DNA helicase and the single-stranded DNA binding protein from the crenarchaeon *Sulfolobus solfataricus*. J. Biol. Chem. *277*, 12118–12127.

Cartwright, R., Tambini, C.E., Simpson, P.J., and Thacker, J. (1998). The XRCC2 DNA repair gene from human and mouse encodes a novel member of the recA/RAD51 family. Nucleic Acids Res. *26*, 3084–3089.

Charlebois, R.L., Lam, W.L., Cline, S.W., and Doolittle, W.F. (1987). Characterization of pHV2 from *Halobacterium volcanii* and its use in demonstrating transformation of an archaebacterium. Proc. Natl. Acad. Sci. USA *84*, 8530–8534.

Chédin, F., Seitz, E.M., and Kowalczykowski, S.C. (1998). Novel homologs of replication protein A in Archaea: implications for the evolution of ssDNA-binding proteins. Trends Biochem. Sci. *23*, 273–277.

Chen, L., Brugger, K., Skovgaard, M., Redder, P., She, Q., Torarinsson, E., Greve, B., Awayez, M., Zibat, A., Klenk, H.P., and Garrett, R.A. (2005). The genome of *Sulfolobus acidocaldarius*, a model organism of the Crenarchaeota. J. Bacteriol. *187*, 4992–4999.

Cline, S.W., and Doolittle, W.F. (1987). Efficient transfection of the archaebacterium *Halobacterium halobium*. J. Bacteriol. *169*, 1341–1344.

Cline, S.W., and Doolittle, W.F. (1992). Transformation of members of the genus Haloarcula with shuttle vectors based on *Halobacterium halobium* and *Haloferax volcanii* plasmid replicons. J. Bacteriol. *174*, 1076-1080.

Conover, R.K., and Doolittle, W.F. (1990). Characterization of a gene involved in histidine biosynthesis in *Halobacterium (Haloferax) volcanii*: isolation and rapid mapping by transformation of an auxotroph with cosmid DNA. J. Bacteriol. *172*, 3244–3249.

Constantinesco, F., Forterre, P., and Elie, C. (2002). NurA, a novel 5′-3′ nuclease gene linked to rad50 and mre11 homologs of thermophilic Archaea. EMBO Rep. *3*, 537–542.

Constantinesco, F., Forterre, P., Koonin, E.V., Aravind, L., and Elie, C. (2004). A bipolar DNA helicase gene, herA, clusters with rad50, mre11 and nurA genes in thermophilic Archaea. Nucleic Acids Res. *32*, 1439–1447.

Contursi, P., Jensen, S., Aucelli, T., Rossi, M., Bartolucci, S., and She, Q. (2006). Characterization of the *Sulfolobus* host-SSV2 virus interaction. Extremophiles *10*, 615–627.

Cubeddu, L., and White, M.F. (2005). DNA Damage Detection by an Archaeal Single-stranded DNA-binding Protein. J. Mol. Biol. *353*, 507–516.

Cubonova, L., Sandman, K., Hallam, S.J., Delong, E.F., and Reeve, J.N. (2005). Histones in crenarchaea. J. Bacteriol. *187*, 5482–5485.

Cui, X., Brenneman, M., Meyne, J., Oshimura, M., Goodwin, E.H., and Chen, D.J. (1999). The XRCC2 and XRCC3 repair genes are required for chromosome stability in mammalian cells. Mutat Res. *434*, 75–88.

Daiyasu, H., Komori, K., Sakae, S., Ishino, Y., and Toh, H. (2000). Hjc resolvase is a distantly related member of the type II restriction endonuclease family. Nucleic Acids Res. *28*, 4540–4543.

Diruggiero, J., Dunn, D., Maeder, D.L., Holley-Shanks, R., Chatard, J., Horlacher, R., Robb, F.T., Boos, W., and Weiss, R.B. (2000). Evidence of recent lateral gene transfer among hyperthermophilic Archaea. Mol. Microbiol. *38*, 684–693.

DiRuggiero, J., Santangelo, N., Nackerdien, Z., Ravel, J., and Robb, F.T. (1997). Repair of extensive ionizing-radiation DNA damage at 95 degrees C in the hyperthermophilic archaeon *Pyrococcus furiosus*. J. Bacteriol. *179*, 4643–4645.

Dolganov, G.M., Maser, R.S., Novikov, A., Tosto, L., Chong, S., Bressan, D.A., and Petrini, J.H. (1996). Human Rad50 is physically associated with human Mre11: identification of a conserved multiprotein complex implicated in recombinational DNA repair. Mol. Cell Biol. *16*, 4832–4841.

Dosanjh, M.K., Collins, D.W., Fan, W., Lennon, G.G., Albala, J.S., Shen, Z., and Schild, D. (1998). Isolation and characterization of RAD51C, a new human member of the RAD51 family of related genes. Nucleic Acids Res. *26*, 1179–1184.

Durr, H., Korner, C., Muller, M., Hickmann, V., and Hopfner, K.P. (2005). X-ray structures of the *Sulfolobus solfataricus* SWI2/SNF2 ATPase core and its complex with DNA. Cell *121*, 363–373.

Dyall-Smith, M., Tang, S.L., and Bath, C. (2003). Haloarchaeal viruses: how diverse are they? Res. Microbiol. *154*, 309–313.

Eder, E., Kutt, W., and Deininger, C. (2001). On the role of alkylating mechanisms, O-alkylation and DNA-repair in genotoxicity and mutagenicity of alkylating methanesulfonates of widely varying structures in bacterial systems. Chem. Biol. Interact *137*, 89–99.

Edmondson, S.P., and Shriver, J.W. (2001). DNA binding proteins Sac7d and Sso7d from *Sulfolobus*. Methods Enzymol *334*, 129–145.

Erauso, G., Stedman, K.M., van de Werken, H.J., Zillig, W., and van der Oost, J. (2006). Two novel conjugative plasmids from a single strain of *Sulfolobus*. Microbiology *152*, 1951–1968.

Esposito, D., and Scocca, J.J. (1997). The integrase family of tyrosine recombinases: evolution of a conserved active site domain. Nucleic Acids Res. *25*, 3605–3614.

Forterre, P., Confalonieri, F., and Knapp, S. (1999). Identification of the gene encoding archeal-specific DNA-binding proteins of the Sac10b family. Mol. Microbiol. *32*, 669–670.

Fricke, W.M., Bastin-Shanower, S.A., and Brill, S.J. (2005). Substrate specificity of the *Saccharomyces cerevisiae* Mus81-Mms4 endonuclease. DNA Repair (Amst) *4*, 243–251.

Furuse, M., Nagase, Y., Tsubouchi, H., Murakami-Murofushi, K., Shibata, T., and Ohta, K. (1998). Distinct roles of two separable *in vitro* activities of yeast Mre11 in mitotic and meiotic recombination. Embo J. *17*, 6412–6425.

Fyodorov, D.V., and Kadonaga, J.T. (2001). The many faces of chromatin remodeling: SWItching beyond transcription. Cell *106*, 523–525.

Gao, Y.G., Su, S.Y., Robinson, H., Padmanabhan, S., Lim, L., McCrary, B.S., Edmondson, S.P., Shriver, J.W., and Wang, A.H. (1998). The crystal structure of the hyperthermophile chromosomal protein Sso7d bound to DNA. Nat. Struct Biol. *5*, 782–786.

Gernhardt, P., Possot, O., Foglino, M., Sibold, L., and Klein, A. (1990). Construction of an integration vector for use in the archaebacterium *Methanococcus voltae* and expression of a eubacterial resistance gene. Mol. Gen. Genet. *221*, 273–279.

Gomes, X.V., and Wold, M.S. (1995). Structural analysis of human replication protein A. Mapping functional domains of the 70-kDa subunit. J. Biol. Chem. *270*, 4534–4543.

Gomes, X.V., and Wold, M.S. (1996). Functional domains of the 70-kilodalton subunit of human replication protein A. Biochemistry *35*, 10558–10568.

Greve, B., Jensen, S., Brugger, K., Zillig, W., and Garrett, R.A. (2004). Genomic comparison of archaeal conjugative plasmids from *Sulfolobus*. Archaea *1*, 231–239.

Grogan, D.W. (1996). Exchange of genetic markers at extremely high temperatures in the archaeon *Sulfolobus acidocaldarius*. J. Bacteriol. *178*, 3207–3211.

Grogan, D.W., and Hansen, J.E. (2003). Molecular characteristics of spontaneous deletions in the hyperthermophilic archaeon *Sulfolobus acidocaldarius*. J. Bacteriol. *185*, 1266-1272.

Guy, C.P., Haldenby, S., Brindley, A., Walsh, D.A., Briggs, G.S., Warren, M.J., Allers, T., and Bolt, E.L. (2006). Interactions of RadB, a DNA repair protein in Archaea, with DNA and ATP. J. Mol. Biol. *358*, 46–56.

Hamiche, A., Sandaltzopoulos, R., Gdula, D.A., and Wu, C. (1999). ATP-dependent histone octamer sliding mediated by the chromatin remodeling complex NURF. Cell *97*, 833–842.

Hamilton-Brehm, S.D., Schut, G.J., and Adams, M.W. (2005). Metabolic and evolutionary relationships among *Pyrococcus* Species: genetic exchange within a hydrothermal vent environment. J. Bacteriol. *187*, 7492–7499.

Hansen, J.E., Dill, A.C., and Grogan, D.W. (2005). Conjugational genetic exchange in the hyperthermophilic archaeon *Sulfolobus acidocaldarius*: intragenic recombination with minimal dependence on marker separation. J. Bacteriol. *187*, 805–809.

Harsojo, Kitayama, S., and Matsuyama, A. (1981). Genome multiplicity and radiation resistance in *Micrococcus radiodurans*. J. Biochem. (Tokyo) *90*, 877–880.

Haseltine, C.A., and Kowalczykowski, S.C. (2002). A distinctive single-strand DNA-binding protein from the Archaeon *Sulfolobus solfataricus*. Mol. Microbiol. *43*, 1505–1515.

Heyer, W.D., Li, X., Rolfsmeier, M., and Zhang, X.P. (2006). Rad54: the Swiss Army knife of homologous recombination? Nucleic Acids Res. *34*, 4115–4125.

Hjort, K., and Bernander, R. (2001). Cell cycle regulation in the hyperthermophilic crenarchaeon *Sulfolobus acidocaldarius*. Mol. Microbiol. *40*, 225–234.

Holmes, M.L., Nuttall, S.D., and Dyall-Smith, M.L. (1991). Construction and use of halobacterial shuttle vectors and further studies on *Haloferax* DNA gyrase. J. Bacteriol. *173*, 3807–3813.

Hopfner, K.P., Karcher, A., Craig, L., Woo, T.T., Carney, J.P., and Tainer, J.A. (2001). Structural biochemistry and interaction architecture of the DNA double-strand break repair Mre11 nuclease and Rad50-ATPase. Cell *105*, 473–485.

Hopfner, K.P., Karcher, A., Shin, D., Fairley, C., Tainer, J.A., and Carney, J.P. (2000a). Mre11 and rad50 from *Pyrococcus furiosus*: cloning and biochemical characterization reveal an evolutionarily conserved multiprotein machine [In Process Citation]. J. Bacteriol. *182*, 6036–6041.

Hopfner, K.P., Karcher, A., Shin, D.S., Craig, L., Arthur, L.M., Carney, J.P., and Tainer, J.A. (2000b). Structural biology of Rad50 ATPase: ATP-driven conformational control in DNA double-strand break repair and the ABC-ATPase superfamily. Cell *101*, 789–800.

Jelinska, C., Conroy, M.J., Craven, C.J., Hounslow, A.M., Bullough, P.A., Waltho, J.P., Taylor, G.L., and White, M.F. (2005). Obligate heterodimerization of the archaeal Alba2 protein with Alba1 provides a mechanism for control of DNA packaging. Structure *13*, 963–971.

Jinks-Robertson, S., Michelitch, M., and Ramcharan, S. (1993). Substrate length requirements for efficient mitotic recombination in *Saccharomyces cerevisiae*. Mol. Cell Biol. *13*, 3937–3950.

Johnson, R.D., and Symington, L.S. (1995). Functional differences and interactions among the putative RecA homologs Rad51, Rad55, and Rad57. Mol. Cell Biol. *15*, 4843–4850.

Johzuka, K., and Ogawa, H. (1995). Interaction of Mre11 and Rad50: two proteins required for DNA repair and meiosis-specific double-strand break formation in *Saccharomyces cerevisiae*. Genetics *139*, 1521–1532.

Kanoksilapatham, W., Gonzalez, J.M., Maeder, D.L., DiRuggiero, J., and Robb, F.T. (2004). A proposal to rename the hyperthermophile *Pyrococcus woesei* as *Pyrococcus furiosus* subsp. *woesei*. Archaea *1*, 277–283.

Kawarabayasi, Y., Hino, Y., Horikawa, H., Jin-no, K., Takahashi, M., Sekine, M., Baba, S., Ankai, A., Kosugi, H., Hosoyama, A., *et al.* (2001). Complete genome sequence of an aerobic thermoacidophilic crenarchaeon, *Sulfolobus tokodaii* strain7. DNA Res. *8*, 123–140.

Kawarabayasi, Y., Sawada, M., Horikawa, H., Haikawa, Y., Hino, Y., Yamamoto, S., Sekine, M., Baba, S., Kosugi, H., Hosoyama, A., *et al.* (1998). Complete sequence and gene organization of the genome of a hyper- thermophilic archaebacterium, *Pyrococcus horikoshii* OT3. DNA Res. *5*, 55–76.

Kawashima, T., Amano, N., Koike, H., Makino, S., Higuchi, S., Kawashima-Ohya, Y., Watanabe, K., Yamazaki, M., Kanehori, K., Kawamoto, T., *et al.* (2000). Archaeal adaptation to higher temperatures revealed by genomic sequence of *Thermoplasma volcanium*. Proc. Natl. Acad. Sci. USA *97*, 14257–14262.

Kelly, T.J., Simancek, P., and Brush, G.S. (1998). Identification and characterization of a single-stranded DNA-binding protein from the archaeon *Methanococcus jannaschii*. Proc. Natl. Acad. Sci. USA *95*, 14634–14639.

Kerr, I.D., Wadsworth, R.I., Cubeddu, L., Blankenfeldt, W., Naismith, J.H., and White, M.F. (2003). Insights into ssDNA recognition by the OB fold from a structural and thermodynamic study of *Sulfolobus* SSB protein. Embo J. *22*, 2561–2570.

Kil, Y.V., Baitin, D.M., Masui, R., Bonch-Osmolovskaya, E.A., Kuramitsu, S., and Lanzov, V.A. (2000). Efficient strand transfer by the RadA recombinase from the hyperthermophilic archaeon *Desulfurococcus amylolyticus*. J. Bacteriol. *182*, 130–134.

Klenk, H.P., Clayton, R.A., Tomb, J.F., White, O., Nelson, K.E., Ketchum, K.A., Dodson, R.J., Gwinn, M., Hickey, E.K., Peterson, J.D., *et al.* (1997). The complete genome sequence of the hyperthermophilic, sulphate-reducing archaeon *Archaeoglobus fulgidus*. Nature *390*, 364–370.

Komori, K., Fujikane, R., Shinagawa, H., and Ishino, Y. (2002). Novel endonuclease in Archaea cleaving DNA with various branched structure. Genes Genet. Syst *77*, 227–241.

Komori, K., and Ishino, Y. (2001). Replication protein A in *Pyrococcus furiosus* is involved in homologous DNA recombination. J. Biol. Chem. *276*, 25654–25660.

Komori, K., Miyata, T., Daiyasu, H., Toh, H., Shinagawa, H., and Ishino a, Y. (2000a). Domain Analysis of an Archaeal RadA Protein for the Strand Exchange Activity. J. Biol. Chem. *275*, 33791–33797.

Komori, K., Miyata, T., Daiyasu, H., Toh, H., Shinagawa, H., and Ishino, Y. (2000b). Domain analysis of an archaeal RadA protein for the strand exchange activity. J. Biol. Chem. *275*, 33791–33797.

Komori, K., Miyata, T., DiRuggiero, J., Holley-Shanks, R., Hayashi, I., Cann, I.K., Mayanagi, K., Shinagawa, H., and Ishino, Y. (2000c). Both RadA and RadB Are Involved in Homologous Recombination in *Pyrococcus furiosus*. J. Biol. Chem. *275*, 33782–33790.

Komori, K., Sakae, S., Shinagawa, H., Morikawa, K., and Ishino, Y. (1999). A Holliday junction resolvase from *Pyrococcus furiosus*: functional similarity to *Escherichia coli* RuvC provides evidence for conserved mechanism of homologous recombination in Bacteria, Eukarya, and Archaea. Proc. Natl. Acad. Sci. USA *96*, 8873–8878.

Kowalczykowski, S.C., Dixon, D.A., Eggleston, A.K., Lauder, S.D., and Rehrauer, W.M. (1994). Biochemistry of homologous recombination in *Escherichia coli*. Microbiol. Rev. *58*, 401–465.

Kowalczykowski, S.C., and Eggleston, A.K. (1994). Homologous pairing and DNA strand-exchange proteins. Annu. Rev. Biochem. *63*, 991–1043.

Krasin, F., and Hutchinson, F. (1977). Repair of DNA double-strand breaks in *Escherichia coli*, which requires *recA* function and the presence of a duplicate genome. J. Mol. Biol. *116*, 81–98.

Kurosawa, N., and Grogan, D.W. (2005). Homologous recombination of exogenous DNA with the *Sulfolobus acidocaldarius* genome: properties and uses. FEMS Microbiol. Lett. *253*, 141–149.

Kuzminov, A. (1999). Recombinational repair of DNA damage in *Escherichia coli* and bacteriophage λ. Microbiol. Mol. Biol. Rev. *63*, 751–813.

Kvaratskhelia, M., Wardleworth, B.N., and White, M.F. (2001). Multiple Holliday junction resolving enzyme activities in the Crenarchaeota and Euryarchaeota. FEBS Lett. *491*, 243–246.

Kvaratskhelia, M., and White, M.F. (2000a). An archaeal Holliday junction resolving enzyme from *Sulfolobus solfataricus* exhibits unique properties. J. Mol. Biol. *295*, 193–202.

Kvaratskhelia, M., and White, M.F. (2000b). Two Holliday junction resolving enzymes in *Sulfolobus solfataricus*. J. Mol. Biol. *297*, 923–932.

Langst, G., and Becker, P.B. (2001). Nucleosome mobilization and positioning by ISWI-containing chromatin-remodeling factors. J. Cell Sci. *114*, 2561–2568.

Langst, G., Bonte, E.J., Corona, D.F., and Becker, P.B. (1999). Nucleosome movement by CHRAC and ISWI without disruption or trans-displacement of the histone octamer. Cell *97*, 843–852.

Lecompte, O., Ripp, R., Puzos-Barbe, V., Duprat, S., Heilig, R., Dietrich, J., Thierry, J.C., and Poch, O. (2001). Genome evolution at the genus level: comparison of three complete genomes of hyperthermophilic Archaea. Genome Res. *11*, 981–993.

Liu, N., Lamerdin, J.E., Tebbs, R.S., Schild, D., Tucker, J.D., Shen, M.R., Brookman, K.W., Siciliano, M.J., Walter, C.A., Fan, W., *et al.* (1998). XRCC2 and XRCC3, new human Rad51-family members, promote chromosome stability and protect against DNA cross-links and other damages. Mol. Cell *1*, 783–793.

Liu, Y., Masson, J.Y., Shah, R., O'Regan, P., and West, S.C. (2004). RAD51C is required for Holliday junction processing in mammalian cells. Science *303*, 243–246.

Low, K.B. (1991). Conjugational methods for mapping with Hfr and F-prime strains. Methods Enzymol *204*, 43–62.

Lucas, S., Toffin, L., Zivanovic, Y., Charlier, D., Moussard, H., Forterre, P., Prieur, D., and Erauso, G. (2002). Construction of a shuttle vector for, and spheroplast transformation of, the hyperthermophilic archaeon *Pyrococcus abyssi*. Appl. Environ. Microbiol. *68*, 5528–5536.

Luo, X., Schwarz-Linek, U., Botting, C.H., Hensel, R., Siebers, B., and White, M.F. (2006). CC1, a novel crenarchaeal DNA binding protein. J. Bacteriol..

Lurz, R., Grote, M., Dijk, J., Reinhardt, R., and Dobrinski, B. (1986). Electron microscopic study of DNA complexes with proteins from the Archaebacterium *Sulfolobus acidocaldarius*. Embo J. *5*, 3715–3721.

Mahillon, J., and Chandler, M. (1998). Insertion sequences. Microbiol. Mol. Biol. Rev. *62*, 725–774.

Makarova, K.S., and Koonin, E.V. (2003). Comparative genomics of Archaea: how much have we learned in six years, and what's next? Genome Biol. *4*, 115.

Malandrin, L., Huber, H., and Bernander, R. (1999). Nucleoid structure and partition in *Methanococcus jannaschii*: an archaeon with multiple copies of the chromosome. Genetics *152*, 1315–1323.

Manzan, A., Pfeiffer, G., Hefferin, M.L., Lang, C.E., Carney, J.P., and Hopfner, K.P. (2004). MlaA, a hexameric ATPase linked to the Mre11 complex in archaeal genomes. EMBO Rep. *5*, 54–59.

Marsh, V.L., Peak-Chew, S.Y., and Bell, S.D. (2005). Sir2 and the acetyltransferase, Pat, regulate the archaeal chromatin protein, Alba. J. Biol. Chem. *280*, 21122–21128.

Martusewitsch, E., Sensen, C.W., and Schleper, C. (2000). High spontaneous mutation rate in the hyperthermophilic archaeon *Sulfolobus solfataricus* is mediated by transposable elements. J. Bacteriol. *182*, 2574–2581.

McAfee, J.G., Edmondson, S.P., Zegar, I., and Shriver, J.W. (1996). Equilibrium DNA binding of Sac7d protein from the hyperthermophile *Sulfolobus acidocaldarius*: fluorescence and circular dichroism studies. Biochemistry *35*, 4034–4045.

McIlwraith, M.J., Hall, D.R., Stasiak, A.Z., Stasiak, A., Wigley, D.B., and West, S.C. (2001). RadA protein from *Archaeoglobus fulgidus* forms rings, nucleoprotein filaments and catalyses homologous recombination. Nucleic Acids Res. *29*, 4509–4517.

Mevarech, M., and Werczberger, R. (1985). Genetic transfer in *Halobacterium volcanii*. J. Bacteriol. *162*, 461–462.

Minton, K.W., and Daly, M.J. (1995). A model for repair of radiation-induced DNA double-strand breaks in the extreme radiophile *Deinococcus radiodurans*. Bioessays *17*, 457–464.

Mira, A., Klasson, L., and Andersson, S.G. (2002). Microbial genome evolution: sources of variability. Curr. Opin. Microbiol. *5*, 506–512.

Mizuuchi, K. (1992). Transpositional recombination: mechanistic insights from studies of mu and other elements. Annu. Rev. Biochem. *61*, 1011–1051.

Muskhelishvili, G., Palm, P., and Zillig, W. (1993). SSV1-encoded site-specific recombination system in *Sulfolobus shibatae*. Mol. Gen. Genet. *237*, 334–342.

Napoli, A., Valenti, A., Salerno, V., Nadal, M., Garnier, F., Rossi, M., and Ciaramella, M. (2005). Functional interaction of reverse gyrase with single-strand binding protein of the archaeon *Sulfolobus*. Nucleic Acids Res. *33*, 564–576.

Napoli, A., Zivanovic, Y., Bocs, C., Buhler, C., Rossi, M., Forterre, P., and Ciaramella, M. (2002). DNA bending, compaction and negative supercoiling by the architectural protein Sso7d of *Sulfolobus solfataricus*. Nucleic Acids Res. *30*, 2656-2662.

Ng, W.V., Kennedy, S.P., Mahairas, G.G., Berquist, B., Pan, M., Shukla, H.D., Lasky, S.R., Baliga, N.S., Thorsson, V., Sbrogna, J., *et al.* (2000). Genome sequence of *Halobacterium* species NRC-1. Proc. Natl. Acad. Sci. USA *97*, 12176-12181.

Nichols, M.D., DeAngelis, K., Keck, J.L., and Berger, J.M. (1999). Structure and function of an archaeal topoisomerase VI subunit with homology to the meiotic recombination factor Spo11. Embo J. *18*, 6177–6188.

Nishino, T., Komori, K., Ishino, Y., and Morikawa, K. (2003). X-Ray and Biochemical Anatomy of an Archaeal XPF/Rad1/Mus81 Family Nuclease. Similarity between Its Endonuclease Domain and Restriction Enzymes. Structure (Camb) *11*, 445–457.

Nunes-Duby, S.E., Kwon, H.J., Tirumalai, R.S., Ellenberger, T., and Landy, A. (1998). Similarities and differences among 105 members of the Int. family of site-specific recombinases. Nucleic Acids Res. *26*, 391–406.

Okada, N., Hamada, M., Ogiwara, I., and Ohshima, K. (1997). SINEs and LINEs share common 3′ sequences: a review. Gene *205*, 229–243.

Oosumi, T., Garlick, B., and Belknap, W.R. (1996). Identification of putative nonautonomous transposable elements associated with several transposon families in *Caenorhabditis elegans*. J. Mol. Evol. *43*, 11–18.

Palm, P., Schleper, C., Grampp, B., Yeats, S., McWilliam, P., Reiter, W.D., and Zillig, W. (1991). Complete nucleotide sequence of the virus SSV1 of the archaebacterium *Sulfolobus shibatae*. Virology *185*, 242–250.

Papke, R.T., Koenig, J.E., Rodriguez-Valera, F., and Doolittle, W.F. (2004). Frequent recombination in a saltern population of *Halorubrum*. Science *306*, 1928–1929.

Pâques, F., and Haber, J.E. (1999). Multiple pathways of recombination induced by double-strand breaks in *Saccharomyces cerevisiae*. Microbiol. Mol. Biol. Rev. *63*, 349–404.

Park, C.H., Bessho, T., Matsunaga, T., and Sancar, A. (1995). Purification and characterization of the XPF-ERCC1 complex of human DNA repair excision nuclease. J. Biol. Chem. *270*, 22657–22660.

Patel, G.B., Nash, J.H., Agnew, B.J., and Sprott, G.D. (1994). Natural and Electroporation-Mediated Transformation of *Methanococcus voltae* Protoplasts. Appl. Environ. Microbiol. *60*, 903–907.

Paull, T.T., and Gellert, M. (1998). The 3′ to 5′ exonuclease activity of Mre 11 facilitates repair of DNA double-strand breaks. Mol. Cell *1*, 969–979.

Paull, T.T., and Gellert, M. (1999). Nbs1 potentiates ATP-driven DNA unwinding and endonuclease cleavage by the Mre11/Rad50 complex. Genes Dev. *13*, 1276-1288.

Peak, M.J., Robb, F.T., and Peak, J.G. (1995). Extreme resistance to thermally induced DNA backbone breaks in the hyperthermophilic archaeon *Pyrococcus furiosus*. J. Bacteriol. *177*, 6316–6318.

Peck, R.F., Dassarma, S., and Krebs, M.P. (2000). Homologous gene knockout in the archaeon *Halobacterium salinarum* with ura3 as a counterselectable marker. Mol. Microbiol. *35*, 667–676.

Peng, X., Holz, I., Zillig, W., Garrett, R.A., and She, Q. (2000). Evolution of the family of pRN plasmids and their integrase-mediated insertion into the chromosome of the crenarchaeon *Sulfolobus solfataricus*. J. Mol. Biol. *303*, 449–454.

Pfeifer, F., Friedman, J., Boyer, H.W., and Betlach, M. (1984). Characterization of insertions affecting the expression of the bacterio-opsin gene in *Halobacterium halobium*. Nucleic Acids Res. *12*, 2489–2497.

Pierce, A.J., Johnson, R.D., Thompson, L.H., and Jasin, M. (1999). XRCC3 promotes homology-directed repair of DNA damage in mammalian cells. Genes Dev. *13*, 2633–2638.

Prangishvili, D., Albers, S.V., Holz, I., Arnold, H.P., Stedman, K., Klein, T., Singh, H., Hiort, J., Schweier, A., Kristjansson, J.K., and Zillig, W. (1998). Conjugation in Archaea: frequent occurrence of conjugative plasmids in *Sulfolobus*. Plasmid *40*, 190–202.

Prangishvili, D., Vestergaard, G., Haring, M., Aramayo, R., Basta, T., Rachel, R., and Garrett, R.A. (2006). Structural and genomic properties of the hyperthermophilic archaeal virus ATV with an extracellular stage of the reproductive cycle. J. Mol. Biol. *359*, 1203–1216.

Pritchett, M.A., Zhang, J.K., and Metcalf, W.W. (2004). Development of a markerless genetic exchange method for *Methanosarcina acetivorans* C2A and its use in construction of new genetic tools for methanogenic Archaea. Appl. Environ. Microbiol. *70*, 1425–1433.

Qian, X., He, Y., Ma, X., Fodje, M.N., Grochulski, P., and Luo, Y. (2006). Calcium stiffens archaeal Rad51 recombinase from *Methanococcus voltae* for homologous recombination. J. Biol. Chem. *281*, 39380–39387.

Rashid, N., Morikawa, M., and Imanaka, T. (1996). A RecA/RAD51 homologue from a hyperthermophilic archaeon retains the major RecA domain only. Mol. Gen. Genet. *253*, 397–400.

Rashid, N., Morikawa, M., Kanaya, S., Atomi, H., and Imanaka, T. (1999). A unique DNase activity shares the active site with ATPase activity of the RecA/Rad51 homologue (Pk-REC) from a hyperthermophilic archaeon. FEBS Lett. *445*, 111–114.

Raymond, W.E., and Kleckner, N. (1993). RAD50 protein of *S.cerevisiae* exhibits ATP-dependent DNA binding. Nucleic Acids Res. *21*, 3851–3856.

Redder, P., She, Q., and Garrett, R.A. (2001). Non-autonomous mobile elements in the crenarchaeon *Sulfolobus solfataricus*. J. Mol. Biol. *306*, 1–6.

Reich, C.I., McNeil, L.K., Brace, J.L., Brucker, J.K., and Olsen, G.J. (2001). Archaeal RecA homologues: different response to DNA-damaging agents in mesophilic and thermophilic Archaea. Extremophiles *5*, 265–275.

Reilly, M.S., and Grogan, D.W. (2001). Characterization of intragenic recombination in a hyperthermophilic archaeon via conjugational DNA exchange. J. Bacteriol. *183*, 2943–2946.

Reiter, W.D., Palm, P., and Yeats, S. (1989). Transfer RNA genes frequently serve as integration sites for prokaryotic genetic elements. Nucleic Acids Res. *17*, 1907–1914.

Rice, G., Stedman, K., Snyder, J., Wiedenheft, B., Willits, D., Brumfield, S., McDermott, T., and Young, M.J. (2001). Viruses from extreme thermal environments. Proc. Natl. Acad. Sci. USA *98*, 13341–13345.

Rice, M.C., Smith, S.T., Bullrich, F., Havre, P., and Kmiec, E.B. (1997). Isolation of human and mouse genes based on homology to REC2, a recombinational repair gene from the fungus *Ustilago maydis*. Proc. Natl. Acad. Sci. USA *94*, 7417–7422.

Richard, D.J., Bell, S.D., and White, M.F. (2004). Physical and functional interaction of the archaeal single-stranded DNA-binding protein SSB with RNA polymerase. Nucleic Acids Res. *32*, 1065–1074.

Robbins, J.B., Murphy, M.C., White, B.A., Mackie, R.I., Ha, T., and Cann, I.K. (2004). Functional analysis of multiple single-stranded DNA-binding proteins from *Methanosarcina acetivorans* and their effects on DNA synthesis by DNA polymerase BI. J. Biol. Chem. *279*, 6315–6326.

Roberts, J.A., Bell, S.D., and White, M.F. (2003). An archaeal XPF repair endonuclease dependent on a heterotrimeric PCNA. Mol. Microbiol. *48*, 361–371.

Roberts, J.A., and White, M.F. (2005a). An archaeal endonuclease displays key properties of both eukaryal XPF-ERCC1 and Mus81. J. Biol. Chem. *280*, 5924–5928.

Roberts, J.A., and White, M.F. (2005b). DNA end-directed and processive nuclease activities of the archaeal XPF enzyme. Nucleic Acids Res. *33*, 6662–6670.

Rosenshine, I., Tchelet, R., and Mevarech, M. (1989). The mechanism of DNA transfer in the mating system of an archaebacterium. Science *245*, 1387–1389.

Sandler, S.J., Satin, L.H., Samra, H.S., and Clark, A.J. (1996). *recA*-like genes from three archaean species with putative protein products similar to Rad51 and Dmc1 proteins of the yeast *Saccharomyces cerevisiae*. Nucleic Acids Res. *24*, 2125–2132.

Sandman, K., Grayling, R.A., Dobrinski, B., Lurz, R., and Reeve, J.N. (1994). Growth-phase-dependent synthesis of histones in the archaeon *Methanothermus fervidus*. Proc. Natl. Acad. Sci. USA *91*, 12624–12628.

Sandman, K., Krzycki, J.A., Dobrinski, B., Lurz, R., and Reeve, J.N. (1990). HMf, a DNA-binding protein isolated from the hyperthermophilic archaeon *Methanothermus fervidus*, is most closely related to histones. Proc. Natl. Acad. Sci. USA *87*, 5788–5791.

Sandman, K., and Reeve, J.N. (2005). Archaeal chromatin proteins: different structures but common function? Curr. Opin. Microbiol. *8*, 656–661.

Sato, T., Fukui, T., Atomi, H., and Imanaka, T. (2003). Targeted gene disruption by homologous recombination in the hyperthermophilic archaeon *Thermococcus kodakaraensis* KOD1. J. Bacteriol. *185*, 210–220.

Schelert, J., Drozda, M., Dixit, V., Dillman, A., and Blum, P. (2006). Regulation of mercury resistance in the crenarchaeote *Sulfolobus solfataricus*. J. Bacteriol. *188*, 7141–7150.

Schild, D., Lio, Y., Collins, D.W., Tsomondo, T., and Chen, D.J. (2000). Evidence for simultaneous protein interactions between human Rad51 paralogs. J. Biol. Chem. *275*, 16443–16449.

Schleper, C., Holz, I., Janekovic, D., Murphy, J., and Zillig, W. (1995). A multicopy plasmid of the extremely thermophilic archaeon *Sulfolobus* effects its transfer to recipients by mating. J. Bacteriol. *177*, 4417–4426.

Schleper, C., Kubo, K., and Zillig, W. (1992). The particle SSV1 from the extremely thermophilic archaeon *Sulfolobus* is a virus: demonstration of infectivity and of transfection with viral DNA. Proc. Natl. Acad. Sci. USA *89*, 7645–7649.

Schleper, C., Roder, R., Singer, T., and Zillig, W. (1994). An insertion element of the extremely thermophilic archaeon *Sulfolobus solfataricus* transposes into the endogenous beta-galactosidase gene. Mol. Gen. Genet. *243*, 91–96.

Schmidt, K.J., Beck, K.E., and Grogan, D.W. (1999). UV stimulation of chromosomal marker exchange in *Sulfolobus acidocaldarius*: implications for DNA repair, conjugation and homologous recombination at extremely high temperatures. Genetics *152*, 1407–1415.

Seitz, E.M., Brockman, J.P., Sandler, S.J., Clark, A.J., and Kowalczykowski, S.C. (1998). RadA protein is an archaeal RecA protein homolog that catalyzes DNA strand exchange. Genes Dev. *12*, 1248–1253.

Seitz, E.M., Haseltine, C.A., and Kowalczykowski, S.C. (2001). DNA recombination and repair in the Archaea. Adv. Appl. Microbiol. *50*, 101–169.

Seitz, E.M., and Kowalczykowski, S.C. (2000). The DNA binding and pairing preferences of the archaeal RadA protein demonstrate a universal characteristic of DNA strand exchange proteins. Mol. Microbiol. *37*, 555–560.

She, Q., Brugger, K., and Chen, L. (2002). Archaeal integrative genetic elements and their impact on genome evolution. Res. Microbiol. *153*, 325–332.

She, Q., Peng, X., Zillig, W., and Garrett, R.A. (2001a). Gene capture in archaeal chromosomes. Nature *409*, 478.

She, Q., Phan, H., Garrett, R.A., Albers, S.V., Stedman, K.M., and Zillig, W. (1998). Genetic profile of pNOB8 from *Sulfolobus*: the first conjugative plasmid from an archaeon. Extremophiles *2*, 417–425.

She, Q., Shen, B., and Chen, L. (2004). Archaeal integrases and mechanisms of gene capture. Biochem. Soc. Trans *32*, 222–226.

She, Q., Singh, R.K., Confalonieri, F., Zivanovic, Y., Allard, G., Awayez, M.J., Chan-Weiher, C.C., Clausen, I.G., Curtis, B.A., De Moors, A., *et al.* (2001b). The complete genome of the crenarchaeon *Sulfolobus solfataricus* P2. Proc. Natl. Acad. Sci. USA *98*, 7835–7840.

Shen, P., and Huang, H.V. (1986). Homologous recombination in *Escherichia coli*: dependence on substrate length and homology. Genetics *112*, 441–457.

Shin, D.S., Pellegrini, L., Daniels, D.S., Yelent, B., Craig, L., Bates, D., Yu, D.S., Shivji, M.K., Hitomi, C., Arvai, A.S., *et al.* (2003). Full-length archaeal Rad51 structure and mutants: mechanisms for RAD51 assembly and control by BRCA2. Embo J. *22*, 4566–4576.

Smith, D.R., Doucette-Stamm, L.A., Deloughery, C., Lee, H., Dubois, J., Aldredge, T., Bashirzadeh, R., Blakely, D., Cook, R., Gilbert, K., *et al.* (1997). Complete genome sequence of *Methanobacterium thermoautotrophicum* deltaH: functional analysis and comparative genomics. J. Bacteriol. *179*, 7135–7155.

Spies, M., Kil, Y., Masui, R., Kato, R., Kujo, C., Ohshima, T., Kuramitsu, S., and Lanzov, V. (2000). The RadA protein from a hyperthermophilic archaeon *Pyrobaculum islandicum* is a DNA-dependent ATPase that exhibits two disparate catalytic modes, with a transition temperature at 75 degrees C. Eur. J. Biochem. *267*, 1125–1137.

Stedman, K.M., She, Q., Phan, H., Holz, I., Singh, H., Prangishvili, D., Garrett, R., and Zillig, W. (2000). pING family of conjugative plasmids from the extremely thermophilic archaeon *Sulfolobus islandicus*: insights into recombination and conjugation in Crenarchaeota. J. Bacteriol. *182*, 7014–7020.

Sugawara, N., Ivanov, E.L., Fishman-Lobell, J., Ray, B.L., Wu, X., and Haber, J.E. (1995). DNA structure-dependent requirements for yeast RAD genes in gene conversion. Nature *373*, 84–86.

Sung, P. (1997). Yeast Rad55 and Rad57 proteins form a heterodimer that functions with replication protein A to promote DNA strand exchange by Rad51 recombinase. Genes Dev. *11*, 1111–1121.

Sung, P., Trujillo, K.M., and Van Komen, S. (2000). Recombination factors of *Saccharomyces cerevisiae*. Mutat Res. *451*, 257–275.

Syvanen, M. (1994). Horizontal gene transfer: evidence and possible consequences. Annu. Rev. Genet. *28*, 237–261.

Tan, T.L., Kanaar, R., and Wyman, C. (2003). Rad54, a Jack of all trades in homologous recombination. DNA Repair (Amst) *2*, 787–794.

Tyson, G.W., Chapman, J., Hugenholtz, P., Allen, E.E., Ram, R.J., Richardson, P.M., Solovyev, V.V., Rubin, E.M., Rokhsar, D.S., and Banfield, J.F. (2004). Community structure and metabolism through reconstruction of microbial genomes from the environment. Nature *428*, 37–43.

Umezu, K., Chi, N.W., and Kolodner, R.D. (1993). Biochemical interaction of the *Escherichia coli* RecF, RecO, and RecR proteins with RecA protein and single-stranded DNA binding protein. Proc. Natl. Acad. Sci. USA *90*, 3875–3879.

Usui, T., Ohta, T., Oshiumi, H., Tomizawa, J., Ogawa, H., and Ogawa, T. (1998). Complex formation and functional versatility of Mre11 of budding yeast in recombination. Cell *95*, 705–716.

Wadsworth, R.I., and White, M.F. (2001). Identification and properties of the crenarchaeal single-stranded DNA binding protein from *Sulfolobus solfataricus*. Nucleic Acids Res. *29*, 914–920.

Wang, G., Kennedy, S.P., Fasiludeen, S., Rensing, C., and DasSarma, S. (2004). Arsenic resistance in *Halobacterium* sp. strain NRC-1 examined by using an improved gene knockout system. J. Bacteriol. *186*, 3187–3194.

Webb, B.L., Cox, M.M., and Inman, R.B. (1997). Recombinational DNA repair: the RecF and RecR proteins limit the extension of RecA filaments beyond single-strand DNA gaps. Cell *91*, 347–356.

Whitaker, R.J., Grogan, D.W., and Taylor, J.W. (2005). Recombination shapes the natural population structure of the hyperthermophilic archaeon *Sulfolobus islandicus*. Mol. Biol. Evol. *22*, 2354–2361.

White, M.F., Giraud-Panis, M.J., Pohler, J.R., and Lilley, D.M. (1997). Recognition and manipulation of branched DNA structure by junction- resolving enzymes. J. Mol. Biol. *269*, 647–664.

Whitehouse, I., Flaus, A., Cairns, B.R., White, M.F., Workman, J.L., and Owen-Hughes, T. (1999). Nucleosome mobilization catalysed by the yeast SWI/SNF complex. Nature *400*, 784–787.

Whitman, W.B., Tumbula, D.L., Yu, J.P., and Kim, W. (1997). Development of genetic approaches for the methane-producing archaebacterium *Methanococcus maripaludis*. Biofactors *6*, 37–46.

Williams, E., Lowe, T.M., Savas, J., and DiRuggiero, J. (2007). Microarray analysis of the hyperthermophilic archaeon *Pyrococcus furiosus* exposed to gamma irradiation. Extremophiles *11*, 19–29.

Woese, C.R., Kandler, O., and Wheelis, M.L. (1990). Towards a natural system of organisms: proposal for the domains Archaea, Bacteria, and Eucarya. Proc. Natl. Acad. Sci. USA *87*, 4576–4579.

Woods, W.G., and Dyall-Smith, M.L. (1997). Construction and analysis of a recombination-deficient (radA) mutant of *Haloferax volcanii*. Mol. Microbiol. *23*, 791–797.

Worthington, P., Hoang, V., Perez-Pomares, F., and Blum, P. (2003). Targeted disruption of the alpha-amylase gene in the hyperthermophilic archaeon *Sulfolobus solfataricus*. J. Bacteriol. *185*, 482–488.

Wu, Y., He, Y., Moya, I.A., Qian, X., and Luo, Y. (2004). Crystal structure of archaeal recombinase RADA: a snapshot of its extended conformation. Mol. Cell *15*, 423–435.

Wu, Y., Qian, X., He, Y., Moya, I.A., and Luo, Y. (2005). Crystal Structure of an ATPase-active Form of Rad51 Homolog from *Methanococcus voltae*: Insights into potassium dependence. J. Biol. Chem. *280*, 722–728.

Xue, H., Guo, R., Wen, Y., Liu, D., and Huang, L. (2000). An abundant DNA binding protein from the hyperthermophilic archaeon *Sulfolobus shibatae* affects DNA supercoiling in a temperature-dependent fashion. J. Bacteriol. *182*, 3929–3933.

Yeats, S., McWilliam, P., and Zillig, W. (1982). A plasmid in the archaebacterium *Sulfolobus acidocaldarius*. Embo J. *1*, 1035–1038.

Zillig, W., Arnold, H.P., Holz, I., Prangishvili, D., Schweier, A., Stedman, K., She, Q., Phan, H., Garrett, R., and Kristjansson, J.K. (1998). Genetic elements in the extremely thermophilic archaeon *Sulfolobus*. Extremophiles *2*, 131–140.

Zivanovic, Y., Lopez, P., Philippe, H., and Forterre, P. (2002). *Pyrococcus* genome comparison evidences chromosome shuffling-driven evolution. Nucleic Acids Res. *30*, 1902–1910.

DNA Repair and DNA Damage Tolerance in Archaeal Bacteria: Extreme Environments and Genome Integrity

7

Takehiko Nohmi, Masami Yamada and Petr Gruz

Abstract

Maintenance of genome integrity is a mechanism central to cellular life. Many Archaeal species live in harsh habitats that are extreme challenges to genome stability. In the habitat at high ambient temperatures, deamination, oxidation and depurination are greatly accelerated and various lesions are supposed to accumulate in the genomic DNA. Thus, the organisms living in such extreme conditions seem to evolve novel strategies for repairing DNA damage and avoiding mutations caused by the lesions. In this chapter, we review mechanisms of DNA repair in archaeal Bacteria and unique properties of archaeal DNA polymerases (Pols) to tolerate DNA damage. In general, archaeal DNA repair proteins are eukaryote-like although many counterparts are missing in the genome sequences. Archaeal B-family DNA Pols, such as *Sso* DNA Pol B1 or *Pfu*, halt DNA replication several basepairs before template uracil or hypoxanthine, deamination products of cytosine or adenine in template DNA, respectively, thereby apparently avoiding mutations (read-ahead mechanism). Fifteen out of 38 archaeal species whose genome sequences have been completely determined seem to possess Y-family DNA Pols, which are specialized to bypass lesions in DNA. Collectively, these molecular features warrant the future investigation on how Archaea accommodate DNA damage inevitably occurring in the extreme harsh environments.

Introduction

Archaea, the third domain of life, is thought to possess pathways involved in DNA replication, repair, recombination and transcription, similar to those of Eukarya, although its morphology is more prokaryotic-like (Olsen and Woese, 1996). Many eukaryotic repair proteins have close homologues in Archaea and thus archaeal proteins have been utilized as structural and biochemical models to reveal the mechanisms of action of the eukaryotic counterparts because of their structural stability and ease of purification. However, DNA-repair pathways in Archaea themselves are not yet well characterized, and several important counterparts are apparently missing in the Archaeal genome, suggesting that Archaea may possess unique mechanisms to protect the genome. This may not be surprising given the environmental challenges faced by hyperthermophilic Archaea living in habitats at extremely high temperatures. Chemical reactions such as hydrolytic deamination are greatly accelerated at high temperatures so that the hyperthermophilic organisms are supposed to be exposed to massive DNA damages. In addition, most of proteins and nucleic acids are denatured at high temperatures. Nevertheless, the mutation frequencies in hyperthermophilic Archaea, i.e. *Sulfolobus acidocaldarius*, are comparable with or lower than those of other microorganisms such as *Escherichia coli* (Grogan *et al.*, 2001). Thus, Archaea seem to have evolved unique and efficient mechanisms to protect their genome from harsh environmental conditions.

In addition to DNA repair, B-family DNA polymerases (DNA Pols) in the hyperthermophic Archaea have a unique property that they halt replication once they encounter uracil (U) or hypoxanthine (HX), which are deamination products of cytosine and adenine, respectively, in the template genome (Fogg *et al.*, 2002; Gruz *et al.*, 2003). Because they stall several basepairs before the lesions, it is called 'read-ahead mechanism' (Greagg *et al.*, 1999). This is regarded as an error avoidance mechanism in hyperthermophilic Archaea because the stalling of DNA replication provides a chance to remove the deaminated bases from the template DNA strands. The molecular feature is unique in hyperthermophic Archaea since DNA Pols from thermophilic eubacteria such as *Taq* DNA Pol do not share similar properties. Besides B-family DNA Pols, which are responsible for chromosome replication, 15 out of 38 archaeal species whose genome sequences have been completed seem to possess Y-family DNA Pols (Ohmori *et al.*, 2001). The Y-family represents a novel family of DNA Pols found in Archaea, Bacteria and Eukarya, and the most remarkable feature is that they can bypass a variety of DNA lesions efficiently (Prakash *et al.*, 2005; Nohmi, 2006). Thus, they are involved in tolerance to DNA damage induced by exogenous and endogenous genotoxic agents. In this chapter, we review unique mechanisms of DNA repair and damage tolerance in Archaea and discuss the future directions.

DNA repair

DNA repair is the molecular mechanism that removes various damages in DNA, thereby ensuring the genome integrity. In fact, all organisms from virus to humans including Archaea are equipped with multiple DNA repair machineries to counteract the genotoxic damage. DNA lesions are induced by exposure to hazardous environmental factors, e.g. sunlight, and endogenous mutagens produced during oxygen consumption and nutritional metabolism, e.g. reactive oxygen species and alkylating agents. Deficit of DNA repair leads to hypersensitivity to toxic and mutagenic effects of genotoxic hazards as has been exemplified by the presence of human genetic diseases such as xeroderma pigmentosum (XP) (Tanaka *et al.*, 1990). The XP patients are sensitive to sun light-induced skin cancer because of the deficit of the ability to remove cyclobutane pyrimidine dimers (CPD) from DNA. In the following section, we view six most important pathways of DNA repair: alkylation repair; photorepair; nucleotide excision repair (NER); base excision repair (BER); mismatch repair (MMR) and homologous recombination (HR) in Archaea (Fig. 7.1, Table 7.1). Of these, alkylation repair and photorepair directly reverse the damages while NER, BER and MMR require multiple steps to remove the damaged or mismatched bases, followed by resynthesis of DNA to fill in the gaps. HR as well as translesion DNA synthesis (TLS), which will be described in detail below, does not remove DNA damage but rescues the stalled DNA replication to complete the duplication of chromosome DNA (damage avoidance). Most Archaea as well as Bacteria and Eukarya have multiple, functionally redundant pathways to combat a variety of damages in DNA induced by environmental stress.

Repair of alkylation damage

DNA repair protein O^6-alkylguanine-DNA alkyltransferase (AGT) is considered to play central roles in the mechanism of cellular resistance to the toxic and mutagenic effects of DNA damage induced by monofunctional alkylating agents (Sedgwick, 2004; Lindahl *et al.*, 1988). AGT operates by the transfer of the offending alkyl groups from the O^6 position of guanine and the O^4 position of thymine in DNA to a cysteine residue at the active site of the protein. This is an irreversible process that results in the stoichiometric inactivation of the protein, the repair of other lesions requiring additional active molecules. The alkylation damage to DNA occurs under a variety of living conditions because alkylating agents including *N*-nitroso compounds are present in the environment and can be generated even endogenously (Taverna and Sedgwick, 1996). The AGT proteins share a highly conserved -PCHRV- amino acid sequence in their active site (Margison *et al.*, 2003). The presence of an AGT gene product or a putative gene has been reported so far in c.a. 100 different species from the three domains of life.

It is shown that extracts of two Crenarchaeota, i.e. *Sulfolobus acidocaldarius* and *Pyrobaculum islandicum*, and two Euryarchaota, i.e. *Pyrococcus furiosus* and *Thermococcus litoralis*, contain the AGT activities (Skorvaga *et al.*,

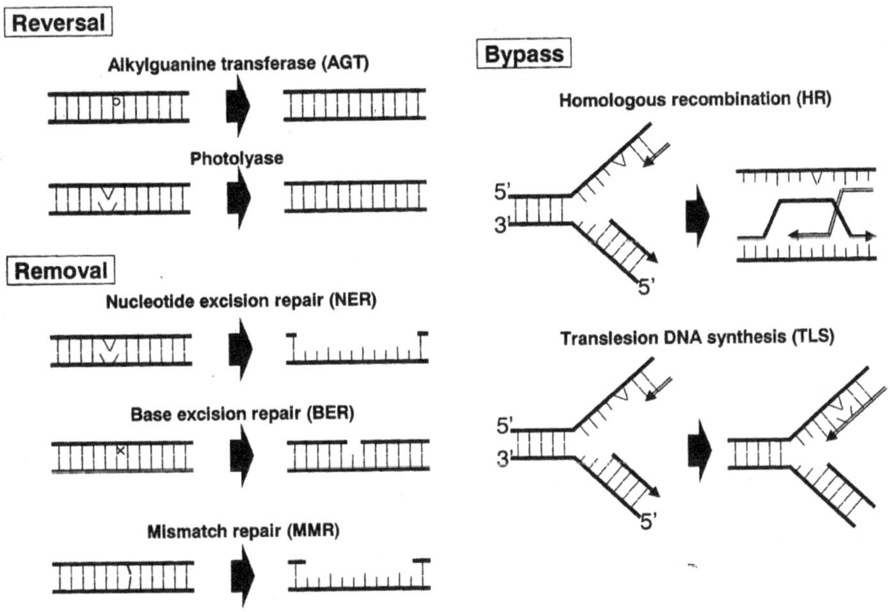

Figure 7.1 Schematic representation of mechanisms of DNA repair. Alkylguanine transferase (AGT) and photolyase directly reverse the damaged bases to correct ones. Nucleotide excision repair (NER), DNA glycosylase involved in base excision repair (BER) and mismatch repair (MMR) remove damaged bases or DNA sequeces containing damaged bases, followed by DNA re-synthesis to fill in the gaps. Homologous recombination (HR) and TLS do not remove damaged bases but assist to resume stalled DNA replication at the damaged sites.

1998). The enzymes seem to be similar to a human AGT because they lack the alkylphosphotriester-DNA alkyltransferase activity seen in some prokaryotes. More over, *P. islandicum* and *Pyrococcus furiosus* have maximal AGT activity at approximately 100°C. These enzymes would thus appear to be able to act on ssDNA (Skorvaga *et al.*, 1998).

AGT in other Euryarchaeota were cloned from *Pyrococcus* sp. KOD1(Leclere *et al.*, 1998), *Archaeoglobus fulgidus* (*Af*-AGT) (Kanugula and Pegg, 2003) and *Ferroplasma acidarmanus* (*Fa*-AGT) (Kanugula *et al.*, 2005). The AGT from KOD1 displays the activity at 90°C for at least 30 min. The enzyme exhibits considerable similarity to the corresponding mammalian, yeast and bacterial enzymes. *Af*-AGT is active in repairing O^6-methylguanine and, at a lower rate, O^4-methylthymine in DNA, and is readily inactivated by the low-molecular-mass inhibitor O^6-benzylguanine. *Fa*-AGT consists of a fusion of the C-terminal active site domain of AGT with an endonuclease V (Endo V) domain. The purified protein repairs O^6-meG in DNA and also cleaves the deaminated bases, i.e. U, HX or xanthine, in a similar manner to *E. coli* Endo V. In this case, two distinct but functional DNA repair proteins co-exist in a single polypeptide chain.

Photoreactivation

Photoreactivation is an efficient and direct repair mechanism for CPDs, which are the major DNA photoproducts induced by the UV component of solar radiation. This mechanism was discovered by the finding that survival of Bacteria irradiated with UV increased after exposure to visible light (Kelner, 1949). The DNA repair process is catalysed by a single enzyme, i.e. photolyase, which monomerizes UV-induced dimers by transfer of blue light energy absorbed by the chromophores (Sancar, 1990; Sancar, 1994). Most photolyases repair CPDs but some repair 6–4 photoproducts (6–4 PPs). Photoreactivation is widely distributed among species, ranging from Bacteria to plants and mammals. The ubiquity and efficient use of visible light emitted by the sun, the same source that induces DNA lesions, indicate that photolyase is an ancient protein that may have played an important role in evolution.

Table 7.1 DNA repair pathways in Archaea

	AGT	PHT	NER	UDG	MIG	AP	OGG	FEN-1	MutS	HR
Crenarchaeota										
Aeropyrum pernix	++	++	+++	++	+++	++	++	+++	++	++
Pyrobaculum aerophilum	++	++	++	+++	+++	++	++	++		++
Pyrobaculum islandicum	+++	++	+	+++		++				+++
Sulfolobus acidocaldarius	+++	+++	++	+++		++	++	++		++
Sulfolobus solfataricus	++	+++	+++	+++		+++	++	++		++
Sulfolobus shibatae				+++						++
Euryarchaeota										
Archaoglobus fulgidus	+++	++	+++	+++		++	+++	+++		++
Ferroplasma acidarmanus	+++		+	+		++	++		++	+
Halobacterium halobium		+++	++	++	++	+++	++	++	++	++
Haloferax		+	++							+++
Methanococcus jannaschii	++		++	+++		++	+++	+++		++
Methanothermobacter thermautotrophicus	++	+++	+++		+++	+++	++	++		++
Pyrococcus furiosus	+++	++	++	+++	+	++	++	+++	++	+++
Pyrocuccus horikoshii	++	++	++	++		++	++	+++	++	++
Pyrocuccus sp. KOD1	+++	++	++	++		++	++	++	++	++
Thermococcus litoralis	+++			+++						

The abbreviations are as follows: AGT – O^6-MeG methyltransferase, PHT – photolyase, NER – nucleotide excision repair (XPF, XPB, XPD), UDG – uracil DNA glycosylase, MIG – mismatch glycosylase, AP – AP endonuclease, OGG – 8-oxoG glycosylase, FEN-1 – flap endonuclease, MutS – mutS, HR – recA. The enzymes characterized biochemically are indicated as +++, enzymes with sequence homology as described in the NCBI GeneBank protein sequence database of March, 2007 are indicated as ++ while proteins with only limited homology are labelled as +.

Among Archaea, photolyases are not common, although the photoreactivation or photolyases are identified in several archaeal species. These include Crenarchaeota, i.e. *Sulfolobus acidocaldarius* (Wood *et al.*, 1997) and *Sulfolobus tokodaii* (Fujihashi *et al.*, 2007), and Euryarchaeota, i.e. *Halobacterium* sp. NRC-1 (McCready and Marcello, 2003) and *Methanothermobacter thermoautotrophicus* (Kiener *et al.*, 1989; Yasui *et al.*, 1994). In addition, several hypothetical genes encoding photolyases are identified in *Haloarcula marismortui* (Baliga *et al.*, 2004c), *Methanopyrus kandleri* (Slesarev *et al.*, 2002) and *Methanosarcina acetivorans* (Galagan *et al.*, 2002). Sso7d and Sac7d are two small (~7 kDa), but abundant, nucleoproteins in *Sulfolobus solfataricus* and *S. acidocaldarius*, respectively (Gao *et al.*, 1998; Robinson *et al.*, 1998). Recently it was demonstrated that Sso7d efficiently repairs the thymine dimer under irradiation conditions, suggesting that Sso7d and Sac7d may be involved in the photoreactivation in the organisms (Tashiro *et al.*, 2006). Two photolyase-like genes, *phr1* and *phr2*, are identified in *Halobacterium NRC-1*, and the *phr2* gene is shown to encode a CPD photolyase (McCready and Marcello, 2003). The photolyase from the *Methanothermobacter thermoautotrophicum* exhibits 37% homology to those of higher eukaryotes and only 15% to microbial photolyases(Yasui *et al.*, 1994)

Nucleotide excision repair

NER is a highly conserved DNA repair pathway able to detect and remove a variety of bulky DNA lesions caused by UV light and environmental mutagens and thereby contributes to the genomic integrity of an organism (Sancar, 1996; Lindahl and Wood, 1999). NER is universal in Bacteria and Eukarya although other mechanisms, i.e. photoreactivation enzymes (see Photoreactivation), DNA glycosylase (see Base excision repair) and endonuclease known as UVDE (Yasui and McCready, 1998), also operate to counteract UV-induced DNA damage in a number of organisms. NER in higher eukaryotes involves the coordinated assembly of a large number of proteins, including the core NER factors XPC-HR23B, TFIIH, XPA, replication protein A (RPA) and two endonucleases XPG and ERCC1-XPF (Aboussekhra *et al.*, 1995). These proteins cooperate to recognize, unravel and excise a 24–32-mer oligonucleotide bearing the DNA lesion prior to filling in the missing gap (Araujo and Wood, 1999). Although Bacteria have a similar overall repair strategy, a much simpler and structurally unrelated multiprotein complex known as UvrABC carries out the same task (Petit and Sancar, 1999). The availability of fully sequenced archaeal genomes has revealed that many Archaea have proteins related to eukaryotic NER factors rather than to the UvrABC bacterial repair system (Aravind *et al.*, 1999; White, 2003; Grogan, 2000). Only certain Euryarchaeota, i.e. *Halobacterium sp.* NRC-1 (Ng *et al.*, 2000), *Methanothermobacter thermoautotrophicus* (Ogrunc *et al.*, 1998), *Haloarcula marismortui* (Baliga *et al.*, 2004a), *Methanosarcina acetivorans* (Galagan *et al.*, 2002) and *Methanosarcina mazei* (Deppenmeier *et al.*, 2002), have Bacteria-type repair proteins, although all of them have eukaryotic-type NER proteins too. It is interesting that Archaea lack damage recognition proteins such as XPA and XPC, which are essential for eukaryotic NER, while they have other eukaryotic-type components, such as XPF/ERCC1, Flap endonuclease, and XPB and XPD helicases (White, 2003). It is worth noting that Crenarchaea have a 'short' form of XPF that lacks the helicase-like domain while Euryarchaeota have a long form of XPF, which includes a helicase-like domain at the N-terminal (Roberts *et al.*, 2003).

In a few species, actual NER activities have been demonstrated. In Crenarchaeota, UV-induced CPDs are efficiently repaired in vivo in the dark in *Sulfolobus solfataricus* (Salerno *et al.*, 2003). In Euryarchaeota, *Methanobacterium thermoautotrophicum* is shown to remove UV-induced 6–4 PPs in the form of 10- to 11-mers by incising 3′- and 5′-sides of the DNA damage (Ogrunc *et al.*, 1998). In addition two structural analyses provide more precise information to this field; XPF homologue in *Aeropyrum pernix* (Newman *et al.*, 2005), and XPB homologue in *Archaeoglobus fulgidus* (AfXPB) (Fan *et al.*, 2006). Interestingly, the DNA damage recognition domain and the flexible thumb motif of AfXPB appear structurally as well as functionally analogous to the MutS mismatch recognition and DNA polymerase thumb domains.

Base excision repair

BER pathway involves the initial recognition of specific types of base damage and/or damage associated basepairing mismatches by a BER DNA glycosylase. An estimated rate of 10^4 damaging events/mammalian cell/day underscores the importance of the BER pathway (Nakamura *et al.*, 1998; Lindahl and Wood, 1999; Lindahl, 1993). After damage recognition, the glycosylase cleaves the damaged or misincorporated base from the DNA backbone, leaving an apurinic/apyrimidinic (AP) site (Lu *et al.*, 2001). These AP sites are cleaved by the major 5′ AP endonuclease, APE1, to leave a 5′ dRp terminus that is then removed by DNA Pol β producing a single nucleotide gap. DNA glycosylases/AP lyases or bifunctional glycosylases, such as endonuclease III (NTH1) and 8-oxoguanine DNA glycosylase (OGG1), besides base excision, cleave DNA at the resultant AP site by β elimination generating a 3′ terminal sugar phosphate that is then removed by APE1 yielding a gap with a 3′ OH (Demple and Harrison, 1994). Ligation will continue either via short-patch (SP) (for one nucleotide) (Dianov *et al.*, 1992) or long-patch (LP) (two or more nucleotides) repair (Frosina *et al.*, 1996; Klungland and Lindahl, 1997). Pol β is the repair-synthesis polymerase of the SP-repair. Pol β is also likely to incorporate the first nucleotide in the LP-repair (Podlutsky *et al.*, 2001) but the elongation step in this pathway is carried out by replicative DNA Pols. Proliferating cell nuclear antigen (PCNA), the sliding clamp for Pol β and flap endonuclease-1 (FEN-1), a structure specific nuclease that excises the displaced oligonucleotide (Pascucci *et al.*, 1999). The ligation step is then operated by the X-ray repair cross complementing 1 (XRCC1)/DNA ligase IIIα complex in the SP-repair. The XRCC1/DNA ligase IIIα complex is dispensable for LP-repair implying that in this pathway the sealing step is performed by DNA ligase I (Cappelli *et al.*, 1997; Sleeth *et al.*, 2004).

Of various BER glycosylases, uracil DNA glycosylase (UDG) have been studied most extensively (Lindahl and Nyberg, 1974; Percival *et al.*, 1989). They are extremely efficient enzymes, which recognize U in ss DNA, in A/U pairs that arise when dUMP is incorporated opposite A during DNA replication, or in G/U mispairs arising through cytosine deamination. Thermostable UDG activities have been identified in Crenarchaeota, i.e. *Sulfolobus solfataricus, S. shibatae, S. acidocaldarius, Thermococcus litoralis, Pyrobaculum islandicum* and P. aerophilum, and Euryarcheota, i.e. *Pyrococcus furiosus, Methanococcus jannaschii* and *Archaeoglobus fulgidus* (Sandigursky and Franklin, 2000; Koulis *et al.*, 1996; Hinks *et al.*, 2002; Chung *et al.*, 2003b). In addition, novel types of UDGs have been identifed in hyperthermophilic Archaea (Pearl, 2000; Sartori *et al.*, 2002; Chung *et al.*, 2003b). One such example is *Pa*-UDGb in *Pyrobaculum aerophilum*, which lacks the polar amino acid at the active site (motif A) and removes HX as well as U in DNA. This enzyme is a functional component of in vitro BER system that consists of *Pa*-UDGb, AP endonuclease IV, DNA Pol B2 and DNA ligase (Sartori and Jiricny, 2003). Besides *Pa*-UDGb, *P. aerophilum* possesses two other UDGs, i.e. *Pa*-UDGa and *Pa*-MIG (mismatch specific DNA glycosylase). *Pa*-UDGa has glutamate at the active site and is more active than *Pa*-UDGb. *Pa*-UDGa appears to be the major UDG in this organism and *Pa*-UDGb is probably responsible for the removal of HX or hydroxymethylU in DNA. *Pa*-MIG can remove T as well as U paired with G so that it may contribute to mismatch correction (Sartori *et al.*, 2002; Yang *et al.*, 2000). MIG activites are also identified in *Aeropyrum pernix* and *Methanobacterium thermoautotrophicum* (Yang *et al.*, 2000; Horst and Fritz, 1996; Fondufe-Mittendorf *et al.*, 2002).

Hypothetical counterparts of OGG1, which removes 8-oxoguanine (8-oxo-G) in DNA, have been identified in the genome of Crenarchaeota, i.e. *Sulfolobus solfataricus* (She *et al.*, 2001) and Euryarchaeota, i.e. *Methanococcus jannaschii* (Gogos and Clarke, 1999), *Archaeoglobus fulgidus* (Chung *et al.*, 2001), *Thermoplasma acidophilum* (Ruepp *et al.*, 2000) and *Thermoplasma volcanicum* (Kawashima *et al.*, 2000). Among them, enzymatic activities against 8-oxo-G in DNA have been demonstrated in *M. jannaschii* and *A. fulgidus* (Gogos and Clarke, 1999; Chung *et al.*, 2001).

FEN-1 has important roles in DNA repair as well as DNA replication and recombination. The enzyme cleaves 5′ flap structure endonucleolytically and has double-strand-specific 5′–3′-exonuclease activity. When working in BER, FEN-1 removes damaged nucleotides after AP endonuclease has incised the 5′ side of the AP site in LP-repair (Klungland and Lindahl, 1997; Kim *et al.*, 1998; Gary *et al.*, 1999). It recognizes a specific DNA structure, independent of the DNA sequence. FEN-1 seems highly conserved in Archaea and Eukaryotes (Matsui *et al.*, 1999; Hosfield *et al.*, 1998; Kaiser *et al.*, 1999; Rao *et al.*, 1998) and the crystal structures have been solved in Euryarchaeota, i.e. *Methanococcus janaschii* (Hwang *et al.*, 1998; Sayers and Artymiuk, 1998; Bae *et al.*, 1999), *Pyrococcus furiosus* (Hosfield *et al.*, 1998), *Archaeoglobus fulgidus* (Chapados *et al.*, 2004) and *Pyrococcus horikoshii* (Matsui *et al.*, 2002) and in Crenarchaeota, i.e. *Aeropyrum pernix* (Collins *et al.*, 2004).

Mismatch repair

The ability to recognize and repair mismatches in DNA after replication has been well documented in many species (Modrich and Lahue, 1996). The most extensively studied general MMR system is the one in the bacterium *E. coli*. It functions with the MutSLH, UvrD, and RecJ proteins through a d(GATC)-specific methylase (dam) methylation-directed pathway and is initiated by the mismatch-stimulated d(GATC) endonuclease activity of MutH, which takes advantage of transient undermethylation of the DNA during replication, and specifically nicks the newly synthesized, that is, unmethylated daughter strand. Subsequently, the RecJ exonuclease selectively degrades the nicked strand, and the mutations are corrected by the action of a DNA Pol, a helicase, and a ligase (Marti *et al.*, 2002).

Hypothetical MutS-like proteins were found in several species of Euryarchaeota, but not in ones of Crenarchaeota, such as *Sulfolobus solfataricus* (She *et al.*, 2001). The MutS-like proteins that have ATPase domain may be divided into two subfamilies, i.e. MutS-I lineage related to MMR and MutS-II lineage related to chromosome aberrations and meiotic crossing-over (Eisen, 1998). The MutS-I lineage includes bacterial MutS and eukaryotic MSH2/6, and the MutS-II includes eukaryotic MSH4 and MSH5. Phylogenetic analysis also shows that the proteins in the MutS-II subfamily are distant and distinct from those involved in MMR (Kawarabayasi *et al.*, 1998; Smith *et al.*, 1997; Robb *et al.*, 2001; Cohen *et al.*, 2003). *Halobacterium* NRC-1 and *Methanosarcina activorans* have both types of MutS homologues as well as a MutL-like protein (Ng *et al.*, 2000).

Unlike MutS and MutL, which are absolutely necessary for MMR, and therefore evolutionarily highly conserved, MutH is significantly less conserved. In fact, no MutH homologue has been identified in either the Eukaryotes or the Archaea. Since it has been revealed that Eukaryotic MMR protein complex, MutLα, is an endonuclease (Kadyrov *et al.*, 2006), it is not surprising that Eukaryote does not possess MutH endonuclease. Although *Halobacterium* NRC-1 does not have MutH endonuclease or a *dam* methylase, it has d(CTAG) methylase (Zim), MutS/L, four RecJ-like exonucleases and the UvrD repair helicase (Ng *et al.*, 2000). However, a d(CTAG)-specific endonuclease was not detected using either sequence- or structure-based annotation methods (Baliga *et al.*, 2004b). *P. furiosus* has detectable enzymatic activities of MutS2 protein, an ATPase activity and a non-mismatch specific DNA-binding activity (Vijayvargia and Biswas, 2002).

Spontaneous mutation frequency in *P. aerophilum* is high, especially in mononucleotide repeats (Fitz-Gibbon *et al.*, 2002). Although this is consistent with the absence of orthologues of MMR proteins, *P. aerophilum* may possess a distinct MMR system. Taken together, it has been, so far, unclear how mismatches generated during DNA replication are corrected in Archaea. It is possible that a novel system that manages mismatch errors operates in Archaea such as *S. solfataricus*, where no MutS/L homologues are identified.

Recombinational repair

DNA double-strand breaks (DSBs) are highly cytotoxic and mutagenic, thus the repair of DSBs in DNA is an essential process in all organisms. DSBs are predominantly repaired by two pathways, i.e. non-homologous end joining and HR. HR conserves genetic identity by using an undamaged chromosome as a template to replace missing or damaged nucleotides. It also serves to underpin genome replication by providing the means to resume replication when the replication machinery is blocked (Seigneur *et al.*, 1998; McGlynn and Lloyd, 2000). Mre11/Rad50 complex plays a key role in DSB repair. The complex participates in the generation of the 3′ ssDNA tails possibly by bridging DNA ends or sister chromatids in the very early steps of HR. RecA family of recombinases entails the reciprocal exchange of the single strands between homologous DNA duplexes by utilizing conserved DNA-binding modules and a common core ATPase domain to form a four-way branched intermediate commonly referred to as the Holliday junction (HJ). In the late step of HR, the HJ is resolved by dual-strand incision across the branch point to release nicked duplexes, then whole procedure completes with sealing by DNA ligase. Enzymes of HR have been unearthed in all three domains of life, i.e. Bacteria, Archaea and Eukarya.

Hyperthermophilic Archaea are reported to be more resistant to ionizing radiation than hyperthermophilic Bacteria or *E. coli* (DiRuggiero *et al.*, 1997). A γ-ray dose allowing 75% survival of *Pyrococcus furiosus* was shown initially to fragment the chromosome into the lengths of about 30 kb, yet full-length chromosomes eventually reappeared *in vivo* after prolonged incubation at 95°C. There is evidence that RadA, the archaeal RecA protein, is induced by DNA damage in thermophilic and mesophilic Archaea (Reich *et al.*, 2001). Although the results imply that effective DSB repair mechanisms, which could be HR, exist in Archaea, very little is known about the enzymes for homologous exchange in Archaea, with the exception of RecA-like strand exchange enzymes (Seitz *et al.*, 1998) and enzymes that cleave HJs *in vitro* found in *P. furiosus* (Komori *et al.*, 1999), or *Sulfolobus solfataricus* (Kvaratskhelia and White, 2000a).

Structural analyses of a Rad50 homologue from *P. furiosus* illuminate the catalytic core of the enzyme, an ATP-binding domain related to the ABC transporter family of ATPases (Hopfner *et al.*, 2000; Hopfner *et al.*, 2001). Its S793R mutation prevents ATP binding and dimerization, and disrupts the communication among the other ATP-binding loops (Moncalian *et al.*, 2004). Crystal structure of Mre11 and Rad50 of *P. furiosus* reveal a dual functional complex consisting of $(Mre11)_2/(Rad50)_2$ heterotetrameric DNA (Hopfner *et al.*, 2001). Putative *recA* and *RAD51* homologues in three archaeans were identified by PCR and referred to as *radA* (Sandler *et al.*, 1996). The same research group amplified fragments of *radA* genes from 11 archaeal species and analysed the phylogenetic relationships (Sandler *et al.*, 1999). RadA proteins that are functional *in vivo* are found in Euryarchaeota, i.e. *Haloferax volcanii* (Woods and Dyall-Smith, 1997) and *P. furiosus* (Komori *et al.*, 2000b; Komori *et al.*, 2000a; Hayashi *et al.*, 1999), and Crenarchaeota, i.e. *Pyrobaculum islandicum* (Spies *et al.*, 2000). Crystal structure is presented for a RadA from *Methanococcus voltae* (Qian *et al.*, 2005). These enzymes share a higher level of sequence and more structural similarity with eukaryotic Rad51 than bacterial RecA (Uemori *et al.*, 1997; DiRuggiero *et al.*, 1997). RadB was also identified as a Rad51-like protein on the basis of conserved ATPase sequences (Rashid *et al.*, 1996). However, RadB from *P. furiosus* is not a recombinase. It does not catalyse strand exchange and does not turn over ATP efficiently (Komori *et al.*, 2000b). Despite the lack of ATPase activities, an ATP-binding activity seems required for RadB to promote survival of *Haloferax volcanii* after UV irradiation. ATP induces pronounced conformational change in RadB (Akiba *et al.*, 2005), which seems to be important for maintaining genome stability in Archaea. RPA is found in *P. furiosus* and is composed of three subunits, i.e. RPA41, RPA14, and RPA32, like the eukaryotic RPA (p70-p14-p32). RPA clearly stimulates a RadA-mediated strand exchange reaction (Komori and Ishino, 2001).

Regarding the late stage, in which the HJ intermediates are processed, the first archaeal HJ resolvase (Hjc, HJ cleavage) is discovered in *P. furiosus* (Komori *et al.*, 1999). Similar resolvases have been purified from *S. solfataricus* (Kvaratskhelia and White, 2000b) and *Methanobacterium thermoautotrophicum* (Bolt *et al.*, 2001). Although Hjc does not resemble any of the known HJ resolvases, it does share significant similarity to restriction endonucleases and exonuclease (Aravind *et al.*, 2000; Kvaratskhelia *et al.*, 2000). Thus, Hjc is an archaeal-specific HJ resolvase. Neither its sequence nor its three-dimensional structure is similar to other known HJ resolvases (Nishino *et al.*, 2001; Komori *et al.*, 2000c). HJ resolvases with distinct properties have been characterized from bacteriophages (T4 endo VII, T7 endo I, RusA and Rap), Bacteria (RuvC), Archaea (Hjc and Hje), yeast (CCE1) and poxviruses (A22R). Thus, the HJ resolvase is quite interesting, from an evolutional point of view (Lilley and White, 2000). The Hjc enzymes target junctions with high specificity but, unlike RuvC and RuvA, they show no sequence-preference for strand cleavage. Hjc from *P. furiosus* has been found to interact with the archaeal equivalent of PCNA and with RadB (Komori *et al.*, 1999; Komori *et al.*, 2000c). The Hje (HJ endonuclease) protein has similar properties to Hjc, including a preference for four-way junction DNA and the ability to resolve both static and mobile HJs (Kvaratskhelia and White, 2000a). Its crystal structure serves more precise information for understanding its function (Nishino *et al.*, 2005). The gene, *hjm* (HJ migration), encodes a protein composed of 720 amino acids. A homology search revealed that Hjm shares sequence similarity with the human PolQ, HEL308, and *Drosophila* Mus308 proteins, which are involved in a DNA repair, whereas no similar sequences were found in Bacteria and yeast (Fujikane *et al.*, 2005).

DNA damage tolerance by archaeal DNA polymerases

Life represents a succession of biological information stored in DNA from ancestors to descendants. Thriving in harsh habitats such as high temperatures, therefore, requires organisms to evolve unique mechanisms to maintain the genome integrity. Temperatures more than 80°C where thermophilic Bacteria optimally grow pose challenges to the genome to be denatured or decomposed. Although denaturation may not be a serious problem for DNA as long as it is covalently closed, heat-induced depurination, oxidation and deamination could have deleterious effects. Depurination leads to AP site, a major cytotoxic lesion, and oxidation generates a variety of DNA lesions such as 8-oxo-G in DNA (Lindahl, 1993). Hydrolytic deamination of exocyclic amino groups of cytosine leads to the formation of U in DNA, and G:U basepairs result in G:C to A:T transitions in a half of the progeny if not repaired (Lindahl and Nyberg, 1974; Pearl, 2000). Although most of the lesions can be removed by DNA repair mechanisms as has been reviewed above, some escape from the repair mechanisms and persist in the template DNA. These residual lesions can block DNA replication and induce mutations or chromosome breaks. To counteract the lesions, cells evolved the mechanisms called DNA damage tolerance to accomplish the chromosome replication. Unlike DNA repair, the damage tolerance mechanism does not remove DNA damage but instead allows DNA Pols proceed beyond the damage and permits the accomplishment of chromosome replication. The mechanisms may also include the behaviour of DNA Pols that stop replication when they encounter spontaneous DNA lesions, thereby giving a chance to cells to remove the lesions before Pols incorporate incorrect dNMPs opposite the lesions. Although DNA damage tolerance mechanisms may include recombination repair in theory, we focus on the mechanisms associated with DNA Pols of Archaea, i.e. read-ahead mechanisms and translesion bypass replication (Lasken *et al.*, 1996; Nohmi, 2006; Fogg *et al.*, 2002). The physiological significance of damage tolerance mechanisms in Archaea is still open to question, but they may be critical to survive in harsh environments and maintain the genome stability. Since Pols involved in damage tolerance are also involved in induction of mutations by incorporation of incorrect dNMP opposite template lesions, they could play a role in evolution or genetic flexibility of archaeal Bacteria living in variable harsh environments. In the following sections, we will see how B-family and Y-family DNA Pols contribute to DNA damage tolerance in Archaea.

Read-ahead mechanism

Deamination of bases in DNA such as cytosine may be the most common genotoxic threat to all living organisms. Deamination of cytosine to U leads to C to T transitions unless repaired and disrupts the succession of genetic information stored in the genome. To counteract the mutagenic potential of U in DNA, Eukarya, Bacteria and plants have a BER pathway (see above) utilizing a UDG or a G:U/T mismatch specific DNA glycosylase. In addition, cells in Bacteria, Eukarya and Archaea possess dUTPase, which converts dUTP to dUMP and pyrophosphate (PPi), to minimize the levels of dUTP in the nucleotide pool (Hogrefe *et al.*, 2002). dUTP is a major source of U in DNA because of their ease of incorporation into DNA by Pols. In fact, it is estimated that 10^4 dUTP is incorporated per DNA replication based on a genome size of 10^{10} basepairs (Savva *et al.*, 1995). In Bacteria and yeast, cells lacking dUTPase exhibit DNA fragmentation and cell death because of the excess excision of U in DNA by the action of UDGs and the following steps of BER (Gadsden *et al.*, 1993). Despite the presence of UDG and dUTPase, elevated temperatures should pose an increase risk of DNA damage in hyperthermophilic Archaea because hydrolytic deamination of U greatly accelerated at high temperatures. However, the spontaneous mutation rate in hyperthermophilic Archaea *Sulfolobus acidocardarium* is reported to be similar to that of *Escherichia coli* (Grogan *et al.*, 2001). Thus, hyperpethermophilic Archaea appear to possess additional mechanisms to protect a stability of the genome from the mutagenic threat of deaminated bases generated at high temperatures.

Interestingly, B-family DNA Pols from hyperthomophilic Archaea such as *Sulfolobus solfataricus* DNA Pol B1 (Pol B1), *Pyrococcus furiosus* DNA Pol (Pfu) and *Thermococcus litoralis* DNA Pol (Vent and Vent exo-), recognize the presence of U in DNA and tightly bind to U-containing oligonucleotides (Lasken *et al.*, 1996). These Pols stall DNA replication several basepairs before template U and thus the recognition and stalling mechanisms seem to contribute to counteract the mutagenic potential of U in DNA because the stalling mechanisms may reduce the chance of misincorporation of adenine opposite template U (Greagg *et al.*, 1999). More interestingly, both Pol B1 and Pfu exo- bind not only to template U but also to template HX, a deamination product of adenine (Gruz *et al.*, 2003). Pfu exo- stalls three to four basepairs before HX as well as U in the template strand, and Pol B1 displays a similar stalling pattern. HX pairs with cytosine during DNA replication and induces A to G transitions if not corrected (Lindahl, 1993). Thus, the stalling behaviour seems to suppress the incorporation of cytosine opposite template HX, thereby reducing the mutagenic potential of HX.

The recognition and stalling mechanism seems unique to archaeal B family DNA Pols because viral B family Pols such as T4 Pol or Pols from hyperthermophilic eubacteria such as *Thermus aquaticus* (Taq) do not stall when the template DNA has U or HX (Greagg *et al.*, 1999). Eukaryotic B-family Pol, i.e. human Pol alpha, does not tightly bind to a U-containing oligonucleotide (Lasken *et al.*, 1996). Several lines of evidence suggest that single molecules of the archaeal Pols scan the template DNA and stall when they encounter U several basepairs ahead of the progress (Asami *et al.*, 2006; Shuttleworth *et al.*, 2004). Thus, this is called 'read-ahead mechanism'. Structural analysis of archaeal B-family Pols suggest that they comprise five domains, i.e. the N-terminal domain, the exonuclease domain and three polymerase active site domains, and also that a pocket in the N-terminal domain is responsible for the discrimination of U from normal DNA bases (Fogg *et al.*, 2002). The pocket seems highly conserved in the N-terminal domains of 15 archaeal B-family Pols including seven euryarchaeal and eight crenarchaeal Pols. It remains to be seen how this pocket accommodates not only U but also HX in template DNA (Savino *et al.*, 2004).

Since archaeal B-family DNA Pols such as *Pfu* or *Vent* are frequently used as PCR enzymes, the inhibitory effects of U are practical problems to achieve long and efficient PCR amplification of DNA. To circumvent the problems, dUTPase is included in the reaction mixture to reduce the levels of dUTP, which are generated during PCR through deamination of dCTP and limit the efficiency of PCR achieved by archaeal B-family DNA Pols (Hogrefe *et al.*, 2002). Pfu dUTPase, which converts dUTP to dUMP and PPi, improves the yield of products amplified by *Pfu* DNA Pol by preventing dUTP incorporation and subsequent inhibition of the Pol by U-containing DNA. It is suggested that archaeal dUTPase may play an important role in preventing dUTP incorporation and inhibition of DNA synthesis by B-family DNA Pols in vivo.

Although it seems plausible that the read-ahead mechanism contributes to the genome stability in hyperthermophilic Archaea, it is an open question of what will happen after the Pols bind to the template U or HX (Fig. 7.2). A hint is that the archaeal Pol binds to UDG (Connolly *et al.*, 2003). It is tempting to speculate, therefore, that HR occurs after the stalling and double-strand DNA is formed where one strand possesses U and the other possesses guanine. UDG excises U from one strand and the resulting single base gap is filled by Pol using a guanine base in the opposite strand as a template. If these events actually occurred, the correct base, i.e. cytosine, would be restored in the original site. Another hint is the interaction between Pol B1 from *S. solfataricus* and Pol Y1 (Dpo4) from the same species (De Felice *et al.*, 2007). Pol Y1 bypasses a variety of DNA lesions including U (Gruz *et al.*, 2001; Boudsocq *et al.*, 2004). It does not stall even when it encounters U or HX during in vitro DNA synthesis (Gruz *et al.*, 2003). Thus, translesion bypass across U or HX by Y-family Pols may occur to continue the DNA replication after removal of stalling Pols at the lesions. The deficit of the bypass pathway is that the translesion reactions induce mutations of C to T as an expense of continuation of chromosome replication. PCNA may be involved in these events because it is a molecular scaffold to recruit several proteins involving in the processing of DNA damage (Gruz *et al.*, 2001; Dionne *et al.*, 2003). In either case, further work is needed to clearly establish the molecular mechanisms after stalling Pols at the lesions in hyperthermophilic Archaea.

DNA lesion tolerance

The need of high fidelity in chromosomal replication imposes significant constrains on the

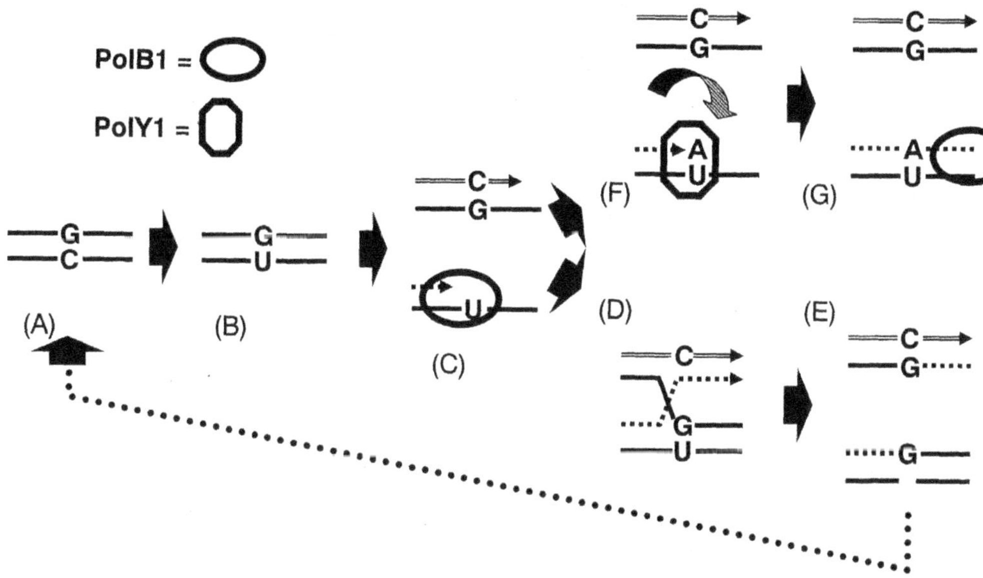

Figure 7.2 Daughter strand gaps by stalling of Pol B1 at template U (or HX) are sealed by HR and/or translesion bypass by Pol Y1. Cytosine in DNA (A) is deaminated by heat, thereby generating G:U mismatch (B). When the strand containing U is copied by Pol B1 before the U is removed, Pol B1 may stop before U (C). ssDNA region downstream of the U can be sealed by HR with sister chromatid (D). UDG may remove the U (E) and provide a chance to regenerate normal G:C basepair (A). However, if Pol B1 is switched to Pol Y1, the Y-family DNA Pol may bypass U by incorporating adenine opposite template U (F). Pol B1 extends the primer strand containing adenine opposite U, thereby generating a mutagenic mispair in DNA (G).

conventional DNA replicases. Whenever they encounter an unrepaired lesion in the template DNA strand, the default replication apparatus would halt. However, most living organisms are equipped with alternative Pols that can tolerate or even correctly accommodate certain abnormalities at the primer-template junction and thus relieve replication blocks. This process termed TLS is faster and energetically less expensive than other DNA reparation mechanisms that do not leave damaged parental strand behind. Depending on DNA lesion context, TLS can result in progeny mutations and that is probably why it is predominantly used under adverse and stressful conditions when an adaptive change in the genotype may be desirable.

According to their primary amino acid sequence similarities signifying common evolutionary origin, Pols are currently classified into seven families, namely A, B, C, D, E, X and Y (Ito and Braithwaite, 1991; Lipps *et al.*, 2003; Ohmori *et al.*, 2001). The enzymes from Y-family are exclusively engaged in the TLS process. They specialize in copying DNA at stalled replication forks, which arise *in vivo* from mismatched or misaligned primer ends usually due to the presence of damage on DNA template. They lack 3′–5′ exonuclease (proofreading) activity and are poorly processive and *error-prone* when copying undamaged DNA resulting often in untargeted mutagenesis (Kim *et al.*, 2001; Kim *et al.*, 1997). The archaeal members of the Y-family Pols have served as great models for solving the three-dimensional structure of these unique enzymes (Ling *et al.*, 2001; Silvian *et al.*, 2001). Their catalytic core consists of finger, palm, thumb and little finger (or wriest or PAD) domains, but the finger and thumb domains are much smaller than in high-fidelity polymerases. An active site cleft consisted of five sequence motifs common to all known Y-family Pols can accommodate DNA and nucleotide triphosphates with relaxed geometric constraints, which result in higher rates of misincorporation and low processivity (Fiala *et al.*, 2007; Ling *et al.*, 2003; Ling *et al.*, 2004b; Zang *et al.*, 2006). Some of these enzymes, however, accurately bypass certain DNA lesions. Examples include accurate bypass across cis-syn thymine-dimer by human Pol eta (Trincao *et al.*, 2001; Washington *et al.*, 2003) and correct incorporation of cytosine opposite template N^2-guanine minor groove adducts (Choi *et al.*, 2006; Jarosz *et al.*, 2006) such as the steroid hormone derived DNA adduct by human Pol kappa (Suzuki *et al.*, 2004). REV1 and Pol zeta can bypass an AP site efficiently *in vitro* (Haracska *et al.*, 2002; Masuda *et al.*, 2002). Although not essential for 'short-term' survival, the Y-family Pols may play important roles in 'long-term' survival (or evolution) by counteracting DNA adducts resulting from radiation and genotoxic byproducts of metabolism in an energetically efficient way (Yeiser *et al.*, 2002).

The archaeal Y-family Pols have been studied mostly as models of their eukaryotic counterparts due to the availability of the three-dimensional structures and feasibility to do co-crystallization studies. The complexes of Pol Y1 (Dpo4) from *S. solfataricus* with benzo[*a*]pyrene (Chandani and Loechler, 2007; Ling *et al.*, 2004b), *cis-syn* thymine dimer (Ling *et al.*, 2003), 8-oxo-G (Zang *et al.*, 2006) and an AP site (Ling *et al.*, 2004a) have been described in great details. The crystal structures also serve as models for the prediction and manipulation of structural properties of other homologous enzymes from low temperature organisms such as the genetic model *E. coli* (Lee *et al.*, 2006). The three dimensional structure of Pol Y1 (Dpo4) was even used in molecular dynamics simulation study to elucidate the catalytic mechanism and kinetics of DNA lesion bypass at atomic resolution (Chandani and Loechler, 2007; Wang *et al.*, 2006). However, it remains to be determined by *in vivo* genetic experiments what the physiological relevance of the Y-family Pols is in Archaea. A pioneering work related to this topic is the disruption of the *dpo4* gene of *S. solfataricus* (P. Blum, unpublished results). The resulting disruptants display enhanced expression of genes involved in protection of oxidative stress, suggesting the roles of the Y-family DNA Pol in the tolerance against oxidative DNA damage.

Apart from the recently heavily studied *Sulfolobus* enzymes, the Y-family Pol homologues are present in numerous but interestingly not all archaeal species. This contrasts with the other two kingdoms of life, i.e. Eukarya and Bacteria, where at least one Y-family member is present. Of the completely sequenced 38 archaeal genomes, there are only 15 having a classified Y-family Pol homologue as described in the current version of the NCBI GeneBank Protein Sequence Database. We have used the program ClustalX (Thompson *et al.*, 1997) to construct an alignment of all the 18 archaeal protein sequences and generated their phylogenetic tree (Fig. 7.3). From this tree we can distinguish four distinct evolutionary branches representing the homologues found in euryarchaeal halobacteria and methanomicrobia as well as Crenarchaeota. Interestingly, the methanomicrobia group splits into two subgroups. One group including *Methanosarcina acetivorans* and *hungate* possesses only one Y-family homologue while another group including *Methanosarcina burtoni, mazei* and *barkeri* harbour two. Although *Picrophilus torridus* is an Euryarchaeota, its Y-family homologue resembles those of Crenarchaeota probably because it originated by a horizontal gene transfer, which is known to occur between this species and Crenarchaea (Futterer *et al.*, 2004). Because the temperate deepwater planktonic Crenarchaeote 4B7 homologue (Beja *et al.*, 2002) is standing apart from all the branches, it may represent a common evolutionary predecessor. It is likely that at least one of the *Methanosarcina mazei* representatives originates by a lateral gene transfer since this species is known to harbour numerous ORFs whose closest homologues are present in the bacterial domain (Deppenmeier *et al.*, 2002) and its natural habitat is rice paddies, a niche shared by other bacterial species such as *Bacillus subtillis* known to harbour numerous Y-family Pol genes (Sung *et al.*, 2003). One of the related species *Methanosarcina acetivorans* with the largest known archaeal genome of 5.8 MB (Galagan *et al.*, 2002) has only one Y-family Pol.

It still remains an open question of how old the Y-family Pols are on evolutionary scale and whether they precede the chromosomal replicases belonging to the B-family. Given the absence of the homologous genes or pseudogenes in about half of the sequenced archaeal genomes, it is tempting to speculate that Y-family Pols are late comers to compensate for genotoxic effects of an aggressive energy metabolism or life under higher radiation exposures. It is conceivable that the tolerated DNA lesions would become gradu-

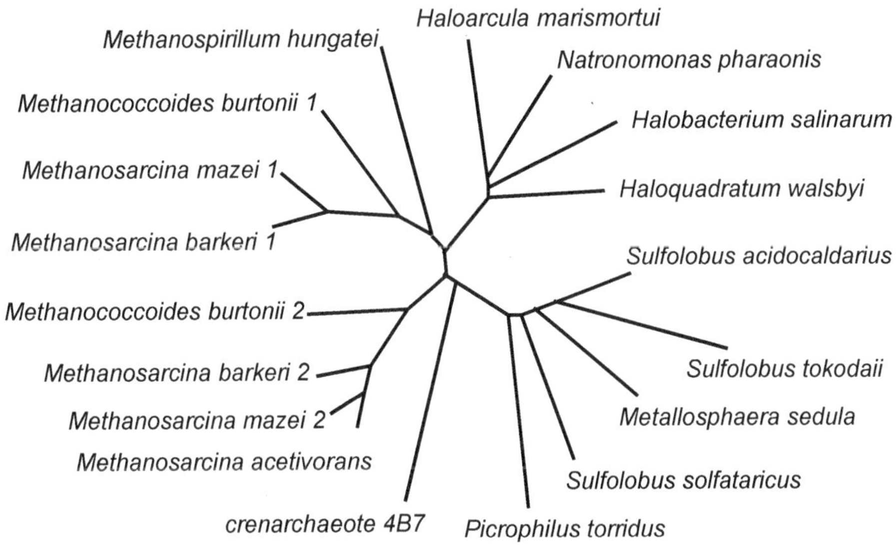

Figure 7.3 Tree showing the phylogeny of Y-family of DNA Pols in archeal species listed in Table 7.2. The tree was drawn using the drawtree program of the PHYLIP phylogeny package based on multiple amino acid sequence alignment generated by the ClustalX program (Thompson *et al.*, 1997). The amino acid sequence alignment is available upon request.

ally diluted under fast growing conditions when symmetric cell division prevails, rather than they would accumulate like in unrepaired genomes of non-dividing cells. In fact, evidence is presented recently that a mechanism may be in place to reduce mutagenesis in non-dividing resting cells by restricting TLS Pol expression to the G2/M phase of cell cycle (Waters and Walker, 2006). Such strategy, i.e. replication first and correctness second, may represent a selective advantage over the opposite strategy, i.e. correctness first and replication second, given that the living environment for Archaea may change rapidly and the most common lesions are bypassed in an *error-free* manner. The obligate symbiont *Nanoarchaeum equitans*, which is deeply branched on an evolutionary tree and probably represents a novel archaeal kingdom, lacks a genuine Y-family homologue despite having an extensive repertoire of other DNA repair enzymes (Waters *et al.*, 2003). Surprisingly, a RecA/RadA recombinase homologue in this species contains a short 53 amino acids segment sharing homology to human Pol iota, a representative of human Y-family Pols (GeneBank accession # AAR39271). The deep water archaeal species like *Pyrococcus* are also lacking Y-family Pols probably because they are well shielded from the surface UV radiation and oxygen. As the major components of archaeal cell membranes are isoprenoid ether-linked glyceroldiethers or tetraethers (De Rosa and Gambacorta, 1988; Itoh *et al.*, 2001), these organisms may be exposed to lower amounts of genotoxic lipid peroxides, which are generated from fatty acid-derived membranes of Bacteria and Eukarya by oxidative metabolism (Chung *et al.*, 2003a). One of the major lipid peroxidation products, trans-4-hydroxy-2-nonenal (Feng *et al.*, 2003; Feng *et al.*, 2004), forms a N^2-dG adduct which is known to be efficiently bypassed in an *error-free* manner by the TLS Y-family Pols (Wolfle *et al.*, 2006; Lone *et al.*, 2007). In fact, at least the DinB homologues, which are the most widespread among all Y-family members, are likely to function during an *error-free* bypass of the N^2-dG minor groove bulky DNA adducts (Jarosz *et al.*, 2006). All the 18 currently sequenced Y-family DNA Pols from Archaea seem to be enzymatically active because they harbour the important catalytic acidic residues in motifs I and III, which bind to the metal Mg^{2+} cofactor. Therefore TLS seems to play an important role in the physiology of certain archaeal species probably by letting them tolerate and become 'immune' against the genotoxic effects of specific DNA lesions resulting either from constant exposure to external radiation or internal reactive by-products of metabolism.

Table 7.2 The archaeal species containing Y-family DNA Pol homologues

Species	Classification	#	Habitat	Metabolism	Genome ref.
Sulfolobus acidocaldarius	*Crenarchaeota Thermoprotei*	1	Acid hot spring, USA Yellowstone Natl. Park, pH 2–3, 80°C	Aerobic, oxidation of elemental sulphur	(Chen *et al.*, 2005)
Sulfolobus tokodaii	*Crenarchaeota Thermoprotei*	1	Beppu hotsprings, Kyushu, pH 2–3, 80°C	Aerobic thermoacidophilic	(Kawarabayasi *et al.*, 2001)
Sulfolobus solfataricus	*Crenarchaeota Thermoprotei*	1	Continental solfatara, Naples, 87°C	Strictly aerobic, produces sulfuric acid	(She *et al.*, 2001)
Crenarchaeote 4B7	*Crenarchaeota Thermoprotei*	1	Temperate water plankton, 200 m deep near the coast of Oregon		(Beja *et al.*, 2002)
Metallosphaera sedula	*Crenarchaeota; Thermoprotei*	1			Unpublished
Picrophilus torridus	*Euryarchaeota; Thermoplasmata*	1	Dry solfataric field in northern Japan, 65°C	Most thermoacidophilic organism, growing on organic substrates, extracellular pH 0, intracellular pH 4.6	(Futterer *et al.*, 2004)
Methanosarcina mazei	*Euryarchaeota; Methanomicrobia*	2	Rice paddies	Fermenting acetate, methylamines and methanol to methane, carbon dioxide and ammonia, metabolically versatile methanogen	(Deppenmeier *et al.*, 2002)
Methanosarcina barkeri	*Euryarchaeota; Methanomicrobia*	2	Freshwater coastal lagoon west of Naples, Italy	Obligate CO_2 reduction with H_2, methyl reduction with H_2, aceticlastic fermentation of acetate or methylotrophic catabolism of methanol, methylated amines and dimethylsulphide, also grows nonmethanogenically with CO, fixes molecular nitrogen, metabolic diversity	(Maeder *et al.*, 2006)
Methanococcoides burtonii	*Euryarchaeota; Methanomicrobia*	2	anoxic and methanesaturated hypolimnion of Ace Lake (Antarctica), 1.7–29.5°C	strictly anaerobic, slightly halophilic, psychrophilic and methylotrophic	Unpublished

Methanosarcina acetivorans	*Euryarchaeota; Methanomicrobia*	1		acetate-utilizing methanogen	(Galagan *et al.*, 2002)
Methanospirillum hungatei	*Euryarchaeota; Methanomicrobia*	1	Anaerobic, freeze-dried granular sludge obtained from an upflow anaerobic sludge bed reactor treating sugar beet wastewater		Unpublished
Halobacterium salinarum	*Euryarchaeota; Halobacteria*	1	High salinity (10 times that of sea water)	Aerobic halophilic chemoorganotroph growing on the degradation products of less halophilic organisms as the salinity reaches near saturation	(Ng *et al.*, 2000)
Haloquadratum walsbyi	*Euryarchaeota; Halobacteria*	1	Spanish solar saltern, thalassic NaCl-saturated environments, sub-lethal conditions of an extremely high $MgCl_2$ concentration and high solar irradiance	Pigmented, synthesize sialic acid and poly-gamma-glutamate capsule, lives in near anoxic conditions due to the extremely high salinity	(Bolhuis *et al.*, 2006)
Natronomonas pharaonis	*Euryarchaeota; Halobacteria*	1	Salt-saturated lakes of pH 11	Grows optimally at pH 8.5 in 3.5 M NaCl, lacks key enzymes of glycolytic pathways but has functional respiration and oxidative phosphorylation, lipoproteins make up one-third of the predicted secretome, harbours blue-light photoreceptor	(Falb *et al.*, 2005)
Haloarcula marismortui	*Euryarchaeota; Halobacteria*	1	Dead Sea	physiologically capable of exploiting diverse environments, large number of environmental response regulators	(Baliga *et al.*, 2004c)

The results are based on the NCBI GeneBank protein sequence database as of March 2007. The number of homologues in each species is ndidicated under # and references are shown for each corresponding genome where available. The information about habitat and metabolism is taken from the EMBL-EBI Integr8 database or directly from the specified literature.

Future perspectives

Hyperthermophilic Archaea are interesting model organisms in terms of finding novel molecular strategies of how cells survive in harsh environments. As has been reviewed here, the replicative DNA Pols have unique molecular features that halt DNA synthesis when they encounter deaminated bases, i.e. U or HX, in template DNA. We expect that more unique mechanisms in DNA repair and damage tolerance will be revealed in Archeaa given the wealth of genome sequence information and ease of crystallization of proteins involved in the processing of DNA damage. However, there are several technical limitations that may hamper the progress of research in this area. The first problem may be the difficulty of cultivation. Many hyperthermophilic Archaea should be cultivated at extremely high temperatures and often require strictly anaerobic conditions and inorganic energy sources. Therefore, the actual genome stabilities such as mutation rates of most of hyperthermophilic Archaea are not experimentally determined. In addition, because of the difficulty to culture, information about chemical analysis of DNA lesions is very rare. We suppose that hyperthermophilic Archaea should have massively damaged bases in DNA such as U, 8-OH-G and AP sites. However, quantitative data concerning the DNA damage are scare and thus it is not sure that hyperthermophilic Archaea protect their genome stability mainly via chemical prevention such as using antioxidants to prevent oxidation of DNA or via efficient DNA repair and damage avoidance mechanisms discussed in this chapter. Collaboration with analytical chemists with experts of Archaea is essential to solve the problem. The second problem may be the lack of genetic approaches in particular gene disruption methods. Only a limited number of Archaea, such as *Halobacterium* species, *Sulfolobus solfataricus*, *Pyrococcus Kodakaraensis*, can be utilized for genetic studies. It is hard to evaluate the physiological roles of the mechanisms even if the proteins involved are crystallized and the fine three-dimensional structures are solved. In this regard, it is desirable to establish the gene knockout and knockin approaches to identify the roles of critical amino acids predicted by structural studies. Nevertheless, recent developments, such as measurement of mutation rates based on DNA instead of phenotypes of living organisms (Bielas and Loeb, 2005), high resolution mass spectrometry equipped with high performance liquid chromatography (Zhu *et al.*, 2003) and RNA-based gene knockdown (Collins and Cheng, 2005), will enable hyperthermophilic Archaea to be better model organisms to establish the relationships between genome integrity and a variety of environmental stresses.

Acknowledgements

This work was supported by grants-in-aid for scientific research from the Ministry of Education, Culture, Sports, Science and Technology, Japan, the Ministry of Health, Labour and Welfare, Japan and the Japan Health Science Foundation. Part of the wok was financially supported by the Budget for Nuclear Research of the Ministry of Education, Culture, Sports, Science and Technology, based on the screening and counselling by the Atomic Energy Commission.

References

Aboussekhra, A., Biggerstaff, M., Shivji, M. K., Vilpo, J. A., Moncollin, V., Podust, V. N., Protic, M., Hubscher, U., Egly, J. M., and Wood, R. D. (1995). Mammalian DNA nucleotide excision repair reconstituted with purified protein components. Cell *80*, 859–868.

Akiba, T., Ishii, N., Rashid, N., Morikawa, M., Imanaka, T., and Harata, K. (2005). Structure of RadB recombinase from a hyperthermophilic archaeon, Thermococcus kodakaraensis KOD1: an implication for the formation of a near-7-fold helical assembly. Nucleic Acids Res. *33*, 3412–3423.

Araujo, S. J. and Wood, R. D. (1999). Protein complexes in nucleotide excision repair. Mutat. Res. *435*, 23–33.

Aravind, L., Makarova, K. S., and Koonin, E. V. (2000). SURVEY AND SUMMARY: holliday junction resolvases and related nucleases: identification of new families, phyletic distribution and evolutionary trajectories. Nucleic Acids Res. *28*, 3417–3432.

Aravind, L., Walker, D. R., and Koonin, E. V. (1999). Conserved domains in DNA repair proteins and evolution of repair systems. Nucleic Acids Res. *27*, 1223–1242.

Asami, Y., Murakami, M., Shimizu, M., Pisani, F. M., Hayata, I., and Nohmi, T. (2006). Visualization of the interaction between archaeal DNA polymerase and uracil-containing DNA by atomic force microscopy. Genes Cells *11*, 3–11.

Bae, K. W., Baek, K. W., Cho, C. S., Hwang, K. Y., Kim, H. R., Sung, H. C., and Cho, Y. (1999). Expression, purification, characterization and crystallization of flap endonuclease-1 from *Methanococcus jannaschii*. Mol. Cells *9*, 45–48.

Baliga, N. S., Bjork, S. J., Bonneau, R., Pan, M., Iloanusi, C., Kottemann, M. C., Hood, L., and DiRuggiero, J. (2004a). Systems level insights into the stress response to UV radiation in the halophilic archaeon Halobacterium NRC-1. Genome Res. *14*, 1025–1035.

Baliga, N. S., Bjork, S. J., Bonneau, R., Pan, M., Iloanusi, C., Kottemann, M. C., Hood, L., and DiRuggiero, J. (2004b). Systems level insights into the stress response to UV radiation in the halophilic archaeon Halobacterium NRC-1. Genome Res. *14*, 1025–1035.

Baliga, N. S., Bonneau, R., Facciotti, M. T., Pan, M., Glusman, G., Deutsch, E. W., Shannon, P., Chiu, Y., Weng, R. S., Gan, R. R., Hung, P., Date, S. V., Marcotte, E., Hood, L., and Ng, W. V. (2004c). Genome sequence of Haloarcula marismortui: a halophilic archaeon from the Dead Sea. Genome Res. *14*, 2221–2234.

Beja, O., Koonin, E. V., Aravind, L., Taylor, L. T., Seitz, H., Stein, J. L., Bensen, D. C., Feldman, R. A., Swanson, R. V., and DeLong, E. F. (2002). Comparative genomic analysis of archaeal genotypic variants in a single population and in two different oceanic provinces. Appl. Environ. Microbiol. *68*, 335–345.

Bielas, J. H. and Loeb, L. A. (2005). Quantification of random genomic mutations. Nat. Methods *2*, 285–290.

Bolhuis, H., Palm, P., Wende, A., Falb, M., Rampp, M., Rodriguez-Valera, F., Pfeiffer, F., and Oesterhelt, D. (2006). The genome of the square archaeon Haloquadratum walsbyi : life at the limits of water activity. BMC. Genomics 7, 169.

Bolt, E. L., Lloyd, R. G., and Sharples, G. J. (2001). Genetic analysis of an archaeal Holliday junction resolvase in Escherichia coli. J. Mol. Biol. *310*, 577–589.

Boudsocq, F., Kokoska, R. J., Plosky, B. S., Vaisman, A., Ling, H., Kunkel, T. A., Yang, W., and Woodgate, R. (2004). Investigating the role of the little finger domain of Y-family DNA polymerases in low fidelity synthesis and translesion replication. J. Biol. Chem. *279*, 32932–32940.

Cappelli, E., Taylor, R., Cevasco, M., Abbondandolo, A., Caldecott, K., and Frosina, G. (1997). Involvement of XRCC1 and DNA ligase III gene products in DNA base excision repair. J. Biol. Chem. *272*, 23970–23975.

Chandani, S. and Loechler, E. L. (2007). Molecular modeling benzo[a]pyrene N2-dG adducts in the two overlapping active sites of the Y-family DNA polymerase Dpo4. J. Mol. Graph. Model. *25*, 658–670.

Chapados, B. R., Hosfield, D. J., Han, S., Qiu, J., Yelent, B., Shen, B., and Tainer, J. A. (2004). Structural basis for FEN-1 substrate specificity and PCNA-mediated activation in DNA replication and repair. Cell *116*, 39–50.

Chen, L., Brugger, K., Skovgaard, M., Redder, P., She, Q., Torarinsson, E., Greve, B., Awayez, M., Zibat, A., Klenk, H. P., and Garrett, R. A. (2005). The genome of Sulfolobus acidocaldarius, a model organism of the Crenarchaeota. J. Bacteriol. *187*, 4992–4999.

Choi, J. Y., Angel, K. C., and Guengerich, F. P. (2006). Translesion synthesis across bulky N2-alkyl guanine DNA adducts by human DNA polymerase kappa. J. Biol. Chem. *281*, 21062–21072.

Chung, F. L., Pan, J., Choudhury, S., Roy, R., Hu, W., and Tang, M. S. (2003a). Formation of trans-4-hydroxy-2-nonenal- and other enal-derived cyclic DNA adducts from omega-3 and omega-6 polyunsaturated fatty acids and their roles in DNA repair and human p53 gene mutation. Mutat. Res. *531*, 25–36.

Chung, J. H., Im, E. K., Park, H. Y., Kwon, J. H., Lee, S., Oh, J., Hwang, K. C., Lee, J. H., and Jang, Y. (2003b). A novel uracil-DNA glycosylase family related to the helix-hairpin-helix DNA glycosylase superfamily. Nucleic Acids Res. *31*, 2045–2055.

Chung, J. H., Suh, M. J., Park, Y. I., Tainer, J. A., and Han, Y. S. (2001). Repair activities of 8-oxoguanine DNA glycosylase from *Archaeoglobus fulgidus*, a hyperthermophilic archaeon. Mutat. Res. *486*, 99–111.

Cohen, G. N., Barbe, V., Flament, D., Galperin, M., Heilig, R., Lecompte, O., Poch, O., Prieur, D., Querellou, J., Ripp, R., Thierry, J. C., Van der, O. J., Weissenbach, J., Zivanovic, Y., and Forterre, P. (2003). An integrated analysis of the genome of the hyperthermophilic archaeon *Pyrococcus abyssi*. Mol. Microbiol. *47*, 1495–1512.

Collins, B. K., Tomanicek, S. J., Lyamicheva, N., Kaiser, M. W., and Mueser, T. C. (2004). A preliminary solubility screen used to improve crystallization trials: crystallization and preliminary X-ray structure determination of Aeropyrum pernix flap endonuclease-1. Acta Crystallogr. D. Biol. Crystallogr. *60*, 1674–1678.

Collins, R. E. and Cheng, X. (2005). Structural domains in RNAi. FEBS Lett. *579*, 5841–5849.

Connolly, B. A., Fogg, M. J., Shuttleworth, G., and Wilson, B. T. (2003). Uracil recognition by archaeal family B DNA polymerases. Biochem. Soc. Trans. *31*, 699–702.

De Felice, M., Medagli, B., Esposito, L., De Falco, M., Pucci, B., Rossi, M., Gruz, P., Nohmi, T., and Pisani, F. M. (2007). Biochemical evidence of a physical interaction between *Sulfolobus solfataricus* B-family and Y-family DNA polymerases. Extremophiles. *11*, 277–282.

De Rosa, M. and Gambacorta, A. (1988). The lipids of archaebacteria. Prog. Lipid Res. *27*, 153–175.

Demple, B. and Harrison, L. (1994). Repair of oxidative damage to DNA: enzymology and biology. Annu. Rev. Biochem. *63*, 915–948.

Deppenmeier, U., Johann, A., Hartsch, T., Merkl, R., Schmitz, R. A., Martinez-Arias, R., Henne, A., Wiezer, A., Baumer, S., Jacobi, C., Bruggemann, H., Lienard, T., Christmann, A., Bomeke, M., Steckel, S., Bhattacharyya, A., Lykidis, A., Overbeek, R., Klenk, H. P., Gunsalus, R. P., Fritz, H. J., and Gottschalk, G. (2002). The genome of Methanosarcina mazei: evidence for lateral gene transfer between Bacteria and Archaea. J. Mol. Microbiol. Biotechnol. *4*, 453–461.

Dianov, G., Price, A., and Lindahl, T. (1992). Generation of single-nucleotide repair patches following excision of uracil residues from DNA. Mol. Cell Biol. *12*, 1605–1612.

Dionne, I., Nookala, R. K., Jackson, S. P., Doherty, A. J., and Bell, S. D. (2003). A heterotrimeric PCNA in the hyperthermophilic archaeon *Sulfolobus solfataricus*. Mol. Cell *11*, 275–282.

DiRuggiero, J., Santangelo, N., Nackerdien, Z., Ravel, J., and Robb, F. T. (1997). Repair of extensive ionizing-radiation DNA damage at 95 degrees C in the hyperthermophilic archaeon *Pyrococcus furiosus*. J. Bacteriol. *179*, 4643–4645.

Eisen, J. A. (1998). A phylogenomic study of the MutS family of proteins. Nucleic Acids Res. *26*, 4291–4300.

Falb, M., Pfeiffer, F., Palm, P., Rodewald, K., Hickmann, V., Tittor, J., and Oesterhelt, D. (2005). Living with two extremes: conclusions from the genome sequence of Natronomonas pharaonis. Genome Res. *15*, 1336-1343.

Fan, L., Arvai, A. S., Cooper, P. K., Iwai, S., Hanaoka, F., and Tainer, J. A. (2006). Conserved XPB core structure and motifs for DNA unwinding: implications for pathway selection of transcription or excision repair. Mol. Cell *22*, 27–37.

Feng, Z., Hu, W., Amin, S., and Tang, M. S. (2003). Mutational spectrum and genotoxicity of the major lipid peroxidation product, trans-4-hydroxy-2-nonenal, induced DNA adducts in nucleotide excision repair-proficient and -deficient human cells. Biochemistry *42*, 7848–7854.

Feng, Z., Hu, W., and Tang, M. S. (2004). Trans-4-hydroxy-2-nonenal inhibits nucleotide excision repair in human cells: a possible mechanism for lipid peroxidation-induced carcinogenesis. Proc. Natl. Acad. Sci. U.S. A *101*, 8598–8602.

Fiala, K. A., Brown, J. A., Ling, H., Kshetry, A. K., Zhang, J., Taylor, J. S., Yang, W., and Suo, Z. (2007). Mechanism of template-independent nucleotide incorporation catalyzed by a template-dependent DNA polymerase. J. Mol. Biol. *365*, 590–602.

Fitz-Gibbon, S. T., Ladner, H., Kim, U. J., Stetter, K. O., Simon, M. I., and Miller, J. H. (2002). Genome sequence of the hyperthermophilic crenarchaeon Pyrobaculum aerophilum. Proc. Natl. Acad. Sci. U.S. A *99*, 984–989.

Fogg, M. J., Pearl, L. H., and Connolly, B. A. (2002). Structural basis for uracil recognition by archaeal family B DNA polymerases. Nat. Struct. Biol. *9*, 922–927.

Fondufe-Mittendorf, Y. N., Harer, C., Kramer, W., and Fritz, H. J. (2002). Two amino acid replacements change the substrate preference of DNA mismatch glycosylase Mig.MthI from T/G to A/G. Nucleic Acids Res. *30*, 614–621.

Frosina, G., Fortini, P., Rossi, O., Carrozzino, F., Raspaglio, G., Cox, L. S., Lane, D. P., Abbondandolo, A., and Dogliotti, E. (1996). Two pathways for base excision repair in mammalian cells. J. Biol. Chem. *271*, 9573–9578.

Fujihashi, M., Numoto, N., Kobayashi, Y., Mizushima, A., Tsujimura, M., Nakamura, A., Kawarabayasi, Y., and Miki, K. (2007). Crystal structure of archaeal photolyase from *Sulfolobus tokodaii* with two FAD molecules: implication of a novel light-harvesting cofactor. J. Mol. Biol. *365*, 903–910.

Fujikane, R., Komori, K., Shinagawa, H., and Ishino, Y. (2005). Identification of a novel helicase activity unwinding branched DNAs from the hyperthermophilic archaeon, *Pyrococcus furiosus*. J. Biol. Chem. *280*, 12351–12358.

Futterer, O., Angelov, A., Liesegang, H., Gottschalk, G., Schleper, C., Schepers, B., Dock, C., Antranikian, G., and Liebl, W. (2004). Genome sequence of Picrophilus torridus and its implications for life around pH 0. Proc. Natl. Acad. Sci. U.S. A *101*, 9091–9096.

Gadsden, M. H., McIntosh, E. M., Game, J. C., Wilson, P. J., and Haynes, R. H. (1993). dUTP pyrophosphatase is an essential enzyme in Saccharomyces cerevisiae. EMBO J. *12*, 4425–4431.

Galagan, J. E., Nusbaum, C., Roy, A., Endrizzi, M. G., Macdonald, P., FitzHugh, W., Calvo, S., Engels, R., Smirnov, S., Atnoor, D., Brown, A., Allen, N., Naylor, J., Stange-Thomann, N., DeArellano, K., Johnson, R., Linton, L., McEwan, P., McKernan, K., Talamas, J., Tirrell, A., Ye, W., Zimmer, A., Barber, R. D., Cann, I., Graham, D. E., Grahame, D. A., Guss, A. M., Hedderich, R., Ingram-Smith, C., Kuettner, H. C., Krzycki, J. A., Leigh, J. A., Li, W., Liu, J., Mukhopadhyay, B., Reeve, J. N., Smith, K., Springer, T. A., Umayam, L. A., White, O., White, R. H., Conway, d. M., Ferry, J. G., Jarrell, K. F., Jing, H., Macario, A. J., Paulsen, I., Pritchett, M., Sowers, K. R., Swanson, R. V., Zinder, S. H., Lander, E., Metcalf, W. W., and Birren, B. (2002). The genome of M. acetivorans reveals extensive metabolic and physiological diversity. Genome Res. *12*, 532–542.

Gao, Y. G., Su, S. Y., Robinson, H., Padmanabhan, S., Lim, L., McCrary, B. S., Edmondson, S. P., Shriver, J. W., and Wang, A. H. (1998). The crystal structure of the hyperthermophile chromosomal protein Sso7d bound to DNA. Nat. Struct. Biol. *5*, 782–786.

Gary, R., Kim, K., Cornelius, H. L., Park, M. S., and Matsumoto, Y. (1999). Proliferating cell nuclear antigen facilitates excision in long-patch base excision repair. J. Biol. Chem. *274*, 4354–4363.

Gogos, A. and Clarke, N. D. (1999). Characterization of an 8-oxoguanine DNA glycosylase from *Methanococcus jannaschii*. J. Biol. Chem. *274*, 30447–30450.

Greagg, M. A., Fogg, M. J., Panayotou, G., Evans, S. J., Connolly, B. A., and Pearl, L. H. (1999). A read-ahead function in archaeal DNA polymerases detects promutagenic template-strand uracil. Proc. Natl. Acad. Sci. U.S. A *96*, 9045–9050.

Grogan, D. W. (2000). The question of DNA repair in hyperthermophilic Archaea. Trends Microbiol. *8*, 180–185.

Grogan, D. W., Carver, G. T., and Drake, J. W. (2001). Genetic fidelity under harsh conditions: analysis of spontaneous mutation in the thermoacidophilic archaeon Sulfolobus acidocaldarius. Proc. Natl. Acad. Sci. U.S. A *98*, 7928–7933.

Gruz, P., Pisani, F. M., Shimizu, M., Yamada, M., Hayashi, I., Morikawa, K., and Nohmi, T. (2001). Synthetic activity of Sso DNA polymerase Y1, an archaeal DinB-like DNA polymerase, is stimulated by processivity factors proliferating cell nuclear antigen and replication factor C.J. Biol. Chem. *276*, 47394–47401.

Gruz, P., Shimizu, M., Pisani, F. M., De Felice, M., Kanke, Y., and Nohmi, T. (2003). Processing of DNA lesions by archaeal DNA polymerases from *Sulfolobus solfataricus*. Nucleic Acids Res. *31*, 4024–4030.

Haracska, L., Prakash, S., and Prakash, L. (2002). Yeast Rev1 protein is a G template-specific DNA polymerase. J. Biol. Chem. *277*, 15546-15551.

Hayashi, I., Morikawa, K., and Ishino, Y. (1999). Specific interaction between DNA polymerase II (PolD) and RadB, a Rad51/Dmc1 homolog, in *Pyrococcus furiosus*. Nucleic Acids Res. *27*, 4695–4702.

Hinks, J. A., Evans, M. C., De Miguel, Y., Sartori, A. A., Jiricny, J., and Pearl, L. H. (2002). An iron–sulfur cluster in the family 4 uracil-DNA glycosylases. J. Biol. Chem. *277*, 16936-16940.

Hogrefe, H. H., Hansen, C. J., Scott, B. R., and Nielson, K. B. (2002). Archaeal dUTPase enhances PCR amplifications with archaeal DNA polymerases by preventing dUTP incorporation. Proc. Natl. Acad. Sci. U.S. A *99*, 596–601.

Hopfner, K. P., Karcher, A., Craig, L., Woo, T. T., Carney, J. P., and Tainer, J. A. (2001). Structural biochemistry and interaction architecture of the DNA double-strand break repair Mre11 nuclease and Rad50-ATPase. Cell *105*, 473–485.

Hopfner, K. P., Karcher, A., Shin, D. S., Craig, L., Arthur, L. M., Carney, J. P., and Tainer, J. A. (2000). Structural biology of Rad50 ATPase: ATP-driven conformational control in DNA double-strand break repair and the ABC-ATPase superfamily. Cell *101*, 789–800.

Horst, J. P. and Fritz, H. J. (1996). Counteracting the mutagenic effect of hydrolytic deamination of DNA 5-methylcytosine residues at high temperature: DNA mismatch N-glycosylase Mig.Mth of the thermophilic archaeon *Methanobacterium thermoautotrophicum* THF. EMBO J. *15*, 5459–5469.

Hosfield, D. J., Mol, C. D., Shen, B., and Tainer, J. A. (1998). Structure of the DNA repair and replication endonuclease and exonuclease FEN-1: coupling DNA and PCNA binding to FEN-1 activity. Cell *95*, 135–146.

Hwang, K. Y., Baek, K., Kim, H. Y., and Cho, Y. (1998). The crystal structure of flap endonuclease-1 from *Methanococcus jannaschii*. Nat. Struct. Biol. 5, 707–713.

Ito, J. and Braithwaite, D. K. (1991). Compilation and alignment of DNA polymerase sequences. Nucleic Acids Res. *19*, 4045–4057.

Itoh, Y. H., Sugai, A., Uda, I., and Itoh, T. (2001). The evolution of lipids. Adv. Space Res. *28*, 719–724.

Jarosz, D. F., Godoy, V. G., Delaney, J. C., Essigmann, J. M., and Walker, G. C. (2006). A single amino acid governs enhanced activity of DinB DNA polymerases on damaged templates. Nature *439*, 225–228.

Kadyrov, F. A., Dzantiev, L., Constantin, N., and Modrich, P. (2006). Endonucleolytic function of MutLalpha in human mismatch repair. Cell *126*, 297–308.

Kaiser, M. W., Lyamicheva, N., Ma, W., Miller, C., Neri, B., Fors, L., and Lyamichev, V. I. (1999). A comparison of eubacterial and archaeal structure-specific 5′-exonucleases. J. Biol. Chem. *274*, 21387–21394.

Kanugula, S., Pauly, G. T., Moschel, R. C., and Pegg, A. E. (2005). A bifunctional DNA repair protein from *Ferroplasma acidarmanus* exhibits O6-alkylguanine-DNA alkyltransferase and endonuclease V activities. Proc. Natl. Acad. Sci. U.S. A *102*, 3617–3622.

Kanugula, S. and Pegg, A. E. (2003). Alkylation damage repair protein O6-alkylguanine-DNA alkyltransferase from the hyperthermophiles Aquifex aeolicus and *Archaeoglobus fulgidus*. Biochem. J. *375*, 449–455.

Kawarabayasi, Y., Hino, Y., Horikawa, H., Jin-no, K., Takahashi, M., Sekine, M., Baba, S., Ankai, A., Kosugi, H., Hosoyama, A., Fukui, S., Nagai, Y., Nishijima, K., Otsuka, R., Nakazawa, H., Takamiya, M., Kato, Y., Yoshizawa, T., Tanaka, T., Kudoh, Y., Yamazaki, J., Kushida, N., Oguchi, A., Aoki, K., Masuda, S., Yanagii, M., Nishimura, M., Yamagishi, A., Oshima, T., and Kikuchi, H. (2001). Complete genome sequence of an aerobic thermoacidophilic crenarchaeon, *Sulfolobus tokodaii* strain7. DNA Res. *8*, 123–140.

Kawarabayasi, Y., Sawada, M., Horikawa, H., Haikawa, Y., Hino, Y., Yamamoto, S., Sekine, M., Baba, S., Kosugi, H., Hosoyama, A., Nagai, Y., Sakai, M., Ogura, K., Otsuka, R., Nakazawa, H., Takamiya, M., Ohfuku, Y., Funahashi, T., Tanaka, T., Kudoh, Y., Yamazaki, J., Kushida, N., Oguchi, A., Aoki, K., and Kikuchi, H. (1998). Complete sequence and gene organization of the genome of a hyper-thermophilic archaebacterium, *Pyrococcus horikoshii* OT3. DNA Res. *5*, 55–76.

Kawashima, T., Amano, N., Koike, H., Makino, S., Higuchi, S., Kawashima-Ohya, Y., Watanabe, K., Yamazaki, M., Kanehori, K., Kawamoto, T., Nunoshiba, T., Yamamoto, Y., Aramaki, H., Makino, K., and Suzuki, M. (2000). Archaeal adaptation to higher temperatures revealed by genomic sequence of Thermoplasma volcanium. Proc. Natl. Acad. Sci. USA 97, 14257–14262.

Kelner, A. (1949). Effect of Visible Light on the Recovery of Streptomyces Griseus Conidia from Ultra-violet Irradiation Injury. Proc. Natl. Acad. Sci. U.S. A 35, 73–79.

Kiener, A., Husain, I., Sancar, A., and Walsh, C. (1989). Purification and properties of *Methanobacterium thermoautotrophicum* DNA photolyase. J. Biol. Chem. *264*, 13880–13887.

Kim, K., Biade, S., and Matsumoto, Y. (1998). Involvement of flap endonuclease 1 in base excision DNA repair. J. Biol. Chem. *273*, 8842–8848.

Kim, S. R., Maenhaut-Michel, G., Yamada, M., Yamamoto, Y., Matsui, K., Sofuni, T., Nohmi, T., and Ohmori, H. (1997). Multiple pathways for SOS-induced mutagenesis in Escherichia coli: an overexpression of dinB/dinP results in strongly enhancing mutagenesis in the absence of any exogenous treatment to damage DNA. Proc. Natl. Acad. Sci. U.S. A *94*, 13792–13797.

Kim, S. R., Matsui, K., Yamada, M., Gruz, P., and Nohmi, T. (2001). Roles of chromosomal and episomal dinB genes encoding DNA pol IV in targeted and untargeted mutagenesis in Escherichia coli. Mol. Genet. Genomics *266*, 207–215.

Klungland, A. and Lindahl, T. (1997). Second pathway for completion of human DNA base excision-repair: reconstitution with purified proteins and requirement for DNase IV (FEN1). EMBO J. *16*, 3341–3348.

Komori, K. and Ishino, Y. (2001). Replication protein A in *Pyrococcus furiosus* is involved in homologous DNA recombination. J. Biol. Chem. *276*, 25654–25660.

Komori, K., Miyata, T., Daiyasu, H., Toh, H., Shinagawa, H., and Ishino, Y. (2000a). Domain analysis of an archaeal RadA protein for the strand exchange activity. J. Biol. Chem. *275*, 33791–33797.

Komori, K., Miyata, T., DiRuggiero, J., Holley-Shanks, R., Hayashi, I., Cann, I. K., Mayanagi, K., Shinagawa, H., and Ishino, Y. (2000b). Both RadA and RadB are involved in homologous recombination in *Pyrococcus furiosus*. J. Biol. Chem. *275*, 33782–33790.

Komori, K., Sakae, S., Daiyasu, H., Toh, H., Morikawa, K., Shinagawa, H., and Ishino, Y. (2000c). Mutational analysis of the *Pyrococcus furiosus* holliday junction resolvase hjc revealed functionally important residues for dimer formation, junction DNA binding, and cleavage activities. J. Biol. Chem. *275*, 40385–40391.

Komori, K., Sakae, S., Shinagawa, H., Morikawa, K., and Ishino, Y. (1999). A Holliday junction resolvase from *Pyrococcus furiosus*: functional similarity to Escherichia coli RuvC provides evidence for conserved mechanism of homologous recombination in Bacteria, Eukarya, and Archaea. Proc. Natl. Acad. Sci. U.S. A *96*, 8873–8878.

Koulis, A., Cowan, D. A., Pearl, L. H., and Savva, R. (1996). Uracil-DNA glycosylase activities in hyperthermophilic micro-organisms. FEMS Microbiol. Lett. *143*, 267–271.

Kvaratskhelia, M., Wardleworth, B. N., Norman, D. G., and White, M. F. (2000). A conserved nuclease domain in the archaeal Holliday junction resolving enzyme Hjc. J. Biol. Chem. *275*, 25540–25546.

Kvaratskhelia, M. and White, M. F. (2000a). An archaeal Holliday junction resolving enzyme from *Sulfolobus solfataricus* exhibits unique properties. J. Mol. Biol. *295*, 193–202.

Kvaratskhelia, M. and White, M. F. (2000b). Two Holliday junction resolving enzymes in *Sulfolobus solfataricus*. J. Mol. Biol. *297*, 923–932.

Lasken, R. S., Schuster, D. M., and Rashtchian, A. (1996). Archaebacterial DNA polymerases tightly bind uracil-containing DNA. J. Biol. Chem. *271*, 17692–17696.

Leclere, M. M., Nishioka, M., Yuasa, T., Fujiwara, S., Takagi, M., and Imanaka, T. (1998). The O6-methylguanine-DNA methyltransferase from the hyperthermophilic archaeon Pyrococcus sp. KOD1: a thermostable repair enzyme. Mol. Gen. Genet. *258*, 69–77.

Lee, C. H., Chandani, S., and Loechler, E. L. (2006). Homology modeling of four Y-family, lesion-bypass DNA polymerases: the case that E. coli Pol IV and human Pol kappa are orthologs, and E. coli Pol V and human Pol eta are orthologs. J. Mol. Graph. Model. *25*, 87–102.

Lilley, D. M. and White, M. F. (2000). Resolving the relationships of resolving enzymes. Proc. Natl. Acad. Sci. U.S. A *97*, 9351–9353.

Lindahl, T. (1993). Instability and decay of the primary structure of DNA. Nature *362*, 709–715.

Lindahl, T. and Nyberg, B. (1974). Heat-induced deamination of cytosine residues in deoxyribonucleic acid. Biochemistry *13*, 3405–3410.

Lindahl, T., Sedgwick, B., Sekiguchi, M., and Nakabeppu, Y. (1988). Regulation and expression of the adaptive response to alkylating agents. Annu. Rev. Biochem. *57*, 133–157.

Lindahl, T. and Wood, R. D. (1999). Quality control by DNA repair. Science *286*, 1897–1905.

Ling, H., Boudsocq, F., Plosky, B. S., Woodgate, R., and Yang, W. (2003). Replication of a cis-syn thymine dimer at atomic resolution. Nature *424*, 1083–1087.

Ling, H., Boudsocq, F., Woodgate, R., and Yang, W. (2001). Crystal structure of a Y-family DNA polymerase in action: a mechanism for error-prone and lesion-bypass replication. Cell *107*, 91–102.

Ling, H., Boudsocq, F., Woodgate, R., and Yang, W. (2004a). Snapshots of replication through an abasic lesion; structural basis for base substitutions and frameshifts. Mol. Cell *13*, 751–762.

Ling, H., Sayer, J. M., Plosky, B. S., Yagi, H., Boudsocq, F., Woodgate, R., Jerina, D. M., and Yang, W. (2004b). Crystal structure of a benzo[a]pyrene diol epoxide adduct in a ternary complex with a DNA polymerase. Proc. Natl. Acad. Sci. U.S. A *101*, 2265–2269.

Lipps, G., Rother, S., Hart, C., and Krauss, G. (2003). A novel type of replicative enzyme harbouring ATPase, primase and DNA polymerase activity. EMBO J. *22*, 2516-2525.

Lone, S., Townson, S. A., Uljon, S. N., Johnson, R. E., Brahma, A., Nair, D. T., Prakash, S., Prakash, L., and Aggarwal, A. K. (2007). Human DNA polymerase kappa encircles DNA: implications for mismatch extension and lesion bypass. Mol. Cell *25*, 601–614.

Lu, A. L., Li, X., Gu, Y., Wright, P. M., and Chang, D. Y. (2001). Repair of oxidative DNA damage: mechanisms and functions. Cell Biochem. Biophys. *35*, 141–170.

Maeder, D. L., Anderson, I., Brettin, T. S., Bruce, D. C., Gilna, P., Han, C. S., Lapidus, A., Metcalf, W. W., Saunders, E., Tapia, R., and Sowers, K. R. (2006). The Methanosarcina barkeri genome: comparative analysis with *Methanosarcina acetivorans* and Methanosarcina mazei reveals extensive rearrangement within methanosarcinal genomes. J. Bacteriol. *188*, 7922–7931.

Margison, G. P., Povey, A. C., Kaina, B., and Santibanez Koref, M. F. (2003). Variability and regulation of O6-alkylguanine-DNA alkyltransferase. Carcinogenesis *24*, 625–635.

Marti, T. M., Kunz, C., and Fleck, O. (2002). DNA mismatch repair and mutation avoidance pathways. J. Cell Physiol *191*, 28–41.

Masuda, Y., Takahashi, M., Fukuda, S., Sumii, M., and Kamiya, K. (2002). Mechanisms of dCMP transferase reactions catalyzed by mouse Rev1 protein. J. Biol. Chem. *277*, 3040–3046.

Matsui, E., Kawasaki, S., Ishida, H., Ishikawa, K., Kosugi, Y., Kikuchi, H., Kawarabayashi, Y., and Matsui, I.

(1999). Thermostable flap endonuclease from the archaeon, *Pyrococcus horikoshii*, cleaves the replication fork-like structure endo/exonucleolytically. J. Biol. Chem. *274*, 18297–18309.

Matsui, E., Musti, K. V., Abe, J., Yamasaki, K., Matsui, I., and Harata, K. (2002). Molecular structure and novel DNA binding sites located in loops of flap endonuclease-1 from *Pyrococcus horikoshii*. J. Biol. Chem. *277*, 37840–37847.

McCready, S. and Marcello, L. (2003). Repair of UV damage in Halobacterium salinarum. Biochem. Soc. Trans. *31*, 694–698.

McGlynn, P. and Lloyd, R. G. (2000). Modulation of RNA polymerase by (p)ppGpp reveals a RecG-dependent mechanism for replication fork progression. Cell *101*, 35–45.

Modrich, P. and Lahue, R. (1996). Mismatch repair in replication fidelity, genetic recombination, and cancer biology. Annu. Rev. Biochem. *65*, 101–133.

Moncalian, G., Lengsfeld, B., Bhaskara, V., Hopfner, K. P., Karcher, A., Alden, E., Tainer, J. A., and Paull, T. T. (2004). The rad50 signature motif: essential to ATP binding and biological function. J. Mol. Biol. *335*, 937–951.

Nakamura, J., Walker, V. E., Upton, P. B., Chiang, S. Y., Kow, Y. W., and Swenberg, J. A. (1998). Highly sensitive apurinic/apyrimidinic site assay can detect spontaneous and chemically induced depurination under physiological conditions. Cancer Res. *58*, 222–225.

Newman, M., Murray-Rust, J., Lally, J., Rudolf, J., Fadden, A., Knowles, P. P., White, M. F., and McDonald, N. Q. (2005). Structure of an XPF endonuclease with and without DNA suggests a model for substrate recognition. EMBO J. *24*, 895–905.

Ng, W. V., Kennedy, S. P., Mahairas, G. G., Berquist, B., Pan, M., Shukla, H. D., Lasky, S. R., Baliga, N. S., Thorsson, V., Sbrogna, J., Swartzell, S., Weir, D., Hall, J., Dahl, T. A., Welti, R., Goo, Y. A., Leithauser, B., Keller, K., Cruz, R., Danson, M. J., Hough, D. W., Maddocks, D. G., Jablonski, P. E., Krebs, M. P., Angevine, C. M., Dale, H., Isenbarger, T. A., Peck, R. F., Pohlschroder, M., Spudich, J. L., Jung, K. W., Alam, M., Freitas, T., Hou, S., Daniels, C. J., Dennis, P. P., Omer, A. D., Ebhardt, H., Lowe, T. M., Liang, P., Riley, M., Hood, L., and DasSarma, S. (2000). Genome sequence of Halobacterium species NRC-1. Proc. Natl. Acad. Sci. U.S. A *97*, 12176-12181.

Nishino, T., Komori, K., Tsuchiya, D., Ishino, Y., and Morikawa, K. (2001). Crystal structure of the archaeal holliday junction resolvase Hjc and implications for DNA recognition. Structure. *9*, 197–204.

Nishino, T., Komori, K., Tsuchiya, D., Ishino, Y., and Morikawa, K. (2005). Crystal structure and functional implications of *Pyrococcus furiosus* hef helicase domain involved in branched DNA processing. Structure. *13*, 143–153.

Nohmi, T. (2006). Environmental stress and lesion-bypass DNA polymerases. Annu. Rev. Microbiol. *60*, 231–253.

Ogrunc, M., Becker, D. F., Ragsdale, S. W., and Sancar, A. (1998). Nucleotide excision repair in the third kingdom. J. Bacteriol. *180*, 5796–5798.

Ohmori, H., Friedberg, E. C., Fuchs, R. P., Goodman, M. F., Hanaoka, F., Hinkle, D., Kunkel, T. A., Lawrence, C. W., Livneh, Z., Nohmi, T., Prakash, L., Prakash, S., Todo, T., Walker, G. C., Wang, Z., and Woodgate, R. (2001). The Y-family of DNA polymerases. Mol. Cell *8*, 7–8.

Olsen, G. J. and Woese, C. R. (1996). Lessons from an Archaeal genome: what are we learning from *Methanococcus jannaschii?* Trends Genet. *12*, 377–379.

Pascucci, B., Stucki, M., Jonsson, Z. O., Dogliotti, E., and Hubscher, U. (1999). Long patch base excision repair with purified human proteins. DNA ligase I as patch size mediator for DNA polymerases delta and epsilon. J. Biol. Chem. *274*, 33696-33702.

Pearl, L. H. (2000). Structure and function in the uracil-DNA glycosylase superfamily. Mutat. Res. *460*, 165–181.

Percival, K. J., Klein, M. B., and Burgers, P. M. (1989). Molecular cloning and primary structure of the uracil-DNA-glycosylase gene from Saccharomyces cerevisiae. J. Biol. Chem. *264*, 2593–2598.

Petit, C. and Sancar, A. (1999). Nucleotide excision repair: from E. coli to man. Biochimie *81*, 15–25.

Podlutsky, A. J., Dianova, I. I., Podust, V. N., Bohr, V. A., and Dianov, G. L. (2001). Human DNA polymerase beta initiates DNA synthesis during long-patch repair of reduced AP sites in DNA. EMBO J. *20*, 1477–1482.

Prakash, S., Johnson, R. E., and Prakash, L. (2005). Eukaryotic translesion synthesis DNA polymerases: specificity of structure and function. Annu. Rev. Biochem. *74*, 317–353.

Qian, X., Wu, Y., He, Y., and Luo, Y. (2005). Crystal structure of Methanococcus voltae RadA in complex with ADP: hydrolysis-induced conformational change. Biochemistry *44*, 13753–13761.

Rao, H. G., Rosenfeld, A., and Wetmur, J. G. (1998). *Methanococcus jannaschii* flap endonuclease: expression, purification, and substrate requirements. J. Bacteriol. *180*, 5406–5412.

Rashid, N., Morikawa, M., and Imanaka, T. (1996). A RecA/RAD51 homologue from a hyperthermophilic archaeon retains the major RecA domain only. Mol. Gen. Genet. *253*, 397–400.

Reich, C. I., McNeil, L. K., Brace, J. L., Brucker, J. K., and Olsen, G. J. (2001). Archaeal RecA homologues: different response to DNA-damaging agents in mesophilic and thermophilic Archaea. Extremophiles. *5*, 265–275.

Robb, F. T., Maeder, D. L., Brown, J. R., DiRuggiero, J., Stump, M. D., Yeh, R. K., Weiss, R. B., and Dunn, D. M. (2001). Genomic sequence of hyperthermophile, *Pyrococcus furiosus*: implications for physiology and enzymology. Methods Enzymol. *330*, 134–157.

Roberts, J. A., Bell, S. D., and White, M. F. (2003). An archaeal XPF repair endonuclease dependent on a heterotrimeric PCNA. Mol. Microbiol. *48*, 361–371.

Robinson, H., Gao, Y. G., McCrary, B. S., Edmondson, S. P., Shriver, J. W., and Wang, A. H. (1998). The hyperthermophile chromosomal protein Sac7d sharply kinks DNA. Nature *392*, 202–205.

Ruepp, A., Graml, W., Santos-Martinez, M. L., Koretke, K. K., Volker, C., Mewes, H. W., Frishman, D., Stocker, S., Lupas, A. N., and Baumeister, W. (2000). The genome sequence of the thermoacidophilic scavenger Thermoplasma acidophilum. Nature *407*, 508–513.

Salerno, V., Napoli, A., White, M. F., Rossi, M., and Ciaramella, M. (2003). Transcriptional response to DNA damage in the archaeon *Sulfolobus solfataricus*. Nucleic Acids Res. *31*, 6127–6138.

Sancar, A. (1994). Structure and function of DNA photolyase. Biochemistry 33, 2–9.

Sancar, A. (1996). DNA excision repair. Annu. Rev. Biochem. *65*, 43–81.

Sancar, G. B. (1990). DNA photolyases: physical properties, action mechanism, and roles in dark repair. Mutat. Res. *236*, 147–160.

Sandigursky, M. and Franklin, W. A. (2000). Uracil-DNA glycosylase in the extreme thermophile *Archaeoglobus fulgidus*. J. Biol. Chem. *275*, 19146-19149.

Sandler, S. J., Hugenholtz, P., Schleper, C., DeLong, E. F., Pace, N. R., and Clark, A. J. (1999). Diversity of radA genes from cultured and uncultured Archaea: comparative analysis of putative RadA proteins and their use as a phylogenetic marker. J. Bacteriol. *181*, 907–915.

Sandler, S. J., Satin, L. H., Samra, H. S., and Clark, A. J. (1996). recA-like genes from three archaean species with putative protein products similar to Rad51 and Dmc1 proteins of the yeast Saccharomyces cerevisiae. Nucleic Acids Res. *24*, 2125–2132.

Sartori, A. A., Fitz-Gibbon, S., Yang, H., Miller, J. H., and Jiricny, J. (2002). A novel uracil-DNA glycosylase with broad substrate specificity and an unusual active site. EMBO J. *21*, 3182–3191.

Sartori, A. A. and Jiricny, J. (2003). Enzymology of base excision repair in the hyperthermophilic archaeon Pyrobaculum aerophilum. J. Biol. Chem. *278*, 24563–24576.

Savino, C., Federici, L., Johnson, K. A., Vallone, B., Nastopoulos, V., Rossi, M., Pisani, F. M., and Tsernoglou, D. (2004). Insights into DNA replication: the crystal structure of DNA polymerase B1 from the archaeon *Sulfolobus solfataricus*. Structure. *12*, 2001–2008.

Savva, R., McAuley-Hecht, K., Brown, T., and Pearl, L. (1995). The structural basis of specific base-excision repair by uracil-DNA glycosylase. Nature *373*, 487–493.

Sayers, J. R. and Artymiuk, P. J. (1998). Flexible loops and helical arches. Nat. Struct. Biol. 5, 668–670.

Sedgwick, B. (2004). Repairing DNA-methylation damage. Nat. Rev. Mol. Cell Biol. 5, 148–157.

Seigneur, M., Bidnenko, V., Ehrlich, S. D., and Michel, B. (1998). RuvAB acts at arrested replication forks. Cell 95, 419–430.

Seitz, E. M., Brockman, J. P., Sandler, S. J., Clark, A. J., and Kowalczykowski, S. C. (1998). RadA protein is an archaeal RecA protein homolog that catalyzes DNA strand exchange. Genes Dev. *12*, 1248–1253.

She, Q., Singh, R. K., Confalonieri, F., Zivanovic, Y., Allard, G., Awayez, M. J., Chan-Weiher, C. C., Clausen, I. G., Curtis, B. A., De Moors, A., Erauso, G., Fletcher, C., Gordon, P. M., Heikamp-de Jong, I., Jeffries, A. C., Kozera, C. J., Medina, N., Peng, X., Thi-Ngoc, H. P., Redder, P., Schenk, M. E., Theriault, C., Tolstrup, N., Charlebois, R. L., Doolittle, W. F., Duguet, M., Gaasterland, T., Garrett, R. A., Ragan, M. A., Sensen, C. W., and Van der, O. J. (2001). The complete genome of the crenarchaeon *Sulfolobus solfataricus* P2. Proc. Natl. Acad. Sci. U.S. A *98*, 7835–7840.

Shuttleworth, G., Fogg, M. J., Kurpiewski, M. R., Jen-Jacobson, L., and Connolly, B. A. (2004). Recognition of the pro-mutagenic base uracil by family B DNA polymerases from Archaea. J. Mol. Biol. *337*, 621–634.

Silvian, L. F., Toth, E. A., Pham, P., Goodman, M. F., and Ellenberger, T. (2001). Crystal structure of a DinB family error-prone DNA polymerase from *Sulfolobus solfataricus*. Nat. Struct. Biol. *8*, 984–989.

Skorvaga, M., Raven, N. D., and Margison, G. P. (1998). Thermostable archaeal O6-alkylguanine-DNA alkyltransferases. Proc. Natl. Acad. Sci. U.S. A *95*, 6711–6715.

Sleeth, K. M., Robson, R. L., and Dianov, G. L. (2004). Exchangeability of mammalian DNA ligases between base excision repair pathways. Biochemistry *43*, 12924–12930.

Slesarev, A. I., Mezhevaya, K. V., Makarova, K. S., Polushin, N. N., Shcherbinina, O. V., Shakhova, V. V., Belova, G. I., Aravind, L., Natale, D. A., Rogozin, I. B., Tatusov, R. L., Wolf, Y. I., Stetter, K. O., Malykh, A. G., Koonin, E. V., and Kozyavkin, S. A. (2002). The complete genome of hyperthermophile Methanopyrus kandleri AV19 and monophyly of archaeal methanogens. Proc. Natl. Acad. Sci. U.S. A *99*, 4644–4649.

Smith, D. R., Doucette-Stamm, L. A., Deloughery, C., Lee, H., Dubois, J., Aldredge, T., Bashirzadeh, R., Blakely, D., Cook, R., Gilbert, K., Harrison, D., Hoang, L., Keagle, P., Lumm, W., Pothier, B., Qiu, D., Spadafora, R., Vicaire, R., Wang, Y., Wierzbowski, J., Gibson, R., Jiwani, N., Caruso, A., Bush, D., Reeve, J. N., and. (1997). Complete genome sequence of *Methanobacterium thermoautotrophicum* deltaH: functional analysis and comparative genomics. J. Bacteriol. *179*, 7135–7155.

Spies, M., Kil, Y., Masui, R., Kato, R., Kujo, C., Ohshima, T., Kuramitsu, S., and Lanzov, V. (2000). The RadA protein from a hyperthermophilic archaeon *Pyrobaculum islandicum* is a DNA-dependent ATPase that exhibits two disparate catalytic modes, with a transition temperature at 75 degrees C. Eur. J. Biochem. *267*, 1125–1137.

Sung, H. M., Yeamans, G., Ross, C. A., and Yasbin, R. E. (2003). Roles of YqjH and YqjW, homologs of the Escherichia coli UmuC/DinB or Y superfamily of DNA polymerases, in stationary-phase mutagenesis and UV-induced mutagenesis of Bacillus subtilis. J. Bacteriol. *185*, 2153–2160.

Suzuki, N., Itoh, S., Poon, K., Masutani, C., Hanaoka, F., Ohmori, H., Yoshizawa, I., and Shibutani, S. (2004). Translesion synthesis past estrogen-derived DNA adducts by human DNA polymerases eta and kappa. Biochemistry *43*, 6304–6311.

Tanaka, K., Miura, N., Satokata, I., Miyamoto, I., Yoshida, M. C., Satoh, Y., Kondo, S., Yasui, A., Okayama, H., and Okada, Y. (1990). Analysis of a human DNA excision repair gene involved in group A xeroderma pigmentosum and containing a zinc-finger domain. Nature *348*, 73–76.

Tashiro, R., Wang, A. H., and Sugiyama, H. (2006). Photoreactivation of DNA by an archaeal nucleoprotein Sso7d. Proc. Natl. Acad. Sci. U.S. A *103*, 16655–16659.

Taverna, P. and Sedgwick, B. (1996). Generation of an endogenous DNA-methylating agent by nitrosation in Escherichia coli. J. Bacteriol. *178*, 5105–5111.

Thompson, J. D., Gibson, T. J., Plewniak, F., Jeanmougin, F., and Higgins, D. G. (1997). The CLUSTAL_X windows interface: flexible strategies for multiple sequence alignment aided by quality analysis tools. Nucleic Acids Res. *25*, 4876–4882.

Trincao, J., Johnson, R. E., Escalante, C. R., Prakash, S., Prakash, L., and Aggarwal, A. K. (2001). Structure of the catalytic core of S. cerevisiae DNA polymerase eta: implications for translesion DNA synthesis. Mol. Cell *8*, 417–426.

Uemori, T., Sato, Y., Kato, I., Doi, H., and Ishino, Y. (1997). A novel DNA polymerase in the hyperthermophilic archaeon, *Pyrococcus furiosus*: gene cloning, expression, and characterization. Genes Cells *2*, 499–512.

Vijayvargia, R. and Biswas, I. (2002). MutS2 family protein from *Pyrococcus furiosus*. Curr. Microbiol. *44*, 224–228.

Wang, Y., Arora, K., and Schlick, T. (2006). Subtle but variable conformational rearrangements in the replication cycle of *Sulfolobus solfataricus* P2 DNA polymerase IV (Dpo4) may accommodate lesion bypass. Protein Sci. *15*, 135–151.

Washington, M. T., Prakash, L., and Prakash, S. (2003). Mechanism of nucleotide incorporation opposite a thymine-thymine dimer by yeast DNA polymerase eta. Proc. Natl. Acad. Sci. U.S. A *100*, 12093–12098.

Waters, E., Hohn, M. J., Ahel, I., Graham, D. E., Adams, M. D., Barnstead, M., Beeson, K. Y., Bibbs, L., Bolanos, R., Keller, M., Kretz, K., Lin, X., Mathur, E., Ni, J., Podar, M., Richardson, T., Sutton, G. G., Simon, M., Soll, D., Stetter, K. O., Short, J. M., and Noordewier, M. (2003). The genome of Nanoarchaeum equitans: insights into early archaeal evolution and derived parasitism. Proc. Natl. Acad. Sci. U.S. A *100*, 12984–12988.

Waters, L. S. and Walker, G. C. (2006). The critical mutagenic translesion DNA polymerase Rev1 is highly expressed during G(2)/M phase rather than S phase. Proc. Natl. Acad. Sci. U.S. A *103*, 8971–8976.

White, M. F. (2003). Archaeal DNA repair: paradigms and puzzles. Biochem. Soc. Trans. *31*, 690–693.

Wolfle, W. T., Johnson, R. E., Minko, I. G., Lloyd, R. S., Prakash, S., and Prakash, L. (2006). Replication past a trans-4-hydroxynonenal minor-groove adduct by the sequential action of human DNA polymerases iota and kappa. Mol. Cell Biol. *26*, 381–386.

Wood, E. R., Ghane, F., and Grogan, D. W. (1997). Genetic responses of the thermophilic archaeon Sulfolobus acidocaldarius to short-wavelength UV light. J. Bacteriol. *179*, 5693–5698.

Woods, W. G. and Dyall-Smith, M. L. (1997). Construction and analysis of a recombination-deficient (radA) mutant of *Haloferax volcanii*. Mol. Microbiol. *23*, 791–797.

Yang, H., Fitz-Gibbon, S., Marcotte, E. M., Tai, J. H., Hyman, E. C., and Miller, J. H. (2000). Characterization of a thermostable DNA glycosylase specific for U/G and T/G mismatches from the hyperthermophilic archaeon Pyrobaculum aerophilum. J. Bacteriol. *182*, 1272–1279.

Yasui, A., Eker, A. P., Yasuhira, S., Yajima, H., Kobayashi, T., Takao, M., and Oikawa, A. (1994). A new class of DNA photolyases present in various organisms including aplacental mammals. EMBO J. *13*, 6143–6151.

Yasui, A. and McCready, S. J. (1998). Alternative repair pathways for UV-induced DNA damage. Bioessays *20*, 291–297.

Yeiser, B., Pepper, E. D., Goodman, M. F., and Finkel, S. E. (2002). SOS-induced DNA polymerases enhance long-term survival and evolutionary fitness. Proc. Natl. Acad. Sci. USA *99*, 8737–8741.

Zang, H., Irimia, A., Choi, J. Y., Angel, K. C., Loukachevitch, L. V., Egli, M., and Guengerich, F. P. (2006). Efficient and high fidelity incorporation of dCTP opposite 7, 8-dihydro-8-oxodeoxyguanosine by *Sulfolobus solfataricus* DNA polymerase Dpo4. J. Biol. Chem. *281*, 2358–2372.

Zhu, H., Bilgin, M., and Snyder, M. (2003). Proteomics. Annu. Rev. Biochem. *72*, 783–812.

Modified Nucleotides in Archaeal RNAs

8

Henri Grosjean, Ramesh Gupta, and E. Stuart Maxwell

Abstract

Modified nucleotides increase the structural and functional diversity of RNA well beyond the four canonical bases. They are abundant in Archaea, Bacteria, and Eukarya and play important roles in regulating RNA function in all three domains of life. In Archaea, both transfer and ribosomal RNAs possess numerous modified nucleosides with some unique to Archaea and others shared with Bacteria and/or Eukarya. The diversity of archaeal nucleoside modifications is matched by the complexity of machineries and biosynthetic pathways that carry out these modifications. In this chapter, we discuss the diversity of nucleosides found in many different archaeal organisms and detail their occurrence in different RNA species. We also examine the enzymes used to accomplish these nucleoside modifications, reviewing both protein-only and RNA-guided ribonucleoprotein enzymes. Throughout this discussion, we provide the reader with additional reference sources that should be helpful for exploring further the modified nucleosides and modification enzymes of Archaea.

Introduction

RNA plays a central role in many cellular processes of all organisms. To accomplish present day functions, nucleotide modification augments the structural and functional diversity of RNAs well beyond the four canonical bases. The utilization of modified bases for RNA structural and functional diversity in Archaea, Bacteria, and Eukarya indicates that the modification of nucleotide bases probably arose from an ancient primordial 'RNA World' before the establishment of the three domains of life. To date, more than 100 modified ribonucleosides have been described (Fig. 8.1). They are found in numerous RNA types with tRNA alone possessing more than 80 modified nucleosides.

Each ribonucleotide modification is catalysed by a specific modification enzyme(s). Only a few modification enzymes have been identified and characterized out of the estimated 150–200 enzymes needed for presently known modifications. Both *in vivo* genetic approaches and *in vitro* assays of cell extracts have been used to characterize these activities and some activities have been characterized using recombinantly expressed enzymes (reviewed in Bujnicki *et al.*, 2004; De Crécy-Lagard, 2004; Björk and Hagervall, 2005; Ofengand and Del Campo, 2005; Andersen and Douthwaite, 2006). The discovery of RNA-guided nucleotide modification by ribonucleoprotein (RNP) complexes has added to the complexity of the nucleotide modification machinery. Large populations of guide RNAs determine the site of nucleotide modification on ribosomal and transfer RNAs while the small set of core proteins bound to each guide RNA catalyse the nucleotide modification reaction (reviewed in Omer *et al.*, 2003; Yu *et al.*, 2005).

Individual modified nucleotides add new dimensions to both the 3D architecture and the function of modified RNA. The majority of modified nucleotides appear to improve the performance of the mature RNA by increasing the RNA's stability and/or function. Often these

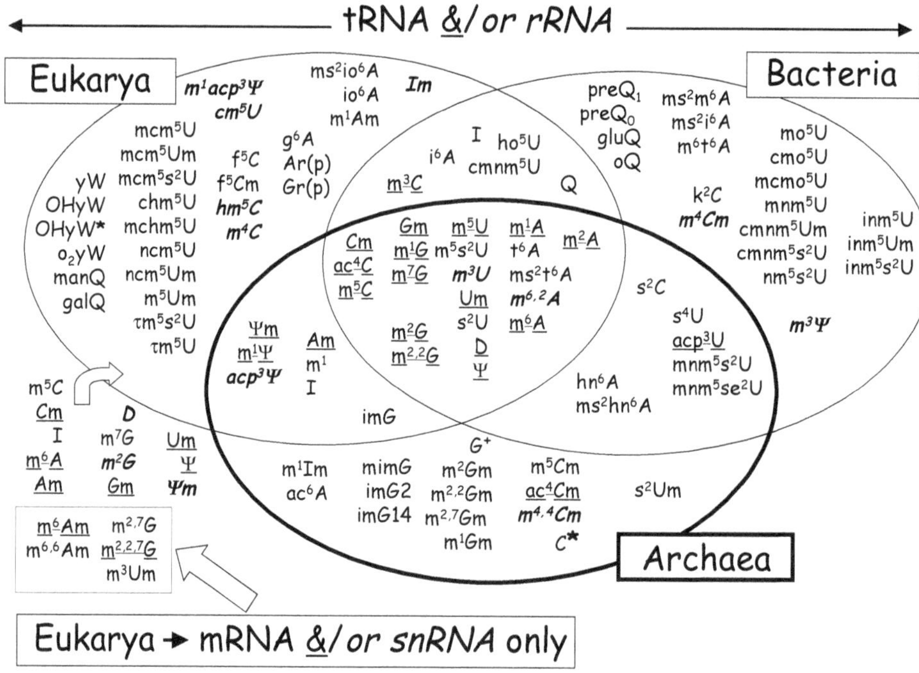

Figure 8.1 Distribution of modified RNA nucleosides in the three domains of life. Compilation of the 109 structurally distinct modified nucleosides derived from the four canonical adenosine (A), guanosine (G), cytosine (C), and uridine (U) nucleosides. Inosine (I) is a deaminated version of A, dihydrouridine (D) is non-aromatic version of U, and pseudouridine (Ψ) is an isomer of U where the usual glycosidic C-N bond is replaced by a C-C bond. All 109 are synthesized by posttranscriptional modification of targeted nucleotides in precursor RNAs during maturation. Enclosed in each circle are the modified nucleotides found in tRNA (cytoplasmic, mitochondrial and chloroplastic) and/or rRNA of the indicated domain of life. Overlapping sectors of the circles designate those modified nucleosides found in two or more domains. Modified nucleosides outside the circles correspond to those found in eukaryal mRNAs and snRNAs. Each presented nucleoside possesses a modification of the base, the ribose, or both base and ribose as indicated. Indicated modifications are the following: a methyl group is abbreviated m; a carboxyl group c; a thiol group s; a seleno group se; an amino group n; an isopentenyl group i; an oxy group o; a hydroxy group h (or OH); a threonylcarbamoyl group t; an acetyl group ac; a 3-amino-3-carboxypropyl group acp; a lysine group k; a methylthio group ms; a hydroxynorvalylcarbamoyl group hn; a methylaminomethyl group mnm, etc. The complete list of the abbreviations for modified nucleosides indicated here can be found at http://library.med.utah.edu/RNAmods/(Limbach *et al.*, 1994). The position of the chemical group on the base is indicated by a superscript number. A repeat of the same number in the abbreviation means that two groups are present on the same base at that numbered position. Two or three numbers after a group mean that particular group is present at those two or three positions of the base. An m or r(p) following the C, U, A, G, I or Ψ signifies a methyl group or a phosphate group located at the 2'-hydroxyl of the ribose. Some hypermodified nucleosides are defined by new symbols such as Q for queuosine, G^+ for archaeosine, imG (or sometimes W) for wyosine, and yW for wybutosine. Wyosine derivatives isowyosine, 4-demethylwyosine and methylwyosine are abbreviated as imG2, imG-14 and mimG, respectively. Only nucleoside C* (archaeal circle) is not conventional and corresponds to an as yet incompletely characterized lysidine-like cytosine (k^2C) derivative as discussed in the text. Designated nucleosides in each circle not underlined signify their occurrence (at least to date) exclusively in tRNA (regular font) or exclusively in ***rRNA (bold italics)***. Designated nucleosides underlined (regular font) indicate modified nucleosides found in both tRNA and rRNA. Modified nucleosides of eukaryal mRNA and snRNA (outside the three circles) are indicated as found only in mRNA (regular font), only in ***snRNA (bold italics)*** and in both mRNA and snRNA (underlined). Five members of this eukaryal group (in the box) are unique to mRNA and/or snRNA, while others (outside the box and circles) are also present in tRNA and/or rRNA. Detailed chemical structures, scientific and common names corresponding to each indicated modified nucleoside can be found at http://library.med.utah.edu/RNAmods/(Limbach *et al.*, 1994). This reference is an excellent source of information for the modified nucleoside Chemical Abstract registration number and index, citation of initial structural characterization and their occurrence, as well as other useful information. A more recent compilation of modified nucleosides found exclusively in rRNA can be found at http://library.med.utah.edu/SSUmods/(McCloskey and Rozenski, 2005). Databases for snoRNAs and sRNAs that are involved in RNA-guided modifications are located at http://people.biochem.

modifications promote the RNA's functional efficiency or its accuracy in catalysis under varying physiological conditions. However, their effects can often be subtle and not easily demonstrated using either *in vivo* or *in vitro* systems (for reviews see Agris, 1996; Hopper and Phizicky, 2003; Anderson and Droogmans, 2005; Douthwaite *et al.*, 2005; Lapeyre, 2005; Suzuki, 2005). In this chapter, we review what is known about RNA modification in Archaea with an emphasis on rRNA and tRNA modifications. Considerable progress has been made since the last comprehensive review in 1985 (Gupta, 1985). The biological or functional consequences of these nucleotide modifications are not explored in this chapter but are discussed elsewhere (Helm, 2006; Grosjean and Oshima, 2007).

Modified nucleosides in archaeal RNAs

Techniques for modified nucleoside detection

The presence of modified nucleosides derived from adenosine, guanosine, cytidine and uridine is a hallmark of all cellular RNAs. Their identification and location in an RNA molecule can often be a difficult but not an impossible task (reviewed in Grosjean *et al.*, 2007). Early identification of modified nucleosides came from tedious RNA sequencing analysis techniques, almost exclusively in tRNA. To date, up to 600 tRNA sequences are available from many different organisms of all three biological domains. Close to 60 or about 10% are archaeal tRNA species (see in Sprinzl and Vassilenko, 2005). In the early days, modified nucleosides in tRNA, as well as those of rRNA, the small nucleolar RNAs (snoRNAs) and other RNAs, were determined by analyzing RNA hydrolysates using paper chromatography. Later, thin-layer cellulose chromatography and more recently high-performance liquid chromatography, eventually coupled with electro-spray mass spectrophotometry, were used for modified nucleoside identification (see for examples Gupta *et al.*, 1976; Gupta and Randerath, 1979; Gupta and Woese, 1980; Pang *et al.*, 1982; Yamaizumi *et al.*, 1982; Buck *et al.*, 1983; Gupta, 1984; Edmonds *et al.*, 1987, 1991; Reddy *et al.*, 1992; Gregson *et al.*, 1993; Kowalak *et al.*, 1994; McCloskey *et al.*, 2001; Noon *et al.*, 2003; Sauerwald *et al.*, 2005; Saikia *et al.*, 2006). The modified nucleosides of purified tRNA and rRNA fragments are now often determined by MALDI-TOF mass spectrometry sequencing (see for examples Kowalak *et al.*, 1993; Kirpekar *et al.*, 2000; Patteson *et al.*, 2001; Hansen *et al.*, 2002; Mengel-Jorgensen and Kirpekar, 2002; Douthwaite and Kirpekar, 2007). Except for classical sequencing techniques, the precise location of a given modified nucleotide within an RNA's primary sequence is not always evident. However, the location of a modified nucleotide can often be inferred based upon its conserved position within the primary sequence of a tRNA or rRNA (see Grosjean *et al.*, 1995b; Ofengand and Bakin, 1997). In the case of pseudouridine (Ψ) and a few other particularly bulky modified nucleosides, their position in an RNA's primary sequence can be determined by primer-extension analysis using reverse transcriptase (reviewed in Motorin *et al.*, 2007). This approach has been particularly successful in analyzing the modified nucleotides of rRNA. An alternative approach is to use the tRNA modification enzymes contained in a cell extract for nucleotide modification and subsequent identification. This *in vitro* assay uses heterologous tRNAs or synthetic T7 RNA polymerase transcripts as substrates (Best, 1978; Constantinesco *et al.*, 1999b; reviewed in Graham and Kramer, 2007; Grosjean *et al.*, 2007). Finally, the presence of 2'-O-ribose methylated nucleotides can be predicted using known box C/D RNAs. The box C/D RNA's guide

Figure 8.1 *continued*

umass.edu/fournierlab/snornadb/main.php (Piekna-Przybylska *et al.*, 2007) and http://lowelab.ucsc.edu/snoRNAdb/. For modified nucleosides of tRNAs, the reader should refer to http://www.tRNA.uni-bayreuth.de/ (Sprinzl and Vassilenko, 2005). Reddy and coworkers (Reddy, 1989; Gu *et al.*, 1998) have compiled a list of snRNAs including their modified nucleosides. Readers should consult http://genesilico.pl/modomics/ (Dunin-Horkawicz *et al.*, 2006) for information about biosynthetic enzymes for nucleotide modifications. Two books contain detailed information concerning the technical approaches used for modified nucleoside determination and analysis (Gott, 2004, 2007).

sequence is complementary to a target RNA and therefore designates the specific target nucleotide for modification (reviewed in Gaspin *et al.*, 2000; Dennis *et al.*, 2001; Omer *et al.*, 2003; also see below).

Occurrence and location of modified nucleosides

Comprehensive analysis of nucleotide modifications is most complete for the tRNAs of two organisms. Sequence analysis of 41 tRNAs in the halophilic archaeon *Haloferax volcanii* (Gupta, 1984, 1986) (Fig. 8.2A) together with 29 tRNA species of the bacterium *Mycoplasma capricolum* (Andachi *et al.*, 1989) has identified the modified nucleotides for nearly all cellular tRNAs of both organisms. Most of the modified nucleotides present in *H. volcanii* tRNAs (both modification type and location) are also found in the few characterized tRNAs of other archaeal organisms, which include *Halobacterium cutirubrum* (12 sequences), *Halococcus morrhuae* (one sequence), *Methanobacterium thermoautotrophicum* (two sequences), *Thermoplasma acidophilum* (two sequences), and *Sulfolobus acidocaldarius* (one sequence). However, as all but five of these tRNAs are derived from mesophilic halophiles, the tRNA modification pattern shown in Fig. 8.2A is not likely representative of the modification pattern of tRNAs belonging to other major families of archaeal organisms. Indeed, the initiator tRNA-Met of *T. acidophilum* and *S. acidocaldarius* contains 1-methyladenosine (m^1A) (see Fig. 8.1 legend for the convention for abbreviating the names of modified nucleosides) at position 58 whereas this modified nucleoside is not present in any halophilic tRNA sequenced thus far. Standard tRNA numbering system (see in Sprinzl and Vassilenko, 2005) is used here. Likewise, 2'-*O*-methylguanosine (Gm) at position 30 and 2'-*O*-methyluridine (Um) at position 54 of *S. acidocaldarius* initiator tRNA-Met is not found in any halophilic tRNA (Sprinzl and Vassilenko, 2005). More recently using techniques other than classical RNA sequencing methods, the existence of new types of modified nucleosides has been revealed. They are unique to archaeal tRNAs and primarily found in thermophilic and hyperthermophilic Archaea (Fig. 8.2B). The modified nucleosides of some archaeal ribosomal RNAs (primarily *H. volcanii*, *Halobacterium halobium*, *Haloarcula marismortui*, and *S. acidocaldarius*) have also been determined. In a few cases, their location within the primary sequence has been determined using HPLC-MS analysis of RNA fragment digests as well as primer-extension/reverse transcription analysis (Bruenger *et al.*, 1993; Ofengand and Bakin, 1997; Noon *et al.*, 1998; Massenet *et al.*, 1999; Kowalak *et al.*, 2000; Hansen *et al.*, 2002; Del Campo *et al.*, 2005). These rRNA modified nucleosides, together with those known from analyses of bacterial and eukaryal rRNAs, can be found in McCloskey and Rozenski, 2005. The occurrence of modified nucleotides in archaeal mRNA has not yet been carefully examined.

As shown in Fig. 8.1, archaeal RNAs possess an extensive set of modified nucleosides. Several of these modified nucleosides are common to all three domains of life, whereas the majority are either unique to Archaea or are found in common with Bacteria and/or Eukarya. Their phylogenetic distribution is compatible with the idea that a few modified nucleosides were present in RNA at the earliest stages of cellular evolution, possibly in the 'RNA World', while the majority of modifications emerged during evolution of the different cell types. The role of these numerous modified nucleosides, usually located in functionally important regions of modern RNA molecules, remains largely unknown. However, in a few cases, they have been shown to play a role in improving and fine tuning RNA function, particularly in response to the diverse, complex, and varying modern cellular contexts in which these RNAs reside (Grosjean, 2005). Improvement of RNA function resulting from new post-transcriptional modifications is dependent upon the progressive acquisition of new nucleoside modification enzymes necessary for the synthesis of these functionally advantageous nucleosides. From recent work with the model organisms *Escherichia coli* and *Saccharomyces cerevisiae*, it is now estimated that the genetic information required to encode all the necessary RNA modification enzymes comprise no less than 1% of the total genome (Hopper and Phizicky, 2003; Björk and Hagervall, 2005; de Crecy-Lagard *et al.*, 2007). This estimate is almost certainly valid for archaeal organisms as well.

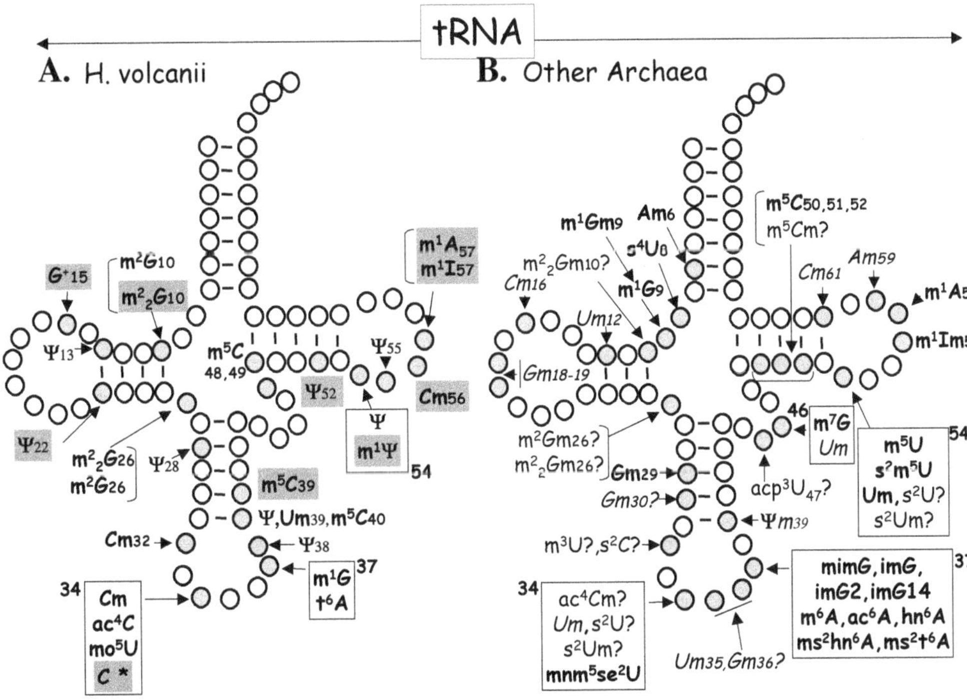

Figure 8.2 Nucleoside modification patterns in archaeal tRNA. (A) Compilation of modified nucleosides characterized in one or more of the 41 fully sequenced tRNAs of *Haloferax volcanii* (Gupta, 1984, 1986). (B) Location of modified nucleosides not present in *H. volcanii* tRNA but present and experimentally verified (bold) or assumed to be present in other archaeal tRNAs (italics or a question mark indicates not experimentally verified). This compilation is assembled from various sources and experimental approaches, which include tRNA sequence analysis, *in vitro* detection of enzymatic activity in a cell extract, HPLC/mass spectrometric analysis of bulk tRNA, and identification of sRNAs belonging to the box C/D RNA-guided RNP machinery which are presumed to guide the 2'-*O*-methylation of tRNA. Designations of modified nucleosides are the same as in Fig. 8.1. The number following a nucleoside indicates the nucleotide position in the tRNA. The conventional tRNA numbering system (Sprinzl and Vassilenko, 2005) is used to indicate tRNA positions. Modified nucleosides in boxes indicate their presence at the indicated position in different tRNAs. Modified nucleosides indicated in a gray background in (A) are uniquely present at that position in archaeal tRNAs and are not found at that position in the more than 450 sequenced bacterial and eukaryal tRNAs. The most characteristic of these are G⁺15, m¹Ψ54, Cm56 and m¹I57, which could be considered as signature modified nucleosides unique to archaeal tRNAs.

Modified nucleosides unique to Archaea

'Doubly' modified nucleosides (xNm)

Analysis of the modified nucleosides in ribosomal and tRNAs from all three domains of life has revealed that 15 nucleosides are unique to Archaea (Fig. 8.1, and for their chemical structures, see Limbach *et al.*, 1995; McCloskey and Rozenski, 2005). Of these 15, nine possess 2'-*O*-methylribose sugars in addition to the modified base. These nine are: 5,2'-*O*-dimethylcytidine (m^5Cm), N^4-acetyl-2'-*O*-methylcytidine (ac^4Cm), 2-thio-2'-*O*-methyluridine (s^2Um), N^2,N^2,2'-*O*-trimethylguanosine (m^2_2Gm or $m^{2,2}$Gm), N^2,7,2'-*O*-trimethylguanosine ($m^{2,7}$Gm), N^2,2'-*O*-dimethylguanosine (m^2Gm), 1,2'-*O*-dimethylinosine (m^1Im), 1,2'-*O*-dimethylguanosine (m^1Gm), and N^4,N^4,2'-*O*-trimethylcytidine (m^4_2Cm or $m^{4,4}$Cm). Seven of these nine (excepting m^4_2Cm in rRNA and probably $m^{2,7}$Gm in mRNA) are found in thermophilic archaeal tRNAs (McCloskey, 1986; Edmonds *et al.*, 1987; McCloskey *et al.*, 1988; Kowalak *et al.*, 1994; McCloskey *et al.*, 2000, 2001). NMR studies have shown that ac^4Cm is exceptionally rigid in conformation owing to the additive nature of the acetylation and methylation modifications which stabilize the 3' *-endo* sugar conformation (Kawai *et al.*, 1992). Both N^4-acetylcytidine (ac^4C) and 2'-*O*-methylcytidine (Cm) are present at position 34 of halobaterial tRNAs (Fig. 8.2A).

This suggests that ac^4Cm may also be present at the same position in thermophilic archaeal tRNAs, possibly stabilizing these tRNAs for a more accurate reading of the genetic code at high temperature. Similar studies of m^5Cm (proposed location at positions 48 – 52 as suggested by the presence of 5-methylcytidine (m^5C) at these sites in other tRNAs) and m^2_2Gm (possibly located at position 10 and/or 26 as N^2,N^2-dimethylguanosine (m^2_2G) is observed at these positions) have shown that these modifications also stabilize the 3′-*endo* sugar conformation (Kawai *et al.*, 1991). It is probable that m^5Cm stabilizes duplex conformation at the junction of the acceptor stem and the T-loop (so named even if ribothymidine is not present, see below) in forming the 3D-core of the tRNA's L-configuration. Modified nucleosides m^2_2Gm, m^1Gm and m^1Im, which cannot form Watson-Crick basepairs when juxtaposed in tRNA's 3D-core, may prevent formation of alternative structures in the G/C-rich archaeal tRNAs, indirectly stabilizing the two interacting loops of the 3D-core (discussed and reviewed in Helm, 2006; Urbonavicius *et al.*, 2006; Grosjean and Oshima, 2007). The location of s^2Um and m^4_2Cm in tRNA and rRNA is currently unknown. However, if s^2Um occurs at position 54 of tRNA, then it could also play a stabilization role as does 5-methyl-2-thiouridine (s^2m^5U at position 54) for *Thermus thermophilus* tRNA (Watanabe *et al.*, 1976; Shigi *et al.*, 2006). If s^2Um is located at wobble position 34 of the anticodon, a more stable codon-anticodon interaction at elevated temperature could be anticipated. Only the possible function of m^4_2Cm in rRNA remains elusive. Finally, nucleosides harbouring a 2′-*O*-methyl group and a modified base are found almost exclusively in archaeal tRNAs. However, the corresponding enzymes or enzymatic systems catalysing these modifications are not necessarily unique to Archaea. Indeed, the modified nucleosides ac^4C, m^5C, m^4C, s^2U, m^2G, m^2_2G, $m^{2,7}G$, m^1G and m^1I, which simply lack the 2′-*O*-methyl modification, are also found in Bacteria and/or Eukarya (Fig. 8.1).

The wyosine family of nucleosides

Several unique nucleosides possessing unmodified ribose sugars are also present in the tRNAs of thermophilic Archaea, but not in the mesophilic halophiles. These include the fluorescent and so called 'minimalist' tricyclic nucleoside 4-demethylwyosine (imG-14) and its monomethylated isowyosine (imG2) and dimethyl methylwyosine (mimG) forms (McCloskey *et al.*, 1987; Zhou *et al.*, 2004). All three are members of the hypermodified wyosine nucleoside family. Members of this family in Eukarya are present exclusively at position 37 of the phenylalanine tRNAs as wybutosine (yW), hydroxywybutosine (OHyW), and peroxywybutosine (o_2yW) (see in Sprinzl and Vassilenko, 2005). However, as we explain below, the uniqueness of imG-14, imG2 and mimG (but not wyosine (imG)) for archaeal tRNAs is only apparent since their corresponding enzymatic machinery has a common evolutionary origin.

Acetyladenosine

An analogous situation probably exists for N^6-acetyladenosine (ac^6A) found in the tRNAs of *Methanopyrus kandleri* (Sauerwald *et al.*, 2005). This newly discovered modified nucleoside is structurally related to N^6-hydroxynorvalinlcarbamoyladenosine (hn^6A), which is located in the corresponding position of tRNAs of the hyperthermophiles *Pyrobaculum islandicum* and *Pyrococcus furiosus*, but not found in the tRNAs of the mesophilic halophiles (Fig. 8.2b) (Reddy *et al.*, 1992). It is also structurally related to the extremely conserved N^6-threonylcarbamoyladenosine (t^6A) nucleoside found at anticodon loop position 37 in tRNAs of Bacteria, Eukarya and Archaea (see in Sprinzl and Vassilenko, 2005). Again, the enzyme catalysing ac^6A formation is unique to thermophilic Archaea and probably belongs to the super family of very conserved enzymes responsible for synthesis of the hypermodified nucleosides hn^6A, t^6A and m^6t^6A (N^6-methyl-N^6-threonylcarbamoyladenosine). These modified nucleosides are found at tRNA position 37 in many organisms from all three domains of life (El Yacoubi and de Crécy-Lagard, personal communication).

Lysidine-related modified cytosine

In Bacteria, the isoleucine codon AUA is read by a tRNA with the anticodon CAU. To accomplish this codon-anticodon interaction, tRNA wobble nucleotide C34 is posttranscriptionnally modified so that tRNA-Ile (CAU) will read only the isoleucine codon AU<u>A</u> and not methionine codon AU<u>G</u>. This modified nucleotide was shown to be lysidine (k^2C), an amino acid-containing nucleoside generated by ATP-dependent lysidine synthase (Soma *et al.*, 2003; reviewed in Grosjean and Bjork, 2004). In Archaea, notably *H. volcanii*, a similar situation appears to exist. Genomic sequencing has indicated that the anticodon for the corresponding tRNA is CAT, suggesting the presence of a modified nucleoside at the wobble position in this organism also (discussed in Marck and Grosjean, 2002). Although a modified nucleoside is found in the tRNA-Ile anticodon at this position, its precise structure has not yet been determined (Gupta, 1984). It is also anticipated that the enzyme catalysing the formation of this modified C in the anticodon wobble position of archaeal tRNA-Ile (as indicated by C* in Figs 8.1 and 8.2A) will be closely related to the one catalysing k^2C in Bacteria.

Archaeosine

Modified nucleoside archaeosine (G^+), first described as X (Gupta and Woese, 1980; Gupta, 1984, 1986), is a 7-deazaguanosine derivative (Gregson *et al.*, 1993). It is found only in archaeal tRNAs (Fig. 8.1) and is located at position 15 (Fig. 8.2A) of the D-loop (so named even if dihydrouridine is not present in the D-loop). In the tRNAs of Bacteria and Eukarya, nucleotide 15 is never modified. Interestingly, G^+ is structurally related to the queuosine (Q) family of deazaguanosine nucleosides, which include 7-aminomethyl-7deazaguanosine ($preQ_1$), epoxyqueuosine (oQ), glutamylqueuosine (gluQ), manosylqueuosine (manQ) and galactosylqueuosine (galQ). These are usually found at wobble position 34 of bacterial and eukaryal tRNAs (see below).

Methylated pseudouridine

One modified nucleoside that is generally considered structurally unique to Archaea is 1-methylpseudouridine ($m^1\Psi$). Here, the C5 atom of the uracil ring is linked to the C1′ of the ribose via a C-C glycosidic bond, and the N1 atom of the uracil forming the original C1′-N1 glycosidic bond is now methylated. This 'doubly' modified nucleoside $m^1\Psi$ is located at tRNA position 54 and found primarily in the tRNAs of Euryarchaeota (Fig. 8.2A). First described in *H. volcanii* as xU (Gupta and Woese, 1980), it was later identified as $m^1\Psi$ (Pang *et al.*, 1982). Nucleotide U54 of tRNAs of all Bacteria (except some Mycoplasmas) and Eukarya (except the initiator tRNA-Met) is nearly always a modified nucleoside, 5-methyluridine (m^5U or ribothymidine (rT)). The absence of m^5U in archaeal tRNAs was initially thought to be a specific feature of this domain (Fox *et al.*, 1980; Gupta and Woese, 1980). However, other studies have revealed the presence of m^5U and its thiolated derivative s^2m^5U in a *Thermococcus* species (Edmonds *et al.*, 1991) and in closely related *Pyrococcus furiosus* (Kowalak *et al.*, 1994). As stated before, s^2m^5U is also present in the tRNAs of the bacterium *Thermus thermophilus*. An S-adenosylmethionine-dependent enzymatic activity for m^5U formation at tRNA position 54 has been reported in a *P. furiosus* extract (Constantinesco *et al.*, 1999b). The corresponding gene for this activity has been recently identified and cloned, and the recombinant protein is shown to catalyse site-specific formation of m^5U in tRNA (Urbonavicius *et al.*, 2007). Interestingly, for the only *S. acidocaldarius* tRNA sequence available, the U at position 54 is Um, while for the tRNAs of *T. acidophilum* position 54 is a Ψ (Kuchino *et al.*, 1982). Thus, while U54 appears to be modified in almost all archaeal tRNAs sequenced to date, the type of modification and consequently the enzymes catalysing these modifications appear to be distinct from organism to organism.

It is noteworthy that the modified nucleoside 3-[3-amino-3-carboxy-propyl]-1-methylpseudouridine ($acp^3m^1\Psi$) has been found in 17S rRNA of the yeast *Saccharomyces carlsbergensis* (position 1191, equivalent *E. coli* 16S rRNA position 966) and 18S rRNA of both HeLa cells and *Drosophila melanogaster* (Brand *et al.*, 1978; Youvan and Hearst, 1981). It is probable that this 'triply' hypermodified nucleoside is present in the small ribosomal RNA of all Eukarya. In yeast, $acp^3m^1\Psi$ biosynthesis begins with the methylation of Ψ at a very early stage of rRNA maturation as evidenced by its presence in yeast 37S pre-rRNA (also Hela cell 45S pre-rRNA). Subsequently, the 3-amino-carboxy-propyl group is attached prior to final cleavage of the

18S precursor into mature 17S. The first methylation step is a nuclear modification while the final step of acp^{3}m$^1\Psi$ formation occurs in the cytoplasm (Brand *et al.*, 1978). Thus, the enzyme necessary for methylation of the cyclic N1 of Ψ is present in Eukarya and is not unique to Archaea as initially thought. However, it remains to be shown whether the eukaryal and archaeal enzymes, both of whose structural genes remain to be identified, are the result of divergent evolution arising from a common ancestral gene or are a consequence of convergent evolution.

Unique modification sites in archaeal tRNAs

Certain modified nucleosides, although chemically not unique to Archaea, are nevertheless found at unique positions in archaeal tRNAs. This is true for the modified nucleotides Ψ22, Ψ52, m^{5}C39, Um39, Um54, m$^2_2$G10, m^{1}I57, Gm30 and Cm56 (Fig. 8.2). Cm56 and m^{1}I57 are unique in this group and are discussed specifically later. Certain modified nucleosides are also noteworthy for their absence in archaeal RNA. These include 2-methyladenosine (m^{2}A), N^6-isopentenyladenosine (i^{6}A) and its derivatives ms^{2}i^{6}A, io^{6}A and ms^{2}io^{6}A which are typically found at anticodon loop nucleotide 37 of bacterial and eukaryal tRNAs. Also absent is the Q nucleoside and its derivatives gluQ, manQ and galQ, as well as inosine (I) and the many modified uridine derivatives typically found at bacterial and eukaryal tRNA wobble position 34 (see Figs 8.1 and 8.2 and Sprinzl and Vassilenko, 2005). Also remarkable and at variance with mesophilic or psychrophilic organisms is the total absence of the thermolabile, 7-methylguanosine (m^{7}G) nucleoside in the tRNAs of thermophiles and hyperthermophiles. Similarly, dihydrouridine (D), which cannot form Watson–Crick basepairs within a double helix due to its non-aromatic character and therefore provides flexibility to the RNA backbone (Dalluge *et al.*, 1996), is totally absent in the tRNAs of hyperthermophiles as well as mesophilic halophiles. This is in direct contrast with its abundance in the tRNAs of mesophilic organisms of bacterial and eukaryal domains, being particularly prominent in the tRNAs of psychrophilic Bacteria (Dalluge *et al.*, 1997; discussed in Grosjean and Oshima, 2007).

RNA modification enzymes and their genes

Our knowledge of RNA modification enzymes is rapidly expanding with the identification and characterization of both genes and encoded modification enzymes. However, most of this information comes from work with classical model organisms such as *E. coli* and *S. cerevisiae* (reviewed in Björk and Hagervall, 2005; Johansson and Byström, 2005; de Crécy-Lagard, 2007). Unfortunately, little is known about RNA maturation in general and RNA modification in particular in the Archaea. This is largely due to the great difficulty in culturing many of these organisms and their low cell yields (especially the extremophiles) which make it difficult to obtain sufficient material for biochemical and enzymatic studies of purified modification enzymes. Only very recently have adequate genetic tools specifically adapted for archaeal organisms become available (reviewed in Allers and Mevarech, 2005; Atomi, 2005). Fortunately, many of these nucleotide modification enzymes are sufficiently conserved in Archaea, Bacteria, and/or Eukarya to allow identification by homology searching of available genome databases. This, coupled with the ability to express these archaeal genes in bacterial expression systems, has often resulted in production of sufficient amounts of purified recombinant proteins for *in vitro* biochemical and enzymatic characterization. Crystal structures of a few archaeal modification enzymes have now been obtained and their detailed structures have been compared with bacterial homologues (see for examples Wang *et al.*, 2000; Ishitani *et al.*, 2002; Ishikawa *et al.*, 2004; Roovers *et al.*, 2004; Hamma *et al.*, 2005; Li and Ye, 2006; Manival *et al.*, 2006; Rashid *et al.*, 2006). Both *in vitro* chemical and enzymatic synthesis of RNA transcripts has been instrumental in assessing the modification activities of these purified enzymes. The availability of completed genome databases from phylogenetically related and diverse organisms (more than 500 at the beginning of 2007) has also allowed comparison of homologous genes (orthologues/paralogues) and the reconstruction of evolutionary histories for some RNA modification enzymes (Anantharaman *et al.*, 2002a). Many questions remain concerning enzyme function. These may be difficult to answer because of their complex and subtle nature.

Also included in future studies will be investigations of cellular location and complex spatial organization with other cellular components as it relates to the RNA maturation pathways in all three domains of life.

Assessment of archaeal RNA modification activities

The first study of archaeal nucleotide modification enzymatic activity was carried out in 1978 (Best, 1978). Using an extract of *Methanococcus vannielii* containing added radiolabelled S-adenosylmethionine as methyl donor and undermethylated *E. coli* tRNA as substrate, the formation of Cm, m^1A, m^1G, m^2G and $m^2{}_2G$ was detected. Notably, no formation of m^7G or m^5U was observed. This pioneering work corroborated independent observations concerning the presence and absence of certain modified nucleosides occurring in native and fully modified *M. vanniellii* tRNA (Gupta and Woese, 1980). From these observations it was concluded that genes encoding certain methyltransferases (specifically for m^7G and m^5U) might not be present in Archaea. We now appreciate that this may be true for certain archaeal organisms but not necessarily true for all Archaea. Indeed, extracts of the hyperthermophilic euryarchaeaote *P. furiosus* catalyse the formation of all nucleotide modifications listed above for *M. vanniellii* as well as m^5U, Am, m^5C, m^1I and Ψ (Constantinesco *et al.*, 1999b). Differences in observed nucleotide modification activities in different cell extracts can result from the type of RNA substrate used. Clearly, synthetic T7 RNA transcripts lacking all modified nucleosides offer a clear advantage over undermethylated, heterologous tRNAs isolated from *E. coli* for detecting and assessing nucleoside modification activities *in vitro* (for details see Grosjean *et al.*, 2007). Most studies now utilize purified recombinant enzymes and synthetic RNA substrates to characterize enzyme activity. To date, about a dozen archaeal nucleotide modification enzymes have been characterized using a variety of assessment methods. They are briefly described below.

Enzymes for mono- and dimethylguanosine formation

The first archaeal RNA modification enzyme to be identified and then investigated *in vitro* was the *P. furiosus* tRNA:$m^2{}_2G26$ methyltransferase (abbreviated Trm1p, Constantinesco *et al.*, 1998). The archaeal gene was easily identified in a homology search using the *TRM1* gene of *S. cerevisiae*, which was also the first gene encoding a tRNA modification enzyme identified in yeast (Ellis *et al.*, 1986). Homologous genes are found in all archaeal and eukaryal genomes thus far sequenced, but not in any bacterial genome except *Aquifex aeolicus*, which probably acquired this gene through lateral transfer during evolution. All belong to the large superfamily of Rossmann-fold containing methyltransferases that use S-adenosylmethionine as methyl donor (Martin and MacMillan, 2002; Schubert *et al.*, 2003). Phylogenetic analysis of the *TRM1* gene is compatible with vertical inheritance in which all archaeal and eukaryal homologues are descended from a common ancestral gene (Anantharaman *et al.*, 2002a).

Both *P. furiosus* and *S. cerevisiae* enzymes are monomeric in solution and catalyse the site-specific formation of mono- and di-methylguanosine ($m^2G/m^2{}_2G$) at position 26 in tRNA (see Fig. 8.2, Edqvist *et al.*, 1994; Constantinesco *et al.*, 1999a). It was shown that the tRNA substrate's sequence and structure around nucleotide G26 positioned at the junction between the D-stem and the anticodon-stem determines the ratio of $m^2G/m^2{}_2G$ formation. Both the archaeal and eukaryal enzymes require the same tRNA structural parameters for methylation, attesting to the conservation of tRNA substrate structure and enzyme recognition during evolution.

Nucleotide m^2G is also present at position 10 of eukaryal tRNA's D-stem. In archaeal tRNAs, this same G10 nucleotide is often found as $m^2{}_2G$, a dimethylguanosine. Genes coding for the site-specific tRNA:m^2G10 or $m^2{}_2G10$ methyltransferases in *Pyrococcus abyssi* (Trm-G10) and *S. cerevisiae* (Trm11p) were found using bioinformatic approaches searching for the presence of S-adenosylmethionine- and RNA-binding motifs as well as phylogenetic sequence conservation. Identification of the putative gene was followed by assessing activity of the recombinant protein *in vitro* (Armengaud *et al.*, 2004; Purushothaman *et al.*, 2005; Urbonavicius *et al.*, 2006). Methylation of tRNA nucleotides G10 (or G26) helps to prevent tRNA misfolding and facilitates the establishment of correctly folded pre-tRNAs by stabilizing the 3-D structure of the interacting D- and T-loops of the core domain (discussed in Steinberg and Cedergren, 1995; Urbonavicius *et al.*, 2006). The presence of this modified nucleoside in tRNA may also stimulate the activity of other tRNA modification enzymes, which require a correctly folded tRNA molecule.

Both the archaeal and eukaryal tRNA methyltransferases display a characteristic Rossmann-fold motif that is responsible for the binding of S-adenosylmethionine. A major difference between the archaeal and eukaryal enzymes is that the archaeal enzyme Trm-G10 functions *in vitro* as a monomer and requires no cofactor other than S-adenosylmethionine, whereas the yeast enzyme Trm11p is strictly dependent upon interaction with another phylogenetically unrelated protein Trm112p. The fact that the N2-exocyclic amine of G10 in archaeal tRNA is usually dimethylated whereas the corresponding nucleotide in eukaryal tRNA is only monomethylated is probably a consequence of the eukaryal enzyme's heteromeric Trm11p/Trm112p structure and its distinct interaction with the tRNA substrate. In contrast to the tRNA:m^2_2G26 methyltransferases (Trm1p) discussed above and which appear to have evolved from a common ancestral gene, the tRNA:G10 methyltransferases (Trm11p and Trm-G10) have evolved independently in the different phyla although fulfilling similar functions. Moreover, the methyltransferases (yeast Trm11p and archaeal Trm-G10) acting upon G10 belong to the cluster of orthologous group of proteins COG1041 (Tatusov *et al.*, 2003). This is distinct from the Trm1p methyltransferases of both yeast and Archaea which act upon G26 (COG1867). This observation indicates that they have emerged through convergent evolution in which different ancestral proteins have evolved to catalyse the same modification reaction on the tRNA substrate but at different nucleotide locations.

Enzymes for 1-methyladenosine formation

Modified nucleoside m^1A is found in the T-loop of many tRNAs from organisms belonging to all three domains of life. For bacterial and eukaryal tRNAs, m^1A is often present at position 58 whereas in archaeal tRNAs it is present at position(s) 58 and/or 57 with the modified m^1A57 being an obligatory intermediate in the biosynthesis of m^1I57 (Fig. 8.2A and below). In yeast, as well as in humans, the site-specific formation of m^1A58 is catalysed by the tetrameric enzyme tRNA:m^1A58 methyltransferase (Anderson *et al.*, 2000). Two identical subunits (Trm6p, alias Gcd14p) possess the Rossmann-fold and the S-adenosylmethionine-binding domain, with the other two identical subunits (Trm61p, alias Gcd10p) required for tRNA substrate binding (Ozanick *et al.*, 2005). The two subunits of the eukaryal tRNA:m^1A58 methyltransferase are phylogenetically related and have certainly evolved from a common ancestral protein (Bujnicki, 2001). In the bacterium *T. thermophilus* and the archaeon *P. abyssi*, the homologous enzyme of the yeast Trm6 enzyme is a homotetramer (TrmI, Droogmans *et al.*, 2003; Roovers *et al.*, 2004). Notable for their variation in substrate specificity, the eukaryal Trm6p/Trm61p and bacterial TrmI enzymes are strictly site-specific for A58. In contrast, the archaeal *P. abyssi* enzyme is region-specific and catalyses the formation of m^1A at two adjacent positions (57 and 58) in the T-loop of certain tRNAs. Also noteworthy is the fact that many tRNAs from *H. volcanii* and other halophilic organisms contain an m^1I (the deaminated form of m^1A, see below) at position 57, but never an m^1A at position 58 (Fig. 8.2A). This clearly demonstrates that tRNA:m^1A57 methyltransferase in this halophilic euryarchaeote is site-specific for position 57 and does not modify adenosine at position 58 as seen for the thermophilic bacterial and hyperthermophilic archaeal enzymes.

As the majority of archaeal, bacterial and eukaryal genomes sequenced to date encode homologues of the methyltransferase Trm6p/TrmI (Cluster of Orthologous Group of proteins COG2519), it is assumed that all present-day tRNA:m^1A methyltransferases acting on the adenine ring N1 at position(s) 57 and/or 58 are vertically inherited. They all appear to emerge from an ancestral precursor gene presumably encoding a tetrameric, prokaryotic-like protein with relaxed specificity for modification of adenosine at positions 57 and 58 (Roovers *et al.*, 2004). Only in the Eukarya has the original prokaryotic-like gene been duplicated and sub-functionalized into two closely related but distinct subunits (Trm6p and Trm61p) with strict site-specific modification of A58, similar to that of the bacterial homologue.

Enzymes for 1-methylinosine formation

Nucleotide m^1I, another 'doubly' modified nucleoside, is exclusively found at T-loop position 57 of many tRNAs in Archaea (Fig. 8.2) (Yamaizumi *et al.*, 1982). In Eukarya, this same modified nucleoside is found at anticodon loop position 37 of only tRNA-Ala (Sprinzl and Vassilenko, 2005). In Archaea, m^1I57 formation proceeds in two steps. In the first reaction, the N1 atom of adenosine is methylated by the TrmI enzyme (see above). Then m^1A57 is deaminated to produce m^1I57 (Grosjean *et al.*, 1995a). This deamination reaction is catalysed by an unknown m^1A-dependent tRNA deaminase. In contrast, these methylation and deamination steps are reversed in the formation of eukaryal tRNA-Ala m^1I37. A37 is first deaminated to produce I37 and then methylation leads to the formation of m^1I37. The deamination reaction is catalysed by the site-specific tRNA:A37 deaminase Tad1p (Gerber *et al.*, 1998), while the methylation step is catalysed by the site-specific tRNA:m^1G37 methyltransferase Trm5p (Björk *et al.*, 2001, see below). Amino acid sequence comparison of eukaryal Trm5p and archaeal TrmI clearly demonstrates that these methyltransferases are phylogenetically unrelated. It is also very likely the eukaryal deaminase Tad1p and the yet to be identified archaeal tRNA:m^1A57 deaminase belong to completely different enzyme families.

Concerning m^1A/m^1I function in both yeast and *T. thermophilus*, this modified nucleoside influences tRNA stability and the corresponding tRNA modification enzyme cannot be deleted without severe growth defects (Droogmans *et al.*, 2003; Kadaba *et al.*, 2004; reviewed in Anderson and Droogmans, 2005). The reason is that 1-methylation of A58 is required for the biosynthesis of s^2m^5U at position 54 in the T-loop of *T. thermophilus* tRNA (Shigi *et al.*, 2006) and this modification is in turn essential for tRNA thermal stability and hence the viability of thermophilic cells at high temperatures.

Enzymes for 1-methylguanosine formation

Nucleoside m^1G is universally present in the anticodon loop at position 37 of many tRNAs of organisms from all three domains of life (Figs 8.1 and 8.2A). This modified nucleoside is critical for maintaining the correct reading frame during mRNA decoding and therefore normal cell growth (Björk *et al.*, 2001). Nucleoside m^1G is also present at tRNA position 9, the junction between the amino acid stem and the D-stem, in several tRNAs of Eukarya and possibly Archaea (Fig. 8.2B), but never in Bacteria (see in Sprinzl and Vassilenko, 2005). Its probable role is to limit the formation of aberrant, alternative pre-tRNA structures during maturation (Helm, 2006), a function distinctly different from the corresponding modification at position 37.

In both *Methanocaldococcus jannaschii* and *P. abyssi*, enzymatic formation of m^1G37 in tRNA is catalysed by the site-specific Trm-G37 enzymes (Christian *et al.*, 2004 and Armengaud, Urbonavicius & Grosjean, unpublished results). The archaeal Trm-G37 enzymes are Rossmann-fold methyltransferases that share significant similarities with the orthologous Trm5p enzymes of *S. cerevisiae* and human (Brulé *et al.*, 2004). They show neither similarity to the bacterial TrmD 'SPOUT' methyltransferases (Anantharaman *et al.*, 2002b; Elkins *et al.*, 2003; Tkaczuk *et al.*, 2007), which also catalyse the formation of m^1G37 (Björk *et al.*, 2001), nor similarity to the Trm10p enzyme which modifies G9 (Jackman *et al.*, 2003). Thus, the non-similarity between the archaeal Trm-G37, eukaryal Trm5p/Trm10p and bacterial TrmD enzymes indicates that they have evolved their common function (methylation of guanosine N1 in tRNAs) by multiple convergence rather than divergence from a common ancestral gene.

Enzymes for archaeosine formation

Archaeosine (G^+) is a modified nucleoside found at position 15 of the D-loop of only archaeal and never bacterial or eukaryal tRNAs. The function of G^+ is probably the stabilization of tRNA 3D-tertiary structure by providing its highly charged imidino side chain for electrostatic interaction with tRNA phosphate groups as well as stacking upon the purine at position 59 (Gregson *et al.*,

1993). Indeed, earlier work has demonstrated that the Watson-Crick pair G15:C48 stacks on purine-59 within the 3D-core, a generally conserved molecular interaction for stabilizing the universally folded L-configuration of all tRNAs (Levitt, 1969; reviewed in Grosjean and Oshima, 2007).

The enzyme archaeosine tRNA-guanine transglycosylase (TGT) catalyses the initial step in G^+ biosynthesis. Recombinant TGT enzymes of *H. volcanii*, *M. jannaschii* and *Pyrococcus horikoshii* have been expressed, isolated, and their enzymatic properties investigated *in vitro* (Watanabe *et al.*, 1997; Bai *et al.*, 2000; Watanabe *et al.*, 2001). The crystal structure of the *P. horikoshii* TGT enzyme bound to its substrate has been solved, revealing that unexpected and important conformational changes are induced in the tRNA substrate upon enzyme binding (Ishitani *et al.*, 2002, 2003). Indeed, to modify nucleotide G15 which is buried within the folded core of the tRNA molecule, the TGT enzyme must first disrupt tRNA folded structure, then position the exposed G15 nucleotide in the enzyme's active site. Catalysis occurs with the exchange (transglycosylation reaction) of the genetically encoded guanine base (G15) and free precursor 7-cyano-7-deazaguanine ($preQ_0$) present in the cytoplasm. Subsequent conversion of the $preQ_0$ nucleoside in tRNA to G^+ requires the net addition of ammonia. Details of this reaction step are not yet available.

From available crystal structures, it appears that the catalytic core of archaeal archaeosine tRNA:G15 transglycosylase (TGT) is very similar to the well characterized bacterial *Zymomonas mobilis* TGT enzyme (Romier *et al.*, 1996). The occurrence of TGT in Archaea and its structural similarity to bacterial TGT was not originally anticipated as Q34 has never been found in the anticodon loop of any archaeal tRNA. Interestingly, the bacterial TGT (a queuosine tRNA:G34 transglycosylase) catalyses the formation of Q at tRNA anticodon position 34 using 7-aminomethyl-7-deazaguanine ($preQ_1$) as the exchanged base. This is in contrast to the archaeal TGT enzyme, which utilizes $preQ_0$ as the exchanged base (Romier *et al.*, 1997; reviewed in Romier *et al.*, 1998). The 'raison d'etre' for the Q34 nucleoside in bacterial tRNA is for increased efficiency and accuracy of mRNA decoding during translation (reviewed in Björk and Hagervall, 2005; Namy *et al.*, 2005).

Both archaeal and bacterial TGTs belong to the same Cluster of Orthologous Group of proteins (COG0343). The primary structural difference between them is the presence of an additional putative RNA-binding domain PUA in the catalytic core of the archaeal enzyme (Aravind and Koonin, 1999; Sabina and Soll, 2006). This difference is likely due to the unique way each enzyme approaches and recognizes the target guanine in tRNA. The G15 of archaeal tRNA is buried within the tRNA core and has to be exposed after binding with the TGT (see above), whereas bacterial anticodon G34 is well exposed naturally. Nevertheless, close inspection of available TGT sequences, including the eukaryal enzymes which catalyse transglycosylation of tRNA G34 for Q formation and utilize yet a third type of 7-deazaguanine derivative for the reaction, reveals a clear phylogenetic link for all TGT enzymes analysed thus far. This suggests that TGT enzymes from all three domains of life exhibit a common folded structure and share a common catalytic mechanism. It is possible that during evolution the different TGT enzymes evolved from a common ancestral precursor. Eukaryal and bacterial enzymes became site-specific for guanine at anticodon position 34 whereas the archaeal enzyme evolved for exclusive modification of guanine at D-loop position 15 and an obviously different function. The archaeal tRNA appears to have acquired the appropriate RNA-binding domain through simple gene fusion during evolution (Sabina and Soll, 2006). Noteworthy is the observation that in solution, the archaeal TGT (position 15 modification) is a homodimer, the bacterial TGT (position 34 modification) is a monomer, and the eukaryal TGT (position 34 modification) is a heterodimer composed of two phylogenetically unrelated subunits (Slany and Muller, 1995; discussed in Garcia and Goodenough-Lashua, 1998).

Protein-only enzymes for methylation at the ribose 2′-hydroxyl position

Transfer RNA often contains 2′-*O*-methylated nucleosides, particularly those tRNAs of the 'hot-loving' thermophiles and hyperthermophiles. The abundance of this particular modification in RNA appears to be directly related to cell growth temperature (Agris *et al.*, 1973; Noon *et al.*, 2003). Methylation of the 2′-hydroxyl group of the ribose sugar limits spontaneous hydrolysis and enzymatic cleavage of the associated phosphodiester bonds. This modification also stabilizes 3′-*endo* ribose conformation, thereby enhancing base stacking interactions and making the RNA helix more rigid (Kawai *et al.*, 1992).

A gene encoding a tRNA ribose 2′-*O*-methyltransferase that methylates invariant nucleotide C56 has been identified in the *P. abyssi* genome (Renalier *et al.*, 2005). This enzyme (aTrm56) belongs to the SPOUT superfamily of S-adenosylmethionine-dependent methyltransferases (COG1303) (Anantharaman *et al.*, 2002b). This modification is one of the few modified nucleosides truly unique to archaeal tRNAs and an orthologous gene is indeed present in all archaeal genomes thus far sequenced, but has not been found in any of the sequenced bacterial or eukaryal genomes. The sole archaeal exception is *Pyrobaculum aerophilum*, a crenarchaeote that lacks the aTm56 gene. Interestingly, it has been recently demonstrated that 2′-*O*-methylation of C56 in this organism is instead mediated by a guide RNA-directed RNP (see box C/D sRNP-guided nucleotide methylation below).

Other Trm56-like enzymes unique to Archaea may exist for ribonucleotide 2′-*O*-methylation of tRNA, rRNA, or possibly other small RNAs. However, box C/D RNP-guided nucleotide methylation is probably the predominant mechanism in Archaea, especially in hyperthermophiles where a large population of guide sRNA genes containing guide sequences complementary to tRNA and rRNA have been identified (see in Gaspin *et al.*, 2000; Dennis *et al.*, 2001; Omer *et al.*, 2003). Some of the 'doubly modified' nucleosides (discussed above and indicated in Figs 8.1 and 8.2) are certainly a result of the two tRNA modification systems (protein-only and box C/D RNA-guided RNP enzymes). Whether these two RNA modification systems work simultaneously or sequentially and whether the first catalysed modification regulates formation of the second modification remains to be determined.

The 2′-*O*-methylation of archaeal tRNA nucleotides C34 and U39 of the intron-containing pre-tRNA-Trp are methylated by a box C/D sRNA-guided mechanism in most, but not all euryarchaeotes. Strikingly, the pre-tRNA-Trp intron possesses the box C/D guide RNA responsible for these two tRNA methylation events (Clouet d'Orval *et al.*, 2001) (see below for details). In several Archaea (primarily Crenarchaeota), introns are sometimes inserted within the mature tRNA at sites different than the typical insertion location between nucleotides 37 and 38 of the anticodon loop. Insertions at these atypical sites in the 'middle' of the pre-tRNA, disrupt folding of the pre-tRNA into its normal cloverleaf and L-shaped structures and necessarily induce alternative folded structures (Marck and Grosjean, 2003). These alternative conformations may well regulate the action or affect the efficiency of modification for other protein-only or RNA-guided modification enzymes.

In contrast, 2′-*O*-methylation of tRNA ribose sugars in the bacterium *E. coli* and the eukaryote *S. cerevisiae*, (but not necessarily rRNA nor other RNAs) appears to be always accomplished using protein-only enzymes. Indeed, the RNA-guided nucleotide modification machinery is not present in Bacteria, but is present in *S. cerevisiae* and does function in methylation of rRNAs and snRNAs (Decatur and Fournier, 2002), but never in tRNA (reviewed in Johansson and Byström, 2005). For example, in *S. cerevisiae* the 2′-*O*-methylation of nucleotides at anticodon loop positions 32 and 34 of several intron-less tRNAs is accomplished using the multisite-specific, Rossmann-fold enzyme Trm7p (Pintard *et al.*, 2002). This enzyme belongs to the same cluster of orthologous group of proteins (COG0293) as *E. coli* RrmJ (a site-specific rRNA:2′-*O*-methyltreansferase which modifies U2552 in 16S rRNA) and yeast fibrillarin (the methylase enzyme of the box C/D RNP complex that extensively modifies rRNA, see below and Feder *et al.*, 2003). Interestingly, the 2′-*O*-methylation of *E. coli* tRNA at anticodon loop position 32 is accomplished by a site-specific RNA:2′-*O*-ribose methyltransferase (TrmJ), which belongs to the minor class of SPOUT family methyltransferases (Purta *et al.*, 2006). However, 2′-*O*-methylation of the nucleotide at position 34 is carried out using the 'sister', site-specific YibK enzyme, also a SPOUT family methyltransferase (Björk G, personal communication). The amino acid sequence of these two bacterial, tRNA-specific 2′-*O*-methyltransferases appears to be completely unrelated to the amino acid sequence of *E. coli* RrmJ (methylates 16S rRNA). Also unrelated are the yeast enzymes Trm7p (methylates the anticodon loop of yeast tRNAs) and fibrillarin (methylates yeast rRNA, see below). This situation is reminiscent of that for the formation of m^1G37 in the tRNAs of Bacteria, Archaea, and Eukarya, as discussed above. These examples illustrate the large potential of individual organisms in the three domains of life to evolve completely different enzymes and modification pathways (creation of new enzymes using convergent evolution) for accomplishing a given nucleotide modification, even at corresponding positions in the same RNA molecule.

Enzymes for isomerization of uridine to pseudouridine

The pseudouridine (5-(β-D-ribofuranosyl) uracil, an isomer of uridine and abbreviated as Ψ) located at position 55 in tRNA is the most conserved tRNA modified nucleoside in all three domains of life. This modification is catalysed by tRNA:pseudouridine synthase TruB in *E. coli* (Nurse *et al.*, 1995) and Pus4p in Eukarya (Becker *et al.*, 1997a), both of which belong to the COG0130 cluster of proteins. Also included in this same protein cluster is Cbf5p, the essential RNA:pseudouridine synthase that is involved in H/ACA RNA-guided pseudouridylation and is responsible for almost all Ψ formation in archaeal and eukaryal rRNAs (see below).

In Archaea, it has been recently demonstrated that two different enzymes (Pus10p and Cbf5) are able to catalyse *in vitro* the site-specific formation of Ψ55 in tRNAs. Pus10p is a newly identified tRNA:pseudouridine synthase that belongs to protein cluster COG1258 and whose corresponding gene is found, without exception, in all sequenced archaeal genomes (Roovers *et al.*, 2006). Notably, both purified recombinant proteins Pus10p and Cbf5p catalyse *in vitro* the formation of Ψ55 but exhibit differences in substrate recognition. Pus10p, considered the probable tRNA:pseudouridine-55 synthase *in vivo*, is able to efficiently modify U55 of a minimalist tRNA stem and loop substrate, similar to both the *E. coli* TruB and *S. cerevisiae* Pus4p enzymes (Gu and Santi, 1991; Becker *et al.*, 1997b). Archaeal Cbf5p alone, a much less efficient enzyme than Pus10p, strictly requires the presence of the 3′ terminal CCA nucleotide triplet within the entire tRNA molecule, probably because it mimics the H/ACA motif of the guide sRNAs that is required for Cbf5p binding (see below). The addition of H/ACA RNP core proteins Gar1 and Nop10 to some extent, improves the efficacy of Ψ55 formation in the intact as well as truncated tRNA, attesting to the fact that the relevant Cbf5p enzyme in the H/ACA RNP is a multi-protein complex (Roovers *et al.*, 2006; Gurha *et al.*, 2007).

TruB, Pus4p, and modern Cbf5p, all of which belong to the same protein cluster COG0130, appear to be direct descendants of a primordial, Cbf5p-like ancestral enzyme. This is supported by the finding that in present day Archaea, Cbf5p may function as a tRNA:Ψ55 synthase, in addition to its primary role as the pseudouridine synthase of the H/ACA RNP. These present day enzymes probably exhibit the relaxed specificity for various cellular RNAs that their ancestral enzymes did early in evolution. After the divergence of Eukarya and Archaea, a gene duplication event in Eukarya may have allowed separation of Cbf5p functions. The primary function of modern Cbf5p as an H/ACA RNP enzyme, is the modification of rRNA within the nucleolus and the nuclear coiled bodies, whereas Pus4p modifies tRNA within the nuclear/cytoplasmic compartments and the mitochondria. Thus in present day Archaea, tRNA and rRNA modification is accomplished independently with two phylogenetically distinct gene products (Cbf5 and Pus10p) catalysing the isomerization of many uridines to pseudouridines in different cellular RNAs.

RNA-guided 2′-*O*-ribose methylation and pseudouridylation

As previously mentioned, two nucleotide modifications in archaeal organisms are accomplished using RNA-guided RNP complexes. Both ribose 2′-O-methylation and the conversion of uridine to pseudouridine are carried out using box C/D and box H/ACA guide RNAs, respectively (reviewed in Dennis *et al.*, 2001; Bachellerie *et al.*, 2002; Terns and Terns, 2002; Omer *et al.*, 2003; Dennis and Omer, 2005). These small non-coding RNAs or sRNAs are associated with specific core proteins to assemble RNP complexes. The box C/D and H/ACA RNAs possess regions of complementarity to target RNAs. Basepairing of the guide and target RNAs determines the specific nucleotide for modification and the associated core proteins carry out the nucleotide methylation and pseudouridylation reactions (Fig. 8.3). The number of protein enzymes needed to carry out these many nucleotide modifications is minimized as all ribose 2′-O-methylations and Ψ formations in targeted RNAs are carried out by common catalytic proteins with the specificity of the modification reactions residing in the myriad guide RNAs themselves. Guide RNAs for nucleotide modification were first identified in Eukarya but subsequently shown to be present in archaeal organisms (Gaspin *et al.*, 2000; Omer *et al.*, 2000). Their absence in Bacteria suggests that these RNA:protein enzymes are evolutionarily ancient complexes, arising before the branching of Archaea and Eukarya some two billion years ago (reviewed in Tran *et al.*, 2004).

Box C/D sRNAs, sRNPs, and ribose 2′-O-methylation

Box C/D sRNAs are present in both Crenarchaeota and Euryarchaeota where they number in the dozens of species (Dennis *et al.*, 2001). A general correlation of increasing sRNA numbers in extreme thermophiles has suggested the increasing importance of 2′-O-ribose methylation for organisms growing in elevated temperatures. The box C/D guide RNAs are characterized by nucleotide boxes C and D as well as C′ and D′ boxes folded into the characteristic box C/D and C′/D′ motifs, respectively (Fig. 8.3). The C′/D′ motif is well conserved in archaeal sRNAs whereas this motif is degenerate, difficult to identify, or even missing in many eukaryal box C/D snoRNAs (Kiss-Laszlo *et al.*, 1998). The archaeal small RNAs or sno-like RNAs (sRNAs) are typically smaller that the eukaryal snoRNAs and the archaeal guide sequences positioned between the box C/D and C′/D′ motifs are small and exhibit a highly conserved distance of 12 nucleotides (Gaspin *et al.*, 2000; Tran *et al.*, 2005). The respective nucleotide box sequences of each motif fold to establish kink-turn or 'K-turn' elements which serve as core protein binding sites (Klein *et al.*, 2001; Kuhn *et al.*, 2002; Goody *et al.*, 2004). Three core proteins bind each motif to establish box C/D and C′/D′ RNP complexes. Ribosomal protein L7, Nop56/58 (Nop5p), and the methylase fibrillarin establish RNP complexes, each capable of directing nucleotide methylation utilizing their respective guide sequences located immediately adjacent and upstream of boxes D and D′ (Fig. 8.3). Site-specific nucleotide methylation is determined using the 'N+5 rule', where the fifth nucleotide of the target RNA upstream of the D or D′ box and within the target RNA:guide RNA duplex is modified (Cavaille *et al.*, 1996; Kiss-Laszlo *et al.*, 1996).

Target RNAs modified by the box C/D sRNP complex include ribosomal RNAs and tRNAs. Additional 'orphan' box C/D sRNAs lacking obvious target RNAs possessing nucleotide sequences complementary to the sRNA's guide sequences have also been identified. Some of the archaeal sRNAs could also function as 'RNA chaperones' to assist in the correct folding of archaeal rRNA and tRNAs as postulated by Dennis and coworkers (Dennis *et al.*, 2001). Interestingly, the 2′-O-methylation of nucleotides C34 and U39 in the tRNA-Trp of most, but not all euryarcheotes, is accomplished using box C/D and C′/D′ motifs encoded within the intron of the pre-tRNA-Trp precursor itself (Clouet d'Orval *et al.*, 2001; Bortolin *et al.*, 2003; Singh *et al.*, 2004). Methylation occurs in the pre-tRNA before splicing of the intron and is accomplished 'in trans' using the intron of another pre-tRNA molecule (Singh *et al.*, 2004). Indeed, an *in vitro* assay, using recombinant proteins and *in vitro* transcribed RNA to assemble the box C/D RNP complex, has shown that either the intron within a pre-tRNA molecule or the excised and circularized intron can carry out these modification reactions. *In vivo*, it is assumed that the excised circular intron is likely responsible for these two tRNA methylation events (Singh

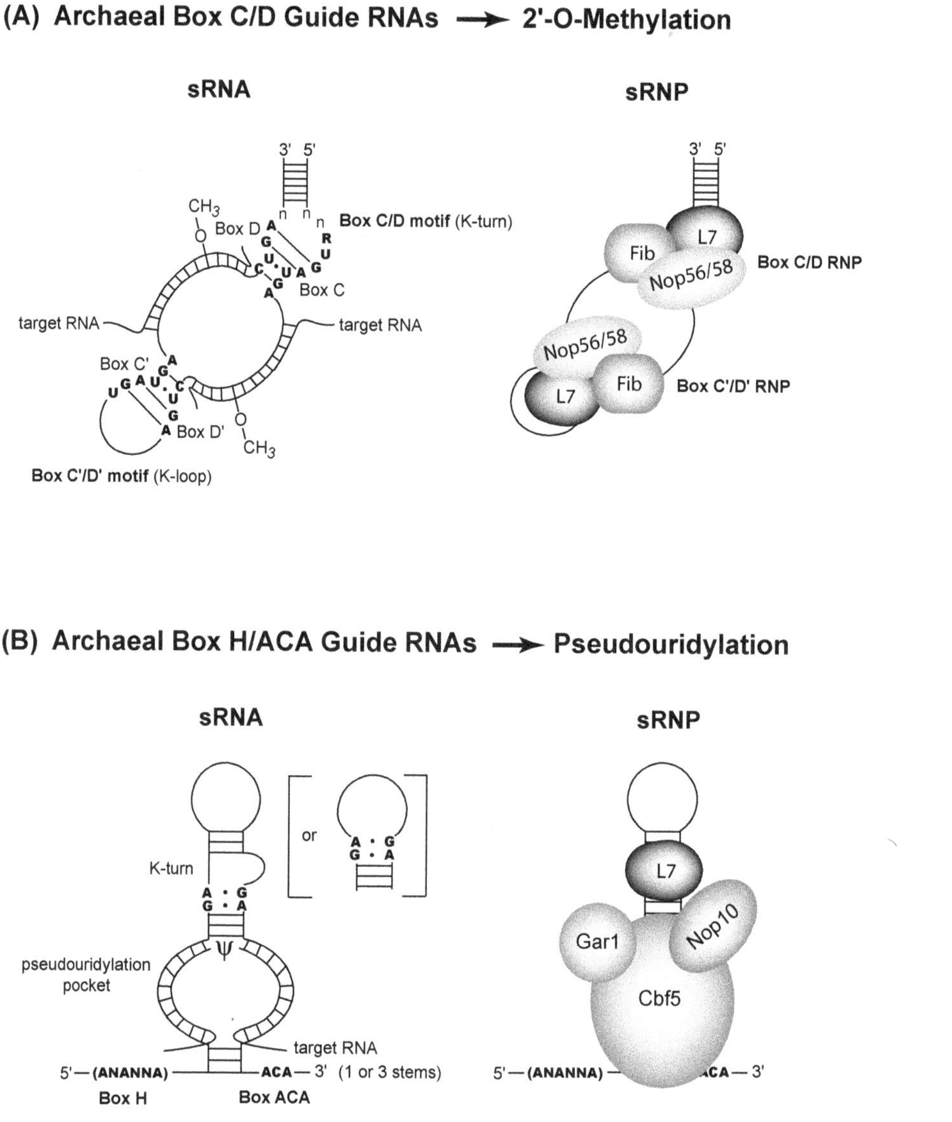

Figure 8.3 Archaeal box C/D and box H/ACA guide sRNAs and sRNP complexes. The characteristic folded secondary structure of archaeal box C/D and box H/ACA sRNAs is shown with target RNAs basepairing to their respective sRNA guide sequences. The site of nucleoside modification on each target RNA is indicated. Illustrated sRNP complexes show the associated sRNP core proteins of each sRNA as well as their known or approximated binding location on each sRNA.

et al., 2004). Furthermore, these reactions occur sequentially with C34 methylation occurring first and required for U39 methylation. In contrast to archaeal tRNA modification, the ribose sugars of eukaryal and bacterial tRNA nucleotides are methylated using protein-only methyltransferase enzymes as discussed above. Strikingly however, some archaeal tRNA 2′-*O*-ribose methylations, such as the conserved archaeal tRNA nucleotide C56 in the T-loop, are accomplished by either box C/D sRNPs or the protein-only enzyme aTrm56, depending on the organism considered (Renalier *et al.*, 2005 and reviewed in Clouet-d'Orval *et al.*, 2005).

In vitro assembly systems that reconstitute catalytically active complexes using *in vitro* transcribed box C/D sRNAs and recombinant core proteins have provided detailed analyses of archaeal box C/D sRNP assembly, structure, and nucleotide methylation function (Wang *et al.*, 2000; Omer *et al.*, 2002; Rashid *et al.*, 2003; Tran *et al.*, 2003). Core protein binding is sequential with L7 first binding the folded K-turns and then followed by the binding of Nop56/58 and fibrillarin. Fibrillarin is a Class I methyltransferase belonging to the superclass of Rossmann-fold enzymes (Feder *et al.*, 2003). S-Adenosylmethionine, the methyl donor, binds a typical nucleotide-binding loop structure within the Rossmann-fold. Efficient box C/D sRNP guided methylation activity is highly dependent upon juxtaposed box C/D and C′/D′ RNPs within a given sRNP complex and requires maintenance of the highly conserved 12 nucleotide spacer/guide regions separating the box C/D and C′/D′ RNPs (Gaspin *et al.*, 2000; Tran *et al.*, 2003; Tran *et al.*, 2005).

Emerging crystal structures of the core proteins are beginning to provide insight into the structural details of the box C/D RNP and elucidate the molecular details of sRNA:core protein interactions and methylation function (Aittaleb *et al.*, 2003; Deng *et al.*, 2004; Hamma and Amare, 2004; Moore *et al.*, 2004; Suryadi *et al.*, 2005). Core protein binding requires changes in sRNA structure or 'RNA remodelling' during sRNP assembly (Gagnon *et al.*, 2006; Omer *et al.*, 2007). Analysis of the Nop56/58-fibrillarin dimer has indicated that Nop56/58 core protein plays a role in stabilizing S-adenosylmethionine binding to fibrillarin (Aittaleb *et al.*, 2004). Unfortunately, a crystal structure of the completely assembled complex is not yet available to answer the numerous remaining questions concerning not only sRNP structure but also methylation function.

Box H/ACA sRNAs, sRNPs, and pseudouridylation

The box H/ACA sRNAs guide the conversion of U to Ψ using the same RNA-guided mechanism as the box C/D RNAs. The fewer number of archaeal H/ACA sRNAs corresponds to the fewer Ψ nucleotides found in archaeal ribosomal and tRNAs (Massenet *et al.*, 1999; Dennis *et al.*, 2001). Archaeal H/ACA sRNAs exhibit a folded hairpin structure possessing a bulged region or 'pseudouridylation pocket' which contains the sRNA's guide sequences (Balakin *et al.*, 1996; Ganot *et al.*, 1997; Bortolin *et al.*, 1999). Typical archaeal H/ACA sRNAs have a single hairpin although some species are composed of three hairpins (Tang *et al.*, 2002; Roshdestvensky *et al.*, 2003). This contrasts with eukaryal H/ACA snoRNAs where all exhibit a bipartite structure consisting of two folded hairpins linked by a hinge region containing the H box. The single hairpin archaeal H/ACA sRNAs often lack the hinge region with included H box. Four core proteins bind archaeal H/ACA sRNAs and include ribosomal protein L7, Nop10, Gar1, and the Ψ synthase, Cbf5p (Watanabe and Gray, 2000; Roshdestvensky *et al.*, 2003; Baker *et al.*, 2005; Charpentier *et al.*, 2005). Ribosomal protein L7 binds a K-turn in the extended H/ACA hairpin and is the only core protein shared by the box C/D and H/ACA sRNAs and only in Archaea. Nhp2p, the eukaryal L7 homologue, is a highly homologous protein but has evolved such that it no longer utilizes a K-turn motif for eukaryal H/ACA snoRNA binding.

Like the box C/D sRNPs, *in vitro* assembly systems for the H/ACA RNAs have been established to study H/ACA RNP assembly, structure, and function (Baker *et al.*, 2005; Charpentier *et al.*, 2005). Similar to the box C/D sRNPs, efficient sRNP assembly involves the temporal binding of the L7, Nop10, and Cbf5p core proteins with the sRNA. This is in contrast to assembly of the eukaryal H/ACA snoRNP complex where the eukaryal core protein homologues function as a trimer for H/ACA snoRNA binding (Wang and Meier, 2004). Emerging NMR and crystal structures are also providing insight into the structure and function of this RNA:protein enzyme (Hamma *et al.*, 2005; Khanna *et al.*, 2006; Manival *et al.*, 2006; Rashid *et al.*, 2006). Most notable is the recent crystal structure of the completely assembled *P. furiosus* complex resolved to 2.3 Å. This structure reveals not only the molecular details of the RNP complex but also shows how a modelled target RNA basepairs to the guide sequence to position the targeted uridine nucleotide within the pseudouridylation pocket for modification (Li and Ye, 2006).

Concluding remarks and future directions

Archaea possess a diverse and extensive array of modified nucleosides. Some of the nucleosides are unique to Archaea while others are shared with either Bacteria or Eukarya. A select few are shared by all three domains of life. The fact that only a few are common to Archaea, Bacteria, and Eukarya is consistent with the idea that these modified nucleosides were present in all cells early in evolution, perhaps even present in the so called 'RNA World'. These observations also suggest that the majority of modified nucleosides found in the modern world have emerged during evolution. Certainly the distinct populations of modified nucleosides found in the kingdoms Crenarcheaota and Euryarchaeota of the archaeal domain are consistent with this notion. The distinct set of nucleosides characteristic of different organisms, even within these same archaeal kingdom, suggests that emergence of new modified nucleosides is an ongoing evolutionary process. This emergence is likely to be in response to changing environmental conditions or cellular growth and survival requirements. These organism-specific needs would explain the apparently recent addition of some modified nucleosides that are unique to a single organism.

Our present knowledge and appreciation of the diversity of the modified nucleoside population is limited by the number of organisms that have been examined in any detail. This limitation includes not only diversity of organisms assessed but also the RNA species examined. Our limited knowledge is in large part a consequence of the complex technical approaches needed to identify and then determine the detailed structure of modified nucleosides as well as their positions within the RNAs. These techniques are certainly more complex and time consuming than simple determination of RNA primary sequence. Hopefully, future developments in experimental techniques including mass spectrometry will aid and speed the identification of new modified nucleosides and their fine structure, which will certainly add to our understanding of the number, distribution, and diversity of the modified nucleoside population. Expanded investigations of both organisms and RNA species promise to expand the number of known modified nucleosides needed and used by organisms in all three domains of life.

The function(s) of modified nucleosides was not explored in detail in this chapter but is no less important. They are typically located within functionally/structurally important regions of the RNA. Modified nucleosides are known to stabilize RNA structure, particularly the 3-D structure of folded RNA, and also help to prevent RNA degradation. They may participate in specific reactions catalysed by a given RNA or stabilize RNA structures critical for catalysis. Specific modifications may well be involved in protein binding, stabilizing or enhancing protein-RNA interactions. It appears that modified nucleosides play important roles in simply improving or fine tuning RNA function to help cells adapt or respond to changes in growth conditions or the general environment. Of course, these changes in growth conditions could well compel cells to evolve new modified nucleosides for new adaptive functions.

The machineries used to catalyse nucleoside modification are themselves diverse both in enzyme types and catalytic approaches. Protein-only and RNA-guided RNPs determine their site of nucleotide modification using the protein and guide RNA components of the enzyme, respectively. Both use proteins for catalysis but protein-only enzymes are diverse in number and approximately correspond to the number of nucleotides they modify. However, a few of them have been shown to modify multiple sites and exhibit even dual-specificity, thus reducing the number of enzymes needed to account for all the known modified nucleotides in RNA. In contrast, the RNA-guided core proteins are very limited in number and it is the large array of guide RNAs whose numbers approximate the number of target nucleotides to be modified. It is striking that different organisms can use these fundamentally different types of enzymes to modify the same nucleotide in a given RNA target.

As discussed earlier, a few cases exist where identical modifications on the same RNA species in a different organism are catalysed by unrelated protein-only enzymes, which are apparently not descended from a common ancestral gene. The existence of such 'isozymes' strongly suggests the unanticipated ability of cells to 'invent' new enzymes by convergent evolution. Likewise, the existence of protein-only and RNA-guided modification enzymes which are able to catalyse identical reactions at the same position of an

RNA molecule, but in different organisms, attests to the ability of cells to select different enzymatic 'solutions' during evolution that accomplish the same catalytic reaction and possibly fulfil the same function.

Protein-only and RNA-guided modification enzymes are fundamentally different in their mechanisms of determining substrate specificity. The co-existence of these two types of enzymes within the same cell points out fundamentally different mechanisms for evolving new enzymes that recognize new nucleotide substrates. Sequence mutation and drift of a guide RNA's guide sequences will naturally evolve a new enzyme that recognizes a new target RNA and hence modifies a new nucleotide. Those new guide RNAs whose modified nucleotides confer a growth or functional advantage to the cell would be maintained whereas those that did not would continue to drift in guide sequence or be lost. During this process, the catalytic proteins would evolve very little. In contrast, protein-only enzymes are more restricted in sequence drift as their substrate recognition lies in the catalytic enzyme itself. Therefore, they may evolve new target nucleotides for modification more slowly.

Another interesting question is the evolutionary relationship of protein-only to RNA-guided modification enzymes and which type emerged first in primitive cells? The absence of guide RNAs in Bacteria indicates the evolution of these RNP enzymes after Archaea and Bacteria emerged from the last universal common ancestor or 'LUCA'. This would suggest that protein-only enzymes preceded RNA-guided modification, or that the RNP enzymes were lost very early at the root of the emerging bacterial phyla. However, the fact that some of the catalytic proteins of the RNA-guided RNP complexes are homologous to protein-only modification enzymes, such as archaeal Pus10p and Cbf5, is consistent with the first scenario. That is, the existence in LUCA of a very limited number of ancestral, simple and probably highly promiscuous protein-only enzymes. Only after fusion with various types of RNA-binding domains, or other motifs needed to improve both enzyme efficacy and specificity, would the very diverse RNA-guided enzymes emerge, able to recognize complementary target substrates and catalyse the modification reaction. This scenario, however, does not exclude the possibility that present-day RNA-guided RNP complexes are remnants of the primordial 'RNA World' and the acquisition of related catalytic proteins from ancestral protein-only enzymes occurred at some stage of evolution via lateral transfer.

Future studies will continue to define the array of modification enzymes, as most are unknown at this time. Assays in cell extracts and testing of cloned gene products for modification activities will identify new enzymes while homology searches of new genomes will reveal new or related enzyme homologues (paralogues and orthologues). These investigations will perhaps suggest how the present day, complex, RNA modification machineries have emerged and progressively evolved. The exact location where these maturation processes occur in cells must also be addressed. As opposed to *in vitro* cell extracts where all cellular architecture has been disrupted, the specific modification of some RNAs may result from their compartmentalization with certain enzymes. Nevertheless, *in vitro* systems using recombinant enzymes will be particularly useful in determining enzyme structure and catalytic mechanisms. Such *in vitro* systems may also help to determine the order of nucleotide modifications in a given RNA and if a particular nucleotide modification is dependent upon prior modification of another. The robust expression of recombinant archaeal enzymes as soluble and properly folded proteins is likely to be important in producing sufficient quantities of protein for both biochemical and biophysical studies. Expression of recombinant enzymes in sufficient quantities will facilitate X-ray crystallographic analysis of enzyme fine structure and yield important details of their catalytic mechanisms and the molecular architecture of their interacting cofactors. Finally, further development of archaeal genetic systems should facilitate not only the examination of nucleotide modification events *in vivo*, but also the biological consequences of specific modifications for RNA function and metabolism as well as their potential roles in cell growth and survival. In this chapter, we have referenced a number of review articles that should help the reader explore many of the fascinating aspects of the archaeal world.

In conclusion, the coming years should be an exciting time in which new modified nucleosides are defined and their roles in cell growth and metabolism for organisms of all three domains of life are further revealed.

Acknowledgements

We thank Valerie de Crécy-Lagard (University of Florida) and Glenn Björk (University of Umea) for providing us with unpublished results, Jef Rozenski (Rega Institute, Belgium) for advice in preparing Fig. 8.1 and Xinxin Zhang for assistance in the preparation of figures. H.G. is Director of Research Emerite at the Universite de Paris-Sud at Orsay, France. This work was supported by NSF Grant MCB 0543741 to E.S.M. and NIH Grant GM55945 to R.G.

References

Agris, P.F. (1996). The importance of being modified: roles of modified nucleosides and Mg2+ in RNA structure and function. Prog. Nucleic Acid Res. Mol. Biol. *53*, 79–129.

Agris, P.F., Koh, H., and Soll, D. (1973). The effect of growth temperatures on the in vivo ribose methylation of *B. stearothermophilus* tRNA. Arch. Biochem. Biophys. *154*, 277–282.

Aittaleb, M., Rashid, R., Chen, Q., Palmer, J., Daniels, C., and Li, H. (2003). Structure and function of archael box C/D sRNP core proteins. Nat. Struct. Biol. *10*, 256-263.

Aittaleb, M., Visone, T., Fenley, M., and Li, H. (2004). Structural and thermodynamic evidence for a stabilizing role of Nop5p in S-adenosyl-L-methionine binding to fibrillarin. J. Biol. Chem. *279*, 41822–41829.

Allers, T., and Mevarech, M. (2005). Archaeal genetics: the third way. Nat. Rev. Genet. *6*, 58–73.

Anantharaman, V., Koonin, E.V., and Aravind, L. (2002a). Comparative genomics and evolution of proteins involved in RNA metabolism. Nucleic Acids Res. *30*, 1427–1464.

Anantharaman, V., Koonin, E.V., and Aravind, L. (2002b). SPOUT: a class of methyltransferases that includes spoU and trmD RNA methylase superfamilies, and novel superfamilies of predicted prokaryotic RNA methylases. J. Mol. Microbiol. Biotechnol. *4*, 71–75.

Andachi, Y., Yamao, F., Muto, A., and Osawa, S. (1989). Codon recognition patterns as deduced from sequences of the complete set of tRNA species in *Mycoplasma capricolum*. Resemblance to mitochondria. J. Mol. Biol. *209*, 37–54.

Andersen, N.M., and Douthwaite, S. (2006). YebU is a m5C methyltransferase specific for 16 S rRNA nucleotide 1407. J. Mol. Biol. *359*, 777–786.

Anderson, J., Phan, L., and Hinnebusch, A.G. (2000). The Gcd10p/Gcd14p complex is the essential two-subunit tRNA(1-methyladenosine) methyltransferase of *S. cerevisiae*. Proc. Natl. Acad. Sci. U.S. A. *97*, 5173–5178.

Anderson, J.T., and Droogmans, L. (2005). Biosynthesis and function of 1-methyladenosine in tRNA. In Fine-Tuning of RNA function by modification and editing, Topics in Current Genetics, vol 12, H. Grosjean, ed. (Berlin-Heidelberg, Springer-Verlag), pp. 122–139.

Aravind, L., and Koonin, E.V. (1999). Novel predicted RNA-binding domains associated with the translation machinery. J. Mol. Evol. *48*, 291- 302.

Armengaud, J., Urbonavicius, J., Fernandez, B., Chaussinand, G., Bujnicki, J.M., and Grosjean, H. (2004). *N2*-methylation of guanosine at position 10 in tRNA is catalyzed by a THUMP domain-containing, S-adenosylmethionine-dependent methyltransferase, conserved in Archaea and Eukaryota. J. Biol. Chem. *279*, 37142–37152.

Atomi, H. (2005). Recent progress towards the application of hyperthermophiles and their enzymes. Curr. Opin. Chem. Biol. 9, 166-173.

Bachellerie, J.P., Cavaille, J., and Huttenhofer, A. (2002). The expanding snoRNA world. Biochimie *84*, 775–790.

Bai, Y., Fox, D.T., Lacy, J.A., Van Lanen, S.G., and Iwata-Reuyl, D. (2000). Hypermodification of tRNA in thermophilic Archaea. Cloning, overexpression, and characterization of tRNA-guanine transglycosylase from *Methanococcus jannaschii*. J. Biol. Chem. *275*, 28731–28738.

Baker, D., Youssef, O., Chastkofsky, M., Dy, D., Terns, R., and Terns, M. (2005). RNA-guided RNA modification: functional organization of the archaeal H/ACA RNP. Genes Dev. *19*, 1238–1248.

Balakin, A., Smith, L., and Fournier, M. (1996). The RNA World of the nucleolus: two major families of small RNAs defined by different box elements with related functions. Cell *86*, 823–834.

Becker, H.F., Motorin, Y., Planta, R.J., and Grosjean, H. (1997a). The yeast gene YNL292w encodes a pseudouridine synthase (Pus4p) catalyzing the formation of psi55 in both mitochondrial and cytoplasmic tRNAs. Nucleic Acids Res. *25*, 4493–4499.

Becker, H.F., Motorin, Y., Sissler, M., Florentz, C., and Grosjean, H. (1997b). Major identity determinants for enzymatic formation of ribothymidine and pseudouridine in the T-psi-loop of yeast tRNAs. J. Mol. Biol. *274*, 505–518.

Best, A.N. (1978). Composition and Characterization of tRNA from *Methanococcus vannielii*. J. Bacteriol. *133*, 240–250.

Björk, G.R., and Hagervall, T.G. (2005). Transfer RNA modification. In *Escherichia coli* and *Salmonella*. Cellular and Molecular Biology, A. Böck, R. Curtis, J.B. Kaper, F.C. Neidhardt, T. Nyström, and C.L. Squires, eds. (Washington DC, ASM Press), pp. http://www.ecosal.org.

Björk, G.R., Jacobsson, K., Nilsson, K., Johansson, M.J., Byström, A.S., and Persson, O.P. (2001). A primordial tRNA modification required for the evolution of life. EMBO J. *20*, 231–239.

Bortolin, M., Bachellerie, J., and Clouet-d'Orval, B. (2003). In vitro assembly and methylation guide activity of an unusual box C/D RNA, cis-acting archaeal pre-tRNA(Trp). Nucleic Acids Res. *31*, 6524–6535.

Bortolin, M., Ganot, P., and Kiss, T. (1999). Elements essential for accumulation and function of small nucleolar RNAs directing site-specific pseudouridylation of ribosomal RNAs. EMBO J. *18*, 457–469.

Brand, R.C., Klootwijk, J., Planta, R.J., and Maden, B.E. (1978). Biosynthesis of a hypermodified nucleotide in *S. carlsbergensis* 17S and *HeLa*-cell 18S ribosomal ribonucleic acid. Biochem. J. *169*, 71–77.

Bruenger, E., Kowalak, J.A., Kuchino, Y., McCloskey, J.A., Mizushima, H., Stetter, K.O., and Crain, P.F. (1993). 5S rRNA modification in the hyperthermophilic Archaea *Sulfolobus solfataricus* and *Pyrodictium occultum*. FASEB J. *7*, 196-200.

Brulé, H., Elliott, M., Redlak, M., Zehner, Z.E., and Holmes, W.M. (2004). Isolation and characterization of the human tRNA-(*N1*-guanine 37) methyltransferase (TRM5p) and comparison to the *E. coli* TrmD protein. Biochemistry *43*, 9243–9255.

Buck, M., Connick, M., and Ames, B.N. (1983). Complete analysis of tRNA-modified nucleosides by high-performance liquid chromatography: the 29 modified nucleosides of *S. typhimurium* and *E. coli* tRNA. Anal. Biochem. *129*, 1–13.

Bujnicki, J.M. (2001). *In silico* analysis of the tRNA:m1A58 methyltransferase family: homology-based fold prediction and identification of new members from Eubacteria and Archaea. FEBS Lett. *507*, 123–127.

Bujnicki, J.M., Droogmans, L., Grosjean, H., Purushothaman, S.K., and Lapeyre, B. (2004). Bioinformatics-guided identification and experimental characterization of novel RNA methyltransferases. In Practical Bioinformatics, vol 15, J.M. Bujnicki, ed. (Berlin, Heidelberg, Springer-Verlag), pp. 139–168.

Cavaille, J., Nicoloso, M., and Bachellerie, J.P. (1996). Targeted ribose methylation of RNA *in vivo* directed by tailored antisense RNA guides. Nature *383*, 732–735.

Charpentier, B., Muller, S., and Branlant, C. (2005). Reconstitution of archaeal H/ACA small ribonucleoprotein complexes active in pseudouridylation. Nucleic Acids Res. *33*, 3133–3144.

Christian, T., Evilia, C., Williams, S., and Hou, Y.M. (2004). Distinct origins of tRNA(m1G37) methyltransferase. J. Mol. Biol. *339*, 707–719.

Clouet d'Orval, B., Bortolin, M.L., Gaspin, C., and Bachellerie, J.P. (2001). Box C/D RNA guides for the ribose methylation of archaeal tRNAs. The tRNA-Trp intron guides the formation of two ribose-methylated nucleosides in the mature tRNA-Trp. Nucleic Acids Res. *29*, 4518–4529.

Clouet-d'Orval, B., Gaspin, C., and Mougin, A. (2005). Two different mechanisms for tRNA ribose methylation in Archaea: a short survey. Biochimie *87*, 889–895.

Constantinesco, F., Benachenhou, N., Motorin, Y., and Grosjean, H. (1998). The tRNA(guanine-26,N2,N2) methyltransferase (Trm1p) from the hyperthermophilic archaeon *Pyrococcus furiosus*: cloning, sequencing of the gene and its expression in E. coli. Nucleic Acids Res. *26*, 3753–3761.

Constantinesco, F., Motorin, Y., and Grosjean, H. (1999a). Characterisation and enzymatic properties of tRNA(guanine 26, *N2*, *N2*)-dimethyltransferase (Trm1p) from *Pyrococcus furiosus*. J. Mol. Biol. *291*, 375–392.

Constantinesco, F., Motorin, Y., and Grosjean, H. (1999b). Transfer RNA modification enzymes from *Pyrococcus furiosus*: detection of the enzymatic activities in vitro. Nucleic Acids Res. *27*, 1308–1315.

de Crécy-Lagard, V. (2004). Finding missing tRNA modification genes: a comparative genomics goldmine. In Practical Bioinformatics, vol 15, J.M. Bujnicki, ed. (Berlin-Heidelberg, Springer-Verlag), pp. 169–190.

de Crécy-Lagard, V. (2007). Identification of genes encoding tRNA modification enzymes via comparative genomics. Methods Enzymol. *425*, 153–183.

de Crécy-Lagard, V., Marck, C., Bochier-Armanet, C., and Grosjean, H. (2007). Comparative rnomics and modomics in mollicutes: prediction of gene function and evolutionary implications. IUMBMB Life, in press.

Dalluge, J.J., Hamamoto, T., Horikoshi, K., Morita, R.Y., Stetter, K.O., and McCloskey, J.A. (1997). Posttranscriptional modification of tRNA in psychrophilic Bacteria. J. Bacteriol. *179*, 1918–1923.

Dalluge, J.J., Hashizume, T., Sopchik, A.E., McCloskey, J.A., and Davis, D.R. (1996). Conformational flexibility in RNA: the role of dihydrouridine. Nucleic Acids Res. *24*, 1073–1079.

Decatur, W.A., and Fournier, M.J. (2002). rRNA modifications and ribosome function. Trends Biochem. Sci. *27*, 344–351.

Del Campo, M., Recinos, C., Yanez, G., Pomerantz, S.C., Guymon, R., Crain, P.F., McCloskey, J.A., and Ofengand, J. (2005). Number, position, and significance of the pseudouridines in the large subunit ribosomal RNA of *Haloarcula marismortui* and *Deinococcus radiodurans*. RNA *11*, 210–219.

Deng, L., Starostina, N., Liu, Z., Rose, J., Terns, R., Terns, M., and Wang, B. (2004). Structure determination of fibrillarin from the hyperthermophilic archaeon *Pyrococcus furiosus*. Biochem. Biophys. Res. Commun. *315*, 726–732.

Dennis, P., and Omer, A. (2005). Small non-coding RNAs in Archaea. Curr. Opin. Microbiol. *8*, 685–694.

Dennis, P.P., Omer, A., and Lowe, T. (2001). A guided tour: small RNA function in Archaea. Mol. Microbiol. *40*, 509–519.

Douthwaite, S., Fourmy, D., and Yoshizawa, S. (2005). Nucleotide methylations in rRNA that confer resistance to ribosome-targeting antibiotics. In Fine-Tuning of RNA function by modification and editing, Topics in Current Genetics, vol 12, H. Grosjean, ed. (Berlin-Heidelberg, Springer-Verlag), pp. 285–307.

Douthwaite, S., and Kirpekar, F. (2007). Identifying modifications in RNA by MALDI mass spectrometry. Methods Enzymol. *425*, 3–20.

Droogmans, L., Roovers, M., Bujnicki, J.M., Tricot, C., Hartsch, T., Stalon, V., and Grosjean, H. (2003). Cloning and characterization of tRNA (m1A58) methyltransferase (TrmI) from *Thermus thermophilus* HB27, a protein required for cell growth at extreme temperatures. Nucleic Acids Res. *31*, 2148–2156.

Dunin-Horkawicz, S., Czerwoniec, A., Gajda, M.J., Feder, M., Grosjean, H., and Bujnicki, J.M. (2006). MODOMICS: a database of RNA modification pathways. Nucleic Acids Res. *34*, D145–149.

Edmonds, C.G., Crain, P.F., Gupta, R., Hashizume, T., Hocart, C.H., Kowalak, J.A., Pomerantz, S.C., Stetter, K.O., and McCloskey, J.A. (1991). Posttranscriptional modification of tRNA in thermophilic Archaea (Archaebacteria). J. Bacteriol. *173*, 3138–3148.

Edmonds, C.G., Crain, P.F., Hashizume, T., Gupta, R., Stetter, K.O., and McCloskey, J.A. (1987). Structural characterization of four ribose-methylated nucleosides from the tRNA of extremely thermophilic Archaebacteria. J. Chem. Soc. Chem. Comm. *1987*, 909–910.

Edqvist, J., Blomqvist, K., and Straby, K.B. (1994). Structural elements in yeast tRNAs required for homologous modification of guanosine-26 into dimethylguanosine-26 by the yeast Trm1p tRNA-modifying enzyme. Biochemistry 33, 9546–9551.

Elkins, P.A., Watts, J.M., Zalacain, M., van Thiel, A., Vitazka, P.R., Redlak, M., Andraos-Selim, C., Rastinejad, F., and Holmes, W.M. (2003). Insights into catalysis by a knotted TrmD tRNA methyltransferase. J. Mol. Biol. *333*, 931–949.

Ellis, S.R., Morales, M.J., Li, J.M., Hopper, A.K., and Martin, N.C. (1986). Isolation and characterization of the *TRM1* locus, a gene essential for the *N2,N2*-dimethylguanosine modification of both mitochondrial and cytoplasmic tRNA in *S. cerevisiae*. J. Biol. Chem. *261*, 9703–9709.

Feder, M., Pas, J., Wyrwicz, L.S., and Bujnicki, J.M. (2003). Molecular phylogenetics of the RrmJ/fibrillarin superfamily of ribose 2'-*O*-methyltransferases. Gene *302*, 129–138.

Fox, G.E., Stackebrandt, E., Hespell, R.B., Gibson, J., Maniloff, J., Dyer, T.A., Wolfe, R.S., Balch, W.E., Tanner, R.S., Magrum, L.J., *et al.* (1980). The phylogeny of prokaryotes. Science *209*, 457–463.

Gagnon, K., Zhang, X., Agris, P., and Maxwell, E.S. (2006). Assembly of the archaeal box C/D sRNP can occur via alternative pathways and requires temperature-facilitated sRNA remodeling. J. Mol. Biol. *362*, 1025–1042.

Ganot, P., Caizergues-Ferrer, M., and Kiss, T. (1997). The family of box H/ACA small nucleolar RNAs is defined by an evolutionarily conserved secondary structure and ubiquitous sequence elements essential for RNA accumulation. Genes Dev. *11*, 941–956.

Garcia, G.A., and Goodenough-Lashua, D.M. (1998). Mechanisms of RNA-modifying and -editing enzymes. In Modification and Editing of RNA, H. Grosjean, and R. Benne, eds. (Washington DC, ASM Press), pp. 135–168.

Gaspin, C., Cavaillé, J., Erauso, G., and Bachellerie, J.P. (2000). Archaeal homologs of eukaryotic methylation guide small nucleolar RNAs: lessons from the *Pyrococcus* genomes. J. Mol. Biol. *297*, 895–906.

Gerber, A., Grosjean, H., Melcher, T., and Keller, W. (1998). Tad1p, a yeast tRNA-specific adenosine deaminase, is related to the mammalian pre-mRNA editing enzymes ADAR1 and ADAR2. EMBO J. *17*, 4780–4789.

Goody, T., Melcher, S., Norman, D., and Lilley, D. (2004). The kink-turn motif in RNA is dimorphic, and metal ion dependent. RNA *10*, 254–264.

Gott, J.M., ed. (2004). RNA Interference, Editing and Modification: Methods and Protocols (Totowa, NJ, Humana Press).

Gott, J.M., ed. (2007). RNA Modification, Methods Enzymol., vol 425 (New York, NY, Academic Press).

Graham, D.E., and Kramer, G. (2007). Identification and characterization of archaeal and fungal tRNA methyltransferase. Methods Enzymol. *425*, 185-209.

Gregson, J.M., Crain, P.F., Edmonds, C.G., Gupta, R., Hashizume, T., Phillipson, D.W., and McCloskey, J.A. (1993). Structure of the archaeal transfer RNA nucleoside G*-15 (2-amino-4,7-dihydro- 4-oxo-7-beta-D-ribofuranosyl-1H-pyrrolo[2,3-d]pyrimidine-5-carboximidamide (archaeosine)). J. Biol. Chem. *268*, 10076-10086.

Grosjean, H. (2005). Fine-tuning of RNA function by modification and editing, Topics in Current Genetics, vol 12 (Berlin-Heidelberg, Springer-Verlag).

Grosjean, H., and Bjork, G.R. (2004). Enzymatic conversion of cytidine to lysidine in anticodon of bacterial isoleucyl-tRNA – an alternative way of RNA editing. Trends Biochem. Sci. *29*, 165–168.

Grosjean, H., Constantinesco, F., Foiret, D., and Benachenhou, N. (1995a). A novel enzymatic pathway leading to 1-methylinosine modification in *Haloferax volcanii* tRNA. Nucleic Acids Res. *23*, 4312–4319.

Grosjean, H., Droogmans, L., Roovers, M., and Keith, G. (2007). Detection of enzymatic activity of tRNA modification enzymes using radiolabelled tRNA substrates. Methods Enzymol. *425*, 57–101.

Grosjean, H., and Oshima, T. (2007). How nucleic acids cope with high temperature. In Physiology and biochemistry of extremophiles, C. Gerday, and N. Glansdorff, eds. (Washington, DC, ASM Press), pp. 39–56.

Grosjean, H., Sprinzl, M., and Steinberg, S. (1995b). Posttranscriptionally modified nucleosides in transfer RNA: their locations and frequencies. Biochimie *77*, 139–141.

Gu, J., Chen, Y., and Reddy, R. (1998). Small RNA database. Nucleic Acids Res. *26*, 160–162.

Gu, X.R., and Santi, D.V. (1991). The T-arm of tRNA is a substrate for tRNA (m5U54)-methyltransferase. Biochemistry *30*, 2999–3002.

Gupta, R. (1984). *Halobacterium volcanii* tRNAs. Identification of 41 tRNAs covering all amino acids, and the sequences of 33 class I tRNAs. J. Biol. Chem. *259*, 9461–9471.

Gupta, R. (1985). Transfer ribonucleic acids of archaebacteria. In The Archaebacteria; the Bacteria: A Treatise on Structure and Function, Vol. 8, C.R. Woese, and R.S. Wolfe, eds. (New York, Academic Press), pp. pp. 311–343.

Gupta, R. (1986). Transfer RNAs of *Halobacterium volcanii*: sequences of five leucine and three serine tRNAs. Syst. Appl. Microbiol. *7*, 102–105.

Gupta, R., and Woese, C.R. (1980). Unusual Modification Patterns in the Transfer RNA of Archaebacteria. Curr. Microbiol. *4*, 245–249.

Gupta, R.C., Randerath, E., and Randerath, K. (1976). A double-labeling procedure for sequence analysis of picomole amounts of nonradioactive RNA fragments. Nucleic Acids Res. *3*, 2895–2914.

Gupta, R.C., and Randerath, K. (1979). Rapid print-readout technique for sequencing of RNA's containing modified nucleotides. Nucleic Acids Res. *6*, 3443–3458.

Gurcha, P., Joardar, A., Chaurasia, P., and Gupta, R., (2007). Differential roles of archaeal box H/ACA proteins in guide RNA-dependent and independent pseudouridine formation. RNA Biol. *4*, 101–109.

Hamma, T., and Amare, A.F.-D. (2004). Structure of protein L7Ae bound to a K-turn derived from an archaeal box H/ACA sRNA at 1.8 A Resolution. Structure *12*, 893–903.

Hamma, T., Reichow, S.L., Varani, G., and Ferré-D'Amaré, R. (2005). The Cbf5-Nop10 complex is a molecular bracket that organizes box H/ACA RNPs. Nature Struct Mol Biol. *12*, 1101–1107.

Hansen, M.A., Kirpekar, F., Ritterbusch, W., and Vester, B. (2002). Posttranscriptional modifications in the A-loop of 23S rRNAs from selected Archaea and Eubacteria. RNA *8*, 202–213.

Helm, M. (2006). Post-transcriptional nucleotide modification and alternative folding of RNA. Nucleic Acids Res. *34*, 721–733.

Hopper, A.K., and Phizicky, E.M. (2003). tRNA transfers to the limelight. Genes Dev. *17*, 162–180.

Ishikawa, I., Sakai, N., Tamura, T., Yao, M., Watanabe, N., and Tanaka, I. (2004). Crystal structure of human p120 homologue protein PH1374 from *Pyrococcus horikoshii*. Proteins *54*, 814–816.

Ishitani, R., Nureki, O., Fukai, S., Kijimoto, T., Nameki, N., Watanabe, M., Kondo, H., Sekine, M., Okada, N., Nishimura, S., *et al.* (2002). Crystal structure of archaeosine tRNA-guanine transglycosylase. J. Mol. Biol. *318*, 665–677.

Ishitani, R., Nureki, O., Nameki, N., Okada, N., Nishimura, S., and Yokoyama, S. (2003). Alternative tertiary structure of tRNA for recognition by a posttranscriptional modification enzyme. Cell *113*, 383–394.

Jackman, J.E., Montange, R.K., Malik, H.S., and Phizicky, E.M. (2003). Identification of the yeast gene encoding the tRNA: m1G methyltransferase responsible for modification at position 9. RNA *9*, 574–585.

Johansson, M.J., and Byström, A.S. (2005). Transfer RNA modifications and modifying enzymes in *S. cerevisiae*. In Fine-Tuning of RNA function by modification and editing, Topics in Current Genetics, vol 12, H. Grosjean, ed. (Berlin-Heidelberg, Springer-Verlag), pp. 87–120.

Kadaba, S., Krueger, A., Trice, T., Krecic, A.M., Hinnebusch, A.G., and Anderson, J. (2004). Nuclear surveillance and degradation of hypomodified initiator tRNA-Met in *S. cerevisiae*. Genes Dev. *18*, 1227–1240.

Kawai, G., Ue, H., Yasuda, M., Sakamoto, K., Hashizume, T., McCloskey, J.A., Miyazawa, T., and Yokoyama, S. (1991). Relation between functions and conformational characteristics of modified nucleosides found in tRNAs. Nucleic Acids Res. Sym. Ser. *25*, 49–50.

Kawai, G., Yamamoto, Y., Kamimura, T., Masegi, T., Sekine, M., Hata, T., Iimori, T., Watanabe, T., Miyazawa, T., and Yokoyama, S. (1992). Conformational rigidity of specific pyrimidine residues in tRNA arises from posttranscriptional modifications that enhance steric interaction between the base and the 2'-hydroxyl group. Biochemistry *31*, 1040–1046.

Khanna, M., Wu, H., Johansson, C., Caizergues-Ferrer, M., and Feigon, J. (2006). Structural study of the H/ACA snoRNP components Nop10p and the 3′ hairpin of U65 snoRNA. RNA *12*, 42–52.

Kirpekar, F., Douthwaite, S., and Roepstorff, P. (2000). Mapping posttranscriptional modifications in 5S ribosomal RNA by MALDI mass spectrometry. RNA *6*, 296-306.

Kiss-Laszlo, Z., Henry, Y., Bachellerie, J.P., Caizergues-Ferrer, M., and Kiss, T. (1996). Site-specific ribose methylation of pre-ribosomal RNA: a novel function for small nucleolar RNAs. Cell *85*, 1077–1088.

Kiss-Laszlo, Z., Henry, Y., and Kiss, T. (1998). Sequence and structural elements of methylation guide snoRNAs essential for site-specific ribose methylation of pre-rRNA. EMBO J. *17*, 797–807.

Klein, D., Schmeing, T., Moore, P., and Steitz, T. (2001). The kink-turn: a new RNA secondary structure motif. EMBO J. *20*, 4214–4221.

Kowalak, J.A., Bruenger, E., Crain, P.F., and McCloskey, J.A. (2000). Identities and phylogenetic comparisons of posttranscriptional modifications in 16 S rRNA from *Haloferax volcanii*. J. Biol. Chem. *275*, 24484–24489.

Kowalak, J.A., Dalluge, J.J., McCloskey, J.A., and Stetter, K.O. (1994). The role of posttranscriptional modification in stabilization of tRNA from hyperthermophiles. Biochemistry *33*, 7869–7876.

Kowalak, J.A., Pomerantz, S.C., Crain, P.F., and McCloskey, J.A. (1993). A novel method for the determination of post-transcriptional modification in RNA by mass spectrometry. Nucleic Acids Res. *21*, 4577–4585.

Kuchino, Y., Ihara, M., Yabusaki, Y., and Nishimura, S. (1982). Initiator tRNAs from archaebacteria show common unique sequence characteristics. Nature *298*, 684–686.

Kuhn, J., Tran, E., and Maxwell, E.S. (2002). Archaeal ribosomal protein L7 is a functional homolog of the eukaryotic 15.5kD/Snu13p snoRNP core protein. Nucleic Acids Res. *30*, 931–941.

Lapeyre, B. (2005). Conserved ribosomal RNA modification and their putative roles in ribosome biogenesis and translation. In Fine-tuning of RNA Function by Modifications and Editing, Topics in Current Genetics, vol 12, H. Grosjean, ed. (Berlin-Heidelberg, Springer-Verlag), pp. 263–284.

Levitt, M. (1969). Detailed molecular model for transfer RNA. Nature *224*, 759–763.

Li, L., and Ye, K. (2006). Crystal structure of an H/ACA box ribonucleoprotein particle. Nature *443*, 302–307.

Limbach, P.A., Crain, P.F., and McCloskey, J.A. (1994). Summary: the modified nucleosides of RNA. Nucleic Acids Res. *22*, 2183–2196.

Limbach, P.A., Crain, P.F., Pomerantz, S.C., and McCloskey, J.A. (1995). Structures of posttranscriptionally modified nucleosides from RNA. Biochimie *77*, 135–138.

Manival, X., Charron, C., Fourmann, J., Godard, F., Charpentier, B., and Branlant, C. (2006). Crystal structure determination and site-directed mutagenesis of the *Pyrococcus abyssi* aCBF5-aNOP10 complex reveal crucial roles of the C-terminal domains of both proteins in H/ACA sRNP activity. Nucleic Acids Res. *34*, 826–839.

Marck, C., and Grosjean, H. (2002). tRNomics: analysis of tRNA genes from 50 genomes of Eukarya, Archaea, and Bacteria reveals anticodon-sparing strategies and domain-specific features. RNA *8*, 1189–1232.

Marck, C., and Grosjean, H. (2003). Identification of BHB splicing motifs in intron-containing tRNAs from 18 Archaea: evolutionary implications. RNA *9*, 1516-1531.

Martin, J.L., and MacMillan, F.M. (2002). SAM (dependent) I AM: the S-adenosylmethionine-dependent methyltransferase fold. Curr. Opin. Struct. Biol. *12*, 783–793.

Massenet, S., Ansmant, I., Motorin, Y., and Branlant, C. (1999). The first determination of pseudouridine residues in 23S ribosomal RNA from hyperthermophilic Archaea *Sulfolobus acidocaldarius*. FEBS Lett. *462*, 94–100.

McCloskey, J.A. (1986). Nucleotide modification in archaebacterial transfer RNA. Syst. Appl. Microbiol. *7*, 247–252.

McCloskey, J.A., Crain, P.F., Edmonds, C.G., Gupta, R., Hashizume, T., Phillipson, D.W., and Stetter, K.O. (1987). Structure determination of a new fluorescent tricyclic nucleoside from archaebacterial tRNA. Nucleic Acids Res. *15*, 683–693.

McCloskey, J.A., Edmonds, C.G., Gupta, R., Hashizume, T., Hocart, C.H., and Stetter, K.O. (1988). Characterization and synthesis of a new nucleoside from archaebacterial transfer RNA: 1,2'-O-dimethylinosine. Nucleic Acids Res. Sym. Ser. *20*, 45–46.

McCloskey, J.A., Graham, D.E., Zhou, S., Crain, P.F., Ibba, M., Konisky, J., Soll, D., and Olsen, G.J. (2001). Post-transcriptional modification in archaeal tRNAs: identities and phylogenetic relations of nucleotides from mesophilic and hyperthermophilic *Methanococcales*. Nucleic Acids Res. *29*, 4699–4706.

McCloskey, J.A., Liu, X.H., Crain, P.F., Bruenger, E., Guymon, R., Hashizume, T., and Stetter, K.O. (2000). Posttranscriptional modification of transfer RNA in the submarine hyperthermophile *Pyrolobus fumarii*. Nucleic Acids Res. Sym. Ser. *44*, 267–268.

McCloskey, J.A., and Rozenski, J. (2005). The Small Subunit rRNA Modification Database. Nucleic Acids Res. *33*, D135-D138.

Mengel-Jorgensen, J., and Kirpekar, F. (2002). Detection of pseudouridine and other modifications in tRNA by cyanoethylation and MALDI mass spectrometry. Nucleic Acids Res. *30*, e135.

Moore, T., Zhang, Y., Fenley, M., and Li, H. (2004). Molecular basis of box C/D RNA-protein interactions; cocrystal structure of archaeal L7Ae and a box C/D RNA. Structure *12*, 807–818.

Motorin, Y., Muller, S., Behm-Ansmant, I., and Branlant, C. (2007). Identification of modified residues in RNAs by reverse transcription-based methods. Methods Enzymol. *425*, 21–53.

Namy, O., Lecointe, F., Grosjean, H., and Rousset, J.-P. (2005). Translational recoding and RNA modifications. In Fine-Tuning of RNA function by modification and editing, Topics in Current Genetics, vol 12, H. Grosjean, ed. (Berlin-Heidelberd, Springer-Verlag), pp. 309–340.

Noon, K.R., Bruenger, E., and McCloskey, J.A. (1998). Posttranscriptional modifications in 16S and 23S rRNAs of the archaeal hyperthermophile *Sulfolobus solfataricus*. J. Bacteriol. *180*, 2883–2888.

Noon, K.R., Guymon, R., Crain, P.F., McCloskey, J.A., Thomm, M., Lim, J., and Cavicchioli, R. (2003). Influence of temperature on tRNA modification in Archaea: *Methanococcoides burtonii* (optimum growth temperature, 23 degrees C) and *Stetteria hydrogenophila* (Topt, 95 degrees C). J. Bacteriol. *185*, 5483–5490.

Nurse, K., Wrzesinski, J., Bakin, A., Lane, B.G., and Ofengand, J. (1995). Purification, cloning and properties of the tRNA Psi-55 synthase from *E. coli*. RNA *1*, 102–112.

Ofengand, J., and Bakin, A. (1997). Mapping to nucleotide resolution of pseudouridine residues in large subunit rRNAs from representative eukaryotes, prokaryotes, archaebacteria, mitochondria and chloroplasts. J. Mol. Biol. *266*, 246-268.

Ofengand, J., and Del Campo, M. (2005). Modified nucleotides of *E. coli* ribosomal RNA. In *Escherichia coli* and *Salmonella*: Cellular and Molecular Biology, A. Böck, R. Curtis, J.B. Kaper, F.C. Neidhardt, T. Nyström, and C.L. Squires, eds. (Washington DC, ASM Press), pp. http://www.ecosal.org.

Omer, A., Lowe, T., Russell, A., Ebhardt, H., Eddy, S., and Dennis, P. (2000). Homologs of small nucleolar RNAs in Archaea. Science *288*, 517–522.

Omer, A., Zago, M., Chang, A., and Dennis, P. (2007). Probing the structure and function of an archaeal C/D-box methylation guide sRNA. RNA *12*, 1708–1720.

Omer, A., Ziesche, S., Ebhardt, H., and Dennis, P. (2002). In vitro reconstitution and activity of a C/D box methylation guide ribonucleoprotein complex. Proc. Nat. Acad. Sci. USA *99*, 5289–5294.

Omer, A.D., Ziesche, S., Decatur, W.A., Fournier, M.J., and Dennis, P.P. (2003). RNA-modifying machines in Archaea. Mol. Microbiol. *48*, 617–629.

Ozanick, S., Krecic, A., Andersland, J., and Anderson, J.T. (2005). The bipartite structure of the tRNA: m1A58 methyltransferase from *S. cerevisiae* is conserved in humans. RNA *11*, 1281–1290.

Pang, H., Ihara, M., Kuchino, Y., Nishimura, S., Gupta, R., Woese, C.R., and McCloskey, J.A. (1982). Structure of a modified nucleoside in archaebacterial tRNA which replaces ribosylthymine, 1-methylpseudouridine. J. Biol. Chem. *257*, 3589–3592.

Patteson, K.G., Rodicio, L.P., and Limbach, P.A. (2001). Identification of the mass-silent post-transcriptionally modified nucleoside pseudouridine in RNA by matrix-assisted laser desorption/ionization mass spectrometry. Nucleic Acids Res. *29*, E49–49.

Piekna-Przybylska, D., Decatur, W., and Fournier, J. (2007). New bioinformatics tools for analysis of nucleotide modifications in eukaryotic rRNA. RNA *13*, 1–8.

Pintard, L., Lecointe, F., Bujnicki, J.M., Bonnerot, C., Grosjean, H., and Lapeyre, B. (2002). Trm7p catalyses the formation of two 2'-*O*-methylriboses in yeast tRNA anticodon loop. EMBO J. *21*, 1811–1820.

Purta, E., van Vliet, F., Tkaczuk, K.L., Dunin-Horkawicz, S., Mori, H., Droogmans, L., and Bujnicki, J.M. (2006). The yfhQ gene of *E. coli* encodes a tRNA:Cm32/Um32 methyltransferase. BMC Mol. Biol. *7*, 23–26.

Purushothaman, S.K., Bujnicki, J.M., Grosjean, H., and Lapeyre, B. (2005). Trm11p and Trm112p are both required for the formation of 2-methylguanosine at position 10 in yeast tRNA. Mol. Cell. Biol. *25*, 4359–4370.

Rashid, R., Aittaleb, M., Chen, Q., Spiegel, K., Demeler, B., and Li, H. (2003). Functional Requirement for Symmetric Assembly of Archaeal Box C/D Small Ribonucleoprotein Particles. J. Mol. Biol. *333*, 295–306.

Rashid, R., Liang, B., Baker, D., Youseff, O., He, Y., Phipps, K., Terns, R., Terns, M., and Li, H. (2006). Crystal structure of a Cbf5-Nop10-Gar1complex and implications in RNA-guided pseudouridylation and dyskeratosis congenita. Mol. Cell *21*, 249–260.

Reddy, D.M., Crain, P.F., Edmonds, C.G., Gupta, R., Hashizume, T., Stetter, K.O., Widdel, F., and McCloskey, J.A. (1992). Structure determination of two new amino acid-containing derivatives of adenosine from tRNA of thermophilic Bacteria and Archaea. Nucleic Acids Res. *20*, 5607–5615.

Reddy, R. (1989). Compilation of small nuclear RNA sequences. Methods Enzymol. *180*, 521–532.

Renalier, M.H., Joseph, N., Gaspin, C., Thebault, P., and Mougin, A. (2005). The Cm56 tRNA modification in Archaea is catalyzed either by a specific 2'-O-methylase, or a C/D sRNP. RNA *11*, 1051–1063.

Romier, C., Meyer, J.E., and Suck, D. (1997). Slight sequence variations of a common fold explain the substrate specificities of tRNA-guanine transglycosylases from the three kingdoms. FEBS Lett. *416*, 93–98.

Romier, C., R., F., and Suck, D. (1998). Structural basis of base exchange by tRNA-guanine transglycosylases. In Modification and Editing of RNA, H. Grosjean, and R. Benne, eds. (Washington DC, ASM Press), pp. 169–200.

Romier, C., Reuter, K., Suck, D., and Ficner, R. (1996). Crystal structure of tRNA-guanine transglycosylase: RNA modification by base exchange. EMBO J. *15*, 2850–2857.

Roovers, M., Hale, C., Tricot, C., Terns, M.P., Terns, R.M., Grosjean, H., and Droogmans, L. (2006). Formation of the conserved pseudouridine at position 55 in archaeal tRNA. Nucleic Acids Res. *34*, 4293–4301.

Roovers, M., Wouters, J., Bujnicki, J.M., Tricot, C., Stalon, V., Grosjean, H., and Droogmans, L. (2004). A primordial RNA modification enzyme: the case of tRNA:m1A methyltransferase. Nucleic Acids Res. *32*, 465–476.

Roshdestvensky, T., Tang, T., Tchirkova, I., Brosius, J., Bachellerie, J.P., and Huttenhofer, A. (2003). Binding of L7Ae protein to the K-turn of archaeal snoRNAs: a shared RNA binding motif for C/D and H/ACA box snoRNAs in Archaea. Nucleic Acids Res. *31*, 869–877.

Sabina, J., and Soll, D. (2006). The RNA-binding PUA domain of archaeal tRNA-guanine transglycosylase is not required for archaeosine formation. J. Biol. Chem. *281*, 6993–7001.

Saikia, M., Dai, Q., Decatur, W.A., Fournier, M.J., Piccirilli, J.A., and Pan, T. (2006). A systematic, ligation-based approach to study RNA modifications. RNA *12*, 2025–2033.

Sauerwald, A., Sitaramaiah, D., McCloskey, J.A., Soll, D., and Crain, P.F. (2005). *N6*-Acetyladenosine: a new modified nucleoside from *Methanopyrus kandleri* tRNA. FEBS Lett. *579*, 2807–2810.

Schubert, H.L., Blumenthal, R.M., and Cheng, X. (2003). Many paths to methyltransfer: a chronicle of convergence. Trends Biochem. Sci. *28*, 329–335.

Shigi, N., Suzuki, T., Terada, T., Shirouzu, M., Yokoyama, S., and Watanabe, K. (2006). Temperature-dependent biosynthesis of 2-thioribothymidine of *Thermus thermophilus* tRNA. J. Biol. Chem. *281*, 2104–2113.

Singh, S.K., Gurha, P., Tran, E.J., Maxwell, E.S., and Gupta, R. (2004). Sequential 2'-*O*-methylation of archaeal pre-tRNA-Trp nucleotides is guided by the intron-encoded but trans-acting box C/D ribonucleoprotein of pre-tRNA. J. Biol. Chem. *279*, 47661–47671.

Slany, R.K., and Muller, S.O. (1995). tRNA-guanine transglycosylase from bovine liver. Purification of the enzyme to homogeneity and biochemical characterization. Eur. J. Biochem. *230*, 221–228.

Soma, A., Ikeuchi, Y., Kanemasa, S., Kobayashi, K., Ogasawara, N., Ote, T., Kato, J., Watanabe, K., Sekine, Y., and Suzuki, T. (2003). An RNA-modifying enzyme that governs both the codon and amino acid specificities of isoleucine tRNA. Mol. Cell *12*, 689–698.

Sprinzl, M., and Vassilenko, K.S. (2005). Compilation of tRNA sequences and sequences of tRNA genes. Nucleic Acids Res. *33*, D139-D140.

Steinberg, S., and Cedergren, R. (1995). A correlation between *N2*-dimethylguanosine presence and alternate tRNA conformers. RNA *1*, 886–891.

Suryadi, J., Tran, E., Maxwell, E.S., and Brown, B. (2005). The crystal structure of the *Methanocaldococcus jannaschii* multi-functional L7Ae RNA-binding protein reveals an induced-fit interaction with the box C/D RNAs. Biochemistry *44*, 9657–9672.

Suzuki, T. (2005). Biosynthesis and function of tRNA wobble modifications. In Fine-Tuning of RNA function by modification and editing, Topics in Current

Genetics, vol 12, H. Grosjean, ed. (Berlin-Heidelberg, Springer-Verlag), pp. 23–69.

Tang, T., Bachellerie, J.P., Rozhdestvensky, T., Bortolin, M., Huber, H., Drungowski, M., Elge, T., Brosius, J., and Huttenhofer, A. (2002). Identification of 86 candidates for small non-messenger RNAs from the archaeon *Archaeoglobus fuldidus*. Proc. Natl. Acad. Sci. USA *99*, 7536–7541.

Tatusov, R.L., Fedorova, N.D., Jackson, J.D., Jacobs, A.R., Kiryutin, B., Koonin, E.V., Krylov, D.M., Mazumder, R., Mekhedov, S.L., Nikolskaya, A.N., *et al.* (2003). The COG database: an updated version includes eukaryotes. BMC Bioinformatics *4*, 41–54.

Terns, M., and Terns, R. (2002). Small nucleolar RNAs: versatile trans-acting molecules of ancient evolutionary origin. Gene Expression *10*, 17–39.

Tkaczuk, K.L., Dunin-Horkawicz, S., Purta, E., and Bujnicki, J.M. (2007). Structural and evolutionary bioinformatics of the SPOUT superfamily of methyltransferases. BMC Bioinformatics *8*, 73–104.

Tran, E., Brown, J., and Maxwell, E.S. (2004). Evolutionary Origins of the RNA-Guided Nucleotide-Modification Complexes: From the Primitive Translation Apparatus? Trends Biochem. Sci. *29*, 343–350.

Tran, E., Zhang, X., Lackey, L., and Maxwell, E.S. (2005). Conserved spacing between the box C/D and C'/D' RNPs of the archaeal box C/D sRNP complex is required for efficient 2'-O-methylation of target RNAs. RNA *11*, 285–293.

Tran, E., Zhang, X., and Maxwell, E.S. (2003). Efficient RNA 2'-O-methylation requires juxtaposed and symmetrically assembled archaeal box C/D and C'/D' RNPs. EMBO J. *22*, 3930–3940.

Urbonavicius, J., Armengaud, J., and Grosjean, H. (2006). Identity elements required for enzymatic formation of N2,N2-dimethylguanosine from N2-monomethylated derivative and its possible role in avoiding alternative conformations in archaeal tRNA. J. Mol. Biol. *357*, 387–399.

Urbonavicius, J., Auxilien, S., Walbott, H., Tachana, K., Golinelli-Pimpaneau, B., Brochier-Armanrt, C., and Grosjean, H. (2007). Acquisition of a bacterial RumA-type tRNA (uracil-54,C5)-methyltransferase by Archaea through an ancient horizontal gene transfer. Mol. Microbiol., in press.

Wang, C., and Meier, T. (2004). Architecture and assembly of mammalian H/ACA small nucleolar and telomerase ribonucleoproteins. EMBO J. *23*, 1857–1867.

Wang, H., Boisvert, D., Kim, K.K., Kim, R., and Kim, S.H. (2000). Crystal structure of a fibrillarin homologue from *Methanococcus jannaschii*, a hyperthermophile, at 1.6 A resolution. EMBO J. *19*, 317–323.

Watanabe, K., Shinma, M., Oshima, T., and Nishimura, S. (1976). Heat-induced stability of tRNA from an extreme thermophile *Thermus thermophilus*. Biochem. Biophys. Res. Commun. *72*, 1137–1144.

Watanabe, M., Matsuo, M., Tanaka, S., Akimoto, H., Asahi, S., Nishimura, S., Katze, J.R., Hashizume, T., Crain, P.F., McCloskey, J.A., *et al.* (1997). Biosynthesis of archaeosine, a novel derivative of 7-deazaguanosine specific to archaeal tRNA, proceeds via a pathway involving base replacement on the tRNA polynucleotide chain. J. Biol. Chem. *272*, 20146-20151.

Watanabe, M., Nameki, N., Matsuo-Takasaki, M., Nishimura, S., and Okada, N. (2001). tRNA recognition of tRNA-guanine transglycosylase from a hyperthermophilic archaeon *Pyrococcus horikoshii*. J. Biol. Chem. *276*, 2387–2394.

Watanabe, Y., and Gray, M. (2000). Evolutionary appearance of genes encoding proteins asociated with box H/ACA snoRNAs: Cbf5p in *Euglena gracilis*, an early diverging eukaryote, and candidate Gar1p and Nop10p homologs in archaebacteria. Nucleic Acids Res. *28*, 2342–2352.

Yamaizumi, Z., Ihara, M., Kuchino, Y., Gupta, R., Woese, C.R., and Nishimura, S. (1982). Archaebacterial tRNA contains 1-methylinosine at residue 57 in T psi C-loop. Nucleic Acids Res. Sym. Ser. *11*, 209–213.

Youvan, D.C., and Hearst, J.E. (1981). A sequence from *Drosophila melanogaster* 18S rRNA bearing the conserved hypermodified nucleoside am psi: analysis by reverse transcription and high-performance liquid chromatography. Nucleic Acids Res. *9*, 1723–1741.

Yu, Y.-T., Terns, R.M., and Terns, M.P. (2005). Mechanisms and functions of RNA-guided RNA modification. In Fine-Tuning of RNA function by modification and editing, Topics in Current Genetics, vol 12, H. Grosjean, ed. (Berlin-Heidelberg, Springer-verlag), pp. 223–262.

Zhou, S., Sitaramaiah, D., Noon, K.R., Guymon, R., Hashizume, T., and McCloskey, J.A. (2004). Structures of two new 'minimalist' modified nucleosides from archaeal tRNA. Bioorg. Chem. *32*, 82–91.

Signal Transduction in the Archaea

9

Peter J. Kennelly

Abstract

Adaptation to changing internal and environmental circumstances requires that living organisms develop mechanisms for monitoring relevant variables and triggering compensatory changes in the functional status of target proteins. The process by which sensory information is received and translated into cellular effects is called signal transduction. Mining of archaeal genomes for homologues of well-known bacterial and eucaryal signal transduction proteins has yielded surprisingly little. While some archaeons acquired copies of the sensor-response machinery that guide the chemotactic swimming of many Bacteria, the only recognizable signal transmission modules conserved throughout the Archaea are the eukaryotic protein kinases, the class II adenylate cyclases, the Pat-like protein acetylases, and several allosteric feedback loops. As the sensor response needs of the Archaea would appear to be, *a priori*, comparable to those of other microorganisms, it appears likely that members of the third domain of life have developed unique paradigms for transducing signals and/or novel ways for exploiting known mechanisms.

Introduction

Signal transduction is a term that, although widely used and universally accepted, is rather amorphously defined. A brief search on the Internet yielded the following descriptions:

1. the movement of signals from outside the cell to inside (http://web.indstate.edu/theme/mwking/signal-transduction.html);
2. any process by which a cell converts one kind of signal or stimulus into another (http://wikipedia.org/wiki/Signal_transduction);
3. signal transduction cascades mediate the sensing and processing of stimuli (http://www.ncbi.nlm.nih.gov/books/bv.fcgi?rid = stryer.chapter.2046);
4. the mechanisms that cells, tissues and organisms use to respond to physiological and environmental stimuli (http://dir.niehs.nih.gov/dirlst/);
5. the process linking the detection of certain kinds of external events to biochemical processes (http://www.cscs.umich.edu/~crshaliz/notebooks/signal-transduction.html);
6. the cascade of processes by which a signal interacts with a receptor, causing a change in the total level or activity of a second messenger or downstream target, and ultimately effecting a change in the functioning of the cell (http://db.yeastgenome.org/cgi-bin/GO/go.pl?goid = 7165).

These descriptions vary significantly in their definitions of both 'signal' and 'transduction'. Definitions 1 and 5 confine the range of eligible signals exclusively to external stimuli, while definition 6 further delimits them to only those species monitored using specific receptors. Definitions 1 and 2 confine the scope of the transduction process to the transmission of information from one location, e.g. external, to another, e.g. internal. On the other hand, definitions 4–6 include the machinery responsible for translating signals into cellular responses, while number 3 cryptically refers to the 'processing' of stimuli.

For the purposes of this chapter, signal transduction will be defined as follows:

> The mechanisms by which cells intervene in molecular processes in response to internal and external cues.

This definition reflects the highly integrated nature of biological sensor-response networks. Cells must monitor and weigh *all* biologically pertinent variables (Fig. 9.1). Therefore, the principal criterion for defining 'signals' should be functional. Classifications based upon location, e.g. external, or mode of action, e.g. receptor-mediated or second messenger-mediated, lend undue importance to secondary and somewhat artificial qualities. Moreover, not all stimuli fall neatly within these latter categories. Cells sense and respond to environmental factors such as temperature, osmolarity, pressure, and pH without resort to prototypic transmembrane receptors such as those which bind hormones. Moreover, the receptors for many 'external' signals, such as diatomic gases, light, and steroid hormones are located in the interior of the cell. By defining signal transduction in explicitly functional terms, it follows that this process encompasses not only the mechanisms by which information is sensed and transmitted, but also the means by which it ultimately is translated into action.

The verb 'intervention' was selected to differentiate the impact of a signal transduction event on the velocity of an enzyme-catalysed reaction from the normal variations that result from shifts in the concentration of the immediate substrates for the reaction being catalysed. The latter constitutes the inevitable, passive consequence of the kinetic behaviour of enzymes, while the former reflects an action – the alteration of one or more of the physical and/or functional properties of a target protein.

The simplest paradigm: sensor-target fusions

A common strategy for rendering a target protein responsive to some type of regulatory signal is to integrate the sensor domain directly into its structure in the form of either a stably associated regulatory subunit or polypeptide fusion (Fig. 9.2). Such sensor-target fusions are sometimes referred to as one-component systems (Ulrich *et al.*, 2005). Their structural simplicity renders them both physically robust and simple to engineer through gene duplication and/or gene fusion events. In the Archaea, sensor-target fusions and stable sensor-target protein complexes are employed to control and coordinate the supply of nucleotides, to balance the production of deoxyribonucleotides, to regulate the biosynthesis of hydrophobic amino acids, and to regulate methanogenesis in response to oxidative stress.

Nucleotide biosynthesis.

The Archaea employ allosterically regulated enzymes as their primary means for sensing and responding to the cellular demand for the nucleotide building blocks of RNA and DNA (Fig. 9.3). While the number of sensor-containing enzymes is relatively small, their strategic loca-

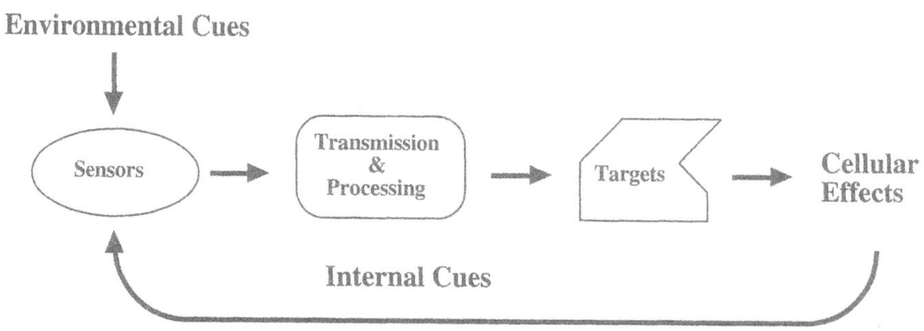

Figure 9.1 The core elements of biological sensor-response.

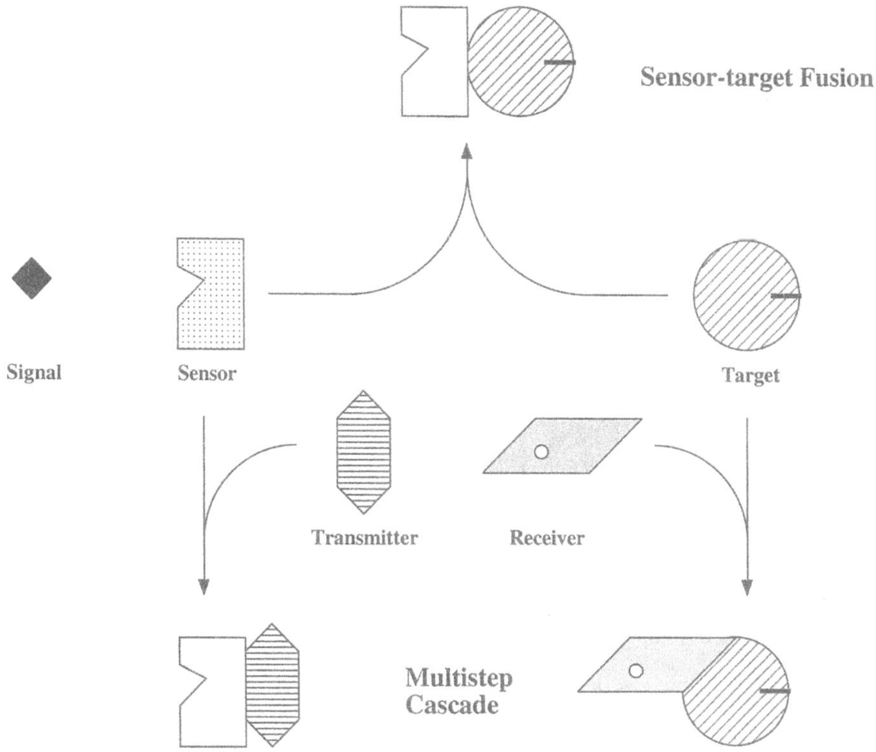

Figure 9.2. Basic route to the engineering of simple signal transduction mechanisms: sensor-target fusions (top) and multistep cascades (bottom).

tions and functional complementarity coalesce to form a sophisticated sensor-response network.

The supply of pyrimidine nucleotides is controlled and coordinated with that of pyrimidine nucleotides through the multivalent allosteric regulation of aspartate transcarbamoylase [ATCase], which catalyses the first committed step in *de novo* pyrimidine biosynthesis. The ATCase from *Pyrococcus abyssi*, for example, is a heterohexamer containing three catalytic subunits complexed with three regulatory subunits (Van Boxstael *et al.*, 2005). The regulatory domains render the catalytic activity of the ATCase sensitive to the levels of the pathway end product CTP (and, with much lower affinity, UTP), as well as the purine nucleotide ATP (Van Boxstael *et al.*, 2003; 2005). Micromolar levels of CTP inhibit ATCase activity, thus reducing the rate of pyrimidine synthesis when the supply of pathway end products threatens to exceed demand. Millimolar concentrations of ATP, on the other hand, stimulate activity to insure that the supply of pyrimidine nucleotides does not lag behind that of their fellow oligonucleotide precursors. The enzyme from *Halobacterium citirubrum* likewise is inhibited by CTP, the only nucleotide tested as a potential effector (Norberg *et al.*, 1973).

Not all archaeal ATCases display such easily rationalized sensor-response behaviour. While the catalytic activity of the ATCase from *Sulfolobus acidocaldarius* was stimulated by either ATP or GTP, as might be expected, the pathway end products UTP or CTP also served as activators at similar, submillimolar, concentrations (Durbecq *et al.* 1999). One potential rationale for this behaviour would be that cells containing high levels of nucleotide triphosphates are primed and ready to engage in an intensive programme of macromolecule biosynthesis potentially leading to cell division. Paradoxically, the ATCase from *Methanocaldococcus jannaschii* reportedly was insensitive to likely nucleotide effectors (Hack *et al.*, 2000). However, this behaviour may reflect some collateral effect of recombinant expression, as this enzyme, like those from *P. abyssi* and *S. solfataricus*, associates with a regulatory subunit to form a stable $\alpha_3\beta_3$ heterohexamer.

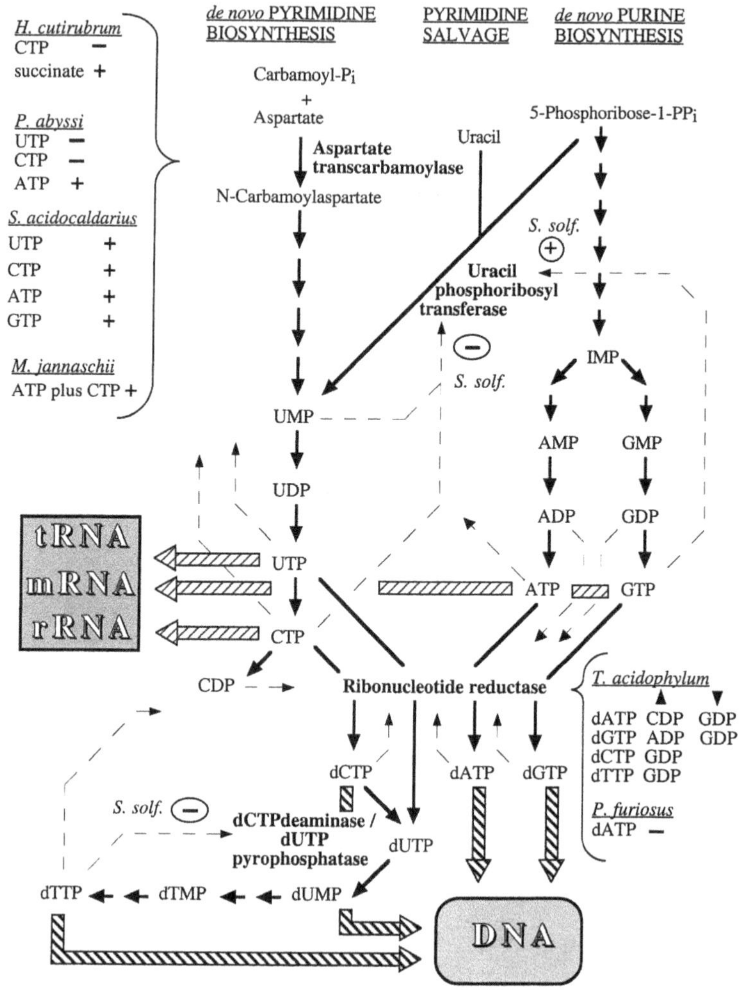

Figure 9.3. Coordination of nucleotide and deoxynucleotide biosynthesis in the Archaea. Shown is a schematic representation of the metabolic pathways responsible for the synthesis of purine (right) and pyrimidine (left) nucleotides. The names of allosterically regulated enzymes are written in bold type. Feedback relationships between products and enzymes are depicted with hatched arrows with the name of the organism in which this behaviour has been observed given in *italic type.* Plus and minus signs indicate whether the compound of interest activates or inhibits the target enzyme. In the case of multivalent regulatory loops targeting aspartate transcarbamoylase and ribonucleotide reductase, the effectors, responses, and organisms are listed to the side. In the case of the latter enzyme, the upward and downward facing arrowheads indicate whether shifts in substrate specificity favour or disfavour the compounds listed below.

In at least one archaeon, *Sulfolobus solfataricus,* management of the supply of pyrimidine nucleotides at the level of *de novo* biosynthesis is complemented by the regulation of the salvage pathway for recovering pyrimidine bases. The critical enzyme in this process, uracil phosphoribosyl transferase, condenses free uracil with phosphoribosyl pyrophosphate to regenerate the pyrimidine nucleotide UMP at the expense of the hydrolysis of two pyrophosphate bonds. It is possible that cytosine can be recycled in this manner as well via chemical deamination to uracil. The activity of the uracil phosphoribosyl transferase from *S. solfataricus* is stimulated by GTP and is inhibited in a synergistic fashion by the combination of UMP and CTP (Arent *et al.,* 2005; Jensen *et al.,* 2005). Consequently, when the supply of purine nucleotides in high, salvage of pyrimidine bases is promoted to help keep pace, while pyrimidine nucleotides themselves feedback inhibit their salvage pathway.

Deoxyribonucleotide biosynthesis.

The deoxyribonucleotide building blocks of DNA are produced via the reduction of nucleotide di(tri)phosphates by the enzyme ribonucleotide reductase. In the archaeon *Thermoplasma acidophylum* the partitioning of flux amongst the four nucleotide bases is achieved by tuning the enzyme's substrate specificity (Eliasson *et al.*, 1999). Binding of the pyrimidine deoxyribonucleotides dCTP or dTTP stimulate the overall catalytic activity of the enzyme somewhat and, more significantly, transform GDP from the least favoured substrate to the most avid. dATP or dGTP, on the other hand, further reduce the enzyme's already low basal activity toward GDP while enhancing activity toward CDP or ADP, respectively. In *Methanococcus jannaschii*, the proportions of the pyrimidine deoxyribonucleotides are further adjusted to meet demand through the allosteric regulation of the bifunctional enzyme dCTP deaminase/dUTP diphosphatase, which catalyses the key steps in the conversion of dCTP and dUTP to dTTP (Li *et al.*, 2003). dTTP inhibits both activities, thus preserving the supply of dCTP when adequate levels of dTTP have accumulated (Huffman *et al.*, 2003).

Presuming that the behaviour of these individual enzymes is representative of their homologues throughout the Archaea, it can readily be seen how, collectively, these sensor-response fusions enable cells to efficiently coordinate the supply nucleotides and deoxyribonucleotides (Fig. 9.3).

Biosynthesis of hydrophobic amino acids

In several species of *Methanococcus*, the biosynthesis of the hydrophobic aliphatic amino acids is autoregulated in response to available supplies through feedback inhibition of the enzyme acetolactate synthase (Xing and Whitman, 1987). Acetolactate synthase catalyses the first committed step in the biosynthesis of these amino acids, the condensation of pyruvate with either a second molecule of pyruvate or 2-oxoglutarate to form carbon dioxide and either acetolactate, the precursor of valine, or acetohydroxybutyrate, the precursor of leucine and isoleucine, respectively. While *M. aeolicus*, *M. voltae*, and *M. maripaludis* utilize a common basic mechanism, each employs a different strategy for assessing and adjusting the pool of available hydrophobic aliphatic amino acids (Xing and Whitman, 1987). The acetolactate synthase from *M. aoelicus*, for example, is inhibited exclusively by valine – which therefore functions as the master regulator for this pathway. *M. voltae* on the other hand, exhibits multivalent sensing capability, as the acetolactate synthase activity in this organism can be shut down when either valine or isoleucine – derivatives of the two potential enzyme products – accumulates. The most sophisticated sensor-response behaviour was exhibited by extracts of *M. maripaludis*, where valine and isoleucine could individually inhibit only a portion of the total acetolactate synthase activity. It is tempting to assume that the combination of the two amino acids, acting either additively on a single enzyme or individually on different isozyme forms, can block biosynthesis when a surfeit of both subgroups of hydrophobic aliphatic amino acids exists.

It is unclear how total flux through the pathway responsible for the *de novo* biosynthesis of the aromatic hydrophobic amino acids is controlled in the Archaea. While the enzyme catalysing the initial step in the canonical pathway, 3-deoxy-D-arabino-2-heptulosonate 7-phosphate synthase [DAHP synthase], is subject to feedback regulation in *Bacteria* and yeast, the only archaeal DAHP synthase characterized to date – that from *P. furiosus*, is reportedly insensitive to phenylalanine, tyrosine, or tryptophan (Schofield *et al.*, 2003). The lack of canonical inhibition may be attributable to its broader substrate specificity and consequent participation in additional metabolic pathways (Schofield at al., 2005). On the other hand, several methanoarchaea lack DAHP synthase (Porat et al, 2004) and instead utilize methylglyoxal and fructose-1,6-bisphospate as their starting material for the biosynthesis of phenylalanine, tyrosine, and tryptophan (Porat *et al.*, 2006).

The latter steps of the core pathway leading to the synthesis of the key precursor chorismate follow the same sequence of reactions as the canonical pathway of the *Bacteria*, as is also the case for the enzymes responsible for converting chorismate to phenylalanine, tyrosine, or tyryptophan. Feedback regulation via sensor-target fusions is prominently featured in the partitioning of flux amongst this trio of potential end products. For example, feedback inhibition of the anthranilate synthases, the enzyme that catalyses the first committed step in the synthesis of tryptophan, by the end product tryptophan has been observed in a broad spectrum of archaeons, including *A. fulgidus* (Byrnes and Vilker, 2004), *M. thermoautotrophicum* (Gast *et al.*, 1994), *S. solfataricus* (Tutino *et al.*, 1997), and *Thermococcus kodakaraensis* (Tang *et al.* 2001).

The other branches of the pathway begin with the conversion of chorismate to prephenate. Here the pathway bifurcates one last time, as prephenate can be converted to phenylpyruvate, the immediate precursor of phenylalanine, by prephenate dehydratase or to 4-hydroxyphenyl pyruvate, the immediate precursor of tyrosine, by prephenate dehydrogenase. In the euryarchaea, both of these enzymes are subject to allosteric inhibition by their respective pathway end products (Jensen *et al.*, 1988; Fischer *et al.*, 1993: Porat *et al.*, 2004; Kleeb *et al.*, 2006).

In addition to short-range, local feedback inhibition, sensor-target fusions are employed to stimulate the *de novo* biosynthesis of phenylalanine when its availability lags behind that of the other hydrophobic amino acids. The prephenate dehydratase of *M. maripaludis*, for example, is activated by tyrosine (Porat *et al.*, 2004), while those from *M. mahii* (Fischer *et al.*, 1993) and *H. vallismortis* (Jensen *et al.*, 1988) can be activated by any one of the following: tyrosine, leucine, methionine, and isoleucine. Paradoxically, while tryptophan activates the prephenate dehydratase from *M. mahii*, as might be expected, it inhibits that from *H. vallismortis*.

M. maripaludis also senses and modulates the *de novo* biosynthesis of tyrosine in response to the levels of *p*-hydroxyphenylacetate. In this methanogen, tyrosine, phenylalanine, and tryptophan can be produced by the carboxylation and transamination of the cognate aryl acids *p*-hydroxyphenylactetate, phenylacetate, and indoleacetate, respectively. It is presumed that synthesis by this route becomes physiologically significant when nutrient-starved anaerobic heterotrophs in the surrounding environment resort to the fermentation of amino acids as a source of energy (Porat *et al.*, 2004). In *M. maripaludis*, *p*-hydroxyphenylacetate functions as an allosteric inhibitor of prephenate dehydrogenase (Porat *et al.*, 2004). Thus, the branch of the *de novo* aromatic amino acid biosynthetic pathway responsible for the synthesis of tyrosine can be inhibited in response either to the accumulation of tyrosine or to exploit the availability of the *p*-hydroxyphenylacetate. As was the case for nucleotide deoxynucleotide synthesis, the emerging picture is that of a network of sensor-target fusion proteins that monitor the availability, not only of their immediate end products, but the entire pool of hydrophobic amino acids.

Glycolysis

The members of the Archaea lack the extensive network of sensor-response fusion proteins for modulating carbon flow through glycolysis characteristic of the *Eucarya* (Verhees *et al.*, 2003). Despite extensive searches, only one glycolytic enzyme – glycerladehyde-3-phosphate dehydrogenase [G3PDH] from *Thermoproteus tenax* (Brunner *et al.*, 1998; 2001) – appears to be allosterically regulated. However, the number of metabolites that G3PDH monitors and responds to is quite impressive. NADPH, NADH, and ATP inhibit the enzyme by reducing its affinity for NAD^+. AMP, ADP, glucose 1-phosphate, and fructose 6-phosphate stimulate activity by increasing the enzyme's affinity for NAD^+. Their effect is further amplified by the concomitant induction of positive cooperativity in the binding of substrate. X-ray crystallographic studies indicate that each of these effectors, both inhibitors and activators, act through a common allosteric binding site (Lorentzen *et al.*, 2004). Moreover, sequence comparisons indicate that the G3P DH's encoded by the crenarachaeons *A. pernix*, *P. furiosus*, *S. solfataricus*, and *S. tokodaii* all contain homologous sensor domains (Lorentzen *et al.*, 2004).

Redox regulation of coenzyme F_{420} levels

Coenzyme F_{420} is a cofactor utilized in the production of methane from CO_2 by methanogenic Archaea. The reduced form, 1,5-dihydro-coenzyme F_{420}, delivers the reducing equivalents used for the reduction 5,10-methenyl-tetrahydromethanopterin to 5-methyl-tetrahydromethanopterin.

The pool of Coenzyme A F_{420} can be rapidly decreased by the enzyme-catalysed addition of AMP, which creates the nonfunctional adduct, coenyme F_{390}. The latter can be converted back to coenzyme F_{420} by the action of the enzyme coenzyme F_{390} hydrolase.

The presence of oxygen is monitored by coenzyme F_{390} hydrolase via a sensor module containing of a pair of closely juxtaposed cysteine residues. Exposure to oxygen triggers the formation of an intramolecular disulphide bond, inactivating the enzyme and thereby reducing the levels of coenzyme F_{420} (Vermeij *et al.*, 1995). When reducing conditions are reestablished, molecular hydrogen reduces the disulphide bond and reactivates the hydrolase. The resulting increase in the coenzyme F_{420} pool is further amplified by the ability of the reduced form of 1,5-dihydro coenzyme F_{420} to inhibit the coenzyme F_{390} synthetase responsible for catalysing the conversion of coenzyme F_{420} to coenzyme F_{390} (Vermeij *et al.*, 1994). Inhibition is competitive, suggesting that the reduced coenzyme acts at the substrate-binding site rather than via a dedicated allosteric site. Other archaeal enzymes whose activity is modulated by the formation and reduction of protein disulphide bonds include an inositiol-1-phosphatase/fructose-1,6-bisphosphatase from *A. fulgidus* (Stieglitz *et al.*, 2003) and an A_1 ATPase from *M. mazei* (Lemker *et al.*, 2003).

Limitations of the sensor-target fusion paradigm

Integrating a receptor directly into the structure of its target protein constitutes a simple means for establishing a communications link between the two, particularly when the signal to be monitored is a small, soluble, rapidly diffusing metabolite. However, the direct, mechanical nature of the signal transmission process in these proteins imposes significant constraints (Fig. 9.4A). With few exceptions, target proteins reside in the interior of the cell and, in some cases, compartmentalized microenvironments therein. On the other hand, it is vitally important for an organism to be aware of the status of its external environment if it is to locate nutrients, escape toxins, identify syntrophic partners, and evade predators.

An obvious solution is to link an extracellular sensor to an intracellular target via a transmembrane domain. While this may work if the target protein acts upon either some membrane associated molecule or upon small, diffusible species; it is difficult to envisage how such a transmembrane protein could stretch and contort itself to interact with supramolecular entities such as the chromosome, glycogen particles, translation complexes, etc. Further challenges in molecular architecture arise when attempting to target more than one protein or in placing a target protein under the control of more than one environmental signal.

In some cases, this dilemma has been resolved by letting the signal come to the target. A handful of environmental cues, such as gases and light, are sufficiently membrane permeable as to render the use of an internal sensor feasible. In some cases, as with the chemotactic sensor for arginine in *H. salinarium*, an internal sensor takes advantage of the presence of an extant membrane transport system (Storch *et al.*, 1999). Placement of environmental sensors inside the cell instead of in direct, immediate contact with the external milieu appears highly counterintuitive. Membrane transit/transport processes may introduce a time lag in the detection of extracellular events or lead to the loss of spatial information. However, internally located environmental sensors do offer some compensatory features, especially for those arhcaeons that reside in extreme niches, by exploiting the ability of the cell membrane to protect receptor proteins against the damaging effects of extreme pH, salinity, etc.

Multistep signal transduction cascades

The limitations of direct, mechanical signal transduction by one-component fusions can be overcome by inserting intermediary trans-acting signal transmission module(s) between sensor and target, thereby forming a multistep signal transduction cascade (Fig. 9.2). By acting intermolecularly, through space instead of via a hard-wired conduit, these intermediary signal transmission modules enable both sensor and target each to be positioned at their optimal locations. Transmission through space is generally achieved in one of two ways. The first is to syn-

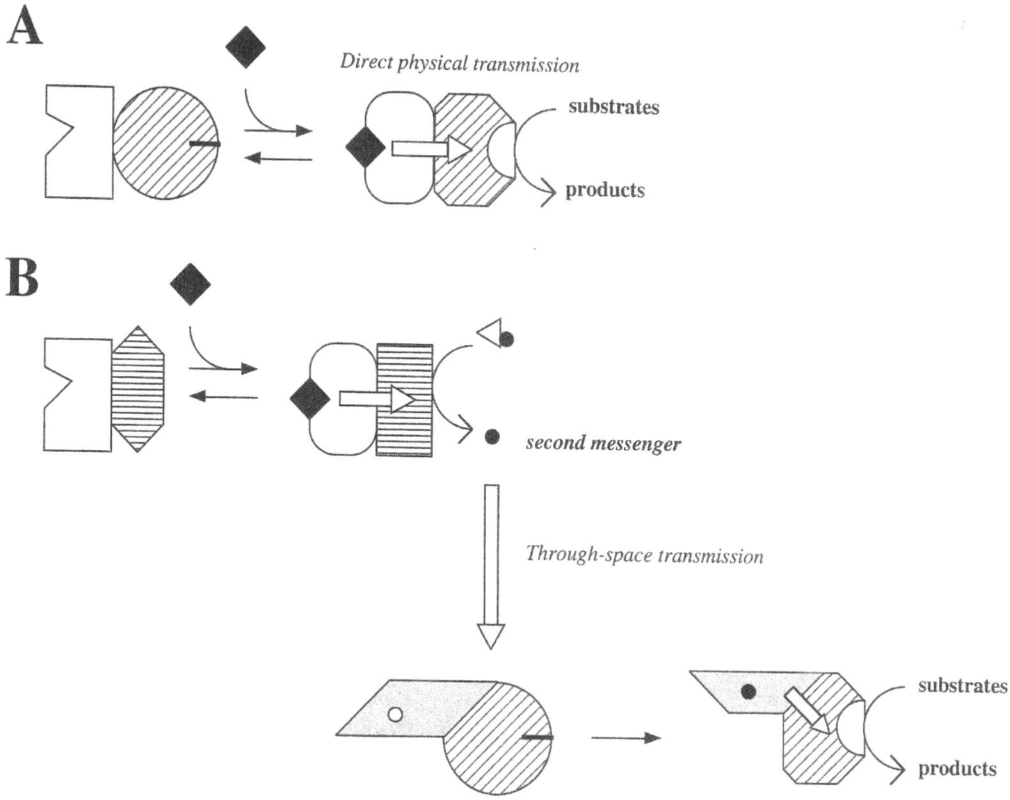

Figure 9.4 Multistep cascades permit through space transmission. Shown are the events that take place when a signal binds to a sensor-target fusion protein (A) or a receptor-transmission fusion protein that produces a second messenger (B). Hollow arrows indicate information transmission events between components. Symbols used include signal, filled diamond; sensor, stippled; target, diagonal cross-hatching; second messenger, solid circle; transmission domain, horizontal cross hatching; receiver domain, light shading.

thesize or release an allosteric effector, or second messenger, that diffuses from a sensor-linked transmitter to a target-linked receiver (Fig. 9.4B). Prototypic second messengers include Ca^{2+}, cyclic nucleotides, polyphosphoinositols, and the component subunits of heterotrimeric G proteins. Branching of the signal to reach multiple targets can readily achieved by simply adding the appropriate receiver domain.

Transmission through space can also be achieved by employing a transmitter module capable of catalysing the covalent modification of the target. Some common regulatory modifications include partial proteolysis (pro-proteins and zymogens), acetylation, oxidation, methylation, and phosphorylation. While this mechanism requires that receiver and transmitter modules come into physical contact, the interaction is transient. Through space transmission can be achieved by having the freely moving target protein encounter the modifying enzyme, undergo modification, and then travel to its site of action in the cell. This is generally the case with two-component cascades. As with second messengers, signal branching is achieved by incorporation of suitable covalent modification site (receiver) on a variety of proteins.

Oftentimes, the covalent modifying protein travels to the target protein rather than the other way around. This is usually accomplished by multiplying the number of steps in the cascade by inserting a dedicated signal transmission protein consisting of a receiver and transmitter domain (Fig. 9.5). Multiplication of the number of steps in a signal transduction cascade also provides additional means for signal branching as well as nodes where signals originating from other receptors can be integrated.

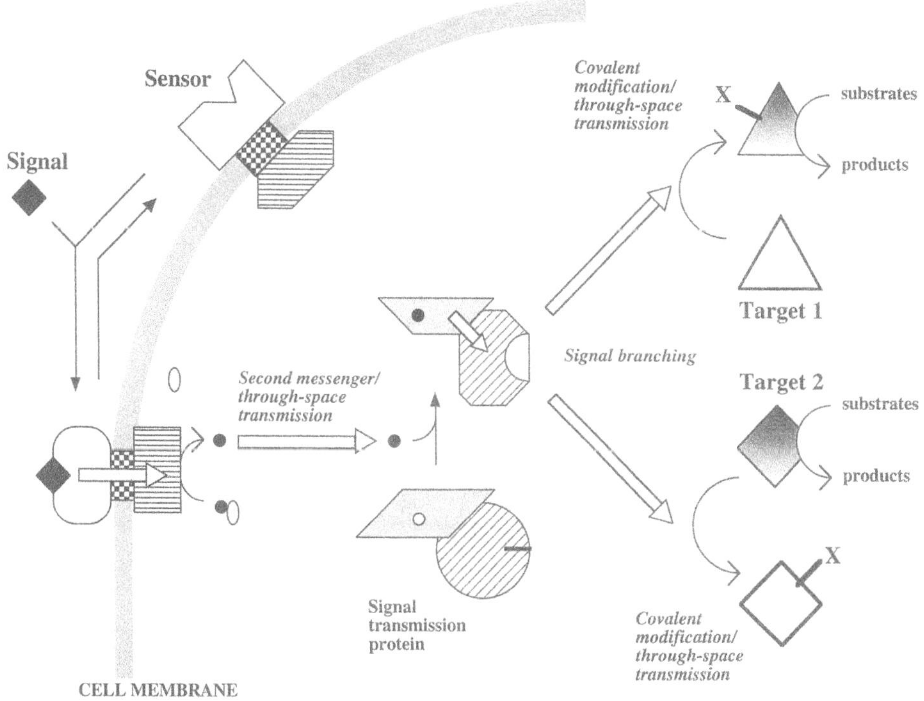

Figure 9.5 Multistep cascades facilitate signal branching. The symbols are the same as those in Fig. 9.4, with the exception that the catalytic domain of the signal transmission protein is indicated by diagonal cross hatching the target proteins are indicated by graduated shading. A bold 'X' indicates the covalent modification of the target proteins. The cell membrane is represented by a stippled band, and the transmembrane domain of the receptor by a checkered pattern.

A census of potential archaeal sensors and signal transmission domains

While the study of archaeal signal transduction processes is just emerging from its infancy, a wealth of information is available concerning sensor-response processes in the members of the Bacteria and Eucarya. As can be seen in Table 9.1, only a handful of these potential signal transmission modules can be identified from inspection of archaeal genome sequences. In fact, the only modules that are generally conserved across the Archaea are the 'atypical' eukaryote-like protein kinases, Pat protein acetyltransferases, and class II adenylate cylcases (Ponting *et al.*, 1999; Galperin, 2005; Marsh *et al.*, 2005). The role of the last named proteins remains cryptic, as the Archaea appear to contain little or no cAMP (Haseltine *et al.*, 1999; Bini and Blum, 2001). In addition the Archaea appear to be devoid of recognizable nucleotide phosphodiesterases or the GGDEF and EAL domains that catalyse the synthesis and degradation, respectively, of the bacterial second messenger cyclic-di-GMP (Galperin, 2005).

The histidine kinases, methyl-accepting chemotaxis proteins [MCP], MCP methylases and demethylases, and response regulators of the two-component regulatory system – while widespread among the euryarchaea – are completely absent from the crenarchaea sequenced to date (Ashby, 2004; 2006). Phyletic analyses indicate that the two-component system is not indigenous to the Archaea, but rather was acquired by an early euryarchaeote from the Bacteria via lateral gene transfer (Koretke *et al.*, 2000). This paradigm can therefore be viewed as an addition or enhancement to the core sensor-response machinery indigenous to the Archaea.

Even less has been gleaned from attempts to mine archaeal genomes for potential extracellular receptors. An estimated 20–30% of archaeal proteins contain predicted (trans)membrane domains (Wallin and Von Heijne, 1998; Arai *et al.*, 2003). However, while several contain potential signal transmission domains such as proteases,

Table 9.1 Census of eucaryal and bacterial signal transmission modules among the Euryarchaea and Crenarchaea

Signal transmission module	Euryarchaea	Crenarchaea
PKN2-like ePK	No	S
Atypical ePK	U	U
PPP-family protein phosphatases	S	S
PPM-family protein phosphatases	No	R
Conventional PTPs	S	S
LMW PTPs	S	S
Cdc25 dual-specific phosphatase	No	No
Two-component system	C	No
Methyl-accepting chemotaxis protein	C	No
SixA histidine phosphatase	S	S
CheZ protein-aspartate phosphatase	No	No
RAP protein-aspartate phosphatase	No	No
CheC protein-aspartate phosphatase	S	No
CheR MCP methyltransferase	C	No
CheB MCP demthylase (deamidase)	C	No
Adenylate cyclase	U	U
Cyclic nucleotide phosphodiesterase	No	No
GGDEF cyclic-di-GMP synthase	No	No
EAL cyclic-di-GMP hydrolase	No	No
EF hand proteins (e.g. calmodulin)	No	No
Pat protein acetylase	U	U
Sir2 protein deacetylase	S	S
SH2 domain	No	No
SH3 domain	No	No
WW domain	No	No
Pleckstrin homology domain	No	No
Caspase	No	No
Ras-like monomeric G protein	No	No
Heterotrimeric G proteins	No	No
Ubiquitin	No	No
Ubiquitin conjugating enzymes	No	No
Phosphatidyl inositol-3 kinases	No	No
Apoptotic NTPases	No	No
Apoptotic HtrA proteases	No	No
Lux-like quorum sensing proteins	No	No

U, virtually ubiquitous; C, commonly encountered; S, sometimes encountered.

MCPs, histidine kinases, and ePKs within their sequences (Taylor and Zhulin, 1999; Galperin *et al.*, 2001; Zhulin *et al.*, 2003; Ulrich *et al.*, 2005; Ashby, 2006), for the most part the signals monitored by their presumed sensor domains remain cryptic. Indeed, when one eliminates the bacterially derived chemo-, aero-, and photoreceptors involved in regulating taxis in many archaeons, the number of unambiguously identified and functionally defined extracellular receptor proteins drops to near insignificance (Kennelly, 2007).

Signal transduction in archaeal chemo-, aero-, and phototaxis

The majority of the euryarchaea contain homologues of the two-component regulatory paradigm that were acquired as complete functional units from the Bacteria via lateral gene transfer (Koretke *et al.*, 2000). The two-component system derives its name from the pair of conserved signal transmission modules linking the environmental sensors for various chemo-, aero-, and photo- tactic cues to their intracellular target, flagellar motor proteins (Hoch, 2000; Bourret and Stock, 2002). The conserved transmitter module is called a histidine kinase, so named because when activated it phosphorylates itself on a conserved histidine residue. The conserved receiver module is an aspartyl autokinase called a response regulator. The response regulator uses the autophosphorylated histidine kinase, rather than ATP, as phosphodonor substrate for catalysing its self-phosphorylation on a conserved aspartate residue. The phosphorylation state of the response regulator is therefore modulated by regulating the availability of its cognate substrate, the phosphorylated form of the histidine kinase.

MCPs are generally found only in those two-component cascades involved in mediating taxis. In chemo- tactic cascades, MCPs serve as a physical link between the ligand-sensing domain of the receptor and its cognate histidine kinase. Methylation and demethlyation of glutamate residues on the MCP, mediated via a response-regulator containing protein demethylase, CheB, serves as a feedback mechanism for adjusting the sensitivity of the receptor (Bourret and Stock, 2002). Consequently, a single receptor-linked sensor-response cascade can work across a wide gradient of chemoattractant or chemorepellant concentration.

By far the best-studied sensor-taxis system among the Archaea is that of *Halobacterium salinarium* (nee *H. halobium*). This extreme halophile is attracted to a variety of chemicals that include D-glucose, acetate, benzoate, indole, $NiSO_4$, histidine, asparagine, leucine, and methionine; it is repelled by phenol (Schimz and Hildebrand, 1979). In addition, *H. salinarium* is both aerotactic (Stoeckenius *et al.*, 1988; Boudko *et al.*, 2003) and phototactic (Hildebrand and Dencher, 1975; Spudich and Stoeckenius, 1979). It migrates towards sources of orange light and away from blue (Stoeckenius *et al.*, 1988). This behaviour is modulated by an estimated seventeen MCP-linked receptors, at least twelve of which are predicted to be membrane proteins (Galperin, 2005). These MCP-linked receptors activate a like number of cognate histidine kinases, thirteen, which serve as phosphodonor substrates for approximately roughly half that number of response regulator proteins.

Phototaxis is mediated by the photoreceptor proteins rhodopsin I and II, each of which employs retinal, a polyisoprenoid prosthetic group containing a series of conjugated double bonds, as its 'antenna' (Spudich, 2006). Stimulation of rhodopsin I, which is sensitive to orange light, modulates taxis toward the source while stimulation of rhodopsin II, which is sensitive to blue light, triggers taxis away from the light source (Bogomolni and Spudich, 1982; Alam *et al.*, 1989; Spudich *et al.*, 1989). The functionally competent form of the photoreceptor is a heterotetramer consisting of two copies of the sensory rhodopsin and two of its cognate MCP, designated HtrI and HtrII, respectively (Chen and Spudich, 2002), which in turn interact with a CheA-like histidine kinase. Intriguingly, Htr II also acts as the chemoreceptor for serine, suggesting that the photoreceptors co-opted an ancient chemoreceptor (Hou *et al.*, 1998).

Oxygen is detected via a receptor, HtrVIII, comprising a six transmembrane segment haem-binding domain, homologous to that of eukaryotic cytochrome c oxidase, fused to an inward-facing MCP (Brooun *et al.*, 1998). A second haem-containing MCP protein, HemAT-Hs, modulates aerophobic behaviour (Hou *et al.* 2000). The haem-binding domain of HemAT-Hs is homologous to that of myoglobin, hence this sensor is predicted to be soluble rather than membrane associated. Since oxidation of the haem iron to the ferric form lowers its affinity for oxygen, it is postulated that its intracellular localization enables HemAT-Hs to integrate streams of information, the oxygen concentration in the surrounding environment and the internal redox state of the organism – thereby favouring taxis away from areas of high oxygen concentration when *H. salinarium* is experiencing oxidative stress.

In addition to the HtrII serine receptor, three other chemoreceptors have been identified to date in *H. salinarium*. The first, Car, is a soluble protein with a fused MCP domain (Storch *et al.*, 1999). As mentioned previously, Car relies of the argninine:ornithine antiporter to deliver arginine to the cell's interior to enable it to perform its sensory role. BasB is activated by sulphur-containing amino acids and works in concert with the MCP BasT (Kokoeva and Oesterheldt, 2000). Similarly, the betaine receptor protein CosB is paired with the MCP CosT (Kokoeva *et al.*, 2002). While BasT and CosT are membrane bound, the sequences of BasB and CosB provide no evidence for a membrane domain. Hence, it has been postulated that the latter are posttranslationally modified by the addition of lipid anchors (Kokoeva *et al.*, 2002).

The sensor-response network that directs the swimming behaviour of *H. salinairum* is organized in a somewhat hierarchal fashion as the photoreceptors exert veto power over chemo- and aero-signalling cascades. Switching the direction of rotation of flagellar motors in response to chemo-, aero-, and phototactic signals in *H. salinarium* requires the presence of the TCA cycle intermediate fumarate (Marwan *et al.*, 1990). Fumarate serves as a switching or potentiating factor whose binding to flagellar motor proteins renders them responsive to CheY, the terminal phosphoprotein of the two-component signalling cascade that becomes activated upon stimulation of the haloarchaeon's chemo-, aero-, and photo- receptors (Montrone *et al.*, 1993). The switch itself is under the control of rhodopsin-containing photoreceptors, whose activation triggers the release of a pool of membrane-bound fumarate molecules (Montrone *et al.*, 1996).

Most, if not all, of the sensor proteins are the targets of modification of glutamate residues to form glutamine. Methylation alters the affinity of the associated receptor for its ligand, thus enabling the organism to adjust the dynamic range of its sensor-response machinery as it moves through a gradient of attractant – a process oftentimes referred to as adaptation (Bourret and Stock, 2002; Baker *et al.*, 2005). While HtrII possesses only a single consensus methylation site, HtrI contains three; suggesting that the latter sensor can be 'tuned' over a wider range (Perazzona and Spudich, 1999). In general, exposure to attractants tends to increase the degree of methylation of the MCPs in *H. salinarium*, while exposure to repellants has the opposite effect (Schmiz, 1981; Alam *et al.*, 1989), indicating that demethylated receptors tend to be the more sensitive. The methylation state of receptor-associated MCPs is mediated via a phosphorylation-dependent feedback loop targeting CheB, whose phosphorylation is dependent upon the availability of the phosphorylated form of the receptor-activated histidine kinase (Bourret and Stock, 2002). CheB homologues are found in virtually every archaeon that encodes two-component chemo-, photo- and/or aerotaxis cascades.

An additional layer of control is provided through the intervention of protein phosphatases acting upon the phosphoproteins of the chemotaxis cascade and receptor adaptation loop. While *Bacteria* harbour a variety of response-regulator phosphatases, the best known of which is the chemotaxis protein CheZ, the Archaea solely contain deduced CheC family protein-aspartyl phosphatases (Szurmant and Ordal, 2004). Some archaeons also contain homologues of SixA, which hydrolyses phosphohistidine *in vitro* (Matsubara and Mizuno, 2000). However, it is unlikely that SixA acts on archaeal histidine kinases as the few archaeons in which it is found, including several crenarchaeons, lack two-component systems (Kennelly, 2007). The signals, if any, modulating the action of archaeal CheC phosphatases are presently unknown.

Genome surveys indicate that the size of the sensory arrays that guide archaeal swimming vary over a wide range (Galperin *et al.*, 2001; Galperin, 2005; Ashby, 2006). While *H. marismortui* and *H. salinarium* each possess seventeen MCPs and thirteen histidine kinases, *P. abyssi* and *P. furiosus* each encode MCPs that act through a single CheA-like histidine kinase. This histidine kinase, in turn, serves as phosphodonor substrate for two response regulators, a CheY-like protein that is predicted to modulate the flagellar motor and a CheB-like protein that modulates receptor sensitivity. Genome surveys also indicate that not all sensors that act through archaeal two-component cascades employ MCPs to bridge receptors with their cognate histidine kinases. *M. acetivorans,* for example, encodes over fifty histidine kinase domains and only two MCPs (Galperin, 2005). The comparable numbers for *M. mazei, H. marismortui,* and *M. thermoautotrophicus* are 33 and 3, 59 and 17, and 16 and none, respectively (Galperin, 2005). It is also noteworthy that two-component cascades play little, if any, role in modulating archaeal gene expression, as the Archaea are virtually devoid of response regulator-transcription factor fusion proteins that are so prolific among the Bacteria (Ashby, 2004).

Protein-serine/threonine/tyrosine phosphorylation

Overview of archaeal protein-serine/threonine/tyrosine kinases

Surveys of archaeal genomes suggest that the phosphorylation of archaeal proteins on serine, threonine, and tyrosine residues is catalysed exclusively by members of the eukaryotic protein kinase [ePK] superfamily (Kennelly, 2003). The number of ORFs encoding deduced ePKs in archaeal genomes typically ranges from three to 12 (Kennelly, 2003; Galperin, 2005). A handful of these possess predicted transmembrane domains suggestive of association with an extracellular receptor (Galperin, 2005).

One of the more striking features of the Archaea is the scarcity of the iconic subfamily of ePKs, the 'typical' ePKs (Fig. 9.6). In fact, typical ePKs are found only among the crenarchaea, while the euryarchaea possess solely atypical ePKs. By contrast, the typical ePK subfamily – which includes such well-known signal transduction proteins as includes the cAMP-dependent protein kinase, protein kinase C, MAP kinase, the insulin receptor kinase, and pp60^{src} – accounts for three-quarters of the eucaryal kinome (Manning *et al.*, 2002). This subfamily also predominates in the Bacteria, which apparently acquired them via lateral gene transfer from the Eucarya (Ponting *et al.*, 1999). It is somewhat ironic that the 'typical' ePK family so strongly associated these latter two phylogenetic domains may have originated in an ancient crenarchaeote.

piD261/Bud32 – the most ancient ePKs?

The piD261/Bud32 protein kinases are the only subfamily of ePKs that are present in virtually every member of the *Eucarya* and the Archaea (Leonard *et al.*, 1998). It therefore appears that the first ePK to be adapted for the purposes of transmitting sensory information was an early piD261/Bud32 protein kinase. The first member of this family to be identified was piD261/

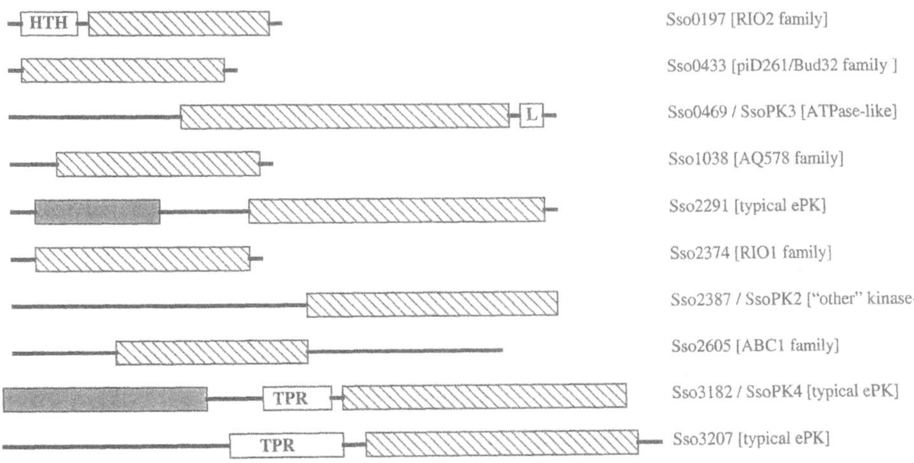

Figure 9.6 Potential ePKs from *S. solfataricus.* Each potential protein kinase is represented by a horizontal line whose length is roughly proportional to the respective length of each polypeptide. The name of the ORF encoding polypeptide is listed at right, with the family of ePKs to which it belongs listed in parentheses. Where relevant, the name of the protein product is also listed. The protein kinase catalytic domain is indicated by the cross hatching area. Predicted transmembrane domains are shaded. The boxes containing the letters HTH, L, and TPR indicate potential helix-turn-helix, leucine zipper, and tetratricopeptide repeat domains.

Bud32 from yeast (Stocchetto *et al.*, 1997). Deletion of the gene encoding this ePK resulted in impaired cell survival and a severe slow-growth phenotype. Complementation with mutationally altered forms of piD261/Bud32 lacking phosphotransferase activity were able to partially rescue this phenotype, suggesting that while its protein kinase activity is important, it does not constitute the only means by which it influences cellular events (Lopreiato *et al.*, 2004). *In vitro*, piD261/Bud32 displayed a marked preference for phosphorylating proteins rich in acidic amino acids, such as casein (Facchin *et al.*, 2002) and glutaredoxin (Lopreiato *et al.*, 2004).

In mammals, the piD261/Bud32 family is represented by the p53-related protein kinase [PRPK] (Abe *et al.*, 2001). While the physiological substrates for both piD261/Bud32 and PRPK remain to be identified, both enzymes display a significant degree of functional conservation. For example, both PRPK and piD261/Bud32 phosphorylate p53 *in vitro*, while expression of PRPK in yeast lacking piD261/Bud32 partially rescues the slow growth phenotype (Facchin *et al.*, 2003). Remarkably, the product of ORF sso0433 exhibits 30% sequence identity and 75% sequence similarity with human PRPK. The high degree of sequence conservation across this vast evolutionary distance implies that the function the piD261/Bud32 ePKs is conserved as well. Superficial support for this notion was supplied by the observation that reombinantly expressed Sso0433, a.k.a. SsoPK5, phosphorylates mammalian p53 *in vitro* (J.D. Haile and P.J. Kennelly, unpublished).

It appears likely that the high degree of conservation of the piD261/Bud32 ePK subfamily reflects a concomitant degree of functional conservation, a supposition supported by the fact the archaeal, yeast, and human members of the family all phosphorylate p53 *in vitro*. It appears likely that the members of this family target proteins involved in gene transcription, since (a) phosphorylation of p53 stimulates its transcription 'activity' (Abe *et al.*, 2001), (b) piD261/Bud32 itself forms complexes with several transcriptional/chromatic-related proteins (Kisseleva-Romanova *et al.*, 2006) and (c) the fundamental importance of gene transcription is consistent with the ubiquitous nature of this protein kinase. One of the puzzling features of the piD261/Bud32 ePKs is its association with a putative protease, Ykr038/Kae1. This protease binds to piD261/Bud32 in yeast two hybrid screens (Lopreiato, 2004), while archaeal homologues of these two proteins are generally encoded by either adjacent ORFs or as a single fusion protein (Lopreiato *et al.*, 2004).

PH1052, a typical ePK

The catalytic domains of the typical ePKs in the crenarchaea display significant homology to the corresponding domain of PKR, a protein kinase that phosphorylates and inactivates translation initiation factor-2α [eIF2α]. In mammals, PKR is activated by double-stranded RNA, thereby preventing invading virions from exploiting the host's intrinsic translational machinery to manufacture viral proteins. Intriguingly, the eIF2α kinase homologue in the archaeon *P. horikoshii*, designated PH1052, phosphorylates eIF2α on the same residue, Ser-51, as does PKR (Tahara *et al.*, 2004). Moreover, archaeal translational initiation factor 2α [aIF2α] contains a serine residue, Ser-48, in a nearly identical position to the phosphoacceptor amino acid in eIF2α. This serine residue could be phosphorylated *in vitro* using either PH1052 or PKR from humans (Tahara *et al.*, 2004). The highly conserved nature of protein translation in the crenarchaea and *Eucarya*, coupled with the obvious physical and functional parallels between PKR and PH1052, strongly suggests that protein translation is regulated by PH1052-mediated phosphorylation in *P. horikoshii* as well as other crenarchaea.

Gene array studies in the crenarchaeote *S. solfataricus* have implicated the phosphorylation of aIF2α in cell cycle regulation (Lundgren and Bernander, 2007). The genome of *S. acidocaldarius* encodes at least two potential aIF2α kinases, designated Saci1193 and Saci1694 [Another member of the sulfolobales, *S. solfataricus*, encodes three (W.K. Ray and P.J. Kennelly, unpublished)]. The message encoding one of these, Saci1193, exhibited strong cyclic variation (Lundgren and Bernander, 2007). In fact, *saci1193* was one of the earliest genes to be induced following release from G_2 arrest. The gene encoding the second typical ePK, *saci1694*, also displayed cyclic variation, albeit of much lower amplitude that trailed that of *saci1193*.

Other ePKs

A handful of additional protein kinases have been characterized from archaeal organisms. Little is known, however, regarding their physiological functions. The first of these was SsoPK2, a member of an unclassified subfamily of atypical ePKs. The ORF encoding SsoPK2, *sso2387*, was identified by homology searches of the *S. solfataricus* genome (Lower and Kennelly, 2003). Topologically, SsoPK2 consists of a C-terminal protein kinase catalytic domain fused to a 300-residue long N-terminal domain(s) of unknown function (Fig. 9.6). Recombinant SsoPK2 displayed sluggish protein-serine phosphotransferase activity toward non-archaeal proteins such as casein, histones, and myelin basic protein. It also catalysed autophosphorylation on serine residues corresponding to the activation loop of typical ePKs.

A second ePK was first identified as a phosphothreonine-containing protein following incubation of a membrane extract of *S. solfataricus* with [^{32}P]ATP (Lower *et al.*, 2004). This fraction was enriched for a protein-threonine kinase activity that had previously been traced to a membrane glycoprotein of undetermined sequence (Lower *et al.*, 2000; Lower and Kennelly, 2002). Mass peptide profiling identified one of the proteins that became radiolabelled as the product of ORF *sso0469*, whose product displayed faint homology to ePKs and ATPases. It also contains a potential leucine zipper near its carboxy terminus (Fig. 9.6). The protein product of ORF *sso0469* exhibited protein-serine kinase activity towards exogenous substrates *in vitro*, but no circumstances could be identified in which it would catalyse its own phosphorylation. Moreover, SsoPK3 was radiolabelled when membrane extracts were incubated with the alternative phosphodonors [^{32}P]GTP or [^{32}P]GDP. Since SsoPK3 is ATP-specific, whereas the membrane-associated glycosylated protein kinase activity exhibited broad nucleotide specificity, it would appear that phosphorylation of SsoPK3 was catalysed by the latter (Lower and Kennelly, 2004). Like SsoPK2, the specific activity of SsoPK3 towards exogenous protein substrates is relatively low. While this may reflect the fact that the phosphoacceptor proteins used for these assays are derived from mammalian, rather than archaeal, organisms; it also remains possible that the observed protein kinase activity is adventitious, and that neither enzyme targets proteins *in vivo*.

The RIO kinases, whose name is an acronym for RIght Open reading frame, are found throughout the Archaea and *Eucarya* (Angemayr *et al.*, 2002). In yeast, RIO1 and RIO2 have been implicated in ribosome biogenesis (Vanrobays *et al.*, 2003) and cell cycle progression (Angemayr *et al.*, 2002). Both RIO1 and RIO2 from the archaeon *A. fulgidus* catalyse their own phosphorylation as well as that of exogenous proteins such as histones and myelin basic protein *in vitro* (LaRonde-LeBlanc *et al.*, 2005a; 2005b). Ironically, while the complete X-ray crystal structures of both *A. fulgidus* RIO kinases have been determined (LaRonde-LeBlanc and Wlodawer, 2004; LaRonde-LeBlanc *et al.* 2005a, 2005b), it remains unclear whether proteins in fact represent the physiological targets for the activity of the RIO kinases.

Archaeal protein-serine/threonine/tyrosine phosphatases

Archaeal genomes generally encode one or more deduced protein-serine/threonine/tyrosine phosphatases drawn from amongst three principle superfamilies: PPP-family, conventional protein-tyrosine phosphatases [cPTP], and low molecular weight PTPs [LMW PTPs] (Kennelly, 2001). All three families appear to be relatively ancient and are found throughout the *Eucarya* and in a large proportion of the *Bacteria*. The archaeal PPP-family protein phosphatases from *S. solfataricus* (Leng *et al.*, 1995), *Methanosarcina thermophila* (Solow *et al.*, 1997), and *Pyrodictium abyssi* (Mai *et al.*, 1998) are all serine/threonine-specific, as is the case for their eucaryal (but not bacterial) homologues (Kennelly, 2001). Only one archaeal PTP has been studied to date, the cPTP *Tk*-PTP from *T. kodakaraensis* (Jeon *et al.*, 2002). While Tk-PTP hydrolysed *p*-nitrophenyl phosphate and free phosphotyrosine, no reports have surfaced as to whether the enzyme acts of protein-bound phosphotyrosine, phosphoserine, or phosphothreonine.

Although all three of these protein phosphatase families are sufficiently ancient to have been indigenous to the Archaea, their distribution follows no easily discerned pattern. The genomes for most contemporary archaeons contain one or two ORFs encoding potential protein-serine/threonine/tyrosine phosphatases (Kennelly, 2003). However, no one protein-serine/threonine/tyrosine phosphatase family is conserved across the Archaea, or even a clear majority of the Archaea. *Aeropyrum pernix* contains only one deduced protein phosphatase, a member of the PPP-family that probably is serine/threonine-specific. *P. horikoshii* contains a single cPTP and *M. jannaschii* contains three, but neither contains an obvious protein-serine/threonine phosphatase (Kennelly, 2003). Hence, it is very likely that the cPTPs in these organisms are dual-specific, i.e. they hydrolyse both protein-tyrosine phosphate and protein-serine/threonine phosphate. Despite their classification as protein-tyrosine phosphatases, many subfamilies of the cPTPs are dual-specific. On the other hand, at least two archaeons – *A. pernix* and *T. volcanium* – lack prototypic protein-tyrosine phosphatases. Does that mean that their PPP-family phosphatases are also dual-specific, or that only select archaeons harbour tyrosine-phosphorylated proteins? *T. volcanium* also holds the distinction of containing the only PPM-family protein-serine/threonine phosphatase among the Archaea, presumably the spoils of a lateral gene transfer event.

One might also ask whether archaeal LMW PTPs might exhibit dual-specific protein phosphatase activity. If so, such behaviour would be unprecedented among this enzyme family. None of the LMW PTPs tested to date show any propensity to dephosphorylate serine- or threonine- phosphorylated proteins (Kennelly, 2001). It may also be noteworthy that while some archaeons have only PPP-family protein phosphatases or cPTPs, none have only LMW PTPs (Kennelly, 2003). In every archaeon in which a potential LMW PTP has been identified, a second predicted protein phosphatase – either a PPP-family protein phosphatase or a cPTP – was also identified. While the number of archaeal genomes represented in these studies remains limited, it lends circumstantial support to the inference that the Archaea have not developed dual-specific variants of the LMW PTP paradigm.

Archaeal phosphoproteins

Thus far, only a handful of proteins have been identified that are subject to covalent modification via the phosphorylation of serine, threonine, or tyrosine residues. In at least two cases, it has been determined that these phosphorylation events alter the functional properties of the protein, as would be expected if they are the target of a sensor-response cascade. Both are from *S. solfataricus*, which has one of the largest complements of potential ePKs among the Archaea (Kennelly, 2003; Galperin, 2005) The regulatory phosphorylation site, Ser-309, on the phosphohexomutase from *S. solfataricus* was identified by mass spectral analysis of peptide digests of proteins in the soluble fraction (Ray *et al.*, 2005). While catalysis involves the formation of a phosphoenzyme intermediate, the catalytic serine residue, Ser-97, is both physically and functionally distinguishable from Ser-309. As modelling studies predicted that Ser-309 resides within the substrate-binding cleft, where the phosphoryl group would presumably form an electrosteric block against substrate binding, it was hypothesized that phosphorylation inhibits catalytic activity. While the protein kinase responsible for the phosphorylation of Ser-309 remains to be identified, mutagenic alteration of this residue to aspartic acid – which also has the potential to serve as an electrosteric block – reduced the catalytic activity of the enzyme to only a few percent that of the dephosphorylated, unmodified phosphohexomutase (Ray *et al.*, 2005). In the case of the D-gluconate dehydrogenase from *S. solfataricus*, it was reported that incubation of the enzyme with alkaline or acid phosphatase led to the release of phosphate from the protein and a concomitant decrease in its specific activity (Kim and Lee, 2005). Hence, it was concluded that phosphorylation activates the enzyme.

Other proteins implicated as the targets of protein phosphorylation in the Archaea include:

1. glycogen synthase from *S. acidocaldarius* (Cardona *et al.*, 2001);
2. Cdc6 from *S. acidocaldarius* (De Felice *et al.*, 2003);
3. Cdc6 from *M. thermoautotrophicum* (Grabowski and Kelman, 2001);

4 methyltransferase activation protein from *M. barkeri* (Daas *et al.*, 1996);
5 a 90 kDa aminopeptidase from *S. solfataricus* (Condo *et al.*, 1998);
6 phenylalanyl-tRNA synthetase from *T. kodakaraensis* (Jeon *et al.*, 2002);
7 phosphohexomutase from *T. kodakaraensis* (Jeon *et al.*, 2002);
8 RNA terminal phosphate cyclase from *T. kodakaraensis* (Jeon *et al.*, 2002).

Phosphorylation of Cdc6 is self-catalysed and, in the case of *M. thermoauotrophicum*, is modulated by association with the minichromosome maintenance protein (Kasiviswanathan *et al.*, 2005). The three-last named proteins, all from *T. kodakaraensis*, are particularly intriguing as they were all identified using a substrate trapping mutant of the *Tk*-PTP and immunoreact with antibodies against phosphotyrosine.

Acetylation of archaeal chromatin proteins

The protein components of archaeal chromatin display a surprising variability. The genomes of most euryarchaea encode histone-like proteins, but archaeal histones are found in only a minority of the creanarchaea (White and Bell, 2002; Sandman and Reeve, 2005). The most widely distributed DNA-binding protein is a uniquely archaeal one, Alba, which is distributed throughout the crenarchaea and in a majority of the euryarchaea (White and Bell, 2002; Sandman and Reeve, 2005). In the crenarchaeon *S. solfataricus*, Alba exists in two forms that differ by the addition of an acetyl group to the side chain of lysine 16 (Bell *et al.*, 2002). Acetylation of Alba decreases its binding affinity roughly 30-fold *in vitro* (Bell *et al.*, 2002), implicating acetylation as a mechanism for regulating the chromatin condensation *in vivo*. Circumstantial evidence supporting this conclusion was provided by the observation that levels of Pat-like acetylase responsible for modifying Alba (Marsh *et al.*, 2005) exhibit clear cell cycle variation (Lundgren and Bernander, 2007).

Alba from *S. solfataricus* binds to and can be deacetylated by the Sir2 deacetylase *in vitro* (Bell *et al.*, 2002), suggesting that Alba modification is physiologically reversible. However, while Alba and the Pat acetylase are fairly ubiquitous among the Archaea, only about half of the members of this domain encode a homologue of Sir2 (White and Bell, 2002). Perhaps other archaeons utilize an as yet unrecognized Alba deacetylase to regenerate the unmodified form. Alternatively, Alba deacetylation may be reversible in only a subset of the Archaea, or Sir2 may not deacetylate Alba *in vivo*. Regardless, the parallels with the histone code in *Eucarya* (Cosgrove and Wolberger, 2005), which involves modification by acetylation, methylation, and phosphorylation, are intriguing. It would not be surprising to discover that archaeal chromatin proteins are subject to a multiplicity of covalent modifications, for which Alba acetylation is the harbinger.

Conclusions

The study of archaeal signal transduction processes is only now emerging from its infancy. Our current state of knowledge thus remains frustratingly fragmentary. Only in a few areas, such as the sensor-response machinery that directs the swimming behaviour of *H. salinarium* or in the internal command and control network that coordinates nucleotide and deoxynucleotide biosynthesis do we have a basic understanding of how the pieces fit together: signal, sensor, signal transmission, target, response. More often than not we have come into possession of a few individual pieces whose positions within the larger signal transduction puzzle can only be guessed at.

Is that all there is?

The human genome contains 1000 ORFs encoding known or potential protein kinases and protein phosphatases, nearly 4% of the entire genome (Manning *et al.*, 2002). *Bacteria* devote 1–2% of their genomes to protein-serine/threonine/tyrosine kinases and phosphatases (Galperin, 2005). Most archaeons allocate less than 0.5% of their genome for these versatile signal transmission modules (Galperin, 2005). The Archaea are devoid of EF-hand proteins and ubiquitin (Ponting *et al.*, 1999); encode only one or two adenylate cyclases (Galprin, 2005), and no cyclic nucleotide phosphodiesterases (Ponting *et al.*, 1999). Thus, we are confronted with the question: Is that all there is? Are the Archaea simply less aware of the events occurring within and without, or have they evolved novel and as yet unrecognized paradigms for sensing and signal transduction?

One can rationalize the spartan nature of archaeal signal transduction networks by appealing to the unique nature of the environmental niches in which they reside. Certainly, one way to explain the patchwork distribution of the various protein-serine/threonine/tyrosine phosphatase families in the Archaea is to invoke reductive evolution. On the other hand, research continues to expand the scope and variety of habitats in which archaeons are known reside and even dominate, and has revealed that many of the Archaea live as members of biologically diverse, interdependent microbial communities. So while archaeal genomics might argue for a limited number of senses, biology suggests that the Archaea face the same requirements to monitor and adapt to their environment as do other microorganisms.

Why not cAMP?; Why so few ePKs?

Given the prolific nature of both cyclic nucleotide second messengers and the typical ePK paradigm in the other two domains of life, it would appear that both represent superior vehicles for signal transmission. Moreover, the Archaea possess the raw material – adenylate cyclases and a range of ePKs families, including the typical ePKs – for exploiting of these molecular paradigms. So why do archaeons make little or no use of cyclic nucleotides? Why haven't their protein-serine/threonine/tyrosine kinase and phosphatase populations – in particular the typical ePK subfamily – expanded to levels comparable to other organisms. Presumably the answers to these questions will become apparent when we discover the physiological role of archaeal adenylate cylcases and learn more about the properties of atypical ePKs.

The future of signal transduction research in the Archaea holds great promise, as the answers to these and other questions have major implications for our understanding of the mechanisms by which these extremophiles interact with their surroundings as well as the origins and development of sensor-response mechanisms across the phylogenetic spectrum.

Acknowledgement

The author wishes to acknowledge the support of grant MCB-0315122 from the National Science Foundation.

References

Abe, Y., Matsumoto, S., Wei, S., Nezu, K., Miyoshi, A., Kito, K., Ueda, N., Shigemoto, K., Hitsumoto, Y., Nikawa, J.-I., and Enomoto, Y. (2001) Cloning and characterization of a p53-related protein kinase expressed in interleukin-2-activated cytotoxic T-cells, epithelial tumor cell lines, and the testes. J. Biol. Chem. *276*, 44003–44011.

Alam, M., Lebert, M. Oesterhelt, D. and Hazelbauer, G.L. (1989). Methyl-accepting chemotaxis proteins in *Halobacterium halobium*. EMBO J. *8* 631–639.

Angermayr, M., Roidl A., and Brandlow W. (2002). Yeast Rio1p is the founding member of a novel subfamily of protein serine kinases involved in control of cell cycle progression. Mol. Microbiol. *44*, 309–324.

Arai, M., Ikeda, M., and Shimizu, T. (2003). Comprehensive analysis of transmembrane topologies in prokaryotic genomes. Gene *304*, 77–86.

Arent, S., Harris P., Jensen, K.F., and Larsen, S. (2005). Allosteric regulation and communication between subunits in uracil phosphoribosyltransferase from *Sulfolobus solfataricus*. Biochemistry *44*, 883–892.

Ashby, M.K. (2004). Survey of the number of two-component response regulator genes in the complete and annotated genome sequences of prokaryotes. FEMS Microbiol. Lett. *231*, 277–281.

Ashby, M.K. (2006). Distribution, structure, and diversity of 'bacterial' genes encoding two-component proteins in the Euryarchaeota. Archaea 2, 11–30.

Baker, M.D., Wolanin, P.M., and Stock, J.B. (2005). Signal transduction in bacterial chemotaxis. BioEssays *28*, 9–22.

Bell, S.D., Botting, C.H., Wardelworth, B.N., Jackson, S.P., and White, M.F. (2002). The interaction of Alba, a conserved archaeal chromatin protein, with Sir2 and its regulation by acetylation. Science *296*, 148–151.

Bini, E., and Blum, P. (2001). Archaeal catabolite repression: A gene regulatory paradigm. Adv. Appl. Microbiol. *50*, 39–367.

Bogomolni, R.A., and Spudich, J.L. (1982) Identification of a third rhodopsin-like pigment in phototactic *Halobacterium halobium*. Proc. Nsatl. Acad. Sci. USA *79*, 6250–6254.

Boudko, D., Yu, H.S., Ruiz, M., Hou, S., and Alam, M. (2003). A time-lapse capillary assay to study aerotaxis in the archaeon *Halobacterium salinarium*. J. Microbiol. Meth. 53, 123–126.

Bourret, R.B., and Stock, A.M. (2002). Molecular information processing: Lessons from bacterial chemotaxis. J. Biol. Chem. *277*, 9625–9628

Brooun, A., Bell, J. Freitas, T., Larsen, R.W., and Alam, M. (1998). An archaeal aeotaxis transducer combines subunit I core structures of eukaryotic cytochrome c oxidase and eubacterial methyl-accepting chemotaxis proteins. J. Bacteriol. *180*, 1642–1646.

Brunner, N.A., Siebers, B., and Hensel, R. (2001). Role of two different glyceraldehydes-3-phosphate dehydrogenases in controlling the reversible Embden-Meyerhof-Parnas pathway in *Thermoproteus tenax*: Regulation on protein and transcript level. Extremophiles 5, 101–109.

Brunner, N.A., Brinkmann, H., Siebers, B., and Hensel, R. (1998). NAD+-dependent glyceraldehyde-3-phosphate dehydrogenase from *Thermoproteus tenax*. J. Biol. Chem. *273*, 6149–6156.

Byrnes, W.M., and Vilker, V.L. (2004). Extrinsic factors potassium chloride and glycerol induce thermostability in recombinant anthranilate synthase from *Archaeoglobus fulgidus*. Extremophiles *8*, 45–462.

Cardona, S., Remonsellez, F., Guiliani, N., and Jerez, C.A.. (2001). The glycogen-bound polyphosphate kinase from *Sulfolobus acidocaldarius* is actually a glycogen synthase. Appl. Environ. Microbiol. *67*, 4773–4780.

Chen, X., and Spudich, J.L. (2002). Demonstration of 2:2 stoichiometry in the functional SRI-HtrI signaling complex in *Halobacterium* membranes by gene fusion analysis. Biochemistry *41*, 3891–3896.

Condo, L., Ruggero, D., Reinhardt, R., and Londel, P. (1998). A novel amino-peptidase associated with the 60 kDa chaperonin in the thermophilic archaeon *Sulfolobus solfataricus*. Mol. Microbiol. *29*, 775–785.

Cosgrove, M.S., and Wolberger, C. (2005). How does the histone code work? Biochem. Cell Biol. *83*, 468–476.

Daas, P.J. H., Wassenaar, R.W., Willemsen, P., Theunissen, R.J., Keltjens, J.T., van der Drift, C., and Vogels, G.D. (1996). Purification and properties of an enzyme involved in the ATP-dependent activation of the methanol:2-mercaptoethanesulfonic acid methyltransferase reaction in *Methanosarcina barkeri*. J. Biol. Chem. *271*, 22339–22345.

De Felice, M., Esposito, L., Pucci, B., Carpentieri, F., De Falco, M., Rossi, M., and Pisani, F.M. (2003). Biochemical characterization of a CDC6-like protein from the crenarchaeon *Sulfolobus solfataricus*. J. Biol. Chem. *278*, 46424–46431.

Durbecq, V., Thia-Toong, T.L., Charlier, D., Villeret, V., Roovers, M., Wattiez, R., Legrain, C., and Glansdorff, N. (1999). Aspartate carbamoyltransferase from the thermoacidophilic archaeon *Sulfolobus acidocaldarius*. Eur. J. Biochem. *264*, 233–241.

Eliasson, R., Pontis, E., Jordan, A., and Reichard, P. (1999). Allosteric control of three B12-dependent (class II) ribonucleotide reductases. J. Biol. Chem. *274*, 7182–7189.

Facchin, S., Sarno, S., Marin, O., Lopreiato, R., Sartori, G., and Pinna, L.A. (2002). Acidophilic nature of yeast PID261/BUD32, a putative ancestor of eukaryotic protein kinases. Biochem. Biophys. Res. Commun. *296*, 1366-1371.

Facchin, S., Lopreiato, R., Ruzzene, M., Marin, O., Sartori, G., Gotz, C., Montenarh, M., Carigani, G., and Pinna, L.A. (2003). Functional homology between yeast PID261/BUD32 and human PRPK: Both phosphorylate p53 and PRPK partially complments piD261/Bud32 deficiency. FEBS Lett. *296*, 61–66.

Fischer, R.S., Bonner, C.A., Boone, D.R., and Jensen, R.A. (1993). Clues from a halophilic methanogen about aromatic amino-acid biosynthesis in Archaebacteria. Arch. Microbiol. *160*, 440–446.

Galperin, M.Y. (2005). A census of membrane-bound and intracellular signal transduction proteins in Bacteria: Bacterial IQ, extroverts and introverts. BMC Microbiol. 5, 35.

Galperin, M.Y., Nikolskaya, A.N., and Koonin, E.V. (2001). Novel domains of the prokaryotic two-component signal transduction system. FEMS Microbiol. Lett. *203*, 11–21.

Gast, D.A., Jenal, U., Wasserfallen, A., and Leisinger, T. (1994). Regulation of tryptophan biosynthesis in *Methanobacterium thermoautotrophicum* Marburg. J. Bacteriol. *176*, 4590–4596.

Grabowski, B., and Kelman, Z. (2001). Autophosphorylation of archaeal Cdc6 homologs is regulated by DNA. J. Bacteriol. *183*, 5459–5464.

Hack, E.S., Vorobyova, T., Sakash, J.B., West, J.M., Macol, C.P., Herve, G., Williams, M.K., and Kantrowitz, E.R. (2000). Characterization of the aspartate transcarbamoylase from *Methanococcus jannaschii*. J. Biol. Chem. *275*, 15820–15827.

Haseltine, C., Montalvo-Rodriguex, R., Bini, E., Carl, A. and Blum,P. (1999). Coordinate transcriptional control in the hyperthermophilic archaeon *Sulfolobus solfataricus*. J. Bacteriol. *181*, 3920–3927.

Hildebrand, E., and Dencher, N. (1975). Two photosystems controlling behavioural responses of *Halobacterium halobium*. Nature. *257*, 46–48.

Hoch, J.A. (2000). Two-component and phosphorelay signal transduction. Curr. Opin. Microbiol. 3, 165–170.

Hou, S., Brooun, A., Yu, H.S., Freitas, T., and Alam, M. (1998). Sensory rhodopsin II transducer HrtII is also responsible for serine chemotaxis in the archaeon *Halobacterium salinarium*. J. Bacteriol. *180*, 1600–1602.

Hou, S., Larsen, R.W., Boudko, D., Riley, C.W., Karatan, E., Zimmer, M., Ordal, G.W., and Alam, M. (2000). Myoglobin-like aerotaxis transducers in Archaea and Bacteria. Nature *403*, 540–544.

Huffman, J.L., Li, H., White, R.H., and Tainer, J.A. (2003). Structural basis for recognition and catalysis by the bifunctional dCTP deaminase and dUTPase from *Methanococcus jannaschii*. J. Mol. Biol. *331*, 885–896.

Jensen, K.F., Arent, S., Larsen, S., and Schack, L. (2005). Allosteric properties of the GTP activated and CTP inhibited uracil phosphoibosyltransferase from the thermoacidophilic archaeon *Sulfolobus solfataricus*. FEBS J. 272, 1440–1453.

Jensen, R.A., d'Amato, T.A., and Hochstein, L.I. (1988). An extreme-halophilic archaebacterium possesses the interlock type of prephenate dehydratase characteristic of Gram-positive eubacteria. Arch. Microbiol. *148*, 365–371.

Jeon, S.-J., Fujiwara, S., Takagi, M., Tanaka, M., and Imanaka, T. (2002). Tk-PTP, a protein-tyrosine phosphatase from the hyperthermophilic archaeon *Thermococcus kodakaraensis* KOD1: Enzymatic characteristics and identification of its substrate proteins. Biochem. Biophys. Res. Commun. *295*, 508–514.

Kasiviswanathan, R., Shin, J.-H., and Kelman, Z. (2005). Interactions between the archaeal Cdc6 and MCM proteins modulate their biochemical properties. Nucleic Acids Res. *33*, 4940–4950.

Kennelly, P.J. (2001). Protein phosphatases: A phylogenetic perspective. Chem. Rev. *101*, 2291–2312.

Kennelly, P.J. (2003). Archaeal protein kinases and protein phosphatases – Insights from genomics and biochemistry. Biochem. J. *370*, 373–389.

Kennelly, P.J. (2007). Sensing, signal transduction, and posttranslational modification. In Archaea: Molecular and Cellular Biology, R. Cavicchioli, ed. (Washington DC: ASM Press), pp. 224–259.

Kim, S., and Lee, S.B. (2005). Identification and characterization of *Sulfolobus solfataricus* D-gluconate dehydratase: A key enzyme in the non-phosphorylated Entner–Doudoroff pathway. Biochem. J. *387*, 271–280.

Kisseleva-Romanova, E., Lopreiato, R., Baudin-Baillieu, A., Rouselle, J.-C., Ilan, L., Hofmann, K., Namane, A., Mann, C., and Libri, D. (2006) Yeast homolog of a cancer-testis antigen defines a new transcription complex. EMBO J. *25*, 3576-3585.

Kleeb, A.C., Kast, P., and Hilvert, D. (2006). A monofunctional and thermostable prephenate dehydratase from the archaeon *Methanocaldococcus jannaschii*. Biochemistry. *45*, 14101–14110.

Kokoeva, M.K., and Oesterhelt, D. (2000). BasT, a membrane-bound transducer protein for amino acid detection in *Halobacterium salinarium*. Mol. Microbiol. *35*, 647–656.

Kokoeva, M.V., Storch, K.F., Klein, C., and Oesterhelt, D. (2002). A novel mode of sensory transduction in Archaea: Binding protein-mediated chemotaxis towards osmoprotectants and amino acids. EMBO J. *21*, 2312–2322.

Koretke, K.K., Lupas, A.N., Warren, P.V., Rosenberg, M., and Brown, J.R. (2000). Evolution of the two-component regulatory system. Mol. Biol. Evol. *17*, 1956–1970.

Krasiviswanathan, R., Shin, J.-H., and Kelman, Z. (2005) Interactions between the archaeal Cdc6 and MCM proteins modulate their biochemical properties. Nucleic Acids Res. *33*, 4940–4950.

LaRonde-LaBlanc, N., and Wlodawer, A. (2004). Crystal structure of A. fulgidus Rio2 defines a new family of serine protein kinases. Structure *12*, 1585–1594.

LaRonde-LeBlanc, N., Guszczynski, T., Copeland, T., and Wlodawer, A. (2005a). Autophosphorylation of *Archaeoglobus fulgidus* Rio2 and crystal structures of its nucleotide-metal ion complexes. FEBS J. *272*, 2800–2810.

LaRonde-LeBlanc, N., Guszczynski, T., Copeland, T., and Wlodawer, A. (2005b). Structure and activity of the atypical serine kinase Rio1. FEBS J. *272*, 3698–3713.

Lemker, T., Gruber, G., Schmid, R., and Muller, V. (2003). Defined subcomplexes of the A1 ATPase from the archaeon *Methanococcus mazei* Go1: Biochemical properties and redox regulation. FEBS Lett. *544*, 206–209.

Leng, J., Cameron, A.J., Buckel, S., and Kennelly, P.J. (1995). Isolation and cloning of a protein-serine/threonine phosphatase from an archaeon. J. Bacteriol. *177*, 6510–6517.

Leonard, C.J., Aravind, L., and Koonin, E.V. (1998). Novel families of putative protein kinases in Bacteria and Archaea: Evolution of the 'eukaryotic' protein kinase superfamily. Genome Res. *8*, 1038–1047.

Li, H., Xu, H., Graham, D.E., and White, R.H. (2003). The *Methanococcus jannaschii* dCTP deaminase is a bifunctional deaminase and diphosphatase. J. Biol. Chem. *278*, 11100–11106.

Lopreiato, R., Facchin, S., Sartori, G., Arrigoni, G., Casonato, S., Ruzzene, M., Pinna, L.A., and Carignani, G. (2004) Analysis of the internaction between piD261/Bud32, an evolutionarily conserved protein kinase of *Saccharomyces cerevisiae*, and the Grx4 glutaredoxin. Biochem. J. *377*, 395–405.

Lorentzen, E., Hensel, R., Knura, T., Ahmed, H., and Pohl, E. (2004). Structural basis of allosteric regulation and substrate specificity of the non-phosphorylating glyceraldehyde 3-phosphate dehydrogenase from *Thermoproteus tenax*. J. Mol. Biol. *341*, 815–828.

Lower, B.H., and Kennelly, P.J. (2002). The membrane-associated protein-serine/threonine kinase from *Sulfolobus solfataricus* is a glycoprotein. J. Bacteriol. *184*, 2614–2619.

Lower, B.H., and Kennelly, P.J. (2003). Open reading frame sso2387 from the archaeon *Sulfolobus solfataricus* encodes a polypeptide with protein-serine kinase activity. J. Bacteriol. *185*, 3436-3445.

Lower, B.H., Bischoff, K.M., and Kennelly, P.J. (2000). The archaeon *Sulfolobus solfataricus* contains a membrane-associated protein kinase that preferentially phosphorylates threonine residues. J. Bacteriol. *182*, 3452–3459.

Lower, B.H., Potters, M.B., and Kennelly, P.J. (2004). A phosphoprotein from the archaeon *Sulfolobus solfataricus* with protein-serine kinase activity. J. Bacteriol. *186*, 463–472.

Lundgren, M., and Bernander, R. (2007). Genome-wide transcription map of an archaeal cell cycle. PNAS. *104*, 2939–2944.

Mai, B., Frey, G., Swanson, R.V., Mathur, E.J., and Stetter, K.O. (1998). Molecular cloning and functional expression of a protein-serine/threonine phosphatase from the hyperthermophilic archaeon Pyrodictium abyssi TAG11. J. Bacteriol. *180*, 4030–4035.

Manning, G., Whyte, D.B., Martinez, R., Hunter, T., and Sudarsanam, S. (2002). The protein kinase complement of the human kinase. Science *298*, 1912–1934.

Marsh, V.L., Peak-Chew, S.Y., and Bell, S.D. (2005). Sir2 and the acetyltransferase, Pat, regulate the archaeal chromatin protein, Alba. J. Biol. Chem. *280*, 21122–21128.

Marwan, W., Schafer, W., and Oesterhelt, D. (1990). Signal transduction in *Halobacterium* depends on fumarate. EMBO J. *9*, 355–362.

Matsubara, M., and Mizuno, T. (2000). The SixA phospho-histidine phosphatase modulates the ArcB phosphorelay signal transduction in *Escherichia coli*. FEBS Lett. *470*, 118–124.

Montrone, M., Oesterhelt, D., and Marwan, W. (1996). Phosphorylation-independent bacterial chemoresponses correlate with changes in the cytoplasmic level of fumarate. J. Bacteriol. *178*, 6882–6887.

Montrone, M., Marwan, W., Grunberg, H., Mubeleck, S., Starostzik, C., and Oesterhelt, D. (1993). Sensory rhodopsin-controlled release of the switch factor fumarate in *Halobacterium salinarium*. Mol. Microbiol. *10*, 1077–1085.

Norberg, P., Kaplan, J.G., and Kushner, D.J. (1973). Kinetics and regulation of the salt-dependent aspartate transcarbamylase of *Halobacterium cutirubrum*, J. Bacteriol. *113*, 680–686.

Oppermann, U.C. T., S. Knapp, V. Bonetto, R. Ladenstein, H. Jornvall. 919980. Isolation and structure of repressor-like proteins from the archaeon *Sulfolobus solfataricus*. FEBS Lett. *432*, 141–144.

Perazzona, B., and Spudich, J.L. (1999). Identification of methylation sites and effects of phototaxis stimuli on transducer methylation in *Halobacterium salinarium*. J. Bacteriol. *181*, 5676–5683.

Ponting, C.P., Aravind, L., Schultz, J., Bork, P., and Koonin, E.V. (1999). Eukaryotic signaling domain homologues in Archaea and Bacteria. Ancient ancestry and horizontal gene transfer. J. Mol. Biol. *289*, 729–745.

Porat, I., Waters, B.W., Teng, Q., and Whitman, W.B. (2004). Two biosynthetic pathways for aromatic amino acids in the archaeon *Methanococcus maripaludis*. J. Bacteriol. *186*, 4940–4950.

Purcarea, C., G. Herve, M.M. Ladjimi, and R. Cunin. (1997). Aspartate transcarbamylase from the deep-sea hyperthermophilic archaeon *Pyrococcus abyssi*: Genetic organization, structure, and expression in *Escherichia coli*. J. Bacteriol. *179*, 4143–4157.

Porat, I., Sieprawska-Lupa, M., Teng, Q., Bohanon, F.J., White, R.H., and Whitman, W.B. (2006). Biochemical and genetic characterization of an early step in a novel pathway for the biosynthesis of aromatic amino acids and p-aminobenzoic acid in the archaeon *Methanococcus maripaludis*. Mol. Microbiol. *62*, 1117–1131.

Ray, W.K., Keith, S.M., DeSantis, A.M., Hunt, J.P., Larson, T.J., Helm, R.F., and Kennelly, P.J. (2005). A phosphohexomutase from the archaeon *Sulfolobus solfataricus* is covalently modified by phosphorylation on serine. J. Bacteriol. *187*, 4270–4275.

Rohlin, L., Trent, J.D., Salmon, K., Kim, U., Gunsalus, R.P., and Liao, J.C. (2005). Heat shock response of *Archaeoglobus fulgidus*. J. Bacteriol. *187*, 6046–6057.

Roosild, T.P., Miller, S., Booth, I.R., and Choe, S. (2002). A mechanism of regulating transmembrane potassium flux through a ligand-mediated conformational switch. Cell. *109*, 781–791.

Rotharmel, T., and Wagner, G. (1995). Isolation and characterization of a calmodulin-like protein from *Halobacterium salinarium*. J. Bacteriol. *177*, 864–866.

Rother, M., Boccazzi, P., Bose, A., Pritchett, M.A., and Metcalf, W.W. (2005). Methanol-dependent gene expression demonstrates that methyl-coenzyme M reductase is essential for *Methanosarcina acetivorans* C2A and allows isolation of mutants with defects in regulation of the methanol utilization pathway. J. Bacteriol. *187*, 5552–5559.

Rudiger, A., Jorgensen, P.L., and Antranikian, G. (1995). Isolation and characterization of a heat-stable pullulanase from the hyperthermophilic archaeon *Pyrococcus woesei* after cloning and expression of its gene in *Escherichia coli*. Appl. Environ. Microbiol. *61*, 567–575.

Rudolph, J., and Oesterhelt, D. (1995). Chemotaxis and phototaxis require a CheA histidine kinase in the archaeon *Halobacterium salinarium*. EMBO J. *14*, 667–673.

Sandman, K., and Reeve, J.N. (2005). Archaeal chromatin proteins: different structures but common function? Curr, Opin. Microbiol. *8*, 656–661.

Schimz, A. (1981). Methylation of membrane proteins is involved in chemosensory and photosensory behavior in *Halobacterium halobium*. FEBS Lett. *125*, 205–207.

Schmiz, A., and Hildebrand, E. (1979). Chemosensory responses of *Halobacterium halobium*. J. Bacteriol. *140*, 749–753.

Schofield, L.R., Patchett, M.L., and Parker, E.B. (2003). Expression, purification, and characterization of 3-deoxy-D-arabino-heptulosonate 7-phosphate synthase from *Pyrococcus furiosus*. Prot. Exp. Purif. *34*, 17–27.

Schofield, L.R., Anderson, B.F., Patchett, M.L., Norris, G.E., Jameson, G.B., and Parker, E.J. (2005). Substrate ambiguity and crystal structure of *Pyrococcus furiosus* 3-Deoxy-D-arabino-heptulosonate-7-phosphate synthase: an ancestral 3-Deoxyald-2-ulosonate-phosphate synthase? Biochemistry, *44*, 11950–11962.

Solow, B., Young, J.C., and Kennelly, P.J. (1997). Gene cloning and expression and characterization of a toxin-sensitive protein phosphatase from the methanogenic archaeon *Methanosarcina thermophila* TM-1. J. Bacteriol. 179, 5072–5075.

Spudich, J.L. (2006). The mulitialented microbial sensory rhodopsins. Trends Microbiol. *14*, 480–487.

Spudich, J.L., and Stoeckenius, W. (1979). Photosensory and chemosensory behavior of *Halobacterium halobium*. Photobiochem. Photobiophys. *1*, 43–53.

Spudich, E.N., Takahashi, T., and Spudich, J.L. (1989). Sensory rhodopsins I and II modulate the methylation/demethylation system in *Halobacterium halobium* phototaxis. Proc. Natl. Acad. Sci. USA *86*, 7746–7750.

Stieglitz, K.A., Seaton, B.A., Head, J.F., Stec, B., and Roberts, M.F. (2003). Unexpected similarity in regulation between an archaeal inositol monophosphatase/fructose bisphosphatase and chloroplast fructose bisphosphatase. Protein Sci. *12*, 760–767.

Stoeckenius, W., Wolff, E.F., and Hess, B. (1988). A rapid population method for action spectra applied to *Halobacterium halobium*. J. Bacteriol. *170*, 2790–2795.

Storch, K.F., Rudolph, J., and Oesterhelt, D. (1999). Car: a cytoplasmic sensor responsible for arginine chemotaxis in the archaeon *Halobacterium salinarum*. EMBO J. 18, 1146-1158.

Szurmant, H., and Ordal, G.W. (2004). Diversity of chemotaxis mechanisms among the Bacteria and Archaea. Microbiol. Mol. Biol. Rev. *68*, 301–319.

Tahara, M., Ohsawa, A., Saito, S., and Kimura, M. (2004). In vitro phosphorylation of initiation factor 2a (aIF2a) from hyperthermophilic archaeon *Pyrococcus horikoshii* OT3. J. Biochem. (Tokyo) *135*, 479–485.

Tang, X.F., Ezaki, S., Atomi, H., and Imanaka, T. (2001). Anthranilate synthase without an LLES motif from a hyperthermophilic archaeon is inhibited by tryptophan. Biochem. Biophys. Res. Commun. *281*, 858–865.

Tutino, M.L., Tosco, A., Marino, G., and Sannia, G. (1997) Expression of *Sulfolobus solfataricus* trpE nd trpG genes in E. coli. Biochem. Biophys. Res. Commun. *230*, 306-310.

Ulrich, L.E., Koonin, E.V., and Zhulin, I.B. (2005). One-component systems dominate signal transduction in prokaryotes. Trends Microbiol. *13*, 52–56.

Van Boxstael, S., Cunin, S., Khan, R., and Maes, D. (2003). Aspartate transcarbamoylase from the hyperthermophilic archaeon *Pyrococcus abyssi*: Thermostability and 1.8A resolution crystal structure of the catalytic subunit complexed with the bisubstrate analogue N-phosphonacetyl-L-aspartate. J. Mol. Microbiol. *326*, 203–216.

Van Boxstael, S., Maes, D., and Cunin, R. (2005). Aspartate transcarbamoylase from the hyperthermophilic archaeon *Pyrococcus abyssi*. Insights into cooperative and allosteric mechanisms. FEBS J. *272*, 2670–2683.

Verhees, C.H., Kengen, S.W. M., Tuininga, J.E., Schut, G.J., Adams, M.W. W., de Vos, W.M., and van der Oost, J. (2003). The unique features of glycolytic pathways in Archaea. Biochem. J. *375*, 231–246.

Vermeij, P., Vinke, E., Keltjens, J.T., and van Der Drift, C. (1995). Purification and properties of coenzyme F390 hydrolase from *Methanobacterium thermoautotrophicum* (strain Marburg). Eur. J. Biochem. *234*, 592–597.

Vermeij, P., Detmers, F.J., Broers, F.J., Keltjens, J.T., and van der Drift, C. (1994). Purification and characterization of coenzyme F390 synthetase from *Methanobacterium thermoautotrophicum* (strain delta H). Eur. J. Biochem. *226*, 185–191.

Wallin, E., and Von Heijne, G. (1998). Genome-wide analysis of integral membrane proteins from eubacterial, archaean, and eukaryotic organisms. Protein Sci. *7*, 1029–1038.

White, M.F., and Bell, S.D. (2002). Holding it together: Chromatin in the Archaea. Trends Genet. *18*, 621–626.

Xing, R., and Whitman, W.B. (1987). Sulfometuron methyl-sensitive and -resistant acetolactate synthases of the archaebacteria *Methanococcus* spp., J, Bacteriol. *169*, 4486–4492.

Zhulin, I.B., Nikolskaya, A.N., and Galperin, M.Y. (2003). Common extracellular sensory domains in transmembrane receptors for diverse signal transduction pathways in Bacteria and Archaea. J. Bacteriol. *185*, 285–294

Functional Genomics of Stress Response in Extremophilic Archaea

10

Sabrina Tachdjian, Keith R. Shockley, Shannon B. Conners, and Robert M. Kelly

Abstract

Given that some members of the Archaea inhabit extreme environments while others occupy more hospitable biological niches, the mechanisms by which this group of microorganisms handles stress are well worth examining for clues to prokaryotic 'disaster' management. The availability of genome sequence information for a number of Archaea has provided a glimpse into the genetic basis for thriving at extremes of temperature, salinity, pH and metal concentrations. Such information has furthermore enabled the application of 'omics' tools to examine the phenotype that results from perturbations that trigger both global and focused survival strategies.

Introduction

Environmental perturbations that result in nucleic acid, protein or membrane damage that significantly alter cellular function can trigger a strategic response designed to improve chances of survival. The so-called 'stress response' comprises a set of mechanisms that, among other things, enables cells to minimize damage under adverse circumstances by repairing or degrading dysfunctional macromolecules, thereby adapting to new environmental conditions (Valentine, 2007). For some microorganisms, such as the extremophilic Archaea, otherwise debilitating conditions are actually optimal. These Archaea thrive at biologically extreme temperatures, salinity, pH, and toxic metals and have been a source of scientific curiosity for decades. But, for such microorganisms, what constitutes 'stress' and what strategies do they invoke when exposed to even more biologically extreme conditions? To date, investigations along these lines have focused on specific genes and proteins as either the weakest link governing survival or as elements of a cellular rescue strategy which seeks to bring order to the cytosol and cellular envelope in the face of destructive forces. Much of the early microbial genome sequencing efforts focused on Archaea because of their small genomes and for clues to evolutionary processes. As a consequence, this facilitated functional genomics efforts to explore global strategies for survival to stress. Such studies have provided complementary information to efforts to understand the role and fate of specific genes and proteins. Among the many interesting findings concerning stress response in Archaea is the realization that, while common elements thread through stress response mechanisms in Archaea, there are also 'customized' features that relate to the type of environmental perturbation encountered. The focus here will be on functional genomics approaches to understand both focused and global strategies to handle stress in extremophilic Archaea.

Archaeal heat shock proteins

Heat shock response mechanisms have been comprehensively examined in Bacteria (Arsene *et al.*, 2000; Hecker *et al.*, 1996; Rosen and Ron, 2002). While some aspects are conserved in Archaea, significant differences exist, especially with respect to families of heat shock proteins (HSPs) (Table 10.1).

Table 10.1 Comparison of archaeal and bacterial HSPs and stress proteins

Archaeal stress proteins	Role in Archaea	Bacterial equivalent	Role in Bacteria
Hsp40*	Binds and tags denatured protein	DnaJ	Binds and tags denatured protein
Hsp70*	Binds Hsp40-protein complex	DnaK	Binds DnaJ–protein complex
Hsp23/GrpE*	- Interacts with Hsp40-protein-Hsp70 complex - ADP/ATP exchange factor	GrpE	– Interacts with DnaJ-protein-DnaK complex – ADP/ATP exchange factor
Type I Hsp60**	Protein folding (Klunker *et al.*, 2003)	GroEL	Protein folding
Type II Hsp60 (also referred to as: Thermosome, ths, rosettasome, TF55, CCT, TRiC, Cpk, MMTS, HHSP)	– Chaperonin activity (Macario and de Macario, 1999; Macario *et al.*, 1999) – Refolding, protein stabilization, protein modification – Structural properties: filament formation, membrane scaffolding (Trent *et al.*, 2003)	none	N/A
Prefoldin	Binds and stabilizes unfolded target polypeptides, delivers them to type II Hsp60	none	N/A
sHsp	Chaperonin: refolding, prevention of protein aggregation	sHsp	Chaperonin: refolding, prevention of protein aggregation
HptX	None characterized – probably proteolytic	HtpX	Degradation of membrane proteins
Proteasome	Protein degradation	Clp	Protein degradation

*Not present in hyperthermophilic Archaea.
**Only present in *Methanosarcinales.*

HSP70, HSP40 and HSP23

These proteins correspond to DnaK, DnaJ and GrpE, respectively, in the bacterial molecular chaperone machine. However, they are not present in all Archaea and are typically absent in hyperthermophiles. For this reason, it has been suggested that DnaK, DnaJ, and GrpE are involved in protein folding at mesophilic temperatures (Cavicchioli *et al.*, 2000). The genes encoding HSP70, HSP40, and HSP23 in Archaea are consistently grouped in an operon akin to the organization found for bacterial heat shock proteins. Whether some Archaea received this group of genes via lateral gene transfer or the operon was initially present in Archaea but was subsequently lost in hyperthermophiles is not clear. However, the lack of bacterial-type intervening sequences in the operon suggests the presence of a completely different regulation mechanism than in Bacteria (Clarens *et al.*, 1995). Information from transcriptional response analysis may shed light on the regulation of these heat shock proteins in the Archaea.

HSP60 (thermosome/rosettasome)

A unique member of the archaeal heat shock protein family is the chaperonin HSP60, also referred to as the 'thermosome' (Phipps *et al.*, 1993) or 'rosettasome' (Kagawa *et al.*, 1995). This protein is similar in size to its bacterial counterpart but otherwise distinct in terms of structure and function. There are actually two types of Hsp60 chaperonin systems: group I HSP60 (GroEL/ES) which is typically bacterial, and group II HSP60 (thermosome/rosettasome), a protein of archaeal origin but related to TCP1 in eukaryotes (Trent *et al.*, 1991). At first, group I HSP60 chaperonins were thought to occur exclusively in Bacteria or in the organelles of eukaryotes, indicative of their bacterial origin. The fact that GroEL/ES was later identified in *Methanococcus mazei* (Klunker *et al.*, 2003), a mesophilic archaeon with significant lateral gene transfer (~20–35%) from Bacteria (Deppenmeier *et al.*, 2002), does not necessarily contradict a bacterial origin of group I chaperonins. The biological role of GroEL/ES as a chaperonin has been thoroughly investigated in Bacteria. While there are *in vitro* indications that group II HSP60 may play a similar role in Archaea (Minuth *et al.*, 1998), its unusual localization in the membrane (as opposed to the cytoplasm) (Trent *et al.*, 2003) and macromolecular self-assembly properties (McMillan *et al.*, 2005) suggest a cytostructural function. For example, the thermosome/rosettasome may contribute to the preservation of cell membrane integrity under heat stress in Archaea, serving as a reinforcing scaffold to minimize leakage into or out of the cytosol. In any case, the thermosome/rosettasome is a predominant component of hyperthermophilic archaeal thermal stress response (Han, 1997; Shockley *et al.*, 2003; Tachdjian and Kelly, 2006).

Prefoldin

Prefoldin (PFD), another multimeric molecular chaperone, interacts with HSP60. Eukaryal PFDs bind selectively to a small subset of proteins, including actin and tubulin, in the process of delivering them to the chaperonin-containing TCP-1 (CCT) (Vainberg *et al.*, 1998). In Archaea, a similar role has been shown for PFD *in vitro*, although high subunit flexibility allows binding to a much wider range of substrates (Martin-Benito *et al.*, 2007). While PFDs from both domains seem to perform the same cellular function, interaction mechanisms between the thermosome and protein substrates diverge to a certain extent. Indeed, structural characterization of the *Pyrococcus horikoshii* PFD (Martin-Benito *et al.*, 2007) has confirmed that the chaperone interacts with the thermosome through its tentacle tips, and that the latter does not penetrate the thermosome cavity to the extent observed for their eukaryal counterpart. Furthermore, binding only occurs on one side of the thermosome at a time (Fig. 10.1). The substrate-binding mechanism also differs, since it does not require encapsulation of denatured proteins as in Eukarya. Instead, contact occurs only through tentacle tips, at a binding site that is close – yet distinct – from the thermosome docking site (Okochi *et al.*, 2004). Because of its relative simplicity (two different subunits compared with six in Eukarya), and lack of substrate specificity, the archaeal PFD can be thought of as a primitive version of the eukaryotic PFD, with high specificity and subunit differentiation arising later in the Eukarya.

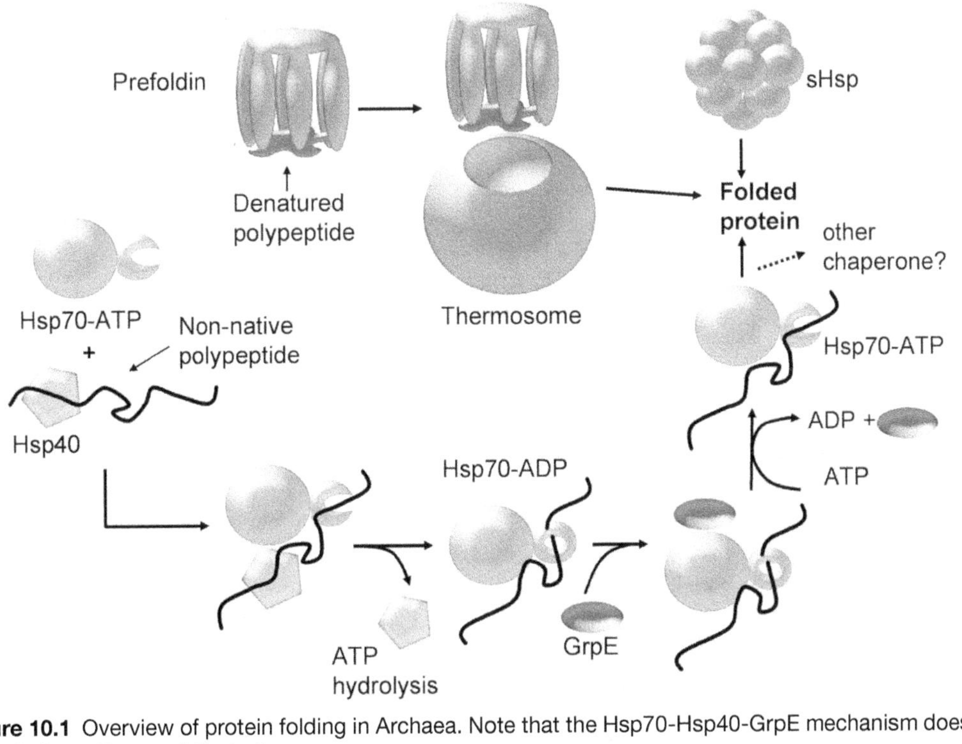

Figure 10.1 Overview of protein folding in Archaea. Note that the Hsp70-Hsp40-GrpE mechanism does not apply to hyperthermophilic Archaea.

Small heat shock proteins (sHSPs)

Small heat shock proteins (sHSPs), typically based on subunits of less than 30~40 kDa, are found in all three domains of life. While common in extremophilic Archaea, only hyperthermophilic versions have been characterized. sHSPs act as chaperones to prevent aggregation of denatured proteins (Ahrman *et al.*, 2007; Jakob *et al.*, 1993; Laksanalamai *et al.*, 2001). The archaeal sHSPs belong to the same family as human alpha-crystallin, a protein component of the eye lens. Study of sHSP regulation in *P. furiosus* first showed that it was heat-shock inducible (Laksanalamai *et al.*, 2001), and led to the identification of heat shock regulator Phr and a cis-acting regulatory sequence (Vierke *et al.*, 2003). Phr is a constitutively expressed repressor that binds the *P. furiosus*-sHSP promoter region, thereby blocking RNA polymerase recruitment. Resolution of the Phr crystal structure, complemented by computational and mutational analyses, showed that the repressor uses a novel DNA-binding mode which may be Archaea-specific (Liu *et al.*, 2007).

The crystal structure of the sHSP in *Methanocaldococcus jannaschii* revealed a 24-subunit homo-oligomeric arrangement into a hollow sphere with multiple 'pores' (Kim *et al.*, 1998a). Comparison with the wheat sHSP structure showed that, while both monomers share low amino-acid sequence homology and have different quaternary structures, monomeric folds are strikingly similar across domains (Laksanalamai and Robb, 2004). The monomer's carboxy- and amino-terminal domains both play a role in self-assembly of the sHSP complex and oligomerzation as well as in chaperoning (Sun and MacRae, 2005), while the central α-crystallin domain is not significantly involved in the overall quaternary structure. Although the chaperone activity of sHSPs is not in doubt, little is known about their mechanism of action *in vivo*. One puzzling fact is that the surface pores are too small to allow entry of the substrate into the sphere's cavity. Hence, it has been proposed that the multimeric form of sHSP performs a cytosolic storage function, while chaperone activity is affected by monomers. Another possibility is for complex assembly to take place after substrate binding, thereby allowing encapsulation of target proteins (Kim *et al.*, 1998b).

Energy-dependent proteases: the proteasome and Lon

Proteolysis is an essential feature of stress response. Not only do intracellular proteases degrade denatured and misfolded proteins (Esser *et al.*, 2004), they also play a regulatory role in ensuring dynamic adjustment of the cytosolic proteome (Maupin-Furlow *et al.*, 2005). In some ways, the 26S proteasome can be viewed as the archaeal counterpart to bacterial protease complexes (e.g. the Clp system). It has a multimeric, stacked ring structure (i.e. two β_7 rings sandwiched between two α_7 rings), with a central cavity in which polypeptide digestion takes place. This cylindrical proteolytic core complex (the 20S proteasome or core particle) is associated with a 19S homo-oligomeric complex of 6 AAA-family ATPase mutlimeric structure, referred to as the proteasome activating nucleotidase (PAN). ATP binding promotes the docking of PAN into the proteasome's α-rings, opening the gate to allow substrate entry. Proteolysis subsequently takes place without ATP hydrolysis, the latter only being required for unfolding globular proteins (Smith *et al.*, 2006). Although the proteasome's presence in the cytosol is maintained under normal and abnormal growth conditions, subunit composition of the 20S core particle appears to be dynamically adjusted in response to environmental conditions and growth phase (Shockley *et al.*, 2003; Kaczowka and Maupin-Furlow, 2003). This phenomenon is thought to influence stability (Madding *et al.*, 2007) and may control gate opening, interactions with AAA regulatory proteins and substrate specificity (Maupin-Furlow *et al.*, 2005).

Another energy-dependent proteolytic complex ubiquitous to all three domains of life is Lon. While Lon protease domain composition varies greatly in complexity from one organism to another, all have in common the presence of a highly conserved Lon domain. Membrane association is also a typical feature of archaeal Lon proteases, as suggested by the presence of signal peptides and/or transmembrane domains (Maupin-Furlow *et al.*, 2005). Several archaeal LonB homologues have been characterized and crystallized, thus shedding additional light the proteolysis mechanism (Besche *et al.*, 2004; Im *et al.*, 2004).

Similar to Lon, HtpX is membrane-bound ATP-dependent protease involved in protein turnover in Archaea. In *E. coli*, HtpX is a membrane-bound protease that most likely plays a role in degradation of denatured membrane proteins (Sakoh *et al.*, 2005). No archaeal version of HtpX has yet been characterized.

Functional genomics studies of stress response in Archaea

The post-genomics era has provided research tools that enable comprehensive analysis of cellular response for the entire cell or organism. These tools have been used to study various types of stress response in the Archaea (Table 10.2).

Thermal stress response

Most functional genomics efforts to date on stress response in Archaea have focused on heat shock. The transcriptional response of the hyperthermophilic archaeon *P. furiosus* to a temperature shift from 90 to 105°C was followed using a targeted cDNA microarray. Results illustrated that *P. furiosus* employs a thermal stress response strategy based on chaperones (HSP60, HSP20), proteolysis, and compatible solute formation (Shockley *et al.*, 2003). In *Archaeoglobus fulgidus*, a whole genome cDNA microarray was used to track transcriptional response over the course of an hour for a temperature shift from 78 to 89°C (Rohlin *et al.*, 2005). Included, among the 14% of the genome that was differentially transcribed following the shift, was the gene encoding a heat shock regulator, *hsr*1. DNA foot-printing analysis revealed a *cis*-binding motif in the promoter region, and showed that HSR1 was responsible for regulating an operon encoding Hsp20 and AAA ATPase cdc48. Thermal stress response of *Methanocaldococcus jannaschii* to temperature shifts from 85 to 65°C and 85 to 95°C showed that protein folding and degradation pathways are mobilized differently during heat shock and cold shock (Boonyaratanakornkit *et al.*, 2005). In contrast to the hyperthermophiles *A. fulgidus*, *P. furiosus*, and *M. jannaschii*, the mesophilic archaeon *Methanosarcina barkeri* genome encodes both HSP70 and HSP60, and seems to induce expression of both sets of genes in the face of thermal stress (Zhang *et al.*, 2006).

Global transcriptional response analysis to thermal shifts can reveal cellular strategies for stress management that go beyond well-characterized folding and proteolysis pathways. For

Table 10.2 Summary of genomics and proteomics studies of stress response in Archaea

Type of stress	Organism	Type of study	References
Temperature	*S. solfataricus*	HS, Transcriptome	(Tachdjian and Kelly, 2006)
	A. fulgidus	HS, Transcriptome	(Rohlin *et al.*, 2005)
	P. furiosus	HS, Transcriptome	(Shockley *et al.*, 2003)
	M. jannaschii	CS, Transcriptome	(Boonyaratanakornkit *et al.*, 2005)
	M. barkeri	HS, Transcriptome	(Zhang *et al.*, 2006)
	Halobacterium NRC-1	HS, Proteome	(Shukla, 2006)
DNA damage (UV, Gamma)	*P. furiosus*	Transcriptome	(Williams *et al.*, 2006)
	Halobacterium NRC-1	UV & Gamma, Transcriptome, LC-MS, systems approach to data analysis	(Baliga *et al.*, 2004; Reich *et al.*, 2001)
Metals	*Halobacterium* NRC-1	Transcriptome, systems approach to data analysis	(Kaur *et al.*, 2006)
	S. solfataricus	Nickel, Transcriptome	(Tachdjian, Stewart, and Kelly unpublished data)
pH	*S. solfataricus*	Acid, Transcriptome	(Tachdjian and Kelly, 2006)
	S. solfataricus	Base, Transcriptome	(Tachdjian & Kelly, unpublished data)
Starvation	*M. mazei*	Nitrogen, Transcriptome	(Veit *et al.*, 2006)
Oxidative	*M. barkeri*	Transcriptome	(Zhang *et al.*, 2006)

instance, in some cases, temperature shifts were found to trigger transcription of genes encoding insertion (IS) elements (Boonyaratanakornkit *et al.*, 2005; Tachdjian and Kelly, 2006) and oxidative stress (Zhang *et al.*, 2006). The basis for heightened expression of IS elements is not clear as yet but a possible role in a stress-induced increase in genome plasticity has been proposed (Tachdjian and Kelly, 2006).

Few studies have utilized advanced proteomics tools for tracking thermal stress response in Archaea. In one example of differential proteomics analysis, heat shock of *Halobacterium salinarum* NRC-1 was studied using a combination of 2D-gel electrophoresis, LC-MS/MS, and MALDI-TOF, showing that a shift from 42 to 49°C triggered the up-regulation – among other genes – of 5 putative HSPs (DnaJ, GrpE, sHSP-1, HSP-5 and sHSP-2) (Shukla, 2006). As proteomics tools become further developed and accessible, they will provide insightful complementary information to transcriptional response analysis.

Global dynamic transcriptional response of P. furiosus to thermal stress

Most reported genome-wide thermal stress response experiments to date have focused on differences in transcript levels before and after heat shock. However, it is clear that heat shock response is a dynamic process that has elements related to growth environment (e.g. media composition) as well as the degree and exposure time of the culture to unfavourable conditions (Pysz *et al.*, 2004). To consider this issue for a hyperthermophilic archaeon, a whole genome cDNA microarray was used to track the dynamic heat shock response (90 to 105°C shift) of *P. furiosus* grown in a 14-litre fermentor on carbohydrate (maltose)- or peptide (tryptone)-based media, both in the presence of elemental sulphur (S°) (Shockley, unpublished data). Samples for transcriptional analysis were obtained prior to heat shock (baseline), when cultures reached 105°C (0 min), and then 5, 30, 60 and 90 min after reaching 105°C. As shown in Fig. 10.2, there is a clear time dependency of thermal stress response. The bulk of differential transcription takes places between 5 and 30 minutes following temperature perturbation, with only small changes noted after 60 minutes. Reflected in the transcriptome dynamics are features that point to both gradual and abrupt cessation of cellular functions. While DNA and RNA metabolism was initially stimulated immediately following temperature shift, perhaps as a biocatalytic response to the

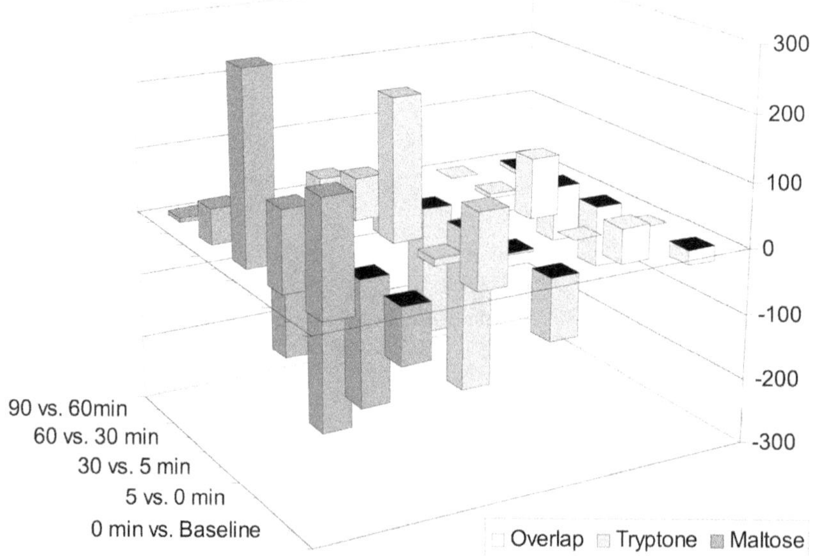

Figure 10.2 Dynamic heat shock response in *P. furiosus* grown on maltose or tryptone. Positive and negative numbers refer to genes up- and down-regulated, respectively. 'Overlap' bars indicate the number of genes responding in the same direction for both substrates.

increased temperature, other cellular functions are affected at longer times. Indeed, ribosomal proteins were down-regulated at the later time points, as were several translation initiation or elongation factors. DNA replication and damage was also apparently an issue, as several genes involved in DNA repair and methylation were transiently induced. Stress response past 60 minutes following temperature shift gave all indications that cellular functions had ceased or been minimized. While heat shock proteins (HSP60, HSP20) are induced immediately after heat shock, transcript levels trailed off at longer times.

Compatible solute biosynthesis

Hyperthermophilic Archaea are known to accumulate compatible solutes in response to thermal stress (Santos and da Costa, 2002); indeed, the thermoprotectant role of di-myo inositol phosphate (DIP), a canonical compatible solute in Archaea, was first identified in *P. furiosus* (Martins *et al.*, 1996). The proposed biosynthesis pathway for DIP in *Pyrococcus woesei* involves two enzymes: L-myo inositol 1-phosphate synthase (encoded by PF1616 in *P. furiosus*) and a putative and yet unidentified DIP synthetase (Scholz *et al.*, 1998). In the dynamic *P. furiosus* heat shock experiment, PF1616 was induced almost immediately on both tryptone and maltose following temperature shift. At longer times, *P. furiosus* transcription of PF1616 decreased, following the trend for many biosynthetic pathways upon prolonged exposure to thermal stress (Shockley, unpublished data).

Response of insertion elements.

In the hyperthermophilic crenarchaeon *S. solfataricus*, insertion (IS) element ORFs are heat shock-inducible (Tachdjian and Kelly, 2006). While there are many fewer IS elements encoded in the *P. furiosus* genome (we have identified 29 thus far) than in *S. solfataricus* (several hundred), they responded significantly to thermal stress. Chromosome shuffling has been proposed to explain the presence of large-scale genomic rearrangements of *P. furiosus* with respect to other *Pyrococcus* species (Zivanovic *et al.*, 2002). As such, it may be that under thermal stress *P. furiosus* is subjected to genome rearrangement, resulting in population diversification.

Cold shock

While the response to stress induced by suboptimal temperatures has, overall, been less studied than the effect of heat shock, this area has recently been gathering interest. This subject is all the more relevant considering that some archaeal species are cold-adapted, and that the way Archaea deal with cold stress appears to be unique among prokaryotes. For instance, cold shock proteins (Csps) are cold-inducible nucleic acid-binding proteins that are widespread in Bacteria, but only a small number of psychrophiles and mesophiles have Csp homologues, which are virtually absent from thermophiles. An archaeal Csp structure has however been solved and the protein has been characterized, indicating a similar role to that played by bacterial Csps (Giaquinto *et al.*, 2007). Proteomic analyses of *Methanococcoides burtonii* cultures grown at 4°C and T_{opt} (23°C) complemented by quantitative PCR revealed an effect of thermal stress on transcriptional regulation, protein folding, and energy generation (Goodchild *et al.*, 2004).

The study of cold shock response has not been limited to psychrophilic Archaea. Indeed, a transcriptional and proteomic study of cold shock (95 to 72°C) in *Pyrococcus furiosus* over the course of 5 hours allowed the identification of two cold-induced proteins or Cip which are widespread in Archaea and seem to be involved in solute binding (Weinberg *et al.*, 2005). The mechanism by which Cip proteins may provide cryoprotection is, however, still not clear.

Toxin–antitoxin (TA) loci as elements of stress response

TA loci contain pairs of consecutive ORFs encoding a toxin and its antidote (antitoxin), which can bind to the former molecule and disables its function as a ribonuclease (Gerdes *et al.*, 2005). The toxin is thought to interfere with the translational machinery by degrading messenger RNAs and/or interacting with the ribosome. TA loci are widespread in microbial genomes (Pandey and Gerdes, 2005). Among the eight types of TA systems identified, the most represented, although least characterized, is the VapBC family. VapC toxins are characterized by the presence of a Pil-N terminus domains (PIN), which functions as a nuclease (Arcus *et al.*, 2004), consistent with a putative role in mRNA degradation. VapBC TA loci expression seems to be heavily regulated and likely plays an important role in stress response (Tachdjian and Kelly, 2006).

DNA damage

DNA damage has been investigated using functional genomics approaches. Through transcriptional profiling, a novel DNA-repair gene cluster was identified in γ-radiation-stressed *P. furiosus*, which seems to be specific to both thermophilic Archaea and Bacteria (Williams *et al.*, 2006). An unexpected outcome of this study was the up-regulation of genes involved in oxygen detoxification and redox homeostasis upon exposure to γ-radiation. Along the same lines, UV-induced DNA damage on the transcriptome in *Halobacterium salinarum* NRC-1 resulted in the identification of 33 UV-responsive proteins, 26 of which could be assigned putative functions (Baliga *et al.*, 2004). The dynamic effect of γ exposure on both the transcriptome and proteome of *H. salinarum* NRC-1 was also investigated, focusing on the correlation between mRNA levels and protein abundance and leading to a systems model for physiological response to γ-radiation (Whitehead *et al.*, 2006).

Transition metals stress

Metal-rich biospheres (acid mine or rock drainage) are mostly dominated by acidophilic Archaea, including members of the *Ferroplasma* genera. However, little is known about archaeal metal resistance and homeostasis mechanisms, although some efforts to study these effects have been reported. The mechanisms of copper resistance in *Ferroplasma acidarmus* Fer1 were examined using proteomics (2D- gels/MALDI-TOF), transcriptional assays (RT-PCR), and genome analyses (Baker-Austin *et al.*, 2005). Two genes, *copZ* and *copB*, were found to be co-expressed and strongly induced in the presence of Cu^{2+}, and most likely play a role in copper efflux. Protein folding and DNA repair functions were also mobilized, showing that copper resistance is a complex phenomenon. The effect of six transition metals on *H. salinarum* NRC1 was modelled using a systems approach to integrate transcriptional data, operon prediction, phylogenetic profiles, chromosomal proximity, protein-DNA interactions, putative functions, and other databases, such as PDB, PFAM, COG and KEGG (Kaur *et al.*, 2006). The study, in which 17 deletion strains were created to probe specific effects, highlighted similarities and differences between global responses to different metals, provided evidence for control of transcriptional regulation by a TRASH domain, and identified ZtnA as a resistance gene against Cu^{2+}, Ni^{2+}, and Co^{2+} stress. A putative global control mechanism mediated by general transcriptional factors and metalloproteins was proposed to explain how excess Mn^{2+} induced Fe deficiency. Mercury resistance in Archaea has best been characterized in *S. solfataricus* by disruption of mercuric reductase MerA and an associated regulator, MerR (Schelert *et al.*, 2004; Schelert *et al.*, 2006). It appears that mercury challenge results in transcription inhibition, which is caused – at least in part – by Hg-mediated inactivation of transcription factor TFB (Dixit *et al.*, 2004).

Nutrient limitation

Nutrient limitation, while not directly resulting in molecular damage, can have a broad impact on archaeal metabolism, as evidenced by the global transcriptional response of *M. mazei* strain Göl to nitrogen limitation (Veit *et al.*, 2006). Promoter analysis of starvation-induced genes showed the presence of a palindromic DNA motif, proposed to be involved in nitrogen control in *M. mazei*. Although not a stress response experiment *per se*, quantitative proteomics and transcriptional analysis of acetate- versus methanol-grown *Methanosarcina acetivorans* revealed a general stress response pattern in the presence of acetate (Li *et al.*, 2007). Several regulatory proteins and two-component systems correlated to substrate type, indicative of the high level of metabolic regulation in *M. acetivorans*.

Comparative genomics approaches for examining archaeal stress response

Insights into how Archaea deal with stress and how extremophilic Archaea have been designed to thrive in biologically restrictive habitats can potentially be gleaned from comparative genomics analysis. For example, the genome sequence of the extreme acidophile *Picrophilus torridus* has been used to examine the basis for growth at pH approaching 0 (Ciaramella *et al.*, 2005; Di Giulio, 2005; Fütterer *et al.*, 2004). The organism's extremely impermeable membrane takes full advantage of the high transmembrane proton gradient, as shown by a relative predilection for secondary transporter over ATP-dependent transport. Surprisingly, the comparison with genomes from other thermoacidophiles, such as close relative *T. acidophilum* and more the distant *S. solfataricus*, revealed that close phylogenetic relationship played a lesser role than ecological proximity in overall genome composition (Fütterer *et al.*, 2004).

The release of the sequence of *Nanoarchaeum equitans* (Waters *et al.*, 2003), a small parasitic archaeon with about 540 genes, provides a basis for understanding the effect of massive reductions in genome content on biological function, including stress response (Brochier *et al.*, 2005; Das *et al.*, 2006; Di Giulio, 2006; Makarova and Koonin, 2005). A key limitation in such approaches is the high proportion of genes annotated as hypothetical proteins in archaeal genomes. Due to the phylogenic distance between Archaea and better-studied Bacteria, the chances of finding sequence-based hits to proteins of known functions are often quite limited.

As functional genomics databases expand, it will become possible to go beyond *in silico* genome sequence analysis and directly compare genome-wide response to particular perturbations as a basis for understanding archaeal stress mechanisms. For example, using whole genome cDNA microarrays for the hyperthermophilic archaeon *P. furiosus* and the hyperthermophilic bacterium *Thermotoga maritima*, we have examined heat shock response from a comparative, functional perspective. Transcriptional response resulting from temperature shifts of genes from these two microorganisms with 60% or higher homology at the amino acid level were clustered side-by-side to assess common features. Fig. 10.3 provides a snapshot of this hierarchical cluster, illustrating a case where utilization of certain genes varies noticeably between Archaea and Bacteria. As can be seen from Fig. 10.4, the magnitude of the transcriptional response to heat shock is greater in *P. furiosus* (several hundreds of responsive genes) than in *T. maritima* (dozens of responsive genes). This may be attributed in part to somewhat different experimental

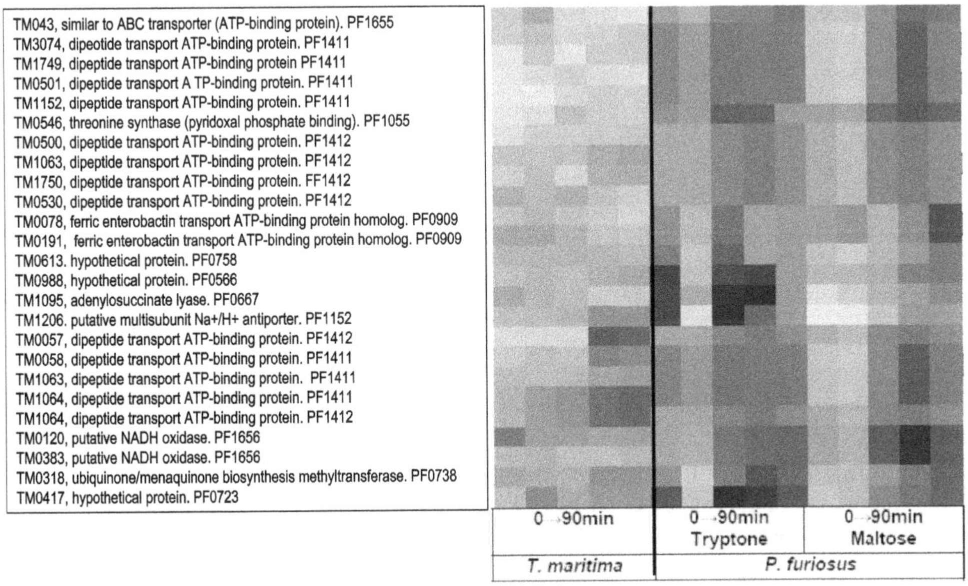

Figure 10.3 Cross-species differential expression hierarchical cluster for Thermotoga maritima and Pyrococcus furiosus genes sharing more than 60% homology. Clustering is based on of log2(fold changes) for each time point with respect to the baseline for a given experiment. Heat shock time points were taken at 0, 5, 30, 60 and 90 min.

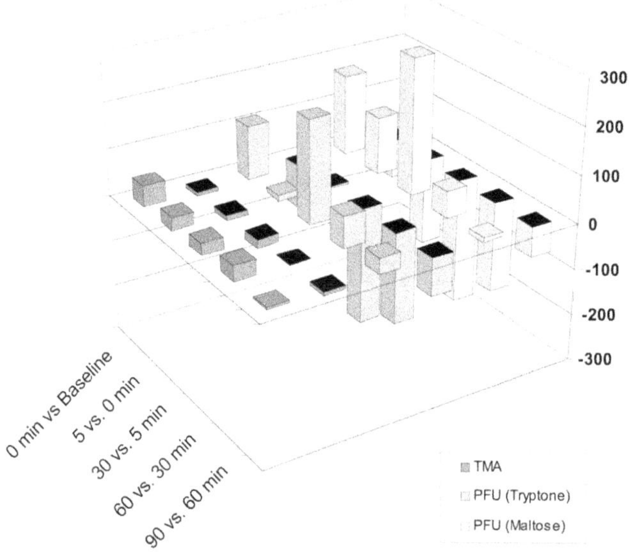

Figure 10.4 Comparison of the magnitude of heat shock response from 80 to 90°C in *T. maritima* and from 90 to 105°C in P. furiosus grown on maltose and tryptone.

conditions; here *T. maritima* was exposed to a shift from 80°C to 90°C, while *P. furiosus* was shifted from 90°C to 105°C. However, when *S. solfataricus* was exposed to a 80°C to 90°C shift approximately one-third (~1000 genes) was differentially transcribed (Tachdjian and Kelly, 2006). Some similarities in heat shock response can be identified beyond the common use of HSPs. For instance, the strong induction of uncharacterized MarR and TetR family regulators has been noticed in both *T. maritima* and *S. solfataricus* (Pysz *et al.*, 2004; Tachdjian and Kelly, 2006). Other similarities include the mobilization of DNA repair proteins as shown by the induction or *recA* genes or homologues in hyperthermophilic Archaea and Bacteria (Pysz *et al.*, 2004; Shockley *et al.*, 2003; Tachdjian and Kelly, 2006).

Summary

Functional genomics efforts to study extremophilic archaeal stress response are in their earliest stages. Nevertheless, it is clear that 'omics' tools will provide unprecedented insights into how this interesting biological phenomenon is initiated and develops in these arguably primitive microorganisms. Such insights will be invaluable in probing the mechanisms by which stress management proceeds in what (to many organisms) are already challenging environments.

Acknowledgements

This work was supported in part by grants from the U.S. National Science Foundation, U.S. Department of Energy and NASA.

References

Ahrman, E., Gustavsson, N., Hultschig, C., Boelens, W.C., and Emanuelsson, C.S. (2007). Small heat shock proteins prevent aggregation of citrate synthase and bind to the N-terminal region which is absent in thermostable forms of citrate synthase. Extremophiles.

Arcus, V.L., Backbro, K., Roos, A., Daniel, E.L., and Baker, E.N. (2004). Distant structural homology leads to the functional characterization of an archaeal PIN domain as an exonuclease. J. Biol. Chem. *279*, 16471–16478.

Arsene, F., Tomoyasu, T., and Bukau, B. (2000). The heat shock response of *Escherichia coli*. Int. J. Food Microbiol. *55*, 3–9.

Baker-Austin, C., Dopson, M., Wexler, M., Sawers, R.G., and Bond, P.L. (2005). Molecular insight into extreme copper resistance in the extremophilic archaeon '*Ferroplasma acidarmanus*' Fer1. Microbiology *151*, 2637–2646.

Baliga, N.S., Bjork, S.J., Bonneau, R., Pan, M., Iloanusi, C., Kottemann, M.C., Hood, L., and DiRuggiero, J. (2004). Systems level insights into the stress response to UV radiation in the halophilic archaeon *Halobacterium* NRC-1. Genome Res. *14*, 1025–1035.

Besche, H., Tamura, N., Tamura, T., and Zwickl, P. (2004). Mutational analysis of conserved AAA+ residues in the archaeal Lon protease from *Thermoplasma acidophilum*. FEBS Lett. *574*, 161–166.

Boonyaratanakornkit, B.B., Simpson, A.J., Whitehead, T.A., Fraser, C.M., El-Sayed, N.M., and Clark, D.S. (2005). Transcriptional profiling of the hyperthermophilic methanarchaeon *Methanococcus jannaschii* in response to lethal heat and non-lethal cold shock. Environ. Microbiol. *7*, 789–797.

Brochier, C., Gribaldo, S., Zivanovic, Y., Confalonieri, F., and Forterre, P. (2005). Nanoarchaea: representatives of a novel archaeal phylum or a fast-evolving euryarchaeal lineage related to Thermococcales? Genome Biol. *6*, R42.

Cavicchioli, R., Thomas, T., and Curmi, P.M. (2000). Cold stress response in Archaea. Extremophiles *4*, 321–331.

Ciaramella, M., Napoli, A., and Rossi, M. (2005). Another extreme genome: how to live at pH 0. Trends Microbiol. *13*, 49–51.

Clarens, M., Macario, A.J., and Conway de Macario, E. (1995). The archaeal dnaK-dnaJ gene cluster: organization and expression in the methanogen *Methanosarcina mazei*. J. Mol. Biol. *250*, 191–201.

Das, S., Paul, S., Bag, S.K., and Dutta, C. (2006). Analysis of *Nanoarchaeum equitans* genome and proteome composition: indications for hyperthermophilic and parasitic adaptation. BMC Genomics *7*, 186.

Deppenmeier, U., Johann, A., Hartsch, T., Merkl, R., Schmitz, R.A., Martinez-Arias, R., Henne, A., Wiezer, A., Baumer, S., Jacobi, C., *et al.* (2002). The genome of *Methanosarcina mazei*: evidence for lateral gene transfer between Bacteria and Archaea. J. Mol. Microbiol. Biotechnol. *4*, 453–461.

Di Giulio, M. (2005). Structuring of the genetic code took place at acidic pH. J. Theor Biol.

Di Giulio, M. (2006). *Nanoarchaeum equitans* is a living fossil. J. Theor Biol. *242*, 257–260.

Dixit, V., Bini, E., Drozda, M., and Blum, P. (2004). Mercury inactivates transcription and the generalized transcription factor TFB in the archaeon *Sulfolobus solfataricus*. Antimicrob Agents Chemother *48*, 1993–1999.

Esser, C., Alberti, S., and Hohfeld, J. (2004). Cooperation of molecular chaperones with the ubiquitin/proteasome system. Biochim. Biophys. Acta *1695*, 171–188.

Fütterer, O., Angelov, A., Liesegang, H., Gottschalk, G., Schleper, C., Schepers, B., Dock, C., Antranikian, G., and Liebl, W. (2004). Genome sequence of *Picrophilus torridus* and its implications for life around pH 0. PNAS *101*, 9091–9096.

Gerdes, K., Christensen, S.K., and Lobner-Olesen, A. (2005). Prokaryotic toxin-antitoxin stress response loci. Nat. Rev. Microbiol. *3*, 371–382.

Giaquinto, L., Curmi, P.M., Siddiqui, K.S., Poljak, A., Delong, E., Dassarma, S., and Cavicchioli, R. (2007). The Structure and Function of Cold Shock Proteins (Csps) in Archaea. J. Bacteriol..

Goodchild, A., Saunders, N.F., Ertan, H., Raftery, M., Guilhaus, M., Curmi, P.M., and Cavicchioli, R. (2004). A proteomic determination of cold adaptation in the Antarctic archaeon, *Methanococcoides burtonii*. Mol. Microbiol. 53, 309–321.

Han, C.J., C.H. Park, and R.M. Kelly (1997). Acquired thermotolerance and 'stressed phase' growth of the extremely thermoacidophilic archaeon *Metallosphaera sedula* in continuous culture. Appl. Environ. Microbiol. *63*, 2391–2396.

Hecker, M., Schumann, W., and Volker, U. (1996). Heat-shock and general stress response in *Bacillus subtilis*. Mol. Microbiol. *19*, 417–428.

Im, Y.J., Na, Y., Kang, G.B., Rho, S.H., Kim, M.K., Lee, J.H., Chung, C.H., and Eom, S.H. (2004). The active site of a lon protease from *Methanococcus jannaschii* distinctly differs from the canonical catalytic Dyad of Lon proteases. J. Biol. Chem. *279*, 53451–53457.

Jakob, U., Gaestel, M., Engel, K., and Buchner, J. (1993). Small heat shock proteins are molecular chaperones. J. Biol. Chem. *268*, 1517–1520.

Kaczowka, S.J., and Maupin-Furlow, J.A. (2003). Subunit topology of two 20S proteasomes from *Haloferax volcanii*. J. Bacteriol. *185*, 165–174.

Kagawa, H.K., Osipiuk, J., Maltsev, N., Overbeek, R., Quaite-Randall, E., Joachimiak, A., and Trent, J.D. (1995). The 60 kDa heat shock proteins in the hyperthermophilic archaeon *Sulfolobus shibatae*. J. Mol. Biol. *253*, 712–725.

Kaur, A., Pan, M., Meislin, M., Facciotti, M.T., El-Gewely, R., and Baliga, N.S. (2006). A systems view of haloarchaeal strategies to withstand stress from transition metals. Genome Res. *16*, 841–854.

Kim, K.K., Kim, R., and Kim, S.H. (1998a). Crystal structure of a small heat-shock protein. Nature *394*, 595–599.

Kim, R., Kim, K.K., Yokota, H., and Kim, S.H. (1998b). Small heat shock protein of *Methanococcus jannaschii*, a hyperthermophile. Proc. Natl. Acad. Sci. USA *95*, 9129–9133.

Klunker, D., Haas, B., Hirtreiter, A., Figueiredo, L., Naylor, D.J., Pfeifer, G., Muller, V., Deppenmeier, U., Gottschalk, G., Hartl, F.U., and Hayer-Hartl, M. (2003). Coexistence of group I and group II chaperonins in the archaeon *Methanosarcina mazei*. J. Biol. Chem. *278*, 33256-33267.

Laksanalamai, P., Maeder, D.L., and Robb, F.T. (2001). Regulation and mechanism of action of the small heat shock protein from the hyperthermophilic archaeon *Pyrococcus furiosus*. J. Bacteriol. *183*, 5198–5202.

Laksanalamai, P., and Robb, F.T. (2004). Small heat shock proteins from extremophiles: a review. Extremophiles *8*, 1–11.

Li, L., Li, Q., Rohlin, L., Kim, U., Salmon, K., Rejtar, T., Gunsalus, R.P., Karger, B.L., and Ferry, J.G. (2007). Quantitative Proteomic and Microarray Analysis of the Archaeon *Methanosarcina acetivorans* Grown with Acetate versus Methanol. J. Proteome Res. *6*, 759–771.

Liu, W., Vierke, G., Wenke, A.K., Thomm, M., and Ladenstein, R. (2007). Crystal Structure of the Archaeal Heat Shock Regulator from *Pyrococcus furiosus*: A Molecular Chimera Representing Eukaryal and Bacterial Features. J. Mol. Biol. *369*, 474–488.

Macario, A.J., and de Macario, E.C. (1999). The archaeal molecular chaperone machine: peculiarities and paradoxes. Genetics 152, 1277–1283.

Macario, A.J., Lange, M., Ahring, B.K., and De Macario, E.C. (1999). Stress genes and proteins in the Archaea. Microbiol. Mol. Biol. Rev. 63, 923–967, table of contents.

Madding, L.S., Michel, J.K., Shockley, K.R., Conners, S.B., Epting, K.L., Johnson, M.R., and Kelly, R.M. (2007). Role of the beta1 subunit in the function and stability of the 20S proteasome in the hyperthermophilic archaeon *Pyrococcus furiosus*. J. Bacteriol. *189*, 583–590.

Makarova, K.S., and Koonin, E.V. (2005). Evolutionary and functional genomics of the Archaea. Curr. Opin. Microbiol. *8*, 586–594.

Martin-Benito, J., Gomez-Reino, J., Stirling, P.C., Lundin, V.F., Gomez-Puertas, P., Boskovic, J., Chacon, P., Fernandez, J.J., Berenguer, J., Leroux, M.R., and Valpuesta, J.M. (2007). Divergent substrate-binding mechanisms reveal an evolutionary specialization of eukaryotic prefoldin compared to its archaeal counterpart. Structure *15*, 101–110.

Martins, L.O., Carreto, L.S., Da Costa, M.S., and Santos, H. (1996). New compatible solutes related to Di-myo-inositol-phosphate in members of the order Thermotogales. J. Bacteriol. *178*, 5644–5651.

Maupin-Furlow, J.A., Gil, M.A., Humbard, M.A., Kirkland, P.A., Li, W., Reuter, C.J., and Wright, A.J. (2005). Archaeal proteasomes and other regulatory proteases. Curr. Opin. Microbiol. *8*, 720–728.

McMillan, R.A., Howard, J., Zaluzec, N.J., Kagawa, H.K., Mogul, R., Li, Y.F., Paavola, C.D., and Trent, J.D. (2005). A self-assembling protein template for constrained synthesis and patterning of nanoparticle arrays. J. Amer. Chem. Soc. *127*, 2800–2801.

Minuth, T., Frey, G., Lindner, P., Rachel, R., Stetter, K.O., and Jaenicke, R. (1998). Recombinant homo- and hetero-oligomers of an ultrastable chaperonin from the archaeon *Pyrodictium occultum* show chaperone activity in vitro. Eur. J. Biochem. *258*, 837–845.

Okochi, M., Nomura, T., Zako, T., Arakawa, T., Iizuka, R., Ueda, H., Funatsu, T., Leroux, M., and Yohda, M. (2004). Kinetics and binding sites for interaction of the prefoldin with a group II chaperonin: contiguous non-native substrate and chaperonin binding sites in the archaeal prefoldin. J. Biol. Chem. *279*, 31788–31795.

Pandey, D.P., and Gerdes, K. (2005). Toxin-antitoxin loci are highly abundant in free-living but lost from host-associated prokaryotes. Nucleic Acids Res. *33*, 966–976.

Phipps, B.M., Typke, D., Hegerl, R., Volker, S., Hoffmann, A., Stetter, K.O., and Baumeister, W. (1993). Structure of a molecular chaperone from a thermophilic archaebacterium. Nature *361*, 475–477.

Pysz, M.A., Ward, D.E., Shockley, K.R., Montero, C.I., Conners, S.B., Johnson, M.R., and Kelly, R.M. (2004). Transcriptional analysis of dynamic heat-shock response by the hyperthermophilic bacterium *Thermotoga maritima*. Extremophiles *8*, 209–217.

Reich, C.I., McNeil, L.K., Brace, J.L., Brucker, J.K., and Olsen, G.J. (2001). Archaeal RecA homologues: different response to DNA-damaging agents in mesophilic and thermophilic Archaea. Extremophiles *5*, 265–275.

Rohlin, L., Trent, J.D., Salmon, K., Kim, U., Gunsalus, R.P., and Liao, J.C. (2005). Heat Shock Response of *Archaeoglobus fulgidus*. J. Bacteriol. *187*, 6046–6057.

Rosen, R., and Ron, E.Z. (2002). Proteome analysis in the study of the bacterial heat-shock response. Mass Spectrom. Rev. *21*, 244–265.

Sakoh, M., Ito, K., and Akiyama, Y. (2005). Proteolytic activity of HtpX, a membrane-bound and stress-controlled protease from *Escherichia coli*. J. Biol. Chem. *280*, 33305–33310.

Santos, H., and da Costa, M.S. (2002). Compatible solutes of organisms that live in hot saline environments. Environ. Microbiol. *4*, 501–509.

Schelert, J., Dixit, V., Hoang, V., Simbahan, J., Drozda, M., and Blum, P. (2004). Occurrence and characterization of mercury resistance in the hyperthermophilic archaeon *Sulfolobus solfataricus* by use of gene disruption. J. Bacteriol. *186*, 427–437.

Schelert, J., Drozda, M., Dixit, V., Dillman, A., and Blum, P. (2006). Regulation of mercury resistance in the crenarchaeote *Sulfolobus solfataricus*. J. Bacteriol. *188*, 7141–7150.

Scholz, S., Wolffa, S., and Hensel, R. (1998). The biosynthesis pathway of di-myo-inositol-1,1'-phosphate in *Pyrococcus woesei*. FEMS Microbiol. Lett. *168*, 37–42.

Shockley, K.R., Ward, D.E., Chhabra, S.R., Conners, S.B., Montero, C.I., and Kelly, R.M. (2003). Heat shock response by the hyperthermophilic archaeon *Pyrococcus furiosus*. Appl. Environ. Microbiol. *69*, 2365–2371.

Shukla, H.D. (2006). Proteomic analysis of acidic chaperones, and stress proteins in extreme halophile *Halobacterium* NRC-1: a comparative proteomic approach to study heat shock response. Proteome Sci. *4*, 6.

Smith, D.M., Benaroudj, N., and Goldberg, A. (2006). Proteasomes and their associated ATPases: A destructive combination. J. Struct Biol. *156*, 72–83.

Sun, Y., and MacRae, T.H. (2005). Small heat shock proteins: molecular structure and chaperone function. Cell Mol. Life Sci. 62, 2460–2476.

Tachdjian, S., and Kelly, R.M. (2006). Dynamic metabolic adjustments and genome plasticity are implicated in the heat shock response of the extremely thermoacidophilic archaeon *Sulfolobus solfataricus*. J. Bacteriol. *188*, 4553–4559.

Trent, J.D., Kagawa, H.K., Paavola, C.D., McMillan, R.A., Howard, J., Jahnke, L., Lavin, C., Embaye, T., and Henze, C.E. (2003). Intracellular localization of a group II chaperonin indicates a membrane-related function. Proc. Natl. Acad. Sci. USA *100*, 15589–15594.

Trent, J.D., Nimmesgern, E., Wall, J.S., Hartl, F.U., and Horwich, A.L. (1991). A molecular chaperone from a thermophilic archaebacterium is related to the eukaryotic protein t-complex polypeptide-1. Nature *354*, 490–493.

Vainberg, I.E. Lewis, S.A., Rommelaere, H., Ampe, C., Vandekerckhove, J., Klein, H.L., and Cowan, N.J. (1998). Prefoldin, a chaperone that delivers unfolded proteins to cytosolic chaperonin. Cell 93, 863–873.

Valentine, D.L. (2007). Adaptations to energy stress dictate the ecology and evolution of the Archaea. Nat. Rev. Microbiol. 5, 316-323.

Veit, K., Ehlers, C., Ehrenreich, A., Salmon, K., Hovey, R., Gunsalus, R.P., Deppenmeier, U., and Schmitz, R.A. (2006). Global transcriptional analysis of *Methanosarcina mazei* strain Go1 under different nitrogen availabilities. Mol. Genet. Genomics *276*, 41–55.

Vierke, G., Engelmann, A., Hebbeln, C., and Thomm, M. (2003). A novel archaeal transcriptional regulator of heat shock response. J. Biol. Chem. *278*, 18–26.

Waters, E., Hohn, M.J., Ahel, I., Graham, D.E., Adams, M.D., Barnstead, M., Beeson, K.Y., Bibbs, L., Bolanos, R., Keller, M., *et al.* (2003). The genome of *Nanoarchaeum equitans:* insights into early archaeal evolution and derived parasitism. Proc. Natl. Acad. Sci. USA *100*, 12984–12988.

Weinberg, M.V., Schut, G.J., Brehm, S., Datta, S., and Adams, M.W. (2005). Cold shock of a hyperthermophilic archaeon: *Pyrococcus furiosus* exhibits multiple responses to a suboptimal growth temperature with a key role for membrane-bound glycoproteins. J. Bacteriol. *187*, 336-348.

Whitehead, K., Kish, A., Pan, M., Kaur, A., Reiss, D.J., King, N., Hohmann, L., DiRuggiero, J., and Baliga, N.S. (2006). An integrated systems approach for understanding cellular responses to gamma radiation. Mol. Syst Biol. *2*, 47.

Williams, E., Lowe, T.M., Savas, J., and Diruggiero, J. (2006). Microarray analysis of the hyperthermophilic archaeon *Pyrococcus furiosus* exposed to gamma irradiation. Extremophiles.

Zhang, W., Culley, D.E., Nie, L., and Brockman, F.J. (2006). DNA microarray analysis of anaerobic *Methanosarcina barkeri* reveals responses to heat shock and air exposure. J. Ind. Microbiol. Biotechnol. *33*, 784–790.

Zivanovic, Y., Lopez, P., Philippe, H., and Forterre, P. (2002). Pyrococcus genome comparison evidences chromosome shuffling-driven evolution. Nucleic Acids Res. *30*, 1902–1910.

Archaeal Antimicrobials: An Undiscovered Country

11

Richard F. Shand and Kathryn J. Leyva

Abstract

Peptide or protein antibiotics have been discovered in all three domains of life, and their production is nearly universal. Bacteriocin and eucaryocin research is well established, while research on archaeocins is still in its infancy. To date, only eight archaeocins (seven halocins and one sulfolobicin) have been partially or fully characterized, but hundreds of archaeocins are believed to exist, especially within the haloarchaea. The prevalence of archaeocins from other members of this domain is unknown simply because no one has looked for them. The discovery of new halocins hinges on recovery and cultivation of haloarchaeal organisms from the environment. For example, samples from a novel hypersaline field site, Wilson Hot Springs, recovered 350 halophilic organisms; preliminary analysis of 75 isolates showed that 48 were archaeal and 27 were bacterial. Significantly, 77% inhibited the growth of at least one other isolate. Inter-domain antagonisms were also present with 43 haloarchaeons inhibiting halophilic members of the domain Bacteria and seven Bacteria antagonized haloarchaeons. Finally, archaeocin research provides excellent opportunities for discovery of novel antibiotics that may have clinical applications.

Introduction

Antimicrobial protein and peptide antibiotics were discovered in the 1920s, beginning with the 'colicins' secreted by *Escherichia coli* (Gratia, 1925). These types of ribosomally synthesized peptides and proteins from the domain *Bacteria* are collectively known as bacteriocins. Similarly, since the early 1960s, antimicrobial peptides and proteins have been discovered in various eukaryotes ranging from protozoa (reviewed in Zhai and Saier, 2000) to plants (reviewed in Broekaert *et al.*, 1997) and to animals (reviewed in Ganz and Lehrer, 1998); these are called eucaryocins (Price and Shand, 2000) to parallel the bacteriocin nomenclature. In 1982, the first report of an antimicrobial substance from the domain Archaea was published (Rodriguez-Valera *et al.*, 1982). At the time, these substances were labelled as bacteriocins, but as more of these antimicrobial compounds were being discovered, they were termed archaeocins to mirror the nomenclature of the other two domains (Price and Shand, 2000). A brief overview of the various antimicrobial protein and peptide antibiotics follows [for more detailed reviews, see Marsh and Goode (1994) for eucaryocins and Riley and Chavan (2007a) for bacteriocins and archaeocins].

Bacteriocins

Almost all bacterial species examined have been shown to produce bacteriocins. It has been postulated that 99% of all bacterial organisms produce and secrete at least one bacteriocin (Klaenhammer, 1988; Riley and Chavan, 2007b). Bacteriocins are plasmid-encoded peptide or protein antibiotics that differ from the more 'traditional' broad spectrum antibiotics in that they have a narrow spectrum of killing activity and are generally active against species closely related to the producer (Riley and Wertz, 2002a; b). Bacteriocins are divided into several categories based on a variety of factors, including the nature of the producing strain, secretion, activity spectrum, extent of post-translational modification, mechanism of killing, and resistance mechanisms. The first and most extensively studied group of bacteriocins is the 'colicins,' named from antagonism studies between strains of *E. coli* (Gratia, 1925). Other categories include Class I bacteriocins (lantibiotics), Class II bacteriocins (nonlantibiotics), and microcins (bacteriocins < 5–10 kDa; reviewed in O'Connor and Shand, 2002). Bacteriocins are also of interest in the food industry. For example, nisin is a bacteriocin produced by lactic acid Bacteria that received FDA approval in 1998 for use as a preservative for pasteurized processed cheese spreads (Chen and Hoover, 2003).

Eucaryocins

Eucaryocins are peptide and protein antibiotics produced by members of the domain *Eucarya*. First discovered in the early 1960s, research in this field has advanced rapidly. As of 2004 (the most recent edition), 895 eucaryocins have been added to the database maintained by Alessandro Tossi (http://www.bbcm.univ.trieste.it/~tossi/pag1.htm) with recent help from the Peptides As Novel Antiinfective Drugs (PANAD) Project. Three dimensional structures of 54 eucaryocins have been elucidated. Eucaryocins have been found in virtually every eukaryotic organism studied, including mammals, amphibians, crustaceans, insects, plants, and protozoans. Generally, eucaryocins are small, heat-resistant cationic peptides that exert their activity by disrupting the membrane of the target cell. Three-hundred-seventy of the eucaryocins in the database are listed as linear peptides and 380 listed as containing disulphide bridges. Many of the eucaryocins have been shown to function as defensins, which comprise a part of the innate immune defences in eukaryotes that possess such immunity (Boman, 1998; Lehrer *et al.*, 1999; Lehrer and Ganz, 1999).

Archaeocins

The term 'archaeocin' was adopted to distinguish peptide and protein antibiotics produced by Archaea from those produced by members of the domain *Bacteria* (Price and Shand, 2000). To date, the only Archaea known to produce archaeocins fall into two phylogenetic groups: euryarchaeal extreme halophiles (haloarchaea) that produce 'halocins' (O'Connor and Shand, 2002; Shand and Leyva, 2007) and the crenarchaeal genus *Sulfolobus*, an aerobic hyperthermophile that produces a 'sulfolobicin' (Prangishvili *et al.*, 2000). Although '[p]roduction of halocin is a practically universal feature of archaeal halophilic rods' (Torreblanca *et al.*, 1994), there appear to be hundreds of different halocins based upon antagonism studies (Meseguer *et al.*, 1986; Torreblanca *et al.*, 1994). However, only a small number of halocins have been discovered, and even fewer have been characterized (see Table 11.1). Halocin protein sequences are unique as they do not match anything in the protein sequence databases (but see Fig. 11.1). Currently, halocin research must take place in the absence of a fully sequenced genome containing a halocin gene, although this may change in the near future with the annotation of a '*halC8*-like' gene in the complete genome for *Haloferax volcanii* (see below for a more complete description). Despite this limitation, all haloarchaea are aerobes and are easy to grow, with typical generation times between 1.5 and 3 hours (Robinson *et al.*, 2005). Detailed protocols for isolating microhalocins are also available (Shand, 2006) as is a complete bibliography of the halocin literature (http://jan.ucc.nau.edu/~shand/bibliography.html).

Archaeocins: halocins and sulfolobicins

Halocins are classified as either peptide (≤ 10 kDa; 'microhalocins') or protein (> 10 kDa) antibiotics produced by members of the archaeal family Halobacteriaceae. To date, all of the known halocin genes are encoded on megaplasmids (> 100 kbp) and possess typical haloarcheal TATA and BRE promoter regions. Halocin transcripts are leaderless and the translated preproteins or preproproteins are most likely exported using the twin arginine translocation (Tat) pathway, as the Tat signal motif (two adjacent arginine residues) is present within the amino terminus (reviewed in Shand and Leyva, 2007). Halocin genes are almost universally expressed at the transition between exponential and stationary phases of growth (Cheung *et al.*, 1997; Shand *et al.*, 1999; Price and Shand,

Table 11.1 Halocin characteristics[a]

Halocin	Producer (source)	Size	GenBank accession No.	Thermal stability	Salt dependent	Activity spectrum[b]	Mechanism	References
C8	*Halobacterium* strain AS7092 (Great Chaidan Salt Lake, China)	~31.1 kDa (prepro-protein; ProC8), 7,427 Da (mature; HalC8)	AY310321	>60 min at 100°C	No	Broad	ND[c]	Li *et al.* (2003), Sun *et al.* (2005)
G1	*Halobacterium* strain GRB (solar saltern, France)	ND	–	ND	ND	Broad	ND	Soppa and Oesterhelt (1989)
H1	*Haloferax mediterranei* M2a (previously Xia3; solar saltern, Spain)	31 kDa	–	<50°C	Yes	Broad	Membrane permeability	Rodriguez-Valera *et al.* (1982), Meseguer *et al.* (1986), Platas (1995), Platas *et al.* (1996), Platas *et al.* (2002)
H2	Haloarchaeon Gla 2.2[d] (solar saltern, Spain)	ND	–	ND	ND	Broad	ND	Rodriguez-Valera *et al.* (1982)
H3	Haloarchaeon Gaa 12 (solar saltern, Spain)	ND	–	ND	ND	Broad	ND	Rodriguez-Valera *et al.* (1982), Meseguer *et al.* (1986)
H4	*Haloferax mediterranei* R4 (solar saltern, Spain)	39.6 kDa (preprotein), 34.9 (mature)	U16389	<60°C	Partially	Broad	Proton flux?	Rodriguez-Valera *et al.* (1982), Meseguer and Rodriguez-Valera (1985), Meseguer and Rodriguez-Valera (1986), Meseguer *et al.* (1986), Meseguer *et al.* (1995), Cheung *et al.* (1997), Shand *et al.* (1999), Perez (2000)
H5	Haloarchaeon Ma 2.20[d] (solar saltern, Spain)	ND	–	ND	ND	Narrow	ND	Rodriguez-Valera *et al.* (1982)

Table 11.1 *continued*

Halocin	Producer (source)	Size	GenBank accession No.	Thermal stability	Salt dependent	Activity spectrum[b]	Mechanism	References
H6/H7	*Haloferax gibbonsii* Ma2.39[e] (solar saltern, Spain)	~3 kDa	–	≤90°C	No	Broad	Na^+/H^+ antiporter inhibitor	Rodriguez-Valera *et al.* (1982), Meseguer *et al.* (1986), Torreblanca *et al.* (1989), Meseguer *et al.* (1995), Alberola *et al.* (1998)
R1	*Halobacterium* strain GN101 (solar saltern, Mexico)	3.8 kDa	–	60°C	No	Broad *Sulfolobus* spp. *M. thermophila*	ND	Rdest and Sturm (1987), Shand *et al.* (1999), Haseltine *et al.* (2001), O'Connor (2002)
S8	Haloarchaeon S8a (Great Salt Lake, UT)	33.9 kDa (preproprotein), 3.58 kDa (mature)	AF276080	≥24 h at boiling[f]	No	Broad *Sulfolobus* spp.	ND	Shand *et al.* (1999), Price and Shand (2000), Haseltine *et al.* (2001)
U1[g]	Uncharacterized haloarchaeon	7,435 Da	–	≥1 week at boiling[f]	No	Broad *Sulfolobus* spp.	ND	Haseltine *et al.* (2001), Duncan (2004)

[a]This table is reprinted with kind permission of Springer Science and Business Media (O'Connor and Shand, 2002).
[b]Activity spectrum refers to inhibition of haloarchaea unless otherwise indicated.
[c]ND: not determined.
[d]Isolates Gla 2.2 and Ma 2.20 were labelled as GLA 22 and MA220, respectively in Rodriguez-Valera *et al.* (1982).
[e]Halocin H6 is produced by *Haloferax gibbonsii* Ma2.39 and was first reported as a 32 kDa protein but is now known to be a microhalocin of about 3 kDa. *Hfx. gibbonsii* Ma2.39 is proprietary and should not be confused with a different halocin-producing strain, *Hfx. gibbonsii* Ma2.38 (ATCC 33595). Halocin H7 is halocin H6, but is produced by a halocin-overproducing mutant of *Hfx. gibbonsii* Ma2.39 called *Hfx. gibbonsii* Alicante SPH7.
[f]This study was done at 2113 metres (7,000 ft); water boils at 93°C at this elevation.
[g]Previously known as HalA4, but recently has been determined not to be produced by strain TuA4. The alphanumeric U1 refers to uncharacterized.

```
HalU1:   DIDITGCSACKYAAGQVCTIGCSAAGGFICGLLGITIPVAGLSsLGfFViI-CTtsADyYSip-DsnaAK...
         □□□□□□□□□□□□□□□□□□□□□□□□□□□□□□□□□□□□□□□□□□□+□□ □□ □ □□  □□ □□++ □   □□
HalC8:   DIDITGCSACKYAAGQVCTIGCSAAGGFICGLLGITIPVAGLSCLG-FVEIVCTV-ADEYSGCGDAV-AKEACNRAGLC

HVO:     DvDvdGCyACgfAvGltCsvGCaAtGaFICGLtGIavPiAGLtCvGFVgIiCdVADalSGCGeAVAeEvCadRtnwC
         □+□+ □□+□□  □+□  □++□□ □ □ □□□□□ □□ +□+□□□+□+□□□ □+□ □□□  □□□□+□□□ □+□  □ ++□
HalC8:   DIDITGCSACKYAAGQVCTIGCSAAGGFICGLLGITIPVAGLSCLGFVEIVCTVADEYSGCGDAVAKEACN-RAGLC

HalU1:   DiDitGCsACkyAaGqvCtiGCsAaGgFICGLlGItiPvAGLsslGfFViI-CttsADyySipdsnaAk...
         □+□+ □□+□□  □+□  □++□□ □ □ □□□□□ □□ +□+□□□+++□ □□ □ □   □□  □      +□
HVO:     DvDvdGCyACgfAvGltCsvGCaAtGaFICGLtGIavPiAGLtcvG-FVgIiCdv-ADalSgcgeavAeevcadrtnwc
```

Figure 11.1 Comparison among the peptide sequences for halocin C8 (HalC8; used as the consensus sequence) to two C8-like peptides, one from Haloferax volcanii strain DS2 (HVO) and one from an uncharacterized haloarchaeon (HalU1). Since HalU1 was sequenced from a purified protein by Edman degradation, three dots at the C-terminus of HalU1 indicate that there may be more amino acids present. In contrast, the HVO peptide sequence was determined from the genome sequence. Key: a = identical amino acides; + = similar amino acids..

2000); the only exception is halocin H1, which is induced during exponential phase (Platas *et al.*, 1996). In contrast, the larger halocin proteins are heat-labile and typically obligately halophilic as they lose their activity (or activity is reduced) when desalted (Perez, 2000; Table 11.1). Table 11.1 summarizes the properties of the characterized or partially characterized halocins; a brief discussion of each halocin is provided in the next sections.

Peptide halocins = microhalocins

Currently, five peptide halocins have been partially or completely characterized at the protein and/or genetic levels: HalS8, HalR1, HalC8, HalH7, and HalU1 (Table 11.1; reviewed in O'Connor and Shand, 2002; Shand and Leyva, 2007). These antimicrobial peptides range from ~3 to 7.4 kDa in molecular mass, consisting of 36 to 76 amino acid residues. Two of the microhalocins (HalS8 and HalC8) are produced by proteolytic cleavage from a larger preproprotein by an unknown mechanism. Microhalocins are hydrophobic peptides that remain active even if desalted and/or stored at 4°C and are fairly insensitive to heat and organic solvents (O'Connor, 2002; Table 11.1). The first microhalocin to be characterized was HalS8, produced by the uncharacterized haloarchaeon S8a isolated from the Great Salt Lake, UT by Penny Amy (Price and Shand, 2000). Halocin R1 (HalR1), produced by *Hbt. salinarum* GN101, isolated from a solar saltern in Guererro Negro, Mexico by Barbara Javor (personal communication), was the second microhalocin to be characterized (Rdest and Sturm, 1987; O'Connor, 2002).

Protein sequence data of the mature peptides HalR1 and HalS8 show that these two microhalocins are remarkably similar to each other, both in molecular mass (3.8 and 3.58 kDa, respectively) and in amino acid sequence; HalR1 is 63% identical (capitalized residues below) and 71% similar to HalS8 (Price and Shand, 2000; O'Connor, 2002; O'Connor and Shand, 2002):

HalR1: lqsNINiNTAAaVILiFNQ...
HalS8: sdcNINsNTAAdVILcFNQ...

HalR1: ...VqvgALCaPTpVsGGgPpP
HalS8: ...VgscALCsPTiV-GG-PvP

The sequence similarity of these two halocins suggests that they might share similar physical properties; this is not the case. HalS8 is extremely stable, even when boiled, desalted, and stored at 4°C while HalR1 can only withstand temperatures at or below 60°C (Table 11.1; O'Connor, 2002; O'Connor and Shand, 2002). While both have a broad activity spectrum, each microhalocin differs in the specific species inhibited (reviewed in O'Connor and Shand, 2002; Shand and Leyva, 2007). It is currently unknown which amino acid residues are responsible for these differences and no information is available regarding the tertiary structure of these, or any other halocins, that may provide insight.

Halocin H7 (HalH7; formerly known as HalH6) produced by *Hfx. gibbonsii* Ma2.39, isolated from a solar saltern near Alicante, Spain (Torreblanca *et al.*, 1986), recently has been identified as a microhalocin (I. Meseguer, personal communication; reviewed in Shand and Leyva,

2007). Although the size of the peptide has been elucidated, unfortunately the gene and protein sequences are proprietary. However, HalH7 has been shown to inhibit the Na^+/H^+ antiporter (Meseguer *et al.*, 1995). This is significant as it not only provides the first specific mechanism of action that can be attributed to any halocin, but it has been shown to inhibit both haloarchaeal and mammalian Na^+/H^+ antiporters (Meseguer *et al.*, 1995; Alberola *et al.*, 1998), with potential clinical applications (see section on Biotechnology applications below).

The last two microhalocins to be characterized are HalU1 and HalC8. HalC8 is produced by *Halobacterium* strain AS7092, isolated from the Great Chaidan Salt Lake, China (Li *et al.*, 2003; Sun *et al.*, 2005). As stated above, HalC8 is processed from a preproprotein, yielding not only HalC8, but a second active protein known as HalI. HalI is cleaved from the amino terminus of the preproprotein and has been shown to impart immunity to HalC8 by the producer strain (Sun *et al.*, 2005). Halocin U1 is produced by an uncharacterized haloarchaeon.

HalU1 and HalC8 show remarkable similarities to one another. Both are thermostable, retain activity after desalting, and exhibit a broad spectrum of activity to other haloarchaea (Table 11.1; O'Connor and Shand, 2002; Li *et al.*, 2003; Duncan, 2004; Sun *et al.*, 2005), but HalU1 also has been shown to inhibit *Solfolobus* spp., a member of the Crenarchaeota (Haseltine *et al.*, 2001). Although both microhalocins are ~7.4 kDa in molecular mass, the 68 amino acids of the peptide HalU1 were determined by Edman degradation of purified protein while the 76 amino acid residues of HalC8 were determined from the HalC8 DNA sequence. HalU1 is 75% identical and 79% similar to HalC8, with the greatest degree of identity at the amino termini (Fig. 11.1; Duncan, 2004). It remains to be determined if HalU1 is cleaved from a preproprotein, or if there is an immunity peptide/protein at the amino terminus.

Recently, the genome sequence for *Hfx. volcanii* strain DS2 was completed and is available on the TIGR website (www.tigr.org). The annotation reveals a '*halC8*-like' gene sequence encoding a peptide with 57% identity and 77% similarity to HalC8 (Fig. 11.1; C.J. Daniels and M. Pohlschröeder, personal communication). Although the open reading frames of *halC8* and the C8-like gene from *Hfx. volcanii* are similar in size (849 vs. 915 bp, respectively), nothing is known about the processing of the protein product from *Hfx. volcanii*. However, there is only a 50% similarity in the alignment of the amino acid residues at the amino terminus of the two proteins (data not shown; C.J. Daniels and M. Pohlschröeder, personal communication). Not surprisingly, the alignment between HalU1 and the HalC8-like peptide from *Hfx. volcanii* shows 43% identity and 58% similarity (Fig. 11.1). Microarray analysis of gene expression has recently been completed for *Hfx. volcanii* cultures grown in complex vs. defined media as well as for cells in exponential vs. stationary phase of growth. Preliminary results reveal that the *Hfx. volcanii* C8-like gene is expressed at very low levels in a complex medium (320 and 130 arbitrary units in exponential vs. stationary phase, respectively) and at low levels in a defined medium (499 and 598 arbitrary units in exponential vs. stationary phase, respectively). The cutoff used for 'significant' expression was 500 arbitrary units (S. Morimoto, A. Sabag-Daigle, R. Nist, K. Kornacker, J. Hemmingsen, and C.J. Daniels, personal communication). The significance of these findings needs to be supported by examining the stability of the mRNA, determining the amount of protein produced, searching for growth conditions that increase expression, and by performing antagonism studies.

Protein halocins

Of the remaining halocins, only two can be classified as protein halocins: HalH1 and HalH4; the molecular masses of the remaining halocins have yet to be elucidated (Table 11.1). Halocin H1 is produced by *Hfx. mediterranei* M2a (formerly strain Xia3), isolated from a solar saltern near Alicante, Spain (Rodriguez-Valera *et al.*, 1982). It is a 31 kDa protein that is heat-labile (Platas, 1995; Platas *et al.*, 1996), loses activity when desalted (Platas *et al.*, 2002), and exhibits a broad range of inhibition within the haloarchaea (Meseguer *et al.*, 1986; Platas *et al.*, 2002). Halocin H1 has yet to be characterized at the protein and genetic levels.

In contrast, HalH4, produced by *Hfx. mediterranei* R4 (ATCC 33500), also isolated from a solar saltern near Alicante, Spain (Rodriguez-Valera *et al.*, 1982) was the first halocin discovered (Rodriguez-Valera *et al.*, 1982) and has been characterized at both the protein and genetic levels (Meseguer and Rodriguez-Valera, 1985; 1986; Cheung *et al.*, 1997; Perez, 2000). The molecular mass of the mature HalH4 protein is 34.9 kDa (359 amino acids), processed from a preprotein of 39.6 kDa (Cheung *et al.*, 1997); the mechanism for processing is unknown. Halocin H4 is an archaeolytic halocin and adsorbs to sensitive *Hbt. salinarum* cells where it may be disrupting membrane permeability (Rodriguez-Valera *et al.*, 1982; Meseguer and Rodriguez-Valera, 1986). However, the specific target site(s) have not been determined (Meseguer *et al.*, 1995).

Sulfolobicins

The archaeocins produced by *Sulfolobus* are entirely different from halocins, since their activity is predominantly associated with the cells and not the supernatant (Prangishvili *et al.*, 2000). Prangishvili *et al.* (2000) were the first to isolate and characterize these proteinaceous toxins, which they called 'sulfolobicins' in keeping with bacteriocin nomenclature. Provisionally, the producer strain has been named '*Sulfolobus islandicus*.' To date, the spectrum of sulfolobicin activity appears to be restricted to other members of the Sulfolobales: the sulfolobicin inhibited *S. solfataricus* P1, *S. shibatae* B12, and six non-producing strains of '*S. islandicus*.' Activity appears to be archaeocidal but not archaeolytic. It does not inhibit *S. acidocaldarius* DSM639, nor does purified sulfolobicin from strain HEN2/2 inhibit *Hbt. salinarum* R1 or *E. coli* (Prangishvili *et al.*, 2000; reviewed in O'Connor and Shand, 2002).

Chemical warfare in a novel hypersaline environment

In the northern Utah desert is a novel hypersaline environment composed of a thick crust of salt (2–10 cm; Fig. 11.2) fed by a group of five geothermal hot springs (68°C) aligned linearly along a transit of less than 1 km (Wilson Hot Springs). The springs maintain a constant supply of water just beneath the crust. Breaking through the crust reveals a carotenoid-rich pink fluid replete in halophilic microorganisms (Fig. 11.3). From a recent survey, we have recovered more than 350 different extreme halophiles based initially upon colony morphology. From this collection, a subset of 75 colonies was purified and characterized (P.J. Polsgrove, B.A. Roberts, M.A. Mishler and R.F. Shand, unpublished; see also http://www4.nau.edu/wilsonhotsprings). Based on domain-specific primers directed against 16S rDNA, 48 were archaeal and 27 were bacterial. We did not recover any extremely halophilic eucaryal organisms in this first subset of 75 purified organisms. Optimum NaCl concentrations, as determined by growth on plates, ranged from 1 to 4 M.

With these 75 purified isolates in hand, an antagonism study was performed. This entailed preparing a top agar lawn of each organism and then spotting small amounts of culture (2.5 μl) of each of the 75 isolates onto each lawn. The 75 plates were incubated at 43°C and surveyed periodically for the presence of zones of inhibition (an antagonism) over a period of 8–10 days (Fig. 11.4). This was done as some zones did not appear until 192 hours. The results were astonishing. Fifty-eight organisms (77%) inhibited at least one other isolate. One isolate (WHSF16, a haloarchaeon) antagonized 33/75 isolates (44%). In addition, there were clear signs of synergy (the opposite of antagonism) where some of the organisms that were spotted grew much better in the presence of specific isolates than on others. Not surprisingly, 49 Archaea antagonized other Archaea, and 8 *Bacteria* antago-

Figure 11.2 A view of the thick salt crust saturated by water from nearby hot springs.

Figure 11.3 Saturated salt brine. The liquid in this figure is a reddish colour due to extreme halophiles that produce carotenoid pigments.

nized other *Bacteria* (intra-domain antagonism). Significantly however, 43 Archaea antagonized other *Bacteria*, and 7 *Bacteria* antagonized other Archaea (inter-domain antagonism) (P.J. Polsgrove, B.A. Roberts, M.A. Mishler and R.F. Shand, unpublished). It is important to note that an antagonism study such as this cannot be done if the organisms cannot be cultured in the laboratory. It follows that regardless of the environment one is examining, an efficient recovery and resuscitation system must be in place. As mentioned above, we recovered hundreds of different extreme halophiles from our field site. This was made possible by the development of an efficient recovery protocol for extreme halophiles (P.J. Polsgrove, L. Fitchett-Hewitt and R.F. Shand, unpublished). The main features of our recovery protocol involved initial resuscitation at low temperature (4°C) in the absence of oxygen and any carbon source, and then plating onto halophile medium that is oligotrophic, containing about one-third the nutrient concentration of the commercial medium R2A.

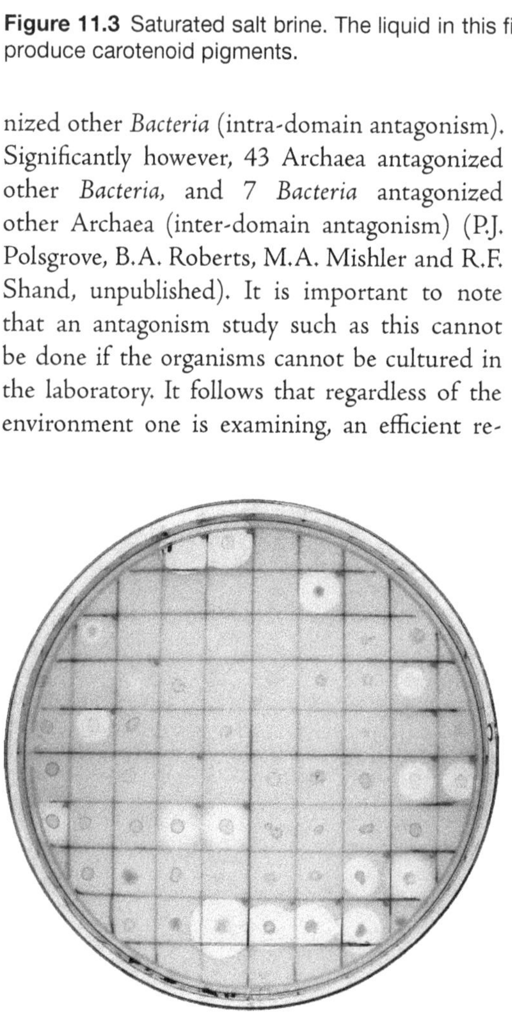

Figure 11.4 A typical antagonism study where one isolate has been challenged by all 75 isolates. Zones of inhibition are clearly visible.

The antagonism study above does not distinguish among zones of inhibition caused by a broad spectrum antibiotic, a narrow spectrum antibiotic (such as a bacteriocin), a metabolic byproduct (such as an acid that would lower the pH), or some other defence molecule (Riley and Chavan, 2007b). Nonetheless, there is an enormous amount of antagonism occurring within this extremely halophilic environment. From an ecological perspective, many view the presence of antimicrobial agents in the environment as a means for reducing competition by suppression or death of another organism (allelopathy) (Kerr, 2007). Clearly some of this is occurring. Typically, antibiotics in the environment are at sublethal concentrations (Hoffman *et al.*, 2007); they can, however, produce profound changes in gene expression. In addition, they are also involved in cell-to-cell signalling (Hoffman *et al.*, 2007). The view that allelopathy in the environment is to reduce competition has begun to change: rather, '...allelopathy may play a critical role in maintaining diversity in these systems.' (Kerr, 2007). This view is supported by the works of Czárán *et al.* (2002), Lenski and Riley (2002) and Kerr *et al.* (2002). Support is also provided by our hypersaline study described above. In the face of substantial antimicrobial activity, the diversity of extreme halophiles is very high. Since we have isolated and purified 75 extreme halophiles and can grow them in the laboratory, we will be able to develop inter- and intra-domain challenge experiments in order to develop hypotheses with respect to the role of antimicrobials in the environment and how they might contribute to maintaining biodiversity.

Finally, we challenged all 75 isolates to a battery of Gram-positive and Gram-negative *Bacteria*. While no Gram-negative *Bacteria* were inhibited, several Gram-positive organisms, including both pathogens and opportunists, were inhibited, but by only two of the 75 isolates, both of which were bacterial and grew optimally on 1 M NaCl. The potential for discovering new antibiotics will rise as we purify and characterize more of the remaining 350 isolates obtained from the Wilson Hot Springs site.

Biotechnology applications of halocins and archaeal antibiotics

Given the amount of antagonism shown at the hypersaline site at Wilson Hot Springs, the fact that *Bacteria* antagonize Archaea and vice versa, and the amount of microbial diversity recovered from the site, the potential for discovering novel antibiotics that may prove beneficial as chemotherapeutic agents against human and/or animal pathogens is vast. Although the mechanisms of action for most halocins have yet to be elucidated, currently one halocin, HalH7, has been shown to inhibit the Na^+/H^+ antiporter in both haloarchaea (Meseguer *et al.*, 1995) and in a dog model (Alberola *et al.*, 1998). In the dog model, treatment with HalH7 after the induction of a myocardial infarct resulted in a reduction in the infarct size and the number of ectopic beats (Alberola *et al.*, 1998). Applications for other halocins will also hinge on the discovery of their mechanisms of action.

Finally, the DNA sequence that encodes immunity to HalC8 may serve as a useful selectable marker especially for haloarchaea that require the highest levels of NaCl for optimal growth (Sun *et al.*, 2005).

Acknowledgements

We thank Dr Chuck Daniels for graciously providing unpublished microarray data from *Hfx. volcanii* DS2 and to Dr Mechtchild Pohlschröeder for providing the sequence alignment for the '*halC8*-like' gene in *Hfx. volcanii*. Special thanks to Peter Polsgrove for providing the unpublished antagonism data from Wilson Hot Springs. Financial support was provided to R.F.S. by the National Institutes of Health grant 1R15AIO75333-01 and the Intramural Grants Program from Northern Arizona University. K.J.L. was supported by the Intramural Grants Program from Midwestern University.

References

Alberola, A., Meseguer, I., Torreblanca, M., Moya, A., Sancho, S., Polo, B., Soria, B., and Such, L. (1998). Halocin H7 decreases infarct size and ectopic beats after myocardial reperfusion in dogs. J. Physiol. *509P*, 148P.

Boman, H.G. (1998). Gene-encoded peptide antibiotics and the concept of innate immunity: an updated review. Scand. J. Immunol. *48*, 15–25.

Broekaert, W.F., Cammue, B.P.A., De Bolle, M.F.C., Thevissen, K., De Samblanx, G.W., and Osborn, R.W. (1997). Antimicrobial peptides from plants. Crit. Rev. Plant Sci. *16*, 297–323.

Chen, H., and Hoover, D.G. (2003). Bacteriocins and their food applications. Compr. Rev. Food Sci. Food Saf. *2*, 82–100.

Cheung, J., Danna, K.J., O'Connor, E.M., Price, L.B., and Shand, R.F. (1997). Isolation, sequence, and expression of the gene encoding halocin H4, a bacteriocin from the halophilic archaeon *Haloferax mediterranei* R4. J. Bacteriol. *179*, 548–551.

Czárán, T.L., Hoekstra, R.F., and Paige, L. (2002). Chemical warfare between microbes promotes biodiversity. Proc. Natl. Acad. Sci. USA *99*, 786–790.

Duncan, J.L. (2004). Haloarchaeal growth physiology, characterization of halocin A4 and cloning *tbp* and *tfb* genes. Masters thesis, Northern Arizona University, Flagstaff AZ.

Ganz, T., and Lehrer, R.I. (1998). Antimicrobial peptides of vertebrates. Curr. Opin. Immunol. *10*, 41–44.

Gratia, A. (1925). Sur un remarquable exemple d'antagonisme entre deux souches de Colibacille. CR Soc. Biol. *93*, 1040–1041.

Haseltine, C., Hill, T., Montalvo-Rodriguez, R., Kemper, S.K., Shand, R.F., and Blum, P. (2001). Secreted euryarchaeal microhalocins kill hyperthermophilic crenarchaea. J. Bacteriol. *183*, 287–291.

Hoffman, L., D'Argenio, D., Bader, M., and Miller, S. (2007). Microbial recognition of antibiotics: ecological, physiological and therapeutic implications. Microbe *2*, 175–182.

Kerr, B., Riley, M.A., Feldman, M.W., and Bohannan, B.J.M. (2002). Local dispersal promotes biodiversity in a real-life game of rock-paper-scissors. Nature *418*, 171–174.

Kerr, B. (2007). The ecological and evolutionary dynamics of model bacteriocin communities. In Bacteriocins: Ecology and Evolution, M.A. Riley and M.A. Chavan, eds. (Heidelberg, Germany: Springer-Verlag), pp. 111–134.

Klaenhammer, T.R. (1988). Bacteriocins of lactic acid Bacteria. Biochimie *70*, 337–49.

Lehrer, R.I., Bevins, C.L., and Ganz, T. (1999). Defensins and other antimicrobial peptides. In Mucosal Immunology, P.L. Ogra, J. Mestecky, M.E. Lamm, W. Strober, J. Bienenstock, and J.R. McGhee, eds. (New York, NY: Academic Press), pp. 89–99.

Lehrer, R.I., and Ganz, T. (1999). Antimicrobial peptides in mammalian and insect host defence. Curr. Opin. Immunol. *11*, 23–27.

Lenski, R.E., and Riley, M.A. (2002). Chemical warfare from an ecological perspective. Proc. Natl. Acad. Sci. USA *99*, 556–558.

Li, Y., Xiang, H., Liu, J., Zhou, M., and Tan, H. (2003). Purification and biological characterization of halocin C8, a novel peptide antibiotic from *Halobacterium* strain AS7092. Extremophiles *7*, 401–407.

Marsh, J., and Goode, J.A. (1994). Antimicrobial Peptides. Ciba Foundation Symposium 186 (Chichester, UK: Wiley).

Meseguer, I., and Rodriguez-Valera, F. (1985). Production and purification of halocin H4. FEMS Microbiol. Lett. *28*, 177–182.

Meseguer, I., and Rodriguez-Valera, F. (1986). Effect of halocin H4 on cells of *Halobacterium halobium*. J. Gen. Microbiol. *132*, 3061–3068.

Meseguer, I., Rodríguez-Valera, F., and Ventosa, A. (1986). Antagonistic interactions among halobacteria due to halocin production. FEMS Microbiol. Lett. *36*, 177–182.

Meseguer, I., Torreblanca, M., and Konishi, T. (1995). Specific inhibition of the halobacterial Na^+/H^+ antiporter by halocin H6. J. Biol. Chem. *270*, 6450–6455.

O'Connor, E.M. (2002). Purification and characterization of microhalocin R1 from *Halobacterium salinarum* GN101. Doctoral dissertation, Northern Arizona University, Flagstaff AZ.

O'Connor, E.M., and Shand, R.F. (2002). Halocins and sulfolobicins: The emerging story of archaeal protein and peptide antibiotics. J. Ind. Microbiol. Biotechnol. *28*, 23–31.

Perez, A.M. (2000). Growth physiology of *Haloferax mediterranei* R4 and purification of halocin H4. Masters thesis, Northern Arizona University, Flagstaff AZ.

Platas, G. (1995). Caracterizacíon de la actividad antimicrobiana de la haloarquea *Haloferax mediterranei* Xia3. Tesis doctoral, Universidad Autónoma de Madrid, Madrid.

Platas, G., Meseguer, I., and Amils, R. (1996). Optimization of the production of a bacteriocin from *Haloferax mediterranei* Xia3. Microbiología *12*, 75–84.

Platas, G., Meseguer, I., and Amils, R. (2002). Purification and biological characterization of halocin H1 from *Haloferax mediterranei* M2a. Int. Microbiol. *5*, 15–19.

Prangishvili, D., Holz, I., Stieger, E., Nickell, S., Kristjansson, J.K., and Zillig, W. (2000). Sulfolobicins, specific proteinaceous toxins produced by strains of the extremely thermophilic archaeal genus *Sulfolobus*. J. Bacteriol. *182*, 2985–2988.

Price, L.B., and Shand, R.F. (2000). Halocin S8: a 36-amino-acid microhalocin from the haloarchaeal strain S8a. J. Bacteriol. *182*, 4951–4958.

Rdest, U., and Sturm, M. (1987). Bacteriocins from halobacteria. In Protein Purification: Micro to Macro, R. Burgess, ed. (New York, NY: Alan R. Liss), pp 271–278.

Riley, M.A., and Chavan, M.A. (2007a). Bacteriocins: Ecology and Evolution (Heidelberg, Germany: Springer-Verlag).

Riley, M.A., and Chavan, M.A. (2007b). Introduction. In Bacteriocins: Ecology and Evolution, M.A. Riley and M.A. Chavan, eds. (Heidelberg, Germany: Springer-Verlag), pp. 1–3.

Riley, M.A., and Wertz, J.E. (2002a). Bacteriocin diversity: ecology and evolutionary perspectives. Biochimie *84*, 357–364.

Riley, M.A., and Wertz, J.E. (2002b). Bacteriocins: Evolution, Ecology, and Application. Annu. Rev. Microbiol. *56*, 117–137.

Robinson, J.L., Pyzyna, B., Atrasz, R.G., Henderson, C.A., Morrill, K.L., Burd, A.M., DeSoucy, E., Fogelman III, R.E., Naylor, J.B., Steele, S.M., Elliott, D.R., Leyva, K.J., and Shand, R.F. (2005). Growth kinetics of extremely halophilic Archaea (Family *Halobacteriaceae*) as revealed by Arrhenius plots. J. Bacteriol. *187*, 923–929.

Rodriguez-Valera, F., Juez, G., and Kushner, D.J. (1982). Halocins: salt-dependent bacteriocins produced by extremely halophilic rods. Can. J. Microbiol. *28*, 151–154.

Shand, R.F. (2006). Detection, quantification and purification of halocins: Peptide antibiotics from haloarchaeal extremophiles. In Extremophiles – Methods in Microbiology, Vol. 35, F.A. Rainey and A. Oren, eds. (Amsterdam: Elsevier/Academic Press), pp. 703–718.

Shand, R.F., and Leyva, K.J. (2007). Peptide and Protein Antibiotics from the Domain Archaea: Halocins and Sulfolobicins. In Bacteriocins: Ecology and Evolution. M.A. Riley and M.A. Chavan, eds. (Heidelberg, Germany: Springer-Verlag), pp. 93–109.

Shand, R.F., Price, L.B., and O'Connor, E.M. (1999). Halocins: Protein antibiotics from hypersaline environments. In Microbiology and Biogeochemistry of Hypersaline Environments, A. Oren, ed. (Boca Raton: CRC Press), pp. 295–306.

Soppa, J., and Oesterhelt, D. (1989). *Halobacterium* sp. GRB: a species to work with? Can. J. Microbiol. *35*, 205–209.

Sun, C., Li, Y., Mei, S., Lu, Q., Zhou, L., and Xiang, H. (2005). A single gene directs both production and immunity of halocin C8 in a haloarchaeal strain AS7092. Mol. Microbiol. *57*, 537–549.

Torreblanca, M., Rodriguez-Valera, F., Juez, J., Ventosa, A., Kamekura, M., and Kates, M. (1986). Classification of non-alkaliphilic halobacteria based on numerical taxonomy and polar lipid composition, and description of *Haloarcula* gen. nov. and *Haloferax* gen. nov. Syst. Appl. Microbiol. *8*, 89–99.

Torreblanca, M., Meseguer, I., and Ventosa, A. (1994). Production of halocin is a practically universal feature of archaeal halophilic rods. Lett. Appl. Microbiol. *19*, 201–205.

Zhai, Y., and Saier, M.H. (2000). The amoebapore superfamily. Biochim. Biophys. Acta *1469*, 87–99.

Index

T

U

V

W

Y

Other Books of Interest